PATTON

A History of
the American Main Battle Tank

Volume I

by
R.P. Hunnicutt

Line Drawings
by
D.P. Dyer

Color Drawing
by
Uwe Feist

Foreword
by
General Donn A. Starry, USA-Retired

E P B M
ECHO POINT BOOKS & MEDIA, LLC

Published by Echo Point Books & Media
Brattleboro, Vermont
www.EchoPointBooks.com

ISBN: 978-1-62654-879-4

Cover image by Uwe Feist

Cover design by Adrienne Núñez,
Echo Point Books & Media

Editorial and proofreading assistance by Christine Schultz,
Echo Point Books & Media

Printed and bound in the United States of America

CONTENTS

Foreword 5
Introduction 6

PART I   THE TRANSITION PERIOD, WORLD WAR II TO KOREA .   7

Planning for the Postwar Army 9
An Interim Tank 10
New Development 32
The Emergency 52

PART II   THE NEW STANDARD MEDIUM TANK 81

Development of the Basic Tank 83
Experimental Modifications 123

PART III   THE ADVENT OF THE MAIN BATTLE TANK 147

The Universal or All Purpose Tank 149
Upgrading the Tank Fleet 219

PART IV   SPECIALIZED ARMOR BASED ON THE PATTON 243

Flame Thrower Tanks 245
Self-Propelled Artillery 255
Antiaircraft Tanks 263
Tank Recovery Vehicles 272
Combat Engineer Vehicles 283
Armored Vehicle Launched Bridges 304
Flotation Devices and Fording Equipment 310
Mine Clearance Equipment.. 323
Training Devices and Special Test Equipment 340

PART V   THE PATTON ON ACTIVE SERVICE 349

U.S. Forces in the United States and Overseas 351
Vietnam 373
Service in Foreign Armies 399
Conclusion 408

PART VI   REFERENCE DATA 411

Color Section 413
Vehicle Data Sheets 421
Weapon Data Sheets 449
References and Selected Bibliography 460
Index 461

ACKNOWLEDGEMENTS

Reviewing the mass of material collected during the research for this book immediately brings to mind the many people who made this project possible. As usual, my friend for many years, Colonel Robert J. Icks provided guidance and suggestions invaluable to the research program. Also, I would like to thank General Donn A. Starry, USA Retired, who agreed to write the foreword for this volume. As he points out, his military career paralleled that of the Patton tank and he saw it from the viewpoint of the user at all stages of its development.

Entire sections of this book would not have been possible without the help of Leon Burg and the Technical Library at the Tank Automotive Command. He found sources of data and photographs that few people knew existed. Similar help was received from R. Paul Ryan of the Ballistics Research Laboratory Library at Aberdeen Proving Ground.

Many of the key people in the various tank development programs took the time to answer questions and to find photographs that were essential in sorting out the history of these vehicles. At the Tank Automotive Command Major General Oscar Decker, Joe Williams, Clifford Bradley, Dan Smith, Clarence Olsen, and Oscar Danielian, all now retired, were particularly helpful. The photographs of the vehicles from the shoot-off to select the cannon for the M60 tank would not have been available without the efforts of Dr. Herbert H. Dobbs, the Technical Director of TACOM. Thanks are also due to Joseph Avesian for finding much of the production data.

At Chrysler, later the Land Systems Division of General Dynamics Corporation, much help was received. Dr. Philip Lett, who entered the M48 development program as a young engineer fresh out of the University of Michigan, arranged access to all of the available material. Many of these items were located by Briggs Jones who provided a very large number of the photographs used in this volume. Other photographs of the early M48 production came from General Motors Corporation and the Ford Motor Company.

John Campbell, John Purdy, and Phyllis Cassler of the Patton Museum at Fort Knox were extremely helpful, particularly in locating many of the early technical manuals. David "Doc" Holliday located and provided prints of numerous photographs from the Armor and Engineer Board. Major Fred Crismon, also at Fort Knox, photographed several vehicles for the book. Royce Taylor of Armor Magazine permitted the use of several items from that publication.

At Bowen-McLaughlin-York Company, James Schroeder answered numerous questions regarding the early development of the M88 recovery vehicle and provided many photographs of this and other BMY projects. The engineer vehicles were a particularly difficult problem which could not have been solved without the help of Major General James Ellis and Elizabeth Slawson of the Engineer School at Fort Belvoir. Material on the DIVAD program came from the Ford Aerospace & Communications Corporation.

As usual in such a research program, many documents were too old for the active files and yet not old enough to be in the National Archives. These were stored in various places and generally were inaccessible. However, Paul Taborn of the Adjutant General's Department managed to track down some of these which were extremely useful. At Teledyne, William Wagner arranged a meeting with Carl Bachle who had been instrumental in the development of the air-cooled tank engines used in the Patton.

Major Geoffrey Tillotson was the source of much useful information, particularly on the development of the British 105mm gun adopted for the M60. He also permitted the use of other material collected during the research for his book on the M48 tank. My thanks also go to Lieutenant Colonel D. Eshel, Retired, for permission to use certain photographs of Israeli vehicles from his excellent publication "Born in Battle/Defence Update International."

Considerable material covering the M48A3 in Vietnam came from Lieutenant Colonel James Loop and Thomas Mayer.

Phil Dyer and Uwe Feist did their usual superb job on the line drawings and the cover painting respectively and Don White helped restore some of the original sketches in the book. The cover painting was based on color photographs taken by Lieutenant Colonel Robert D. Hurt III who, as a captain, commanded Company H, 2nd Squadron, 11th Armored Cavalry Regiment in Vietnam. They were located through the efforts of Colonel James Leach who commanded the Regiment at that time.

It is impossible to recall everyone who helped with a research program such as this. However, others who made a significant contribution were as follows: Brigadier General Charles Adsit, Major General A. H. Anderson, Robert Barnowske, David H. Bartle, Steve Bern, Gary Binder, James W. Burdick, Michael Clear, Chris Delwiche, Clement Detloff, Colonel Claude Donovan, Lieutenant General Jean Engler, John L. Everett, George Forsyth, Leon S. Gallant, Donald F. Hays, Colonel Thomas Huber, Edward R. Jackovich, Ron Johnson, Marvin H. Kabakoff, Tim Kubica, James A. Kronner, Jacques Littlefield, Colonel Charles D. McGaw, Captain Dwight McLemore, Colonel Samuel L. Myers, Jr, J. D. Neu, Colonel Jimmy Pigg, Walter Spielberger, Bruce Stainitis, R.A. Stauss, Regina Strother, Major General John F. Thorlin, Alfred B. Wilkinson, and John McCartney

FOREWORD

by
General Donn A. Starry, U.S. Army (Retired)

While Richard Hunnicutt's 1971 book on the Pershing tank series and his 1978 book on the Sherman tank series were both widely acclaimed, this book on the Patton tank series will appeal to an even wider and more enthusiastic audience. For the Patton tank fleet, in all its variants, spans, and indeed overarches, the active military careers of all today's tank and cavalry soldiers in the U.S. Army.

As a newly commissioned lieutenant, I trained at Fort Knox in 1949 in the Pershing, and served a tour in a Pershing equipped unit in Germany. Towards the end of the tour, in 1952, our Pattons—M47s arrived to replace our tired M26s. Returning to Europe for duty in 1960 I found myself in a Patton equipped unit—M48A1s; then our M60s arrived, followed closely by M60A1s. In the short span of four years in the same unit we worked our way through two tank fleets and well into the third—all Pattons.

In Vietnam it was comforting to join a unit equipped with Pattons—M48A3s this time. It was further comforting to report, in 1973, to command Fort Knox knowing that while there were many new faces among the soldiers, the tried and true faces in the motor parks were Patton tanks—M60s, M60A1s, M60A2s and, later, M60A3s. Returning later to Germany to command V Corps, it was homecoming; for the tank parks were lined with Pattons, the tanks in which I'd served as a battalion commander in that same Corps. It was on my watch as Commander, Training and Doctrine Command that we began to usher in the Abrams tank fleet, training and deploying the first units to the active Army. And so as I left the Army's active ranks after thirty-five years of commissioned service, it was an Army which had been equipped with Pattons—in many variations, for virtually all of my active service.

In our own, and many other armies as well, Patton tanks will continue to do yeoman work until well on to the the turn of the century.

Sturdy and dependable, sometimes cantankerous and balky, but always most responsive to good handling and proper care and maintenance, and ever deadly in battle; Patton tanks taught the tank soldiers of my own and several succeeding generations as well, all we know about tanks. The lessons of those Patton tanks were ever present, indeed overriding, as we deliberated about and experimented with design and performance characteristics needed in the M1 Abrams tank.

As the career of General Patton himself has cast a long shadow over armor tactics, training, leadership and equipment design in our Army, so the tank that bears his distinguished name has, in its own way, taught generations of tank and cavalry soldiers in our army their important lessons about tactical maneuver, care and maintenance, and other critical elements of tank culture. In so doing it showed the way to design of a successor tank that promises to do all for following generations of tank soldiers what the Patton has done for the rest of us. Richard Hunnicutt's final judgment that "the Patton must be rated as one of the most important tank designs of all time," is, if anything, understated. It was and continues to be an old, tried and tested, faithful friend.

INTRODUCTION

The development of the medium and main battle tank in the United States after World War II followed two major paths. The first was the design and modification of the Patton tank and its product improved descendant, the M60 series. The second important line of development was devoted to the various experimental vehicles or concepts which were intended to replace the Patton. Obviously, there was considerable interaction between the two areas of work with many items developed for the new experimental vehicles being applied as product improvements to the production tanks.

The objective of this book is to trace the development history of the Patton and its various modifications. These include the M60 series as well as other vehicles based on the tank chassis or those using many components from that chassis. The completely new experimental tank designs such as the T95 and the MBT70 are mentioned only in regard to their effect on the Patton tank and its descendants. The new tanks resulting from the T95 and MBT70 projects were expected to be in service by the early 1960s and 1970s respectively. However, development problems, excessive costs, and changing user requirements delayed these programs and resulted in their eventual cancellation. In the first case, the product improved Patton, redesignated as the M60, replaced the T95 as the new main battle tank. Approximately ten years later Congress terminated the MBT70/XM803 program considering the new tank to be too expensive and complicated. Thus a further improved version of the M60 was required to soldier on until a new main battle tank could be developed to meet the Army's requirements. The third try was successful with the production of the new tank designated as the M1 and named after General Creighton Abrams. In addition to these major projects, numerous other concepts were studied during this period as possible replacements for the Patton. A detailed history of these parallel development programs will be covered in a future volume.

Space limitations permit only a brief outline of the Patton's operational history. This is restricted to important events such as the first use of the various vehicles by United States troops and their employment in operations of historical significance. It is also intended to illustrate the effect of operational experience on the modification of the existing tanks as well as the course of future development programs. A more complete coverage of the Patton's battle record may be found in the histories of the various conflicts in which it was employed. The operations of the M48A3 in Vietnam are described in the official publication "Mounted Combat in Vietnam" which is quoted in this volume. Another excellent illustrated history of these same actions can be found in Simon Dunstan's "Vietnam Tracks" also published by Presidio Press. These and other references are included in the bibliography.

The development of the Patton and the M60 series is not yet complete with numerous modifications still being made both in the United States and in foreign countries. Some of these are still classified and can not be included. Also, the basic chassis retains the potential for future upgrading in many areas which will enable it to maintain its position as a first line main battle tank for many years to come. The major emphasis in this book has been on the development and modification of these vehicles in the United States. Foreign modifications are briefly covered when they provide a significant improvement to the tank's effectiveness or when they influence future United States development. However no attempt is made to provide an exhaustive coverage of such foreign variants.

Since the potential for modification and improvement of these vehicles is far from exhausted, this book must be regarded as an interim history. Possible future developments in power trains, running gear, armament, armor, signature reduction and other protective measures will, no doubt, provide a wealth of material for yet another volume to complete the story.

R. P. Hunnicutt
Belmont, California
March 1984

PART I

THE TRANSITION PERIOD, WORLD WAR II TO KOREA

Following the end of World War II, military expenditures were drastically reduced in the United States as the country rapidly returned to a peacetime economy. The large tank contracts were cancelled and most of the later production vehicles were placed in storage for possible future use. The advent of the atomic bomb had changed the concept of future wars and most of the money available was allocated to the development of nuclear weapons and their delivery systems. The Air Force and the Navy competed for funds to support strategic weapons such as the B36 intercontinental bomber and the super aircraft carrier. With no strategic role in the nuclear weapons race, the Army's share of the budget was extremely small. No funds were available for new tank production and very little for research and development. These limited resources were allocated to component development and concept studies of future armored vehicles.

Planning for the postwar period started prior to the end of hostilities. On 2 January 1945, the Army Ground Forces (AGF) Equipment Review Board convened in Washington, D.C. to consider the requirements of the postwar Army. This study included all equipment used by the ground troops as well as any required by the Air Force for direct support of ground operations. The Board's report, dated 20 June 1945, recommended the development of light, medium, and heavy tanks with maximum weights of 25, 45, and 75 tons respectively as well as a 150 ton super heavy tank for experimental studies.

As proposed, the 45 ton medium tank had armor 8 inches thick on the front with 3 inches on the sides. The main armament was to be a 3 inch gun capable of penetrating 8 inches of homogeneous armor at 30 degrees obliquity at a range of 1000 yards. This weapon was to be stabilized in both azimuth and elevation and equipped with an optical range finder. Development also was requested for a radar range finder and a lead computing sight. The radar was to incorporate automatic identification equipment (IFF) similar to that used in aircraft. An automatic loader was specified for the main gun although the full five man crew was retained. The Board also noted that special power plants should be developed for armored vehicles with consideration given to multifuel engines and gas turbines.

Artist's concept of the 45 ton medium tank proposed by the Army Ground Forces Equipment Review Board. Note the coaxial .30 and .50 caliber machine guns as well as the .30 caliber blister guns on the turret. The bow machine guns were to be operated by remote control.

A few months after the original Equipment Review Board submitted its report, a far more senior board was appointed to review the postwar equipment situation. The new board consisted of one Brigadier General and three Lieutenant Generals with General Joseph W. Stilwell as president. Although officially designated as the War Department Equipment Review Board, it was usually referred to by the name of its president. Convened on 1 November 1945, the Stilwell Board submitted its report on 19 January 1946. As far as armored vehicles were concerned, the new report generally concurred with that submitted earlier by the AGF Board, although it did drop the request for the 150 ton super heavy tank study. It also recommended that the development of tank destroyers and towed antitank guns be terminated. Like its predecessor, the Stilwell Board emphasized the importance of component development. This included more powerful weapons designed specifically for tank use as well as special armor to defeat both shaped charge and kinetic energy projectiles. The report also recommended the development of special tank engines and improved running gear. Postwar tank development in the United States followed these recommendations until the appearance of the Army Equipment Development Guide in December 1950.

When production ended in October 1945, Chrysler (above) and General Motors (below) had produced a total of 2212 M26 Pershings including the ten T26E1 pilot tanks.

AN INTERIM TANK

At the end of hostilities in the Fall of 1945, the latest standard tank in the U.S. Army was the M26 General Pershing. Although it was the final production tank of the war, the M26 did not fully reflect the latest design concepts. Its greatest deficiency was the relatively low power of the Ford GAF engine. The 500 horsepower (hp) output which successfully drove the 35 ton Sherman was inadequate for the 45 ton Pershing. However, with more than 2000 M26s in the inventory and no funds for new production, the obvious source of supply for a postwar tank was a modernization program for the Pershing.

The new air-cooled V-12 gasoline engine then under development at Continental Motors Corporation offered the best answer to the power plant problem. Conceived as the "ideal" tank engine, development had been initiated by the wartime Tank Engine Committee. Cylinder design and general construction were based on the results of a development program using two 2-cylinder test engines.

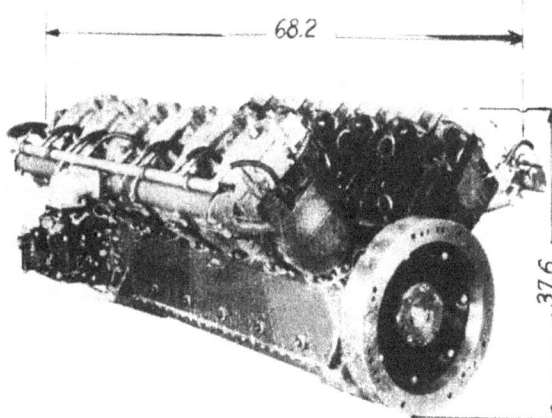

68.2

37.6

Although poor in quality, the two views above show the original AV-1790-1 engine without its fans and oil coolers. The fins on the individual air-cooled cylinders are clearly visible.

Three full size engines were authorized by Ordnance Committee action on 22 July 1943. They were designated as the AV-1790-1 reflecting the air-cooled V design with a displacement of approximately 1790 cubic inches. The characteristic sheet dated 11 February 1946 rated this engine at 740 gross horsepower at 2600 revolutions per minute (rpm). The first engine completed was used for preliminary engineering tests. Serial number 2 was installed in a modified M26 and number 3 was used for 500 hour dynamometer testing. The modified Pershing was designated as the M26E2 and the AV-1790-1 was installed with the new General Motors CD-850-1 cross drive transmission.

Referred to as the cross drive because of its transverse installation in the tank chassis, the CD-850-1 combined the functions of a transmission, a steering control, and vehicle brakes. As a transmission, it used two hydraulically selected gear ranges forward and one in reverse driving through a single phase hydraulic torque converter with part of the energy transmitted through a mechanical path bypassing the torque converter. The power from the mechanical and hydraulic paths was applied equally to both output shafts except when steering. For steering, all of the mechanical power was applied to one side providing the required speed difference between the tracks. The driver steered the tank using a single lever referred to as a wobble stick which operated a hydraulically actuated controlled differential. Moving the wobble stick to one side with the transmission in neutral caused the tracks to move opposite directions pivoting the tank in its own length. Built-in disc brakes were operated mechanically by a pedal in the drivers' compartment. Dual controls were provided allowing either the driver or the assistant driver to operate the vehicle.

The CD-850-1 transmission is shown below (top view) and at the left (side view).

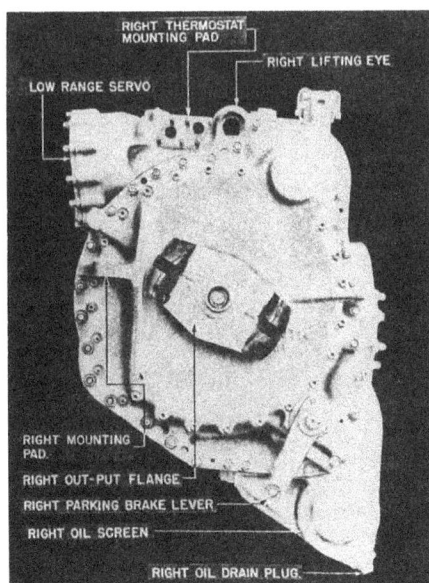

11

TOTAL PACKAGE WEIGHT
WITH COOLING SYSTEM—
8000 LBS.

25 IN.
CONTROLLED
DIFFERENTIAL

31.5 IN.
TORQMATIC
TRANSMISSION

45.5 IN.
GAF ENGINE

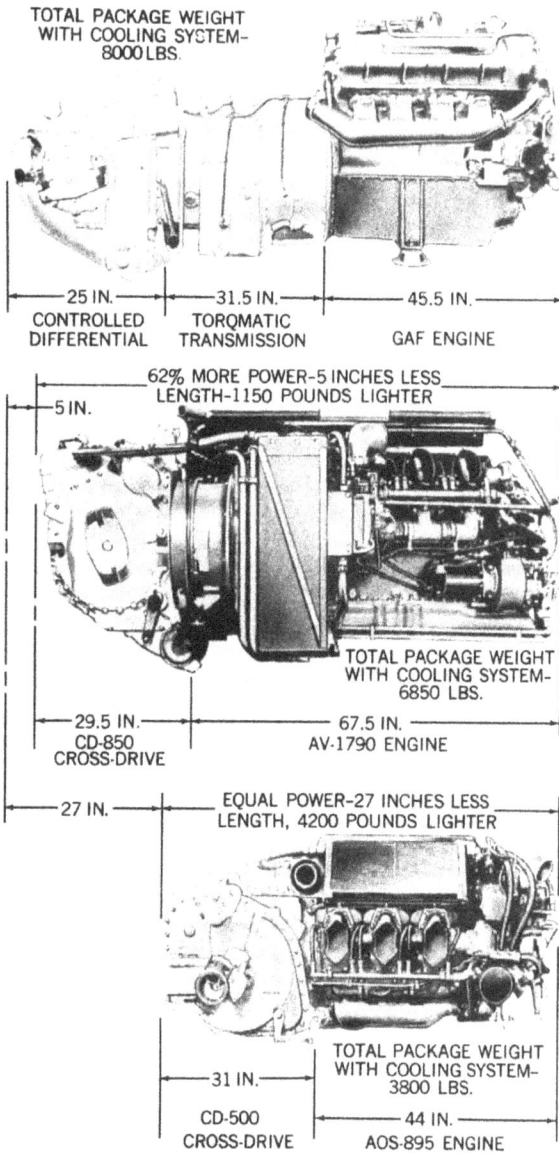

At the left, the new power package is compared with those from the Pershing (top) and the light tank T41E1 (bottom). The later model AV-1790-5B engine appears in this illustration.

62% MORE POWER—5 INCHES LESS LENGTH—1150 POUNDS LIGHTER

5 IN.

TOTAL PACKAGE WEIGHT
WITH COOLING SYSTEM—
6850 LBS.

29.5 IN.
CD-850
CROSS-DRIVE

67.5 IN.
AV-1790 ENGINE

27 IN.

EQUAL POWER—27 INCHES LESS LENGTH, 4200 POUNDS LIGHTER

TOTAL PACKAGE WEIGHT
WITH COOLING SYSTEM—
3800 LBS.

31 IN.
CD-500
CROSS-DRIVE

44 IN.
AOS-895 ENGINE

The combination of the AV-1790-1 and the CD-850-1 resulted in a power package shorter in overall length than the original Ford GAF engine, torqmatic transmission, and controlled differential in the M26. After testing at Detroit, the M26E2 was shipped to Aberdeen Proving Ground, Maryland, arriving in May 1948.

The original plans for a modernized Pershing included improved armament such as the 90mm gun T54 tested in the M26E1. Later tests, also in the M26E1, replaced the 90mm gun with the high powered 3 inch gun T98. This weapon was identical ballistically with the Navy's 3 inch 70 caliber anti-aircraft gun. However, plans for more powerful armament were cancelled when the specifications for the modernized M26 were revised at a meeting in the Spring of 1948 between the Ordnance Department and the Army Field Forces. The final decision was to retain the 90mm gun M3 with the addition of a bore evacuator. The modified gun was redesignated as the 90mm gun M3A1 and a single baffle muzzle brake replaced the double baffle model on the earlier weapon. The fire control system was improved by the installation of a M83 telescope with two moveable reticles in a cant correcting mount.

Below, the 3 inch gun T98 in mount T136 is installed in the M26E1 tank during tests to select armament for the modernized medium tank.

USA 3012ro82

This is T40 serial number 4 (above and below) during tests at Fort Knox. The machine guns have not been mounted in this vehicle. Note the square transmission inspection holes in the rear armor.

Ten tanks were authorized for construction using fiscal year 1948 funds under the designation medium tank T40. These vehicles were powered by the AV-1790-3 engine using the CD-850-1 transmission at first and later the CD-850-2.

The AV-1790-3 was an improved version of the earlier AV-1790-1. Design changes in the camshafts and manifolds increased the power output of the AV-1790-3 to 810 gross horsepower at 2800 rpm. Ignition on the later engines was provided by four magnetos with two spark plugs fitted in each cylinder. Two magnetos were used for each bank of six cylinders, each firing one set of spark plugs. The AV-1790-1 also differed from the later engines in that it was a dry sump model and used battery ignition instead of the magnetos.

The CD-850-2 cross drive transmission incorporated a polyphase hydraulic torque converter replacing the single phase unit in the CD-850-1. The later model torque converter provided better efficiency at high speeds by operating as a fluid coupling after the point of 1:1 torque multiplication was reached.

The new power package required design changes in the top and rear of the M26 tank hull. Armor grills permitted cooling air to be sucked into the engine compartment along the sides of the top deck. The air passed over the fuel tanks, up between the engine cylinders, through the fans, and was ejected from the vehicle through the center grills. Engine exhaust pipes extended sideways from the top center of the rear deck to mufflers mounted on each fender. Three square armor covered holes in the lower part of the vertical rear hull plate provided access to the cross drive transmission.

On the T40, the suspension system was modified to include a small track tension idler between the rear road wheel and the sprocket on each track. Sprung by a small torsion bar, it was intended to maintain track tension during turns and when maneuvering over rough terrain thus reducing track throwing tendencies.

T40 serial number 2 arrived at Aberdeen on 2 August 1949 for engineering tests. It was followed on 8 September by serial number 3. The latter was loaded to a simulated combat weight of 95,800

pounds and subjected to an extended endurance test on the various courses at the Proving Ground. Nine of the ten T40 tanks were used in test programs at Aberdeen, Fort Knox, and Detroit Arsenal. The tenth tank, registration number 30162865, was retained at Detroit for conversion into the armored engineer vehicle T39. Eight of the nine test vehicles were later returned to the Arsenal for conversion into medium tanks M46A1. The remaining tank was converted to simulate the weight of the medium tank T42 and used to evaluate the 500 hp engine and transmission for that vehicle. It was later utilized by the Cadillac Division of General Motors during the further development of the power package for the light tank T41E1.

Item 32312 of the Ordnance Technical Committee Minutes (OCM or later OTCM) dated 30 July 1948 standardized the modernized Pershing as the medium tank M46. This same action reclassified the M26 as Limited Standard. It was originally intended that the Pershings would be declared obsolete, but since those tanks in the hands of the troops would have to be maintained with an adequate supply of spare parts, the reclassification was changed to Limited Standard. However, any further issue of the M26 to the troops required authorization by the Department of the Army. An initial production run of 800 M46s was included in the 1949 budget and it was estimated that 1215 additional M26s were available for conversion during the 1950 fiscal year. All of these plans were to be modified by unexpected events in the Far East. OCM 32312 also nicknamed the M46 the General Patton in honor of General George S. Patton, Jr.

T40 serial number 3 is shown here at Aberdeen Proving Ground on 3 October 1949. The instruments mounted near the gun travel lock are part of the test equipment. In the view below, the armor grills are open revealing the engine and transmission.

14

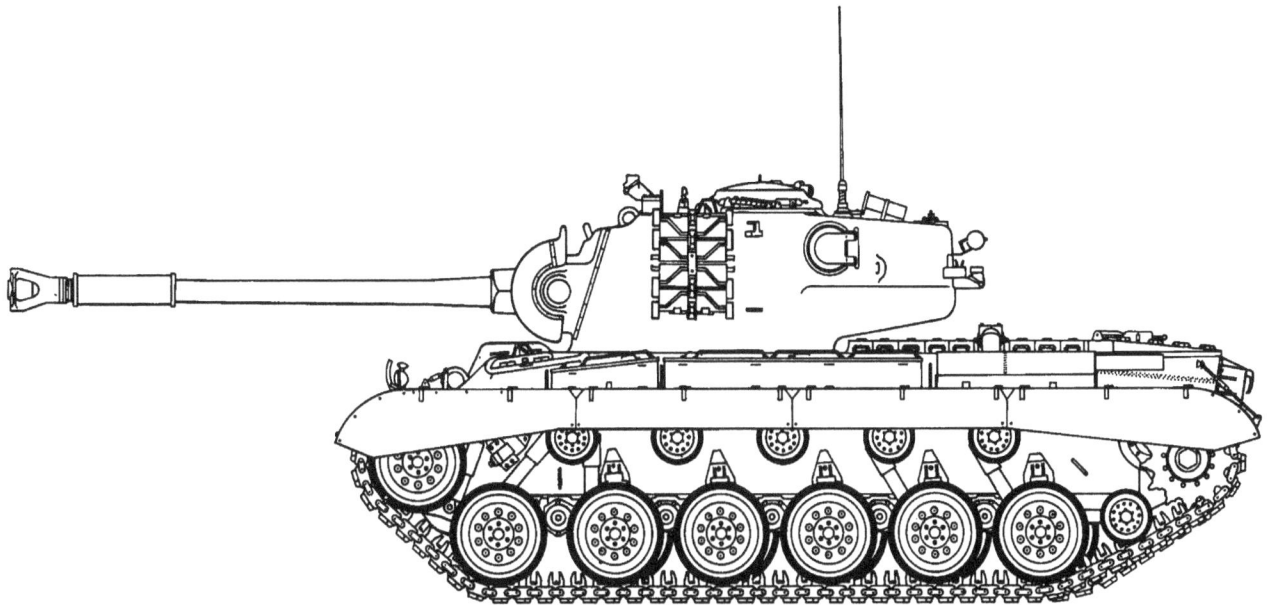

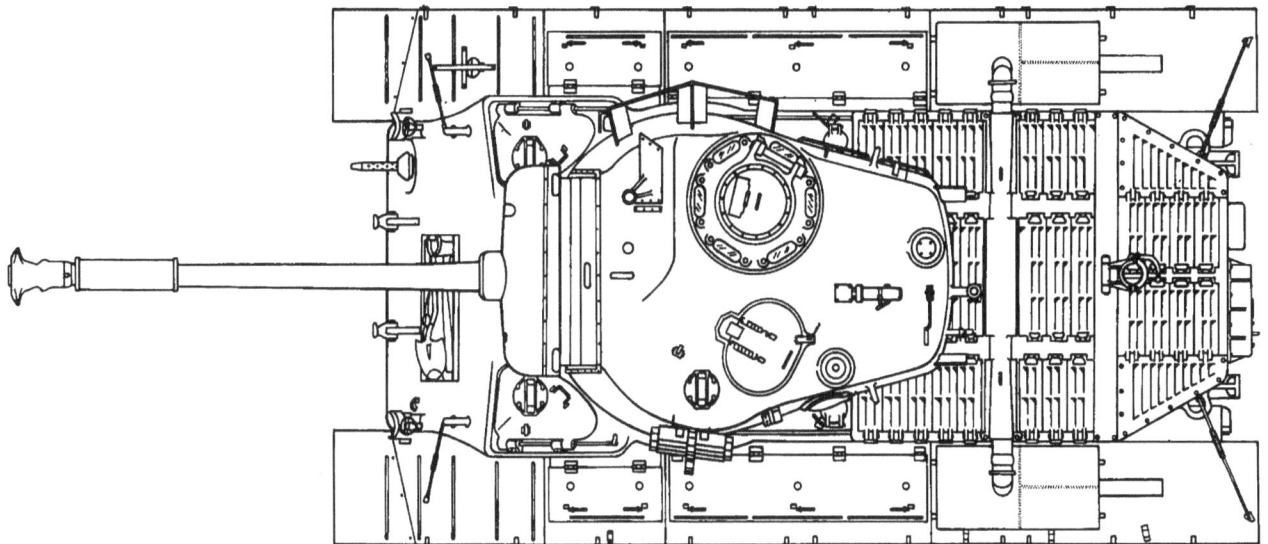

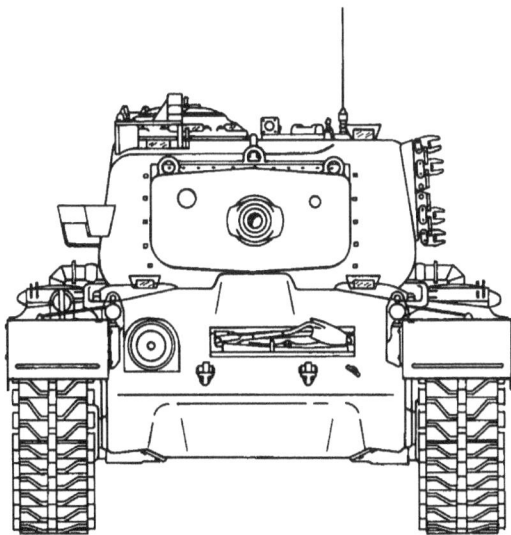

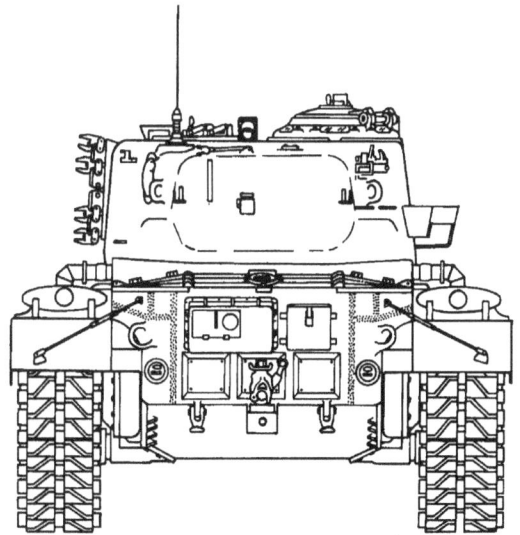

Scale 1:48

Medium Tank T40

The first of the production M46 tanks arrived at Aberdeen Proving Ground in early November 1949 and immediately entered the endurance testing program. Like the T40, the M46 retained the turret and forward hull structure of the M26 and was armed with the 90mm gun M3A1 fitted with a bore evacuator and a single baffle muzzle brake. Both the T40 and the M46 could be easily identified from the M26 and M26A1 by the small track tension idlers between the last road wheels and sprockets as well as the engine mufflers on each rear fender. The production M46 differed from the ten T40s by having three circular access hole covers in the vertical rear armor compared to the square covers on the earlier tank. The production M46 was powered by the Continental AV-1790-5 engine, later designated as the AV-1790-5A after the appearance of the slightly modified AV-1790-5B. The production model cross drive transmission was designated as the CD-850-3.

Below is M46 serial number 31, also at Aberdeen. This photograph was dated 20 February 1950. None of these early tanks were fitted with muffler shields.

The tank shown here is representative of the later M46 conversions. It does not have the stowage rack on the right side of the turret and it is fitted with shields over the exhaust mufflers. Note the designations; adjusting idler wheel and compensating idler wheel. Later this terminology was changed to compensating idler wheel and track tension idler wheel respectively.

With further modifications, it became the CD-850-4 in the later production tanks. The CD-850-4 was similar to the CD-850-3 except that a split hydraulic power path replaced the hydraulic-mechanical power path in the earlier transmission. The final drive reduction gear ratio was changed from 3.95:1 to 4.47:1 in the later vehicles.

After the outbreak of the Korean War in June 1950, drastic changes occurred in the tank production plans. Although the M46 program was in full gear, the M26s available for conversion would not begin to meet the Army's tank requirements. Many Pershings in the hands of the troops were rushed to Korea to help stem the enemy advance, further reducing the numbers available. It was not desirable to initiate new production of the M46 since many of its features were considered obsolete. However, the need for immediate production did not permit sufficient time to develop a completely new design. To solve the problem, an emergency program designed a new production tank combining the proven hull and chassis of the M46 with the new turret and armament of the experimental tank T42.

By this time, proving ground tests and troop use had introduced numerous modifications to the M46. These changes were applied to the late production M46 as well as to the new tank design which was now designated as the medium tank M47. They included a better oil cooling system, a new instrument panel and hull wiring, improved brakes, and a new fire extinguisher system. The power package consisting of the AV-1790-5B engine and the CD-850-4 transmission was interchangeable between the M46(New) and the M47. With 360 of the M46(New) tanks scheduled for production by 1 April 1951, they were redesignated as the medium tank M46A1 to prevent confusion between the two versions at some future date.

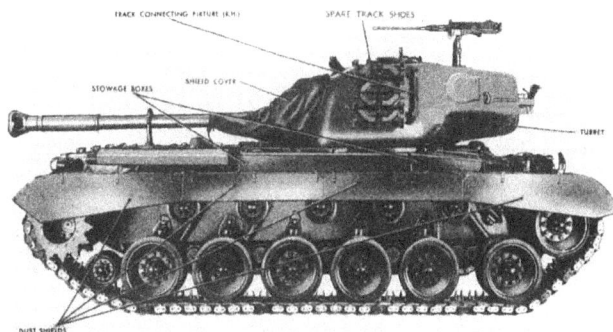

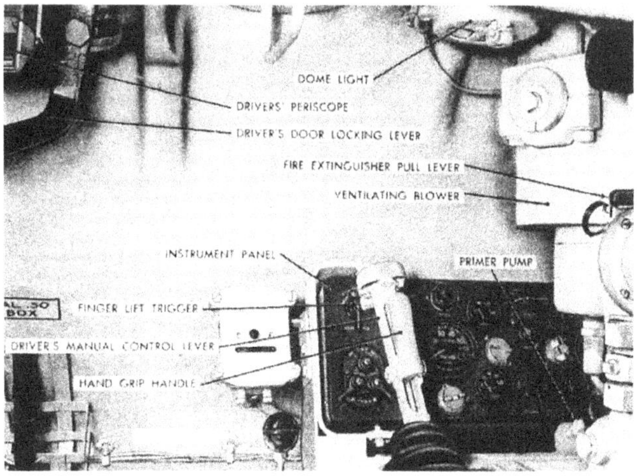

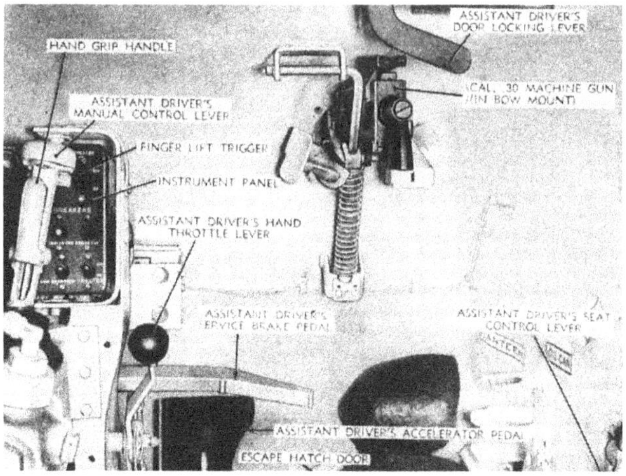

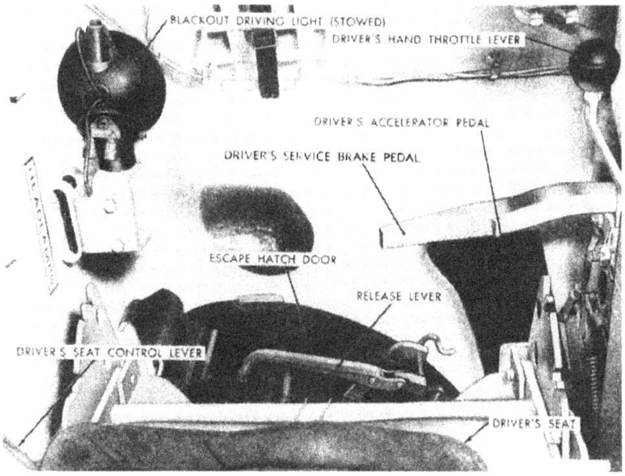

The driver's controls appear at the left and top left with the assistant driver's station directly above. Each is provided with a wobble stick manual control lever and the other necessary controls to operate the tank.

The two views at the lower left compare the drivers' instrument panels in the M46 (upper) and M46A1 (lower). The transmission linkage can be seen below with a view of driver's hatch door at the bottom.

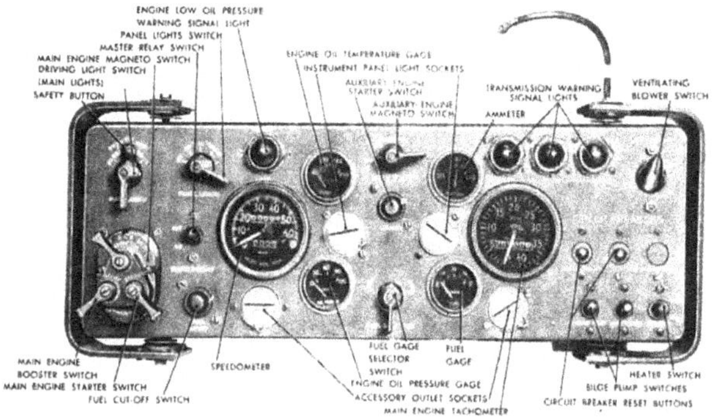

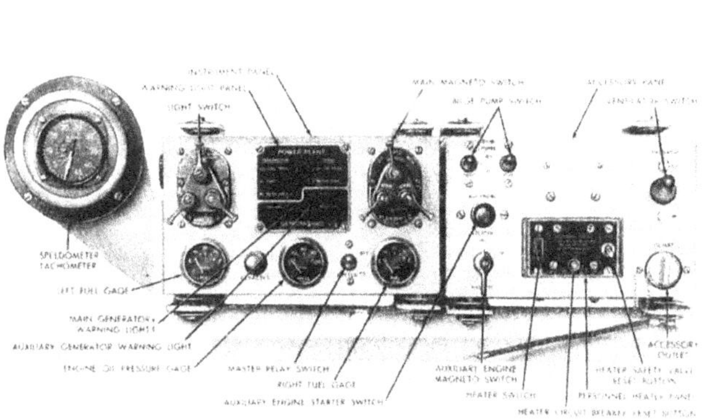

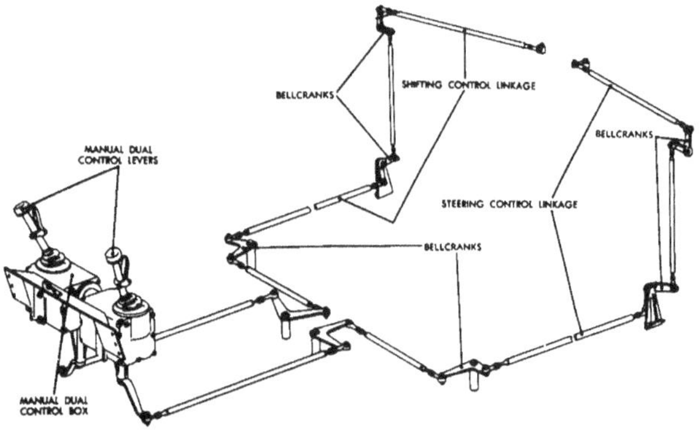

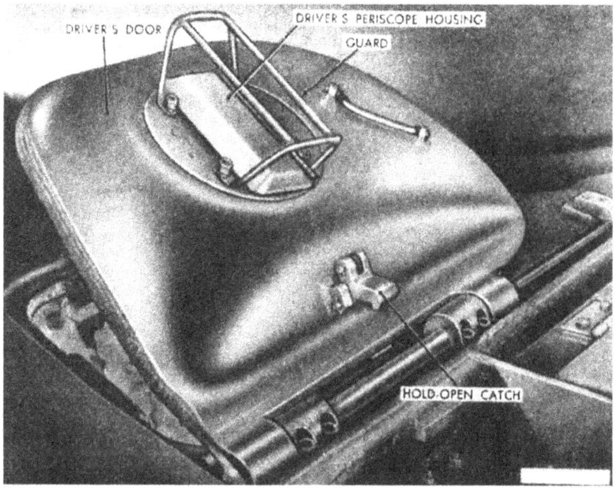

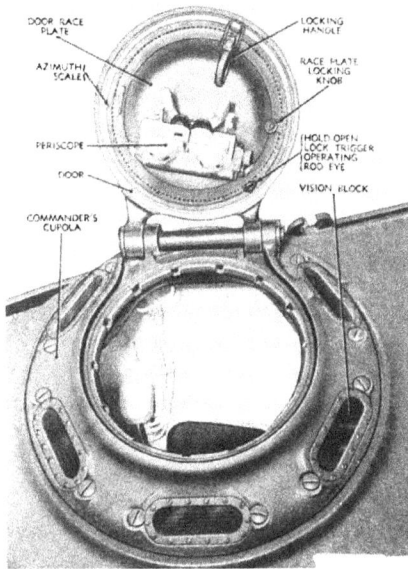

DOOR RACE PLATE
AZIMUTH SCALE
PERISCOPE
DOOR
COMMANDER'S CUPOLA
LOCKING HANDLE
RACE PLATE LOCKING KNOB
HOLD OPEN LOCK TRIGGER OPERATING ROD EYE
VISION BLOCK

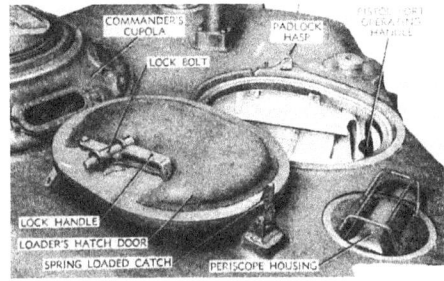

COMMANDER'S CUPOLA
LOCK BOLT
PADLOCK HASP
PISTOL PORT OPERATING HANDLE
LOCK HANDLE
LOADER'S HATCH DOOR
SPRING LOADED CATCH
PERISCOPE HOUSING

At the left is the turret vision cupola (upper) and the seats for the commander and gunner (lower). The empty side ammunition racks are folded up. Above is the loader's hatch locked in the open position. A plan view of the turret is below.

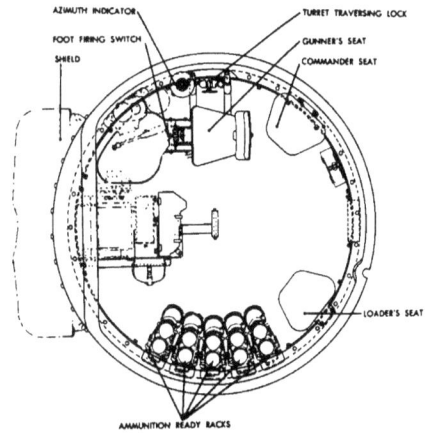

AZIMUTH INDICATOR
FOOT FIRING SWITCH
SHIELD
TURRET TRAVERSING LOCK
GUNNER'S SEAT
COMMANDER SEAT
LOADER'S SEAT
AMMUNITION READY RACKS

TURRET SWITCH BOX
COMMANDER'S SEAT
GUNNER'S SEAT
COMMANDER'S SEAT CONTROL HANDLE
AMMUNITION RACKS (FOLDED UP)

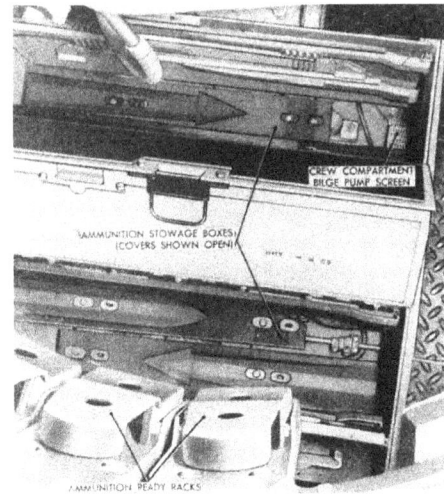

CREW COMPARTMENT BILGE PUMP SCREEN
AMMUNITION STOWAGE BOXES (COVERS SHOWN OPEN)
AMMUNITION READY RACKS

Above, the empty ammunition ready racks are stowed in the folded position. The gunner's controls can be seen at the left and below.

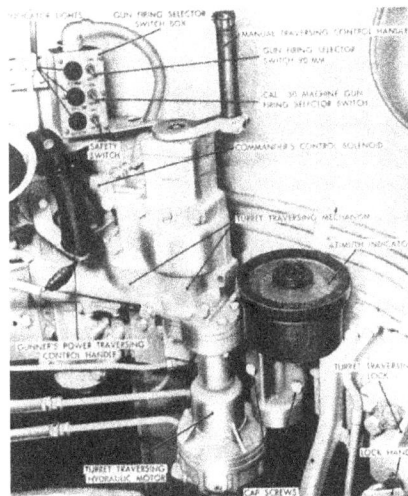

INDICATOR LIGHTS
GUN FIRING SELECTOR SWITCH BOX
MANUAL TRAVERSING CONTROL HANDLE
GUN FIRING SELECTOR SWITCH 90 MM
CAL .50 MACHINE GUN FIRING SELECTOR SWITCH
SAFETY SWITCH
COMMANDER'S CONTROL SOLENOID
TURRET TRAVERSING MECHANISM
AZIMUTH INDICATOR
GUNNER'S POWER TRAVERSING CONTROL HANDLE
TURRET TRAVERSING LOCK
TURRET TRAVERSING HYDRAULIC MOTOR
LOCK HANDLE
CAP SCREWS

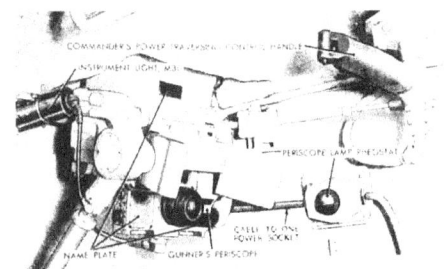

COMMANDER'S POWER TRAVERSING CONTROL HANDLE
INSTRUMENT LIGHT M33
PERISCOPE LAMP RHEOSTAT
CASES HOLDING POWER SOCKET
NAME PLATE
GUNNER'S PERISCOPE

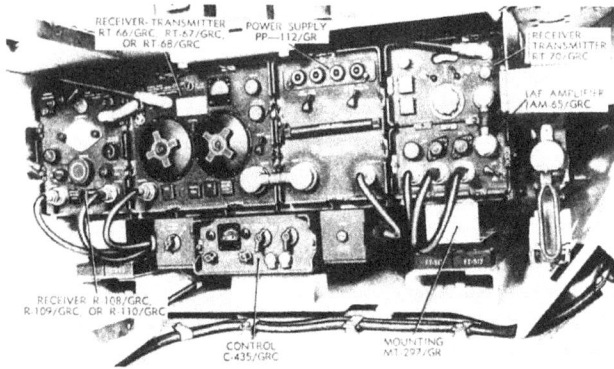

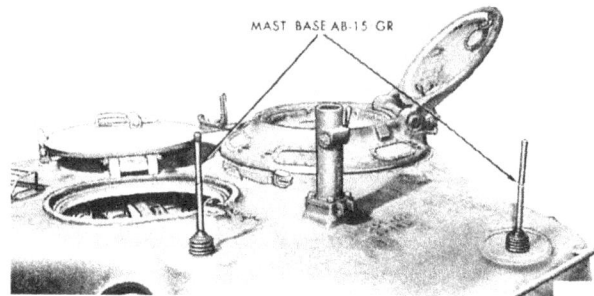

The installation of the radio set AN/GRC-3,-4,-5,-6,-7, or -8 appears at the left above. At the right above are the mast bases for the radio antennas on the turret roof. Below are alternate radio sets SCR-528 (left) and AN/VRC-3 (right). Note the inside view of the closed pistol port in the latter photograph.

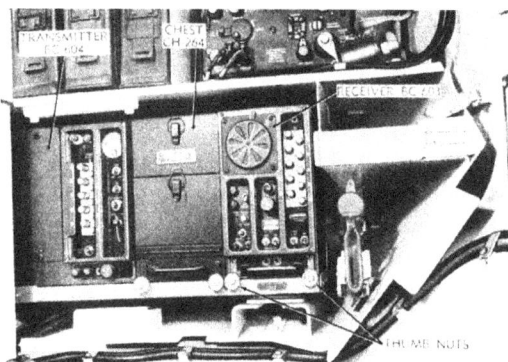

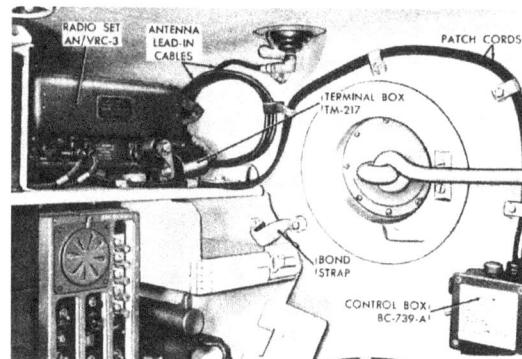

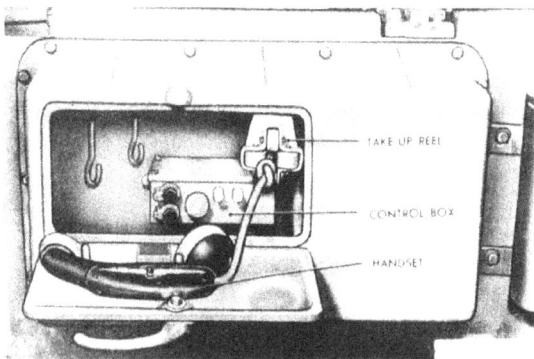

The external interphone control box on the rear hull plate is shown above open and closed. Below are views of the hull floor 90mm ammunition boxes open and closed. Note the hull side wall ammunition racks. Both views are looking toward the rear of the tank.

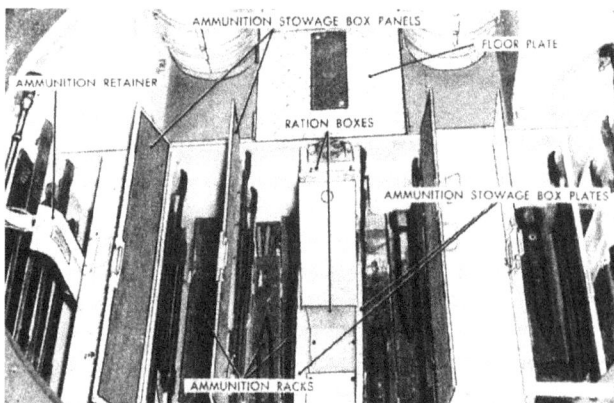

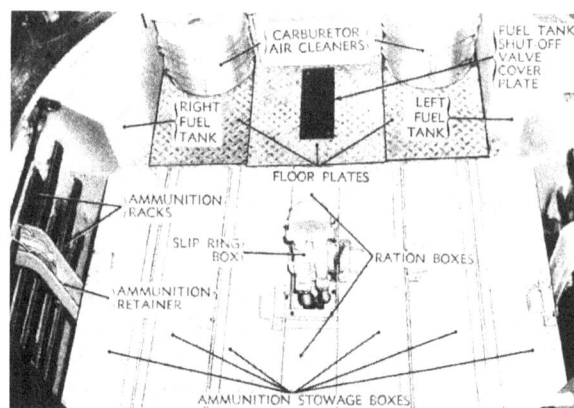

20

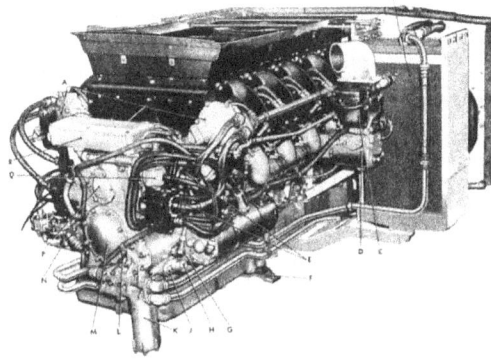

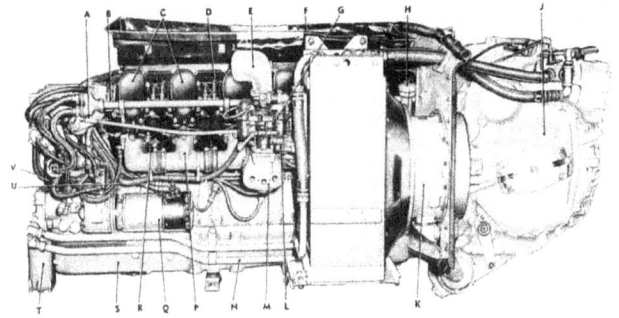

A—Camshaft drive gear housing covers.	J—Oil filter by-pass valve.
B—Oil cooler core.	K—Engine oil filter.
C—Carburetor.	L—Oil cooler by-pass valve.
D—Hot-spot manifold.	M—Accessory drive gear cover.
E—Right magnetos.	N—Main fuel line quick disconnect.
F—Engine mounting bracket.	P—Accessory driven gear cover.
G—Engine oil temperature gage sending unit.	Q—Tachometer sending unit.
H—Oil pressure control valve.	R—Engine wiring junction box.

A—Throttle linkage.	L—Carburetor degasser.
B—Ignition harness.	M—Hot spot manifold.
C—Valve covers.	N—Engine oil cooler lines.
D—Engine (right cylinder bank).	P—Intake manifold.
E—Carburetor inlet elbow.	Q—Generator.
F—Engine top shroud.	R—Cylinder oil drain lines.
G—Carburetor.	S—Oil pan.
H—Engine oil filler cap.	T—Engine oil filter.
J—Transmission (left side).	U—Fuel pump to carburetor fuel line.
K—Oil cooler fan housing.	V—Right—lower magneto.

The AV-1790-5A engine of the M46 above can be compared with the AV-1790-5B in the M46A1 below. The diagonal stiffener on the oil cooler housing of the AV-1790-5B is an easy point of identification.

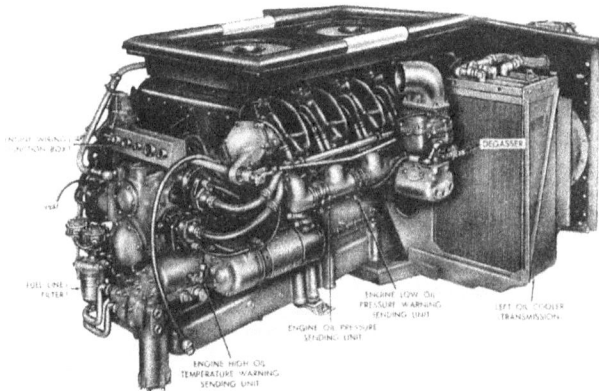

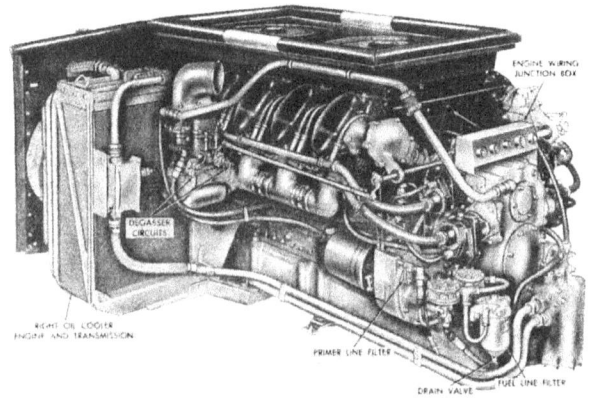

At the left is a rear view of the CD-850-3 cross drive transmission installed in the early M46s. Below is the CD-850-4 fitted to the late M46s and M46A1s.

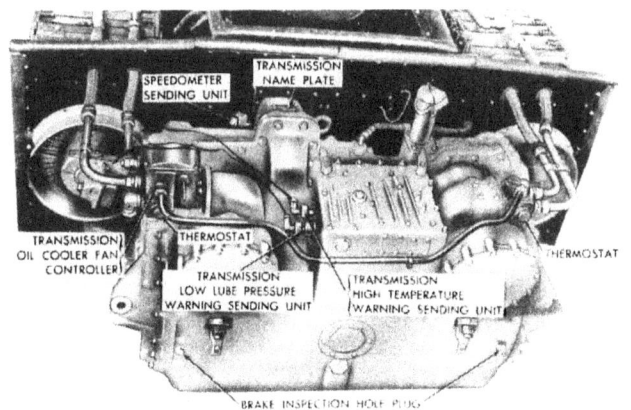

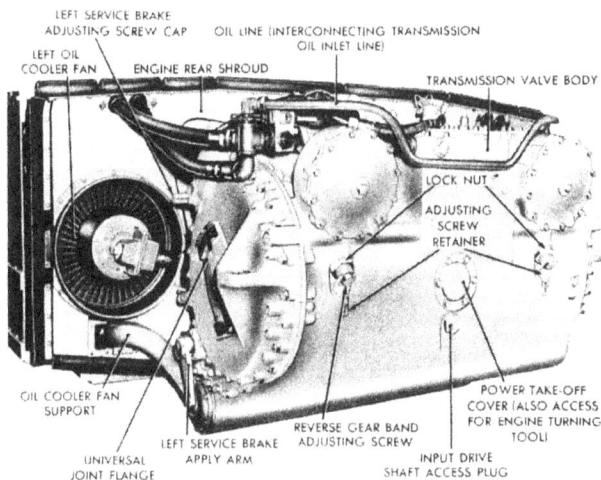

21

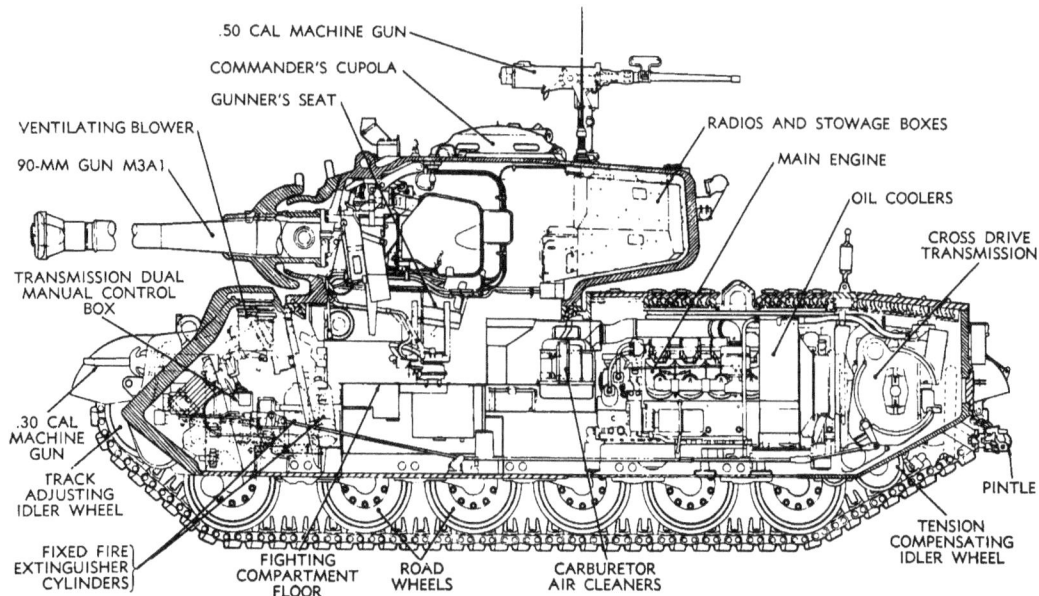

Labels in top cutaway diagram:
.50 CAL MACHINE GUN
COMMANDER'S CUPOLA
GUNNER'S SEAT
VENTILATING BLOWER
90-MM GUN M3A1
RADIOS AND STOWAGE BOXES
MAIN ENGINE
OIL COOLERS
CROSS DRIVE TRANSMISSION
TRANSMISSION DUAL MANUAL CONTROL BOX
.30 CAL MACHINE GUN
TRACK ADJUSTING IDLER WHEEL
FIXED FIRE EXTINGUISHER CYLINDERS
FIGHTING COMPARTMENT FLOOR
ROAD WHEELS
CARBURETOR AIR CLEANERS
PINTLE
TENSION COMPENSATING IDLER WHEEL

The compact power package obtainable with the AV-1790 engine and the CD-850 cross drive transmission is obvious in the sectional view of the M46 above.

A—Commander's cupola.
B—Center battery cover.
C—Exhaust pipe housing.
D—Lifting loops.
E—Engine compartment rear cross beam.
F—Gun traveling lock.
G—Rear grille door support.
H—Speedometer sending unit.
J—Exhaust pipe clamp.
K—Left side batteries.
L—Air cleaner pipe.
M—Center battery cover hold-down clamp.

A—Engine compartment rear cross beam.
B—Oil cooler tie-rods.
C—Engine cooling fan vanes.
D—Engine cooling fans.
E—Engine top shroud.
F—Engine top shroud air shield.
G—Engine oil filler cap.
H—Engine and transmission oil coolers.
J—Engine top should shroud plates.
K—Left side batteries.
L—Fuel gage sending units.
M—Lifting eyes.
N—Auxiliary engine and generator.

The engine compartment is shown above with the grill doors open (left) and with the front and intermediate deck sections removed (right). Below, details of the suspension and tracks can be seen at the left and at the right is a view of the tank hull from the bottom rear.

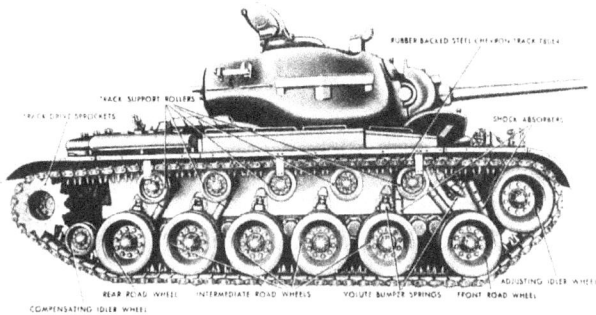
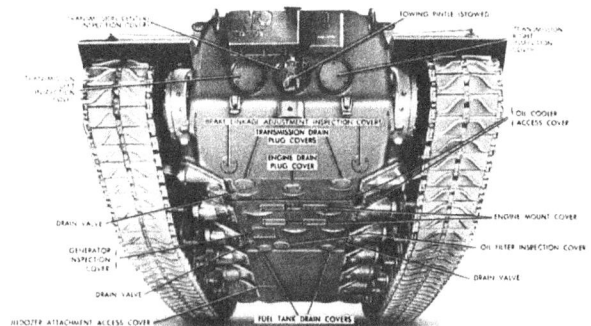

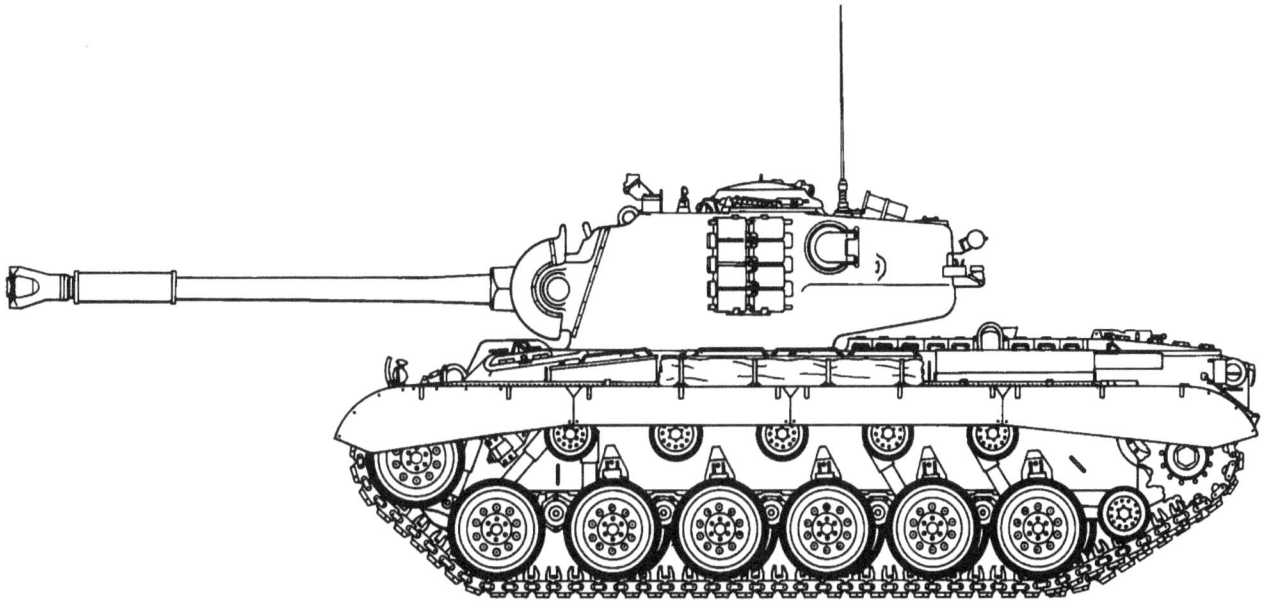

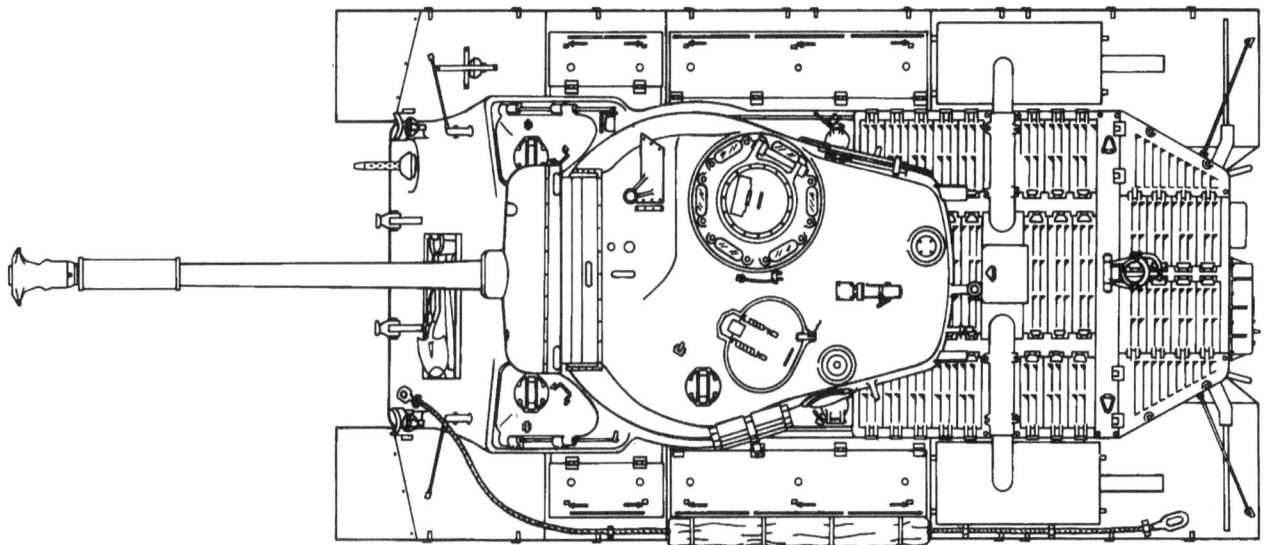

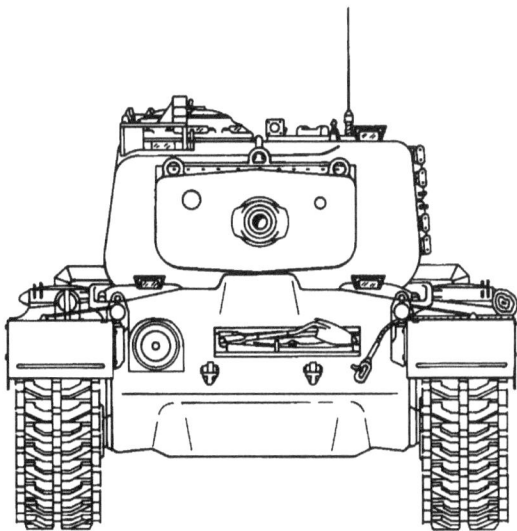

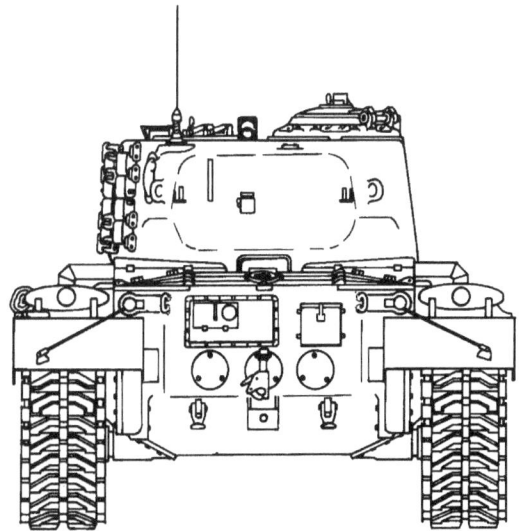

Medium Tank M46

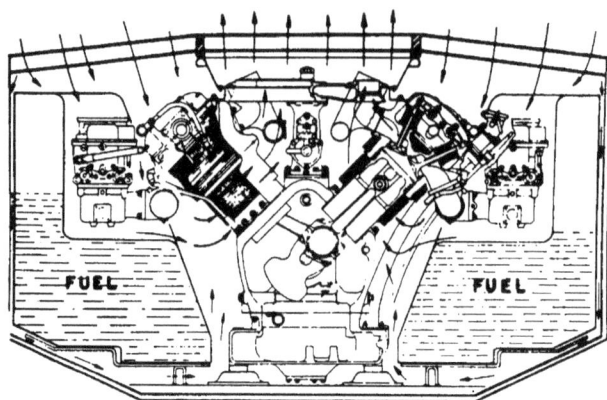

The M46 at the left was photographed during winter tests near Devil's Lake, North Dakota in January 1951. The grill covers were installed to restrict the cooling air flow during cold weather operations. The drawing above shows the normal air flow around the AV-1790 engine.

The tanks at the right and below are typical of the later production M46. Note the shields over the exhaust mufflers. Both tanks are probably M46A1s, but there are no external points of identification.

Above, an M46 is unloaded from the S.S. Coe Victory at a South Korean port on 12 August 1950. Note the early version of the exhaust muffler shield on this tank.

When the North Korean Army pushed south across the 38th parallel in late June 1950, their Soviet built T34/85 tanks rapidly overran both the South Korean Army and the United Nations Forces sent to bolster the defense. Desperate measures to strengthen the retreating troops included the shipment of old tanks rebuilt in Japan as well as new tank battalions rushed over from the United States. Among the new units unloading at Pusan, Korea on 8 August 1950 was the 6th Tank Battalion equipped with the M46 General Patton. These were the first of the new tanks to go into battle, fighting alongside the old Shermans and Pershings. By the end of 1950, many M46s were arriving in Korea. Some of these were lost, still on their railway flatcars, during the retreat from the north after the Chinese entered the war.

The aerial photograph above shows four Patton tanks abandoned on their railway cars near P'yongyang, Korea during the retreat from the north. They were subsequently destroyed by the U.S. Air Force.

At the right, tank crews of C Company, 64th Tank Battalion check their M46s before moving into action on 12 February 1951 near Anyang-ni, Korea.

Above, the crew poses with their tiger marked M46 early in 1951. At the left, tank 63 from C Company, 64th Tank Battalion has been knocked out by a U.S. rocket captured by the Chinese. The bottom photograph shows the hole in the side armor which penetrated and ignited a fuel tank. This action occurred in early June 1951.

The high power to weight ratio of the Patton compared to the Pershing was a great advantage in the mountainous terrain of Korea and the great agility provided by the cross drive transmission was extremely valuable. During the breakthrough to Chipyong-ni on 15 February 1951, Task Force Crombez, led by Colonel Marcel G. Crombez, placed its Pattons in the lead. These tanks from D Company, 6th Tank Battalion made effective use of their 90mm guns and the pivot capability of the M46 allowed them to turn completely around in the narrow defiles along the attack route.

Reports from Korea indicated the need for battlefield illumination to permit tank operations at night. The first attempt to solve the problem considered an updated version of a World War II design which mounted a searchlight inside a special turret on the medium tank M4A3. Designated as the

An M46 from the 1st Marine Division is being used above as artillery on 23 November 1951. The tank is positioned on a slope to obtain elevations greater than the 20 degrees available in the gun mount. Below, the commander and loader of another marine tank stand in their hatches during operations in January 1952.

Above a 1st Marine Division M46 is engaged in "bunker busting" during the Korean winter of 1951-52.

Below, another M46 carries its load of marines across the Pukchon River during training exercises in Korea.

Above, a 1st Marine Division tank in Korea is fitted with the 18 inch diameter searchlight. This photograph was dated 27 August 1952. At the right, another marine tank is equipped with the searchlight during operations near Chang-Dan, Korea on 1 October 1952.

searchlight tank T52, it was armed with a lightweight 75mm gun coaxial with the searchlight. Production plans for the special searchlight tank were scrapped after tests at Fort Knox indicated that unarmored 18 inch searchlights could be used effectively when mounted above the cannon on standard tanks. In the dark such searchlights were extremely difficult to hit and the cost of an occasional loss was insignificant compared to the expense of producing a special searchlight tank. The 18 inch searchlights also could be installed on any standard tank providing great operational flexibility.

The M46 also was fitted with an infrared searchlight and suitable viewers for night operations. This installation (at right) was referred to as the Leaflet II continuing the nomenclature originally used for the World War CDL project.

Above, two Pattons await orders to move up to positions supporting marine infantry in Korea on 20 November 1952. A tiger marked M46 (at left) repaired by the 2nd Ordnance Maintenance Company is taken for a test run in Korea on 10 May 1952.

As the war continued, increasing numbers of Pattons were introduced into both the Marine Corps and Army units fighting in Korea. Many of these would remain after the armistice to guard the uneasy truce. Since the terms of the armistice agreement restricted the introduction of new weapons into Korea, the Patton remained in service here far longer than elsewhere. On 14 February 1957, OTCM 36468 included the M46 and M46A1 on a list of vehicles reclassified as obsolete. However, it also specified that U.S. forces in Korea would continue to use those tanks on hand and that spare parts sufficient to maintain them for at least five years would be retained in Korea or Japan. It is interesting to note that on the same list of vehicles declared obsolete was the 76mm gun M4A3 medium tank with the horizontal volute spring suspension, thus finally ending the Sherman's 15 year career in the U.S. Army.

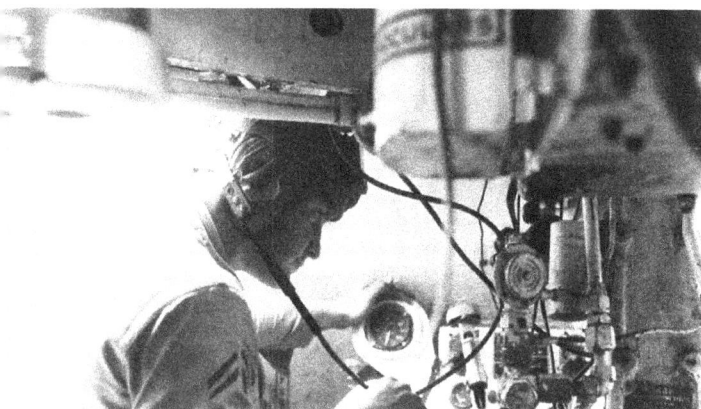

The driver's compartment in the Patton is shown at the left. Note the instrument panel characteristic of the M46A1 in this photograph.

Above, an M46 tankdozer from D Company, 1st Marine Tank Battalion is fitted with a screen constructed from heavy wire fencing. This screen was intended to protect the tank from rocket launched antitank weapons.

Below, a marine Patton is in action supporting the Turkish Brigade attached to the U.S. Army 25th Division near Chongdong, Korea on 5 July 1953. Note the basket installed on the turret to salvage the empty brass cartridge cases from the 90mm rounds.

The light tank T37 (above) was armed with the 76mm gun T94. With a muzzle velocity of 2600 feet/second, its performance was similar to the 76mm gun in the World War II Sherman tank. This weapon was replaced by the higher velocity 76mm gun T91 in the later light tank T41 (bottom of page). Turret mounted .30 caliber blister machine guns were installed on both tanks.

NEW DEVELOPMENT

In response to the recommendations of the Stilwell Board, a design for a new light tank was initiated in July 1946. Intended as a long range project, the new vehicle was to take advantage of the lessons learned during World War II and to incorporate the latest ideas in tank design. With only limited funds available, the light vehicle was the only tank development recommendation of the Stilwell Board implemented at that time. Designated as the light tank T37, the project moved slowly and by September 1948 it was modified to carry a more powerful gun and a new fire control system. The modified vehicle was designated as the light tank T41.

By the Summer of 1948 the world situation had grown increasingly dangerous. In June, Berlin was isolated by the blockade of all surface transportation routes forcing the western powers to supply the city by air. In the Far East, the rapidly deteriorating situation in China foreshadowed the collapse of the Nationalist government the following Spring. Faced with these conditions, the program had been started to modernize the inventory of Pershings. However, their numbers were limited and they were based on a design at least five years old. A new medium tank project was obviously overdue. The progress in the development of the T37 indicated that the basic design might be applied to a new medium tank. The interior dimensions and power pack of the light tank would be retained, but the new vehicle would have heavier armor and a 90mm gun in a new turret.

The T42 mock-up is shown above after modification. The .30 caliber blister machine guns have been removed from the turret. Note the large housing for the early T37 range finder originally considered for installation. At the bottom of the page is the 90mm gun T119 selected for the T42.

A conference at Detroit Arsenal on 21 September 1948 outlined the military characteristics of the new medium tank. These original specifications were published in OTCM 32529 dated 2 December 1948 which assigned the designation medium tank T42. The OTCM called for a tank weighing approximately 36 tons with better armor protection than the medium tank M46 and equivalent armament. The 90mm gun was to be stabilized in elevation and the turret in azimuth with a provision for an automatic loader and a stereoscopic range finder. A concentric recoil system for the cannon conserved turret space by replacing the usual recoil cylinders with a single hollow cylinder surrounding the gun barrel. The main gun mount also carried a direct sight telescope as well as a .50 caliber coaxial machine gun. A .30 caliber machine gun was mounted in a blister on each side of the turret. Like the T37, the inside diameter of the turret ring was 69 inches continuing the use of that ring size on medium tanks which dated from the introduction of the Sherman in 1941.

Approval was given for the construction of a T42 wooden mock-up in March 1949. Review of this

mock-up at meetings in October and December 1949 resulted in changes and a further revision of the specifications. The ground contact length was increased from 122 inches to 127 inches and the tread was narrowed from 115 inches to 111 inches. The turret ring diameter was enlarged to 73 inches and the turret mounted blister machine guns were eliminated. The direct sight telescope was deleted in favor of a gunner's periscope as an auxiliary sighting device. The stereoscopic range finder was retained as the primary sighting equipment for the gunner, but the lead computer and panoramic sight were dropped.

The British liaison officer objected to the original specification of a 90mm gun equivalent to that in the M46. Such a gun was considered inadequate for a new medium tank. He pointed out that the British 20 pounder was a much more powerful weapon. However, when finally approved, the military characteristics were revised specifying the 90mm gun T119. This new weapon was required to penetrate 11.1 inches of homogeneous armor at 30 degrees obliquity at 1000 yards range using armor piercing discarding sabot (APDS) ammunition. The

The difference between the T24 cartridge case on the 3000 feet/second AP-T M318A1(T33E7) at the left and the M19 case on the M71 high explosive round above is not obvious. However, the shoulder on the T24 is about 0.55 inches further forward than that on the M19 preventing it from chambering in the earlier 90mm guns.

chamber volume of the T119 gun was the same as that of the 90mm gun M3A1, but it was designed for a maximum pressure of 47,000 pounds per square inch (psi) compared to 38,000 psi for the earlier weapon. A muzzle velocity of 3000 feet per second (ft/sec) was attained with the 24 pound armor piercing shot AP-T T33E7. Although the T119 could safely fire all rounds used in the M3A1 gun, the reverse was not true. To prevent accidental use of the high pressure rounds in the earlier gun, the shoulder of the new T24 cartridge case was slightly longer (had less taper) than the earlier M19 case and thus would not chamber in the 90mm guns of the M1, M2, and M3 series. However, the M19 case could be used in the T119 gun.

The light tank power package retained by the T42 consisted of the Continental AOS-895 gasoline engine and the General Motors CD-500 cross drive transmission. As its designation indicated, the engine was an air-cooled, opposed, supercharged design with a displacement of approximately 895 cubic inches. Using many of the standard components of the Continental air-cooled engine family, the AOS-895 developed 500 hp at 2800 rpm. Its six cylinders were interchangeable with those on the AV-1790 in the M46. The CD-500 transmission was similar to the CD-850 described earlier, but it was designed for use with engines ranging from 375 hp to 500 hp.

Below is the AOS-895 air-cooled engine. A cutaway drawing of the CD-500 transmission appears at the right along with a sketch of the drive train.

Concern was expressed during the meetings on the T42 about the performance which could be expected with only a 500 hp engine and it was recommended that all possible means of increasing the power to weight ratio be explored. To determine the performance characteristics prior to the production of a T42 pilot, the first medium tank T40, registration number 30162856, was converted for test purposes by installing the AOS-895 engine with the CD-500 transmission. Its weight was then adjusted

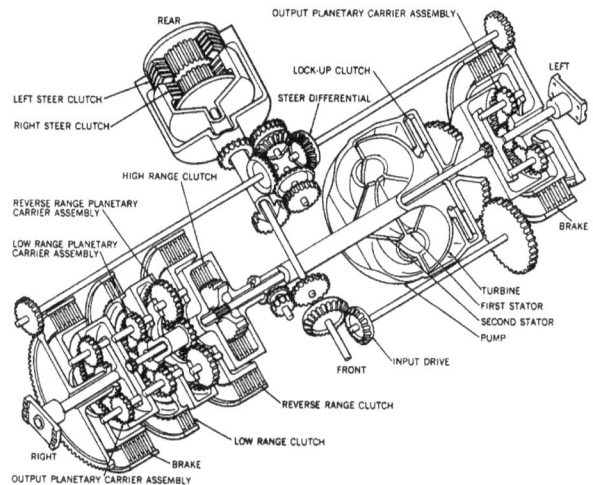

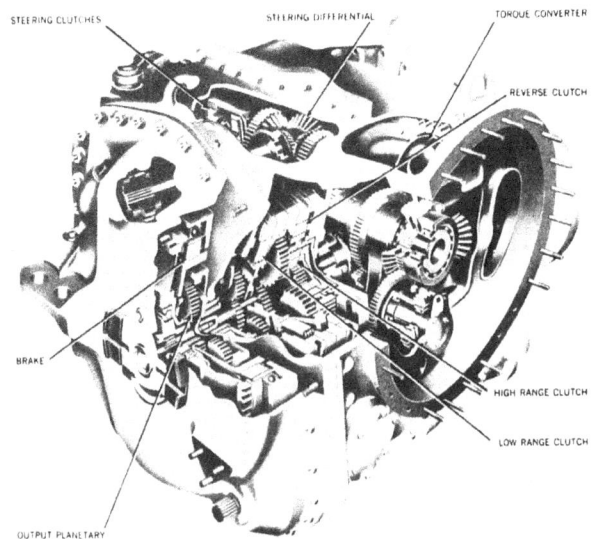

The first pilot T42, registration number 30163669, is shown in the three photographs on this page. In the view above, the welded marking NON-BALLISTIC CASTING, NOT FOR COMBAT USE can be seen on the glacis plate and the turret side. The 90mm gun is still fitted with the single baffle muzzle brake. Note the double pin T80E6 tracks.

to equal that of the T42. This simulated T42 arrived at Aberdeen in June 1950 and was tested in comparison with the late model M4A3. Although it was equal in performance to the Sherman, it did not meet the design estimates, further reinforcing the arguments that the T42 was underpowered.

On 7 November 1950, the Ordnance Committee introduced a change in nomenclature for U.S. tanks. It was concluded that the weight bracket method of classifying tanks as light, medium, or heavy was no longer practical due to changing trends in development and employment. Under the new nomenclature, the tanks were designated as light, medium, or heavy gun tanks referring to the caliber of their main armament. With this system, the light tank T41, the medium tank T42, and the heavy tank T43 became the 76mm gun tank T41, the 90mm gun tank T42, and the 120mm gun tank T43 respectively.

Six T42 pilot tanks were authorized for construction and the first of these, registration number 30163669, arrived at Aberdeen for test on 30 December 1950. By this time, the Korean War was six months old and the crash program for a new production medium tank was in full swing. Although the Army Field Forces had rejected the T42 for the production program, Ordnance continued the development in the hope that it would result in a lighter and more economical future medium tank.

The single baffle muzzle brake on T42 pilot number 1 (above) has been replaced with a cylindrical blast deflector. The pistol port is visible in the turret left side wall and single pin T95 tracks have been installed. On pilot number 3 (below), registration number 30163671, the pistol port has been deleted and the pioneer tool rack has been relocated from the glacis plate to the left side of the turret bustle.

T42 number 3 is shown here during tests at Fort Knox. Shields were installed over the exhaust mufflers on this as well as the other late pilot tanks.

INSTALLATION, TURRET ELECTRICAL - 7520057 □
INSTALLATION, TURRET ELECTRICAL - 7572500 □
INSTALLATION, GUN & TURRET CONTROL AND STABILIZER (TURRET) - 7531799 □
INSTALLATION, HYDRAULIC ELEVATING AND TRAVERSING - 7401455 ○
TURRET ASSEMBLY - 7537161

INSTALLATION, TOP DECK GRILLE - 7527750 □
INSTALLATION, POWER PLANT - 7527957 □
INSTALLATION, ENGINE ELECTRICAL - 7526520 □
SCREW - 19?? (40)

MOUNT, COMBINATION GUN - 7536700 □
MOUNT, COMBINATION GUN - 7352873 □
INSTALLATION, HULL ELECTRICAL - 7527708 □
INSTALLATION, PERSONNEL HEATER - 7528464 □
INSTALLATION, FUEL TANKS & LINES - 7528167 □
HULL MACHINING & ASSEMBLY - 7527088 □

INSTALLATION, BILGE PUMPS & DOME CONTROL - 7520068
SEAT(?) ARRANGEMENT, HULL - 7527008 □
WINDOW ARRANGEMENT, HULL - 7528951 □
INSTALLATION, SUSPENSION - 7557183
CONTROLS INSTALLATION, ENGINE, STEERING & SHIFTING & ACCEL - 7527040

PLATFORM ASSEMBLY - 7537314 □
ELEVATION ASSEMBLY - 7401365 ○
INSTALLATION, PPU GENERATOR - 7528166
INSTALLATION, GUN & TURRET CONTROL AND STABILIZER (HULL) - 7520038 □

□   STABILIZED
○   UNSTABILIZED

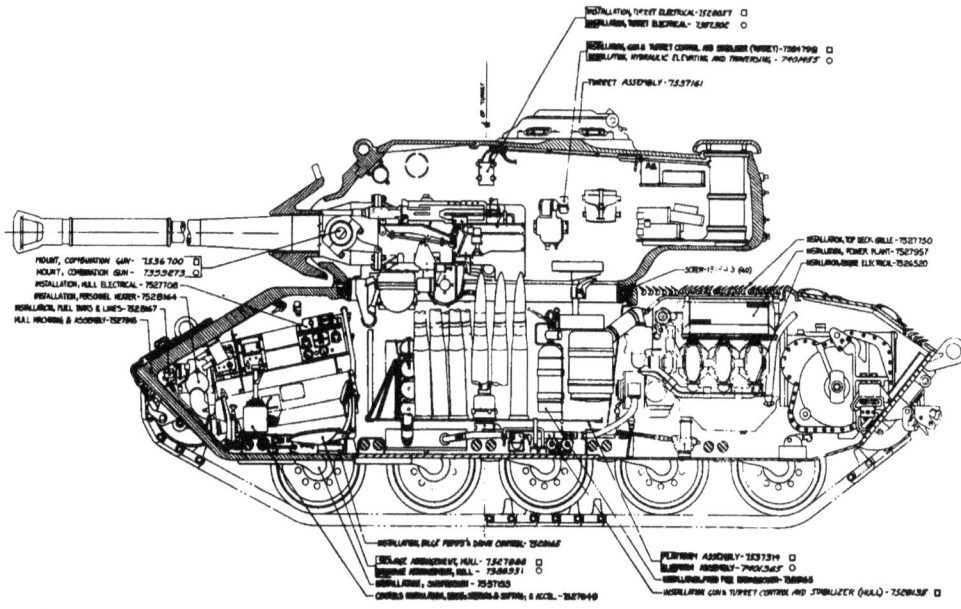

The internal arrangement of the T42 can be seen in the sectional drawing above. Note the compact engine and transmission which helped minimize the size and weight of the tank.

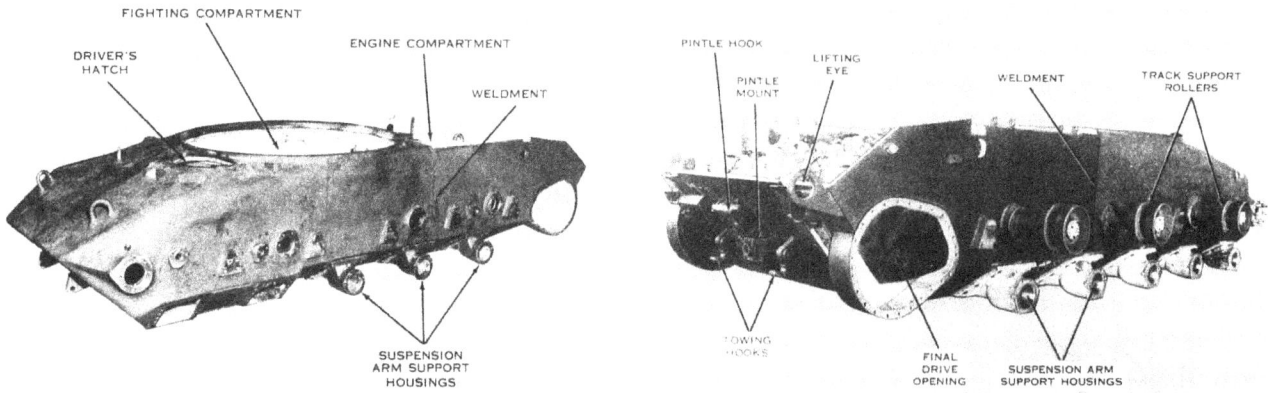

FIGHTING COMPARTMENT
DRIVER'S HATCH
ENGINE COMPARTMENT
WELDMENT
SUSPENSION ARM SUPPORT HOUSINGS

PINTLE HOOK
PINTLE MOUNT
LIFTING EYE
WELDMENT
TRACK SUPPORT ROLLERS
TOWING HOOKS
FINAL DRIVE OPENING
SUSPENSION ARM SUPPORT HOUSINGS

Above, the welded assembly of castings and rolled plates used to fabricate the hull is obvious. Details of the lights and horn installed on the hull front are shown below. The group mounted at the right front appears at the left and the group on the driver's side is at the right.

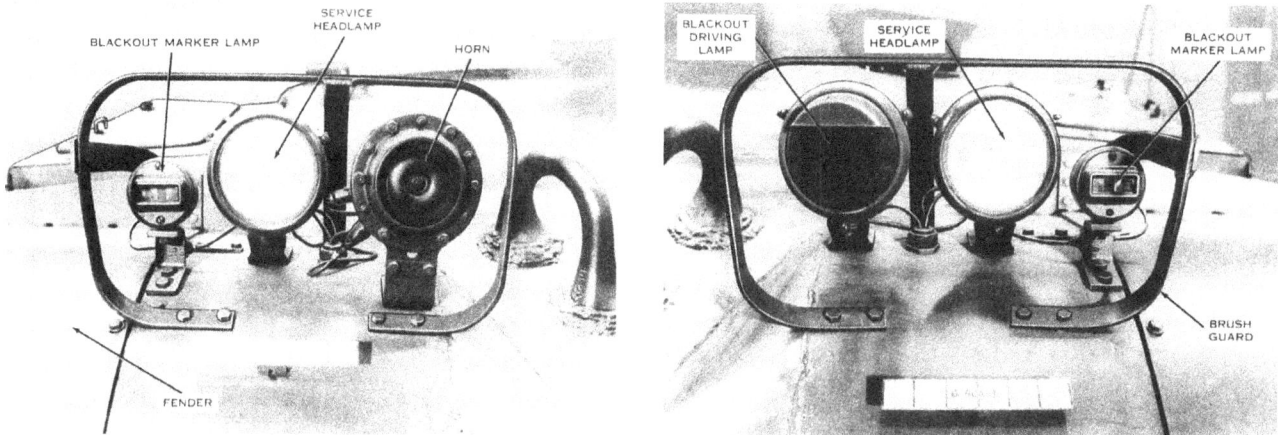

BLACKOUT MARKER LAMP
SERVICE HEADLAMP
HORN
FENDER

BLACKOUT DRIVING LAMP
SERVICE HEADLAMP
BLACKOUT MARKER LAMP
BRUSH GUARD

38

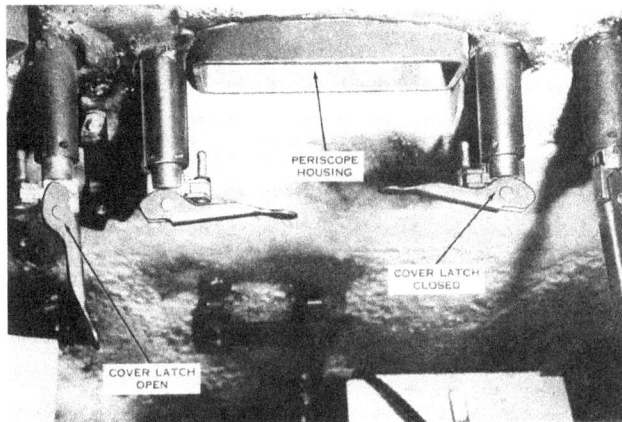

The driver's periscope latches are shown above with the periscopes removed. At the top right is an exterior view of the driver's hatch with the cover closed. The turret has been removed in this photograph and the 90mm ammunition rack alongside the driver can be seen in the lower right-hand corner.

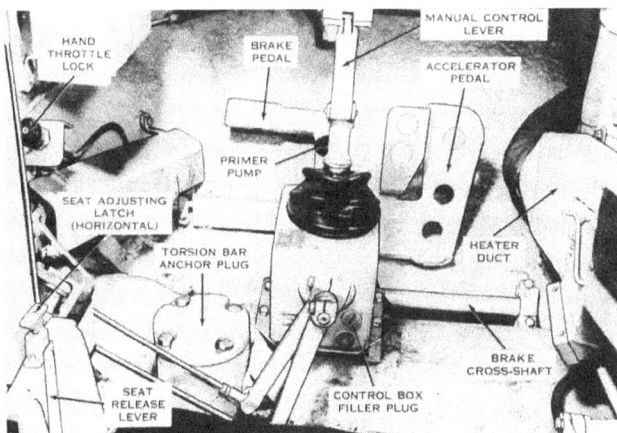

The T42 hull was assembled from two sections. The forward part was a homogeneous armor steel casting and the rear was a welded assembly of rolled armor plate. The two halves were joined by a vertical weld at the rear of the fighting compartment. The upper front of the hull casting was 4.0 inches thick at an angle of 60 degrees from the vertical. The vehicle was manned by a crew of four consisting of the tank commander, gunner, loader, and driver. The single driver's position was located in the left hull front with a 36 round 90mm ammunition rack on the right replacing the usual assistant driver.

The driver's manual control lever (wobble stick) and other controls appear above. Below are views of the instrument panel (left) and the 36 round 90mm ammunition rack in the right front of the hull (right).

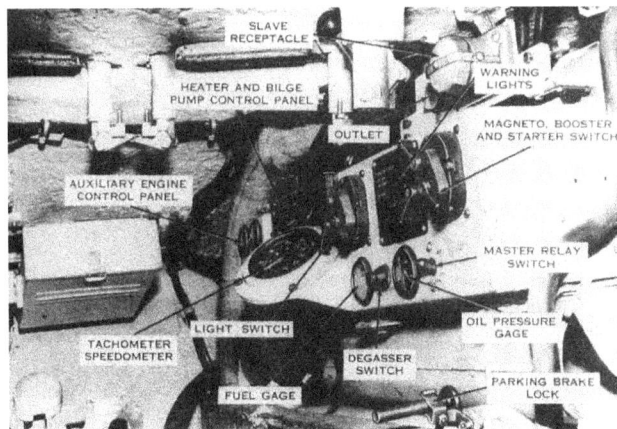

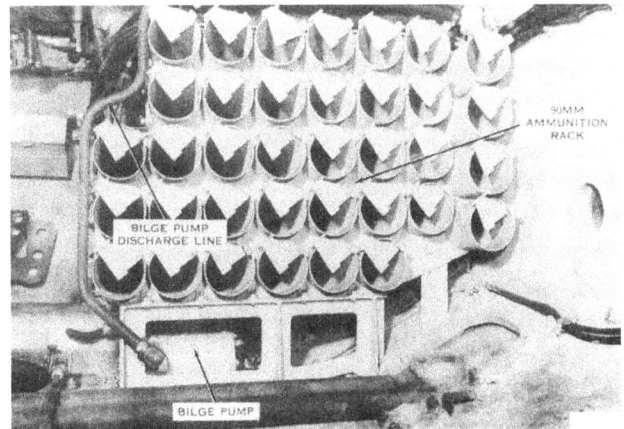

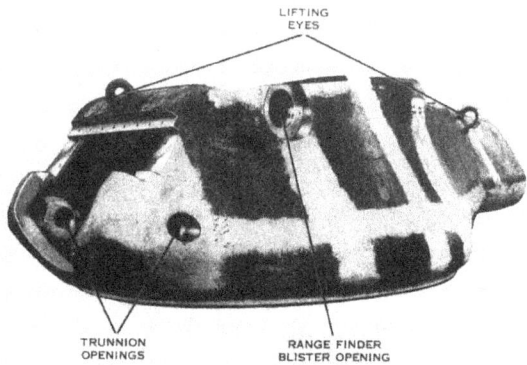

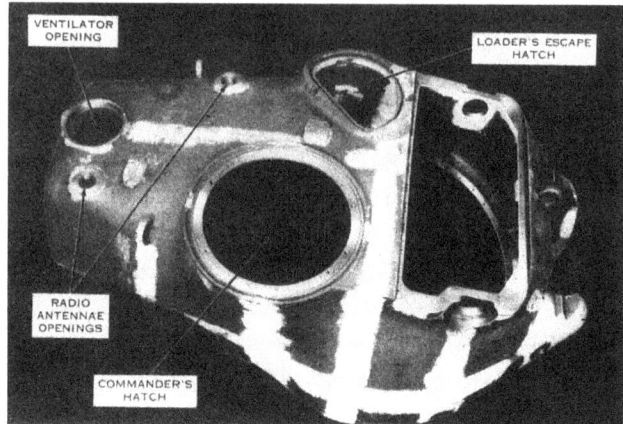

Details of the turret casting are visible above before installation of the various fittings. The turret top is shown in the two views below after the roof sections, hatches, and other equipment have been installed.

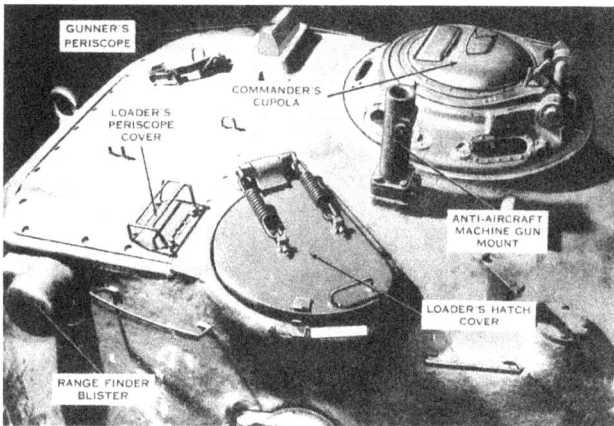

On the pilot tanks the cast armor steel turret mounted the T119 90mm gun coaxially with either a .30 caliber or a .50 caliber machine gun. A .50 caliber antiaircraft machine gun was pedestal mounted on the turret roof adjacent to the tank commander's vision cupola. The loader was provided with a flat oval shaped hatch in the left side of the turret roof as well as a pistol port in the left turret wall. The pistol port was deleted on the later pilots. The gunner was located in the usual position forward of the tank commander in the right front of the turret. The gunner's primary sight was the T41 stereoscopic range finder with a T35 periscopic sight as a backup device. Although designed to incorporate an IBM stabilization system, the pilots were delivered without such equipment. Twenty-four rounds of 90mm ammunition were stowed on the turret platform bringing the total carried to 60 rounds. The standard vision cupola was installed on the pilots pending the development of an improved model. Several experimental cupolas incorporating a .50 caliber machine gun mount were installed on the T42 mock-up and some also were tested on the pilot tanks.

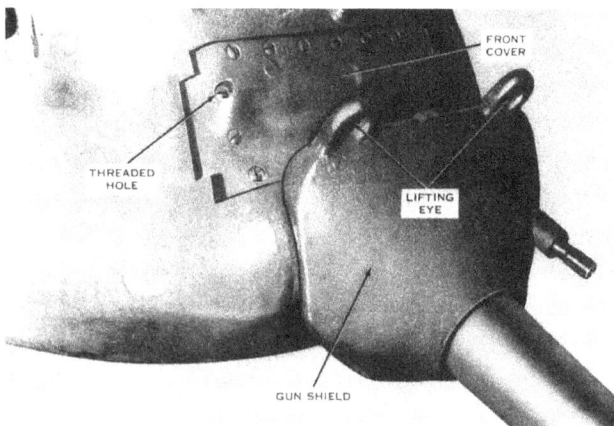

At the left is a view of the turret front and the gun shield. Note the protrusion of the long .50 caliber coaxial machine gun barrel.

40

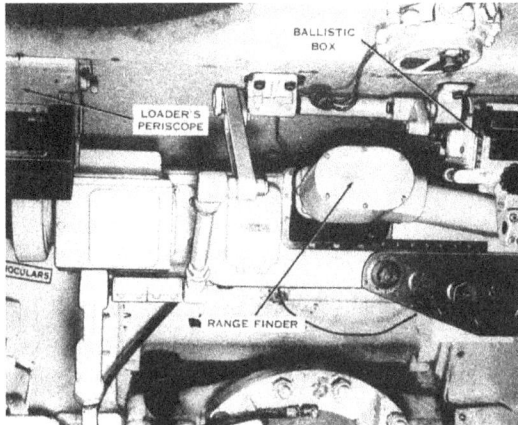

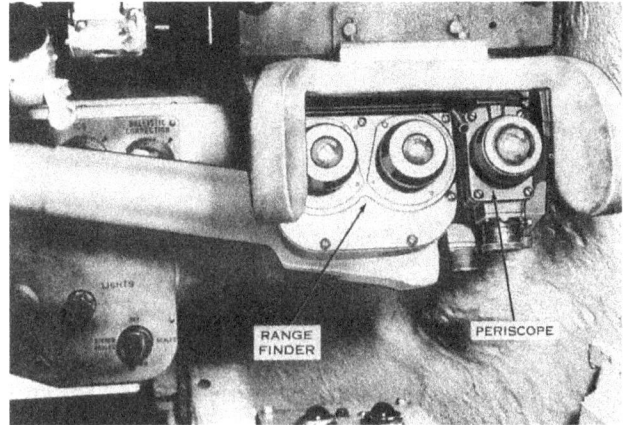

Above is the T46E2 range finder extending from the gunner's station (right) to the loader's position (left). The eyepieces for the range finder and the gunner's periscope are combined under a single brow pad. Note that the coaxial machine gun is not fitted in the left photograph. Below, another view of the loader's station (left) shows the .50 caliber coaxial machine gun installed with the flexible ammunition chute, booster and ready box. The gunner's controls are below at the right.

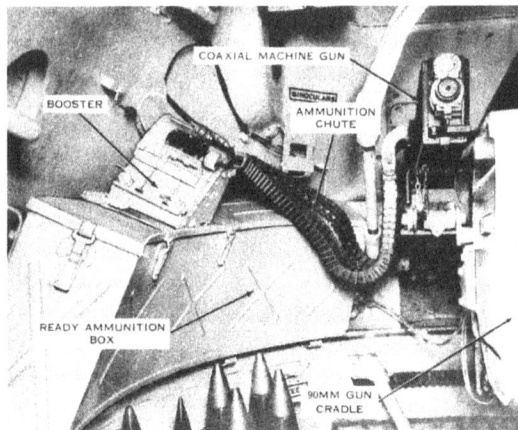

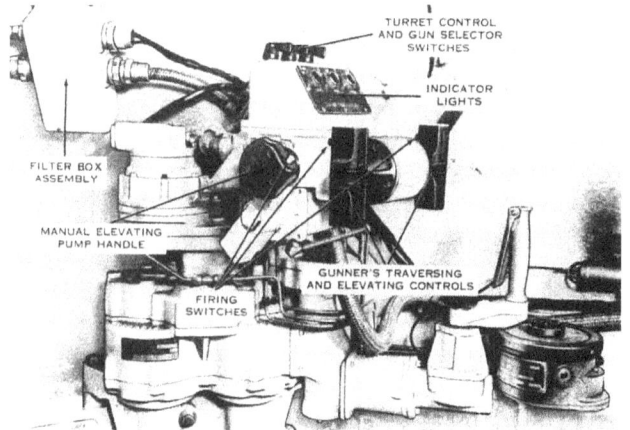

Below are views of the 90mm ammunition racks on the front and left side of the turret platform (left) and the engine compartment bulkhead with the turret removed (right). The torsion bars from the number three road wheels are at the bottom of the latter photograph and the fixed fire extinguisher cylinder appears at the left.

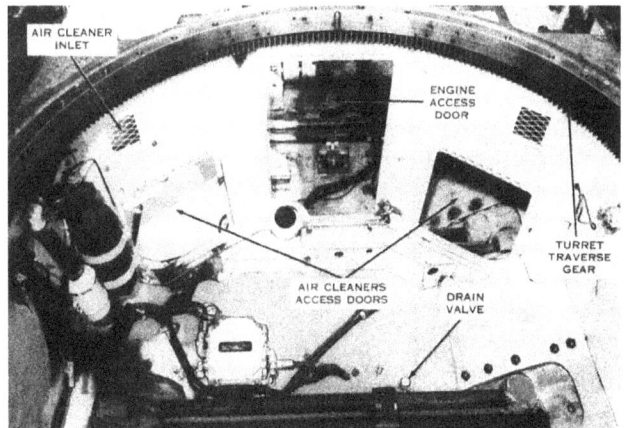

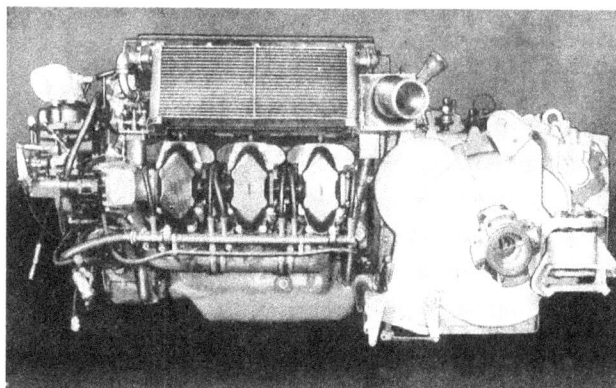

The power pack, consisting of the AOS-895 engine and the CD-500 transmission, is shown above removed from the tank. At the left below is a view of the accessory end of the AOS-895 engine.

The pilot T42s were fitted with the AOS-895-3 engine and the CD-500-3 transmission. The CD-500-3 was approximately 500 pounds lighter than the earlier models primarily due to the substitution of aluminum for steel components. Like the CD-850, it had two speed ranges forward, but there was a lockup feature in the high range. The lockup which

automatically engaged at about 23 miles/hour and disengaged at approximately 21 miles/hour provided a direct mechanical drive eliminating hydraulic losses and reducing fuel consumption. The tank was steered by a wobble stick control similar to that in the M46. During experiments at the Proving Ground, the wobble stick was replaced by a steering wheel. The performance of the pilot T42 with the later power package met the OTCM specifications exceeding the results obtained with the converted T40. However, the Army Field Forces still considered the T42 underpowered. By this time their interest had shifted to the M47 and T48 programs which were already well advanced.

The torsion bar suspension in the T42 consisted of five dual road wheels on each side with the front and rear torsion bar springs somewhat heavier than the three center bars. The 24 inch wide T95 single pin tracks were carried by three track support rollers on each side. An adjustable idler was installed at the front of each track. Double pin 23 inch wide T80E6 or T84E1 tracks also were installed at various times during the test program.

At the left below, the power plant is installed in the T42 hull. The fan on top of the air-cooled engine can be clearly seen in this photograph. The view at the right shows how the shields were fitted over the fender mounted exhaust mufflers.

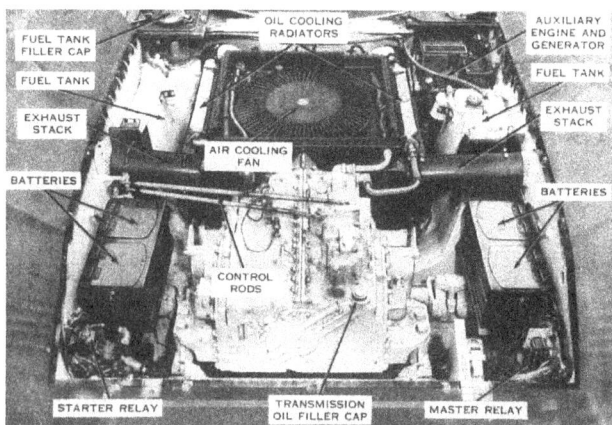

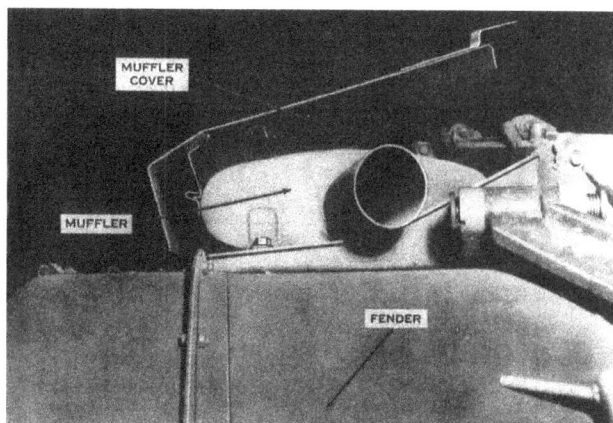

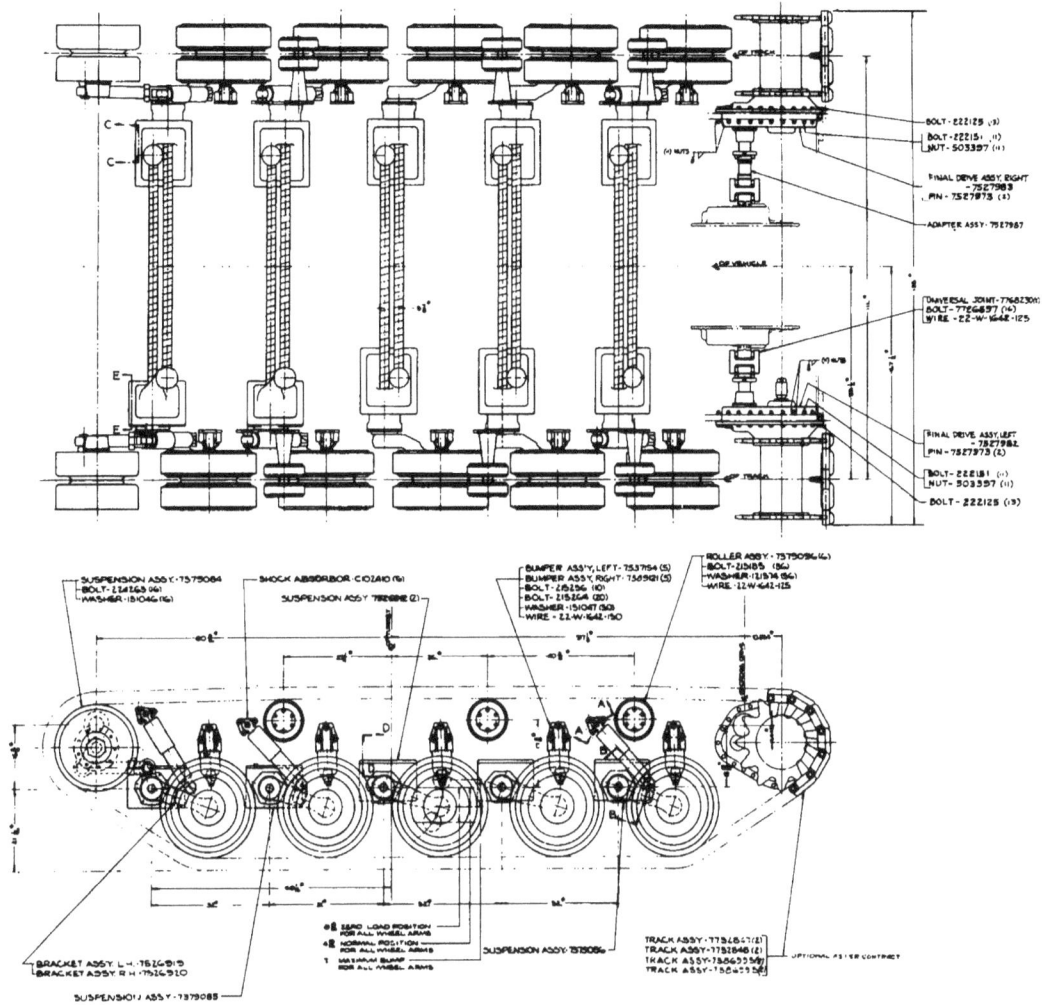

The layout of the torsion bar suspension system on the T42 is illustrated by the drawing above. Note the offset between the road wheels on opposite sides necessary to provide clearance for the torsion bars. The road wheels on the left side were 5-7/8 inches ahead of those on the right.

Below, the torsion bars can be seen (left) in the empty engine compartment and the installation of the road wheel arms and other exterior suspension components is clearly visible (right) with the road wheels removed.

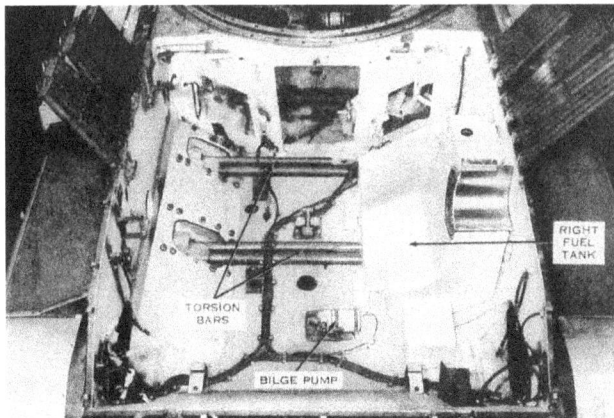

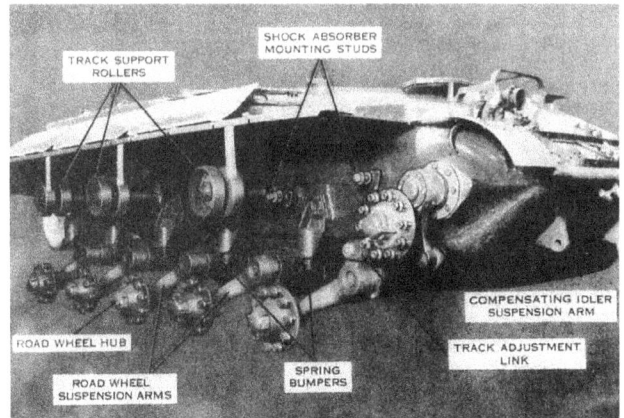

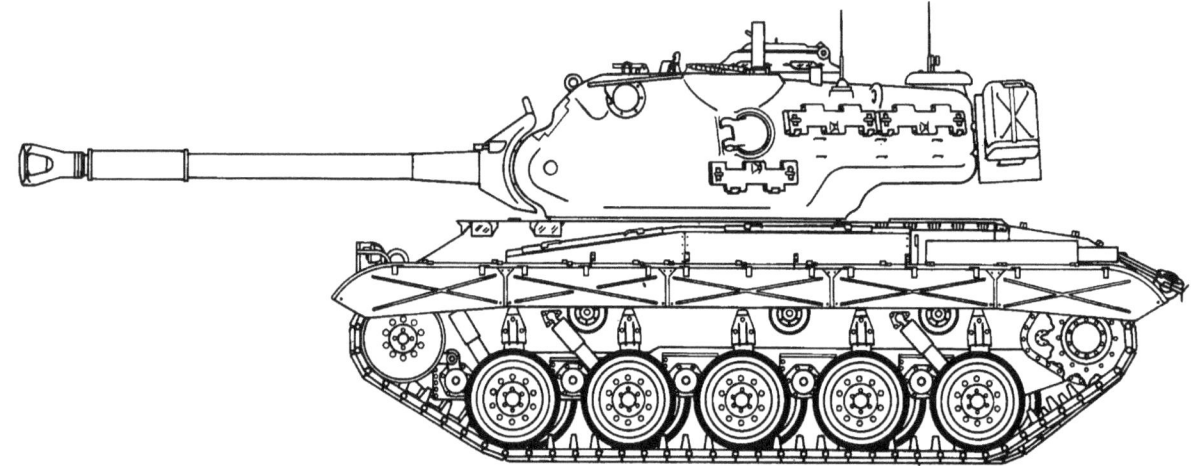

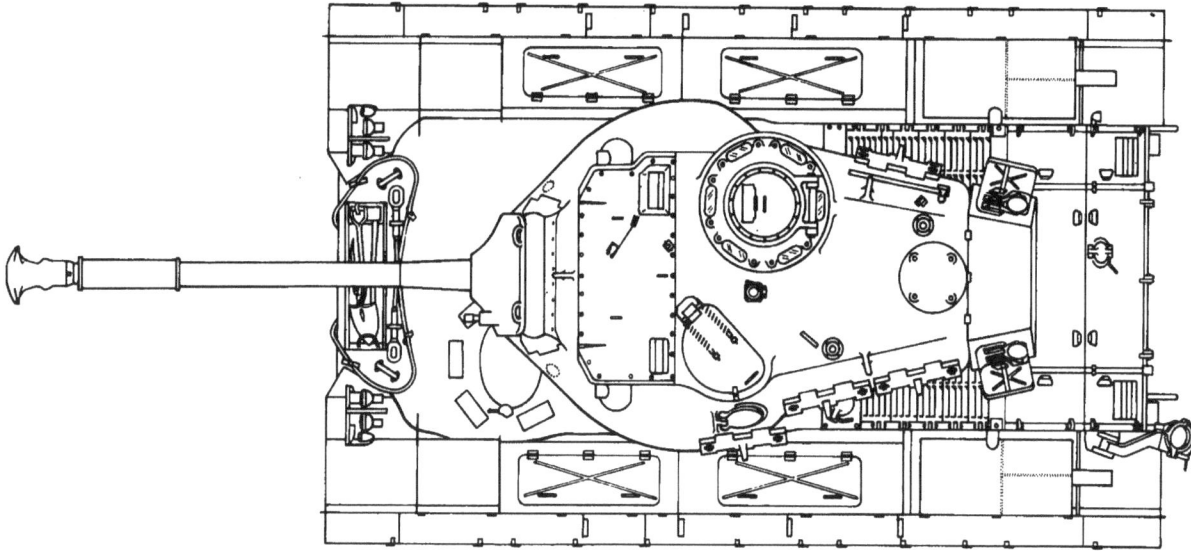

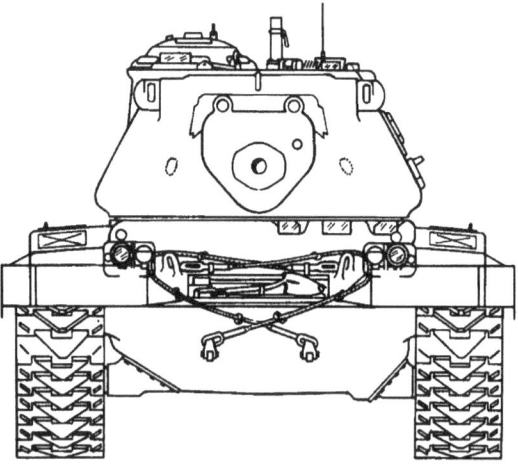

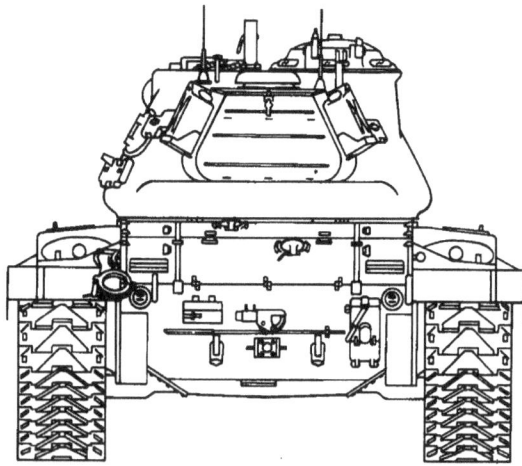

Scale 1:48

90mm Gun Tank T42

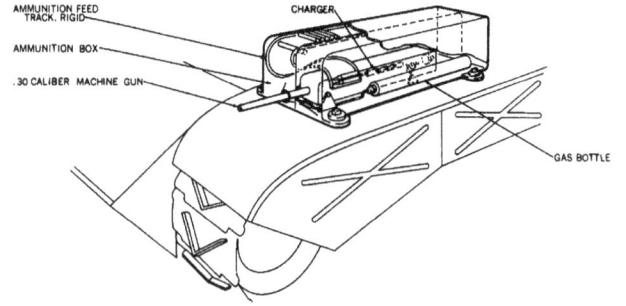

The T42 at the left is fitted with two .30 caliber fender machine guns during tests at Aberdeen Proving Ground. The general arrangement of this installation is shown in the drawing above.

Although the turret blister guns were eliminated early in the design program, it remained desirable to provide additional machine gun firepower to compensate for the lack of a bow gunner. One solution to this problem was the installation of fender mounted gun container kits. Each kit consisted of an armored box containing a .30 caliber M1919A4 machine gun, 680 rounds of ammunition, a pneumatic charger, a firing solenoid, and a compressed air bottle. The driver controlled the guns from a fire control box. The guns were charged separately, but they were fired from a single switch. Tests at Aberdeen indicated that the fender gun kits were unsatisfactory for firing at specific targets, but that they could be used for general area fire if the terrain was uniform. The Proving Ground recommended further development to incorporate elevation adjustments to compensate for pitching of the vehicle.

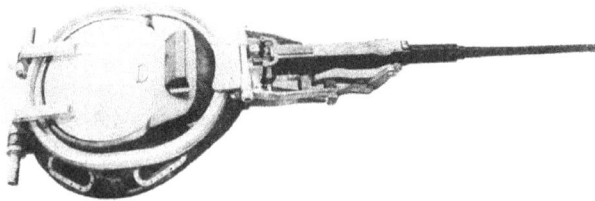

A mock-up of an experimental machine gun mount for the tank commander's cupola is shown above. In the two photographs below, it is installed on the T42 turret mock-up.

The experimental .50 caliber machine gun mount, shown in mock-up form at the left, is illustrated below during tests on the T42 at Aberdeen.

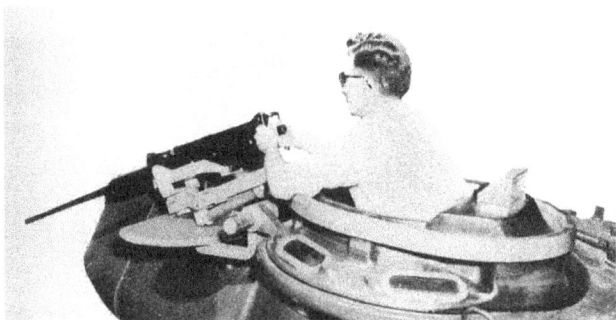

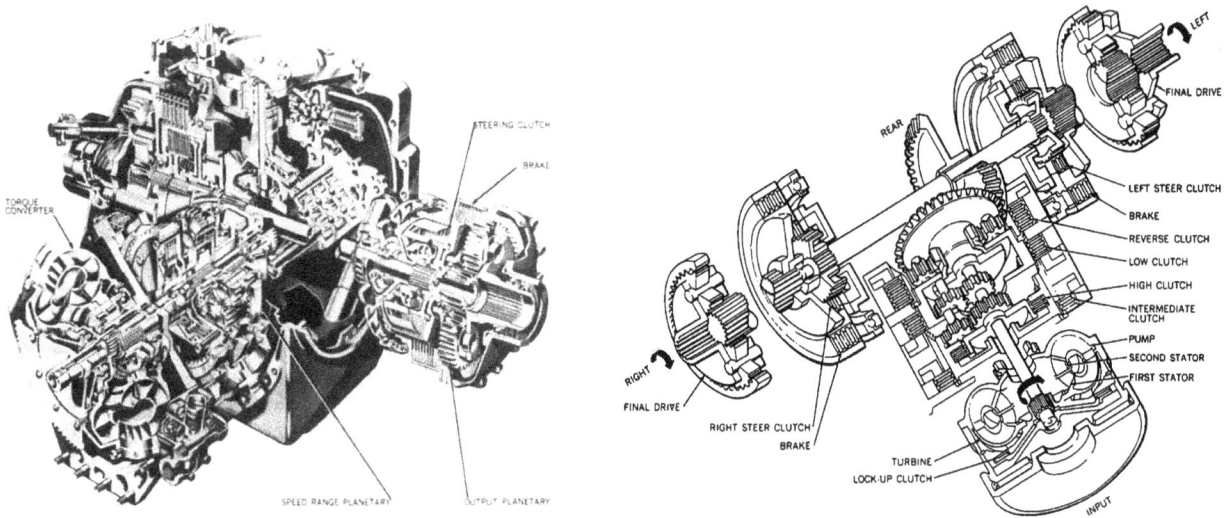

The cutaway view (top left) gives the general arrangement of the XT type transmission and the sectional sketch (top right) shows the components in the XT-500. Note the clutch-brake steering in the schematic (below right) of the XT-500. T42, serial number 2, shows the modifications to the hull necessary to accommodate the XT-500 transmission.

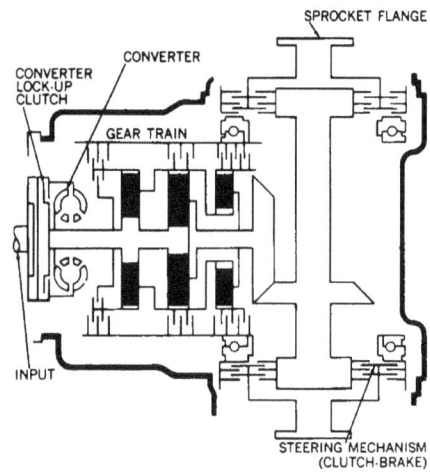

Below, the wobble stick in the second pilot T42 has been replaced by a steering wheel during tests at Aberdeen. At the left, the wheel is in the upper position for driving with the hatch open. At the right, it has been rotated to the lower position for operation with the hatch closed.

The first pilot T42 was constructed from soft steel castings and plate instead of armor. Early in the test program at Aberdeen, this tank was badly damaged by fire. A pin came out of the final drive quick disconnect fitting releasing the universal joint which promptly ripped a hole in the fuel tank. The gasoline sprayed up onto the hot exhaust crossover pipes and the tank was engulfed in flames. Fortunately, pilot number 2 had arrived at the Proving Ground on 17 April 1951 and after number 1 was partially destroyed, it was used to continue the automotive test program. Later, this same tank was modified to permit the installation of the XT-500-1 transmission. The required changes to the engine compartment resulted in a vertical rear hull plate. The XT-500 was designed to be more efficient than the original cross drive transmission. It consisted of a single stage multiphase hydraulic torque converter with a lockup clutch and a 3-speed plus reverse planetary gear train. The torque converter end of the transmission was connected to the engine with a clutch-brake steering mechanism at each side on the output end. In addition to the improved efficiency, the relatively simple XT-500 was lower in cost with approximately 60 per cent fewer parts than the CD-500.

Pilot number 3 was shipped to Fort Knox for evaluation by the Army Field Forces. It was later returned to Detroit and eventually, it was fitted with an oscillating turret to become the 90mm gun tank T69. Pilot number 4 also went to Aberdeen and number 5 was assigned to the United States Marine Corps. T42 number 6, registration number 30163673, arrived at Aberdeen in December 1952 where it was used to evaluate additional experimental components. These included such items as a glowmeter instrument panel, an hydraulically operated driver's seat, rubber snubber springs, and the amplidyne fire control system.

Although production had been ruled out for the T42, it was proposed during the Spring of 1953 that the program be continued and expanded to develop a lighter and more economical medium tank. New features studied for application to the T42 included a cast steel elliptical hull and a flat track suspension. The latter eliminated the track return rollers and supported the track on top of the large road wheels. At that time, it was intended to designate the improved vehicle as the 90mm gun tank T87. However, a conference in May 1953 rejected these proposals and dropped the T42 program completely. This was officially confirmed by Ordnance Committee action in October 1954.

In the original military characteristics for the T42, the Ordnance Committee specified that automatic loading would be included if feasible. Preliminary attempts to design an automatic loader for the T42 turret were unsuccessful primarily because of the limited space available and the relative motion between the mechanical loader mounted in the turret and the gun breech. Further studies at Rheem Manufacturing Company indicated that it would be practical to design an automatic loader for the 90mm gun if it was installed in an oscillating or ball type trunnion mounted turret. With this arrangement, the entire turret moved in both elevation and azimuth and the only movement between the gun and turret was due to recoil and counterrecoil. With the gun breech only moving back and forth always along the same line, the design of the loader was greatly simplified.

Under a new contract, Rheem prepared design drawings and built mock-ups of the oscillating turret and the automatic loader. OTCM 33597, dated 10 January 1951, listed the characteristics of the proposed turret and recommended the development and manufacture of a pilot turret mounting a T139 type 90mm gun suitable for installation on the T42 tank. Although full scale work began in May 1951, design difficulties and problems in obtaining government furnished equipment delayed completion of the pilot until early 1955. The turret was installed on the third pilot T42, registration number 30163671, and the complete vehicle was designated as the 90mm gun tank T69. Like pilot number 2, this particular chassis had been modified and the rear hull had been rebuilt to accommodate the XT-500 transmission.

The T42 continued to serve in various test programs. Below at the left, the 1st Cavalry Division conducts frozen rice paddy mobility tests near Wangchon, Korea on 18 January 1963. At the right, the T69 turret is installed on the chassis of the third pilot T42. The latter photograph was dated 25 January 1955.

This is the 90mm gun tank T69 at Aberdeen Proving Ground during the test program in the Summer of 1955. Details of the oscillating turret can be clearly seen in these photographs. Note the open turret roof in the view at the bottom right.

The long nosed cast turret body on the T69 was mounted on trunnions and supported by an armored yoke assembly which rotated on the 73 inch turret ring. It was armed with the 90mm gun T178. This weapon was essentially the 90mm gun T139 mounted upside down with the mounting lugs modified to permit installation of the recoil mechanism in the forward part of the turret. A coaxial .30 caliber machine gun was mounted to the left of the main armament.

The automatic loading equipment utilized an 8 round cone-shaped magazine mounted on the bottom of the turret. A revolving spider inside the magazine was reloaded manually. Any one of three types of ammunition could be chosen by setting the selector switch. A slot in the top of the magazine permitted the loading mechanism to withdraw the round which was then rammed into the chamber. The cyclic ramming rate varied from 33 rounds/minute with

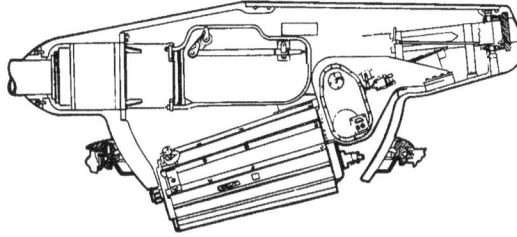

Above is a sectional view of the T69 turret and automatic loader. The revolving 8 round cone-shaped magazine is at the bottom of the drawing.

a single type of ammunition to 18 rounds/minute when three types of ammunition were used. After firing, the ejected case passed along the top of the rammer assembly, through the ejection chute, and out of the automatically opened ejection port in the turret bustle. This port also closed automatically when the gun returned to battery, however, a manual control was provided for emergency use. In addition to the 8 rounds of 90mm ammunition in the magazine, 32 were carried in the front hull rack. In the T42, this rack contained 36 rounds, but in the T69, the close proximity of the automatic loader and the turret ring prevented access to four spaces in the rack. A control panel with colored lights indicated the round type location and the interlock functioning. The human loader located on the left side of the turret was responsible for round selection and magazine replenishment. He was provided with a vision periscope and a flat circular hatch in the turret roof. The gunner was located in the right front of the turret using a T35 periscopic sight. The Cadillac Gage

The T69 turret interior is viewed above through the open turret roof. The T46E2 range finder is visible as well as the gunner's controls. Below at the left is the loader's station and the control panel for the automatic loader. This panel was relocated from the gunner's position (below right) during the test program.

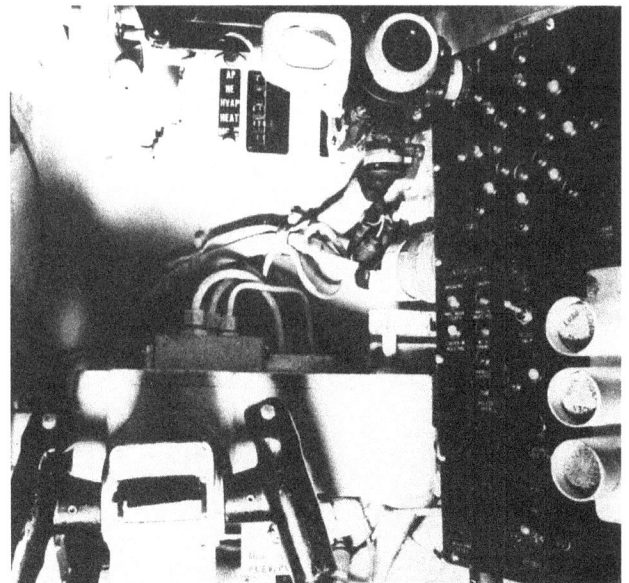

49

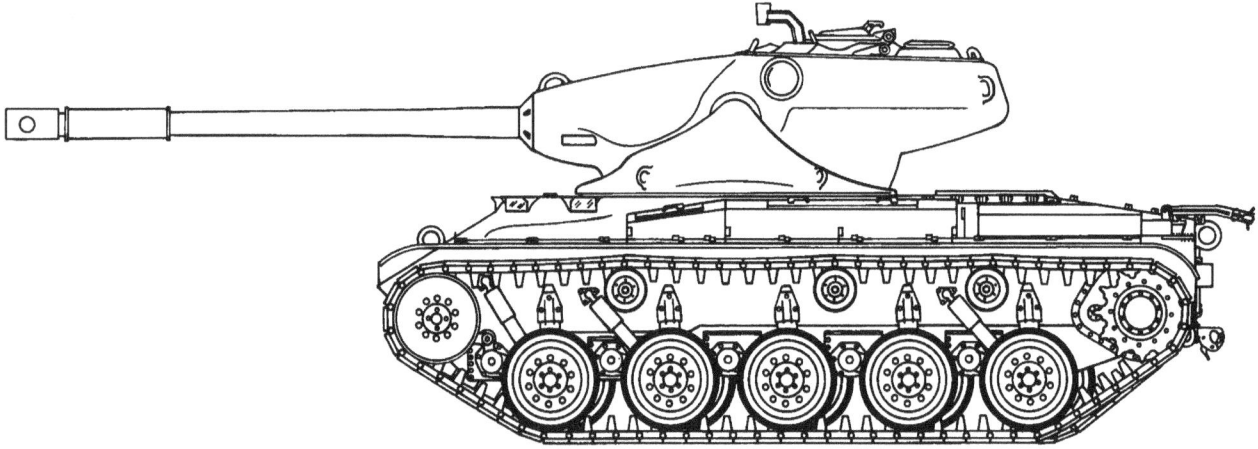

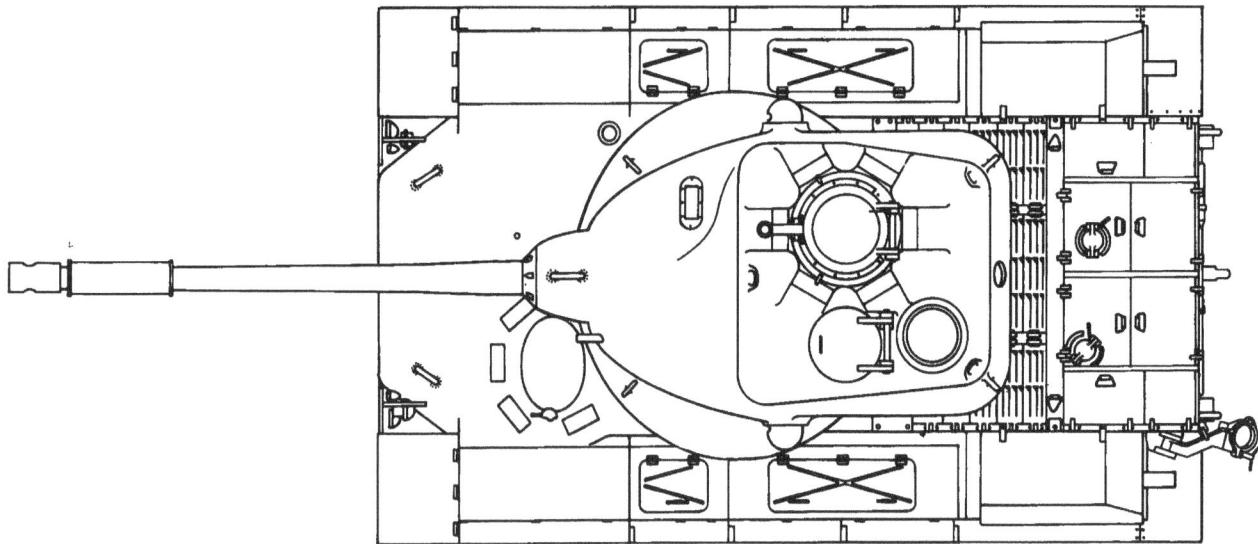

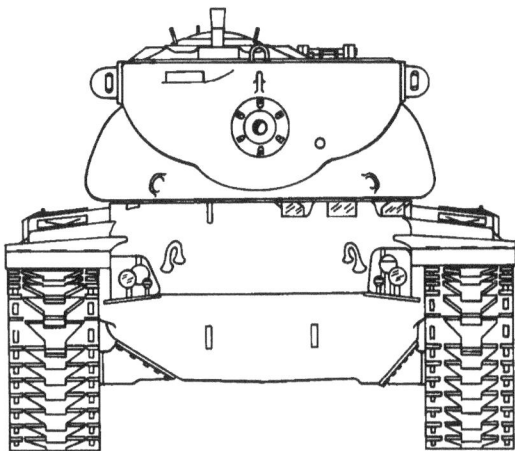

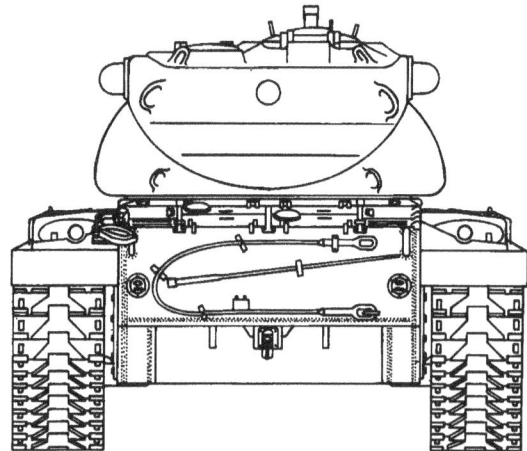

Scale 1:48

90mm Gun Tank T69

The side view above shows the turret roof in the open position. The turret is at approximately half the maximum elevation. At full elevation, the turret bustle would touch the rear deck.

gun control system was installed in front of the gunner. To the rear of the gunner, the tank commander had a T46E2 range finder and six periscopes arranged around his hatch. A .50 caliber antiaircraft machine gun was mounted in the International Harvester Company rotating hatch. An hydraulically operated access cover formed a large portion of the turret roof. This arrangement allowed easy servicing and removal of the automatic loader and provided an emergency exit from the turret. A control on the loader's side of the turret raised and lowered the hinged roof.

The T69 was under test at Aberdeen from June 1955 through April 1956. During this time a high rate of component failure prevented a thorough study of automatic ammunition handling in the oscillating turret. Although considered unsatisfactory for service, much valuable information was obtained for future turret designs. In addition to the pilot tank, six ballistic turrets were manufactured and shipped to Aberdeen for evaluation. By this time, interest had shifted to later design concepts and OTCM 36729 terminated the T69 project on 11 February 1958.

The interphone amplifier appears in the rear of the turret (below left) and the ejection port for empty cartridge cases can be seen in the closed position in the turret bustle wall. The .50 caliber machine gun is mounted on the commander's hatch (below right). No periscopes have been installed in the six openings around the hatch.

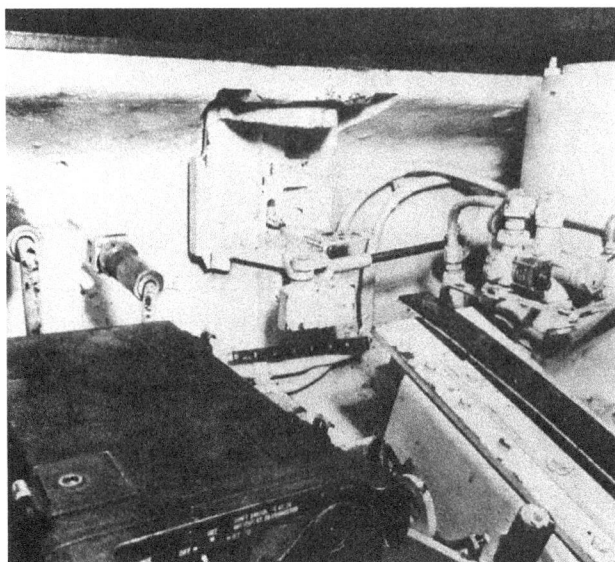

Above is an artist's concept of the new tank with the T42 turret installed on the hull of the M46, dated October 1950.

THE EMERGENCY

The outbreak of the Korean War in June 1950 brought the Army's tank shortage into sharp focus. At the highest levels of government, it was feared that Korea might be the opening battle of a new global war. Particular concern also was felt about the situation in Europe. A little more than a year had elapsed since the United States had committed itself to defend western Europe as part of the new North Atlantic Treaty Organization (NATO). Modern tanks in large numbers would be required to meet these worldwide commitments. The stock of Pershings available for conversion to the new M46 was far too limited to fill this requirement and even the M46 did not reflect the latest design concepts. A new tank was needed which could be put into immediate production as the stock of M26s was depleted.

In late June, Colonel Joseph M. Colby at Detroit Arsenal recommended that the medium tank T42 be released for immediate production along with the light tank T41 and the heavy tank T43. However, the Army Field Forces would not accept the T42 without extensive tests to disprove their contention that it was underpowered. With the emergency situation in Korea, the selection of a new production tank could not be delayed. At the same time, it was not desirable to initiate large scale new production of the M46 with many of its features rapidly approaching obsolescence. In July, it was suggested that the problem be solved by installing the T42 turret with its improved armament and fire control system on the proven M46 chassis. A meeting at Detroit Arsenal on 12 September resulted in an agreement between the Army Field Forces, the Corps of Engineers, and Ordnance for the production of an interim tank based on that combination.

On 1 November 1950, OTCM 33485 assigned the designation 90mm gun tank M47 to the new production vehicle even though it had not been standardized. This unusual procedure was authorized since quantity production had been ordered and early standardization was expected. By using the M47 designation at the outset, the confusion resulting from numerous later changes on drawings and other documents could be avoided.

The military characteristics of the M47 were outlined in OTCM 33805, dated 27 January 1951, and they were approved on 19 July. The basic M46 hull

The single M46E1 is shown above. In the view at the left, the muffler shields have not been installed and the muzzle brake is rotated about 90 degrees from its normal position.

was retained with the front armor modified to increase the slope of the 4 inch front plate to an angle of 60 degrees from the vertical thus equaling the protection on the T42. The rotoclone blower in the upper front hull of the M46 was eliminated, improving the contour of the armor in this area. It was no longer required since the T42 turret was fitted with a ventilation blower in the turret bustle. The bow gunner and his .30 caliber machine gun were retained in the right front hull. The turret ring diameter was increased to 73 inches to match the T42 turret. An improved commander's cupola was installed on the production turret and the pistol port was eliminated from the left side wall. Since the assistant driver or bow gunner was retained in the

In order to obtain test data prior to the availability of the first M47, a pilot tank was assembled at Detroit. This vehicle used an M46 hull without the modifications to the front armor. A T42 turret was installed with the early commander's vision cupola and pistol port. OTCM 33805 also designated this pilot vehicle as the medium tank M46E1 and it was shipped to Aberdeen Proving Ground for test in March 1951.

M47, the 71 rounds of 90mm ammunition were arranged similar to those in the M46. Eleven rounds were carried in the turret basket ready racks with 12 additional rounds located on the hull side walls, six on each side. The remaining 48 rounds were stowed in racks below the turret basket floor.

Below, the M46E1 is under test at Aberdeen with the muzzle brake properly installed. Note that this is an early T42 turret with the pistol port in the left turret wall.

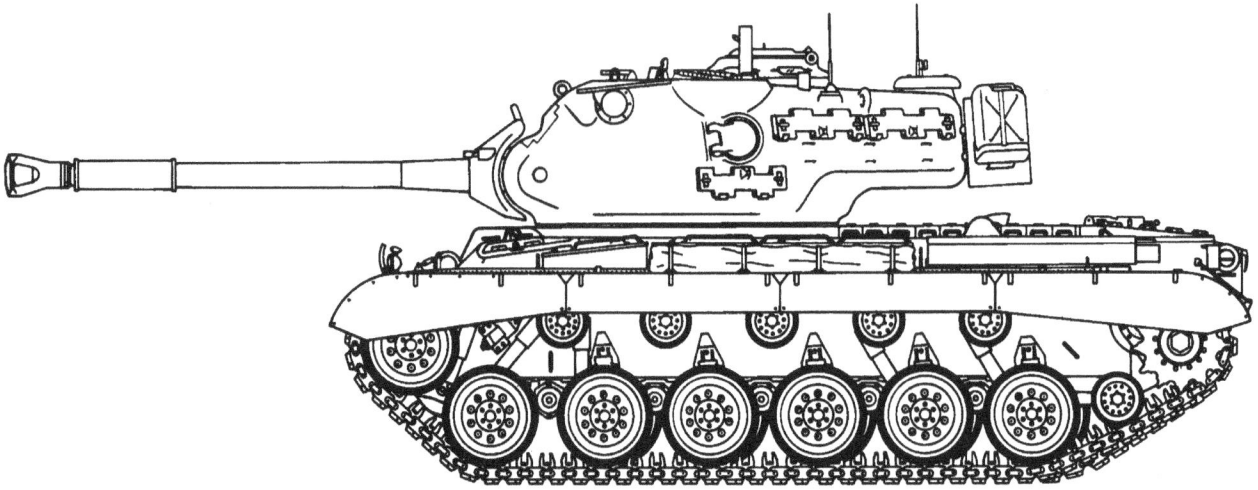

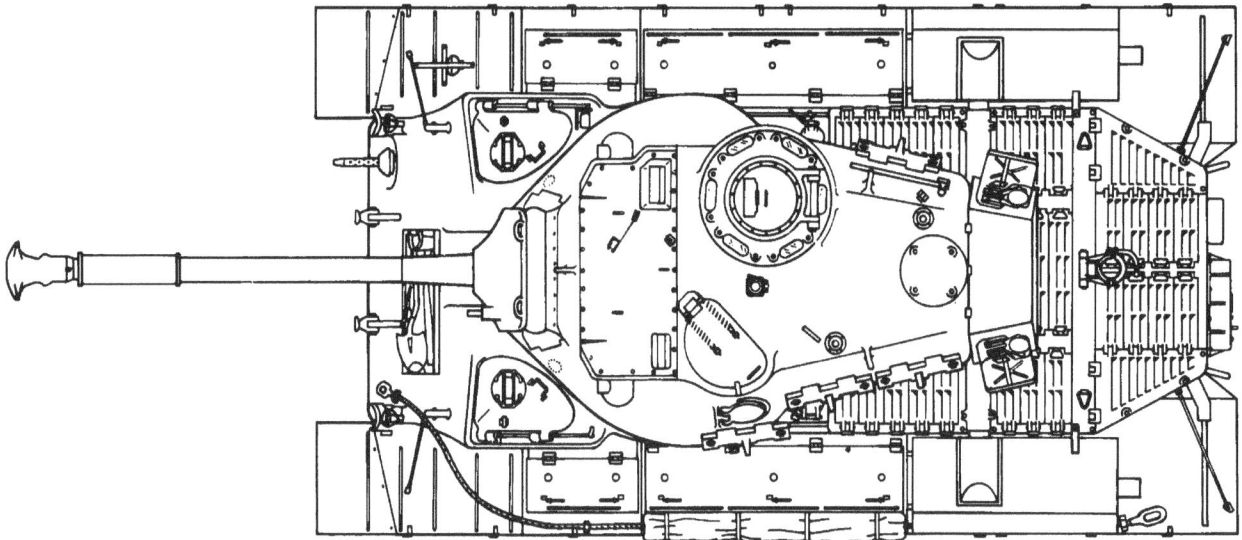

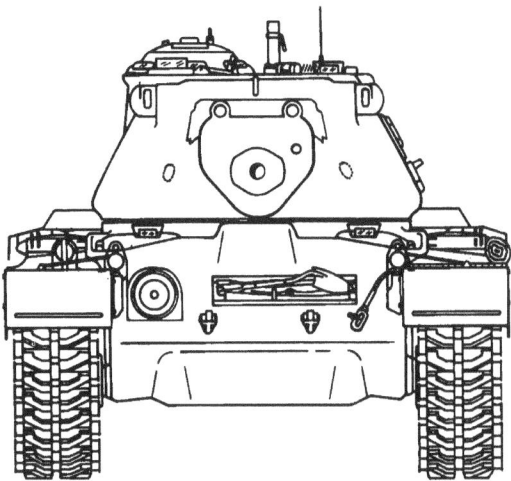

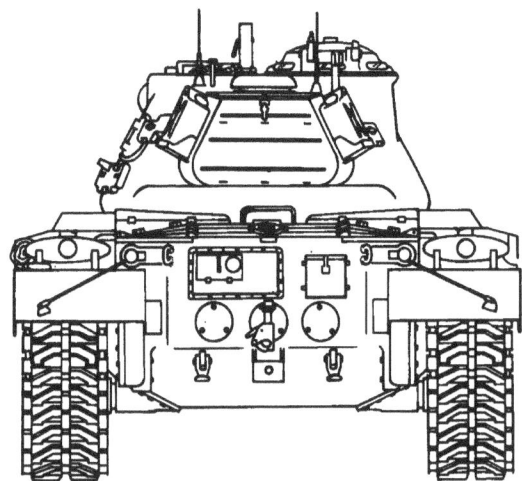

Scale 1:48

Medium Tank M46E1

Above, an early M47 has been delivered without a range finder and the ports for the instrument have been fitted with cover plates. Note the original location of the pedestal mount for the .50 caliber machine gun on the turret roof.

Production of the M47 started in June 1951 at Detroit Arsenal and somewhat later at the American Locomotive Company. At that time, Detroit Arsenal was under the direct control of the Ordnance Department. Ordnance continued to run the Arsenal until the Summer of 1952. Chrysler Corporation then took over the production operations under contract, resuming the role they had played during World War II.

Production tanks were shipped to Aberdeen and the Army Field Forces Board in late July 1951. The first M47 to arrive at the Proving Ground was serial number 6, registration number 30164213. Engineering and endurance tests began at once to determine the characteristics of the vehicle and to evaluate its mobility and mechanical reliability. The tests ran from 27 August 1951 to 7 August 1952. During this period of time, seven other M47s were being evaluated at Aberdeen. Five of these tanks were engaged in track qualification work and the other two were used for inspection control tests. With eight tanks available, much valuable data were obtained on the component failure rates for the new vehicle.

In addition to the new turret and modified front hull, the M47 had its suspension changed from that on the original M46 by eliminating the second and fourth track return rollers on each side. These production tanks also were equipped with the improved AV-1790-5B engine and the CD-850-4 transmission. The driver's controls were similar to those in the M46A1, but no controls were provided

The two views at the right show the early M47 without the range finder. Note the muzzle brake originally installed on the 90mm gun. This was replaced by the cylindrical blast deflector on the production tanks.

for the assistant driver or bow gunner. Although the same single baffle muzzle brake used on the T42 was originally specified for the 90mm gun T119E1, it was soon replaced on the production tanks by a cylindrical blast deflector. Since the production of the T41 range finder lagged behind that of the tanks, the first M47s were delivered with flat plates installed over the ports for the range finder blisters. The range finders were to be installed when they became available. Prior to that time, T35 periscopic sights were provided for the gunner and the tank commander.

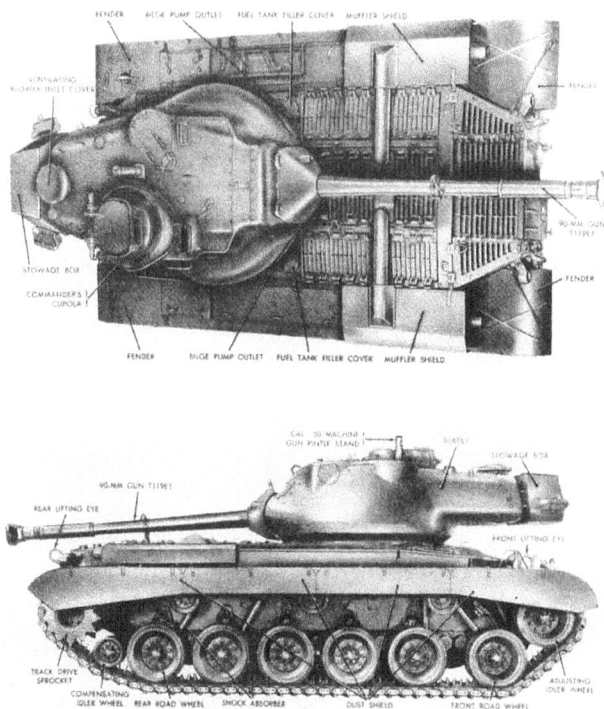

M47, serial number 25, is under evaluation at Fort Knox. The long barreled .50 caliber coaxial machine gun protrudes from the gun shield. Note the early location of the commander's .50 caliber machine gun pedestal mount. The range finder has not been installed on this tank and the infrared headlights have not been fitted.

A later production M47 is shown here during test at Fort Knox. The range finder is installed and the .50 caliber machine gun pedestal mount has been increased in height and relocated on the turret roof. Infrared headlights are installed with new brush guards for both light groups.

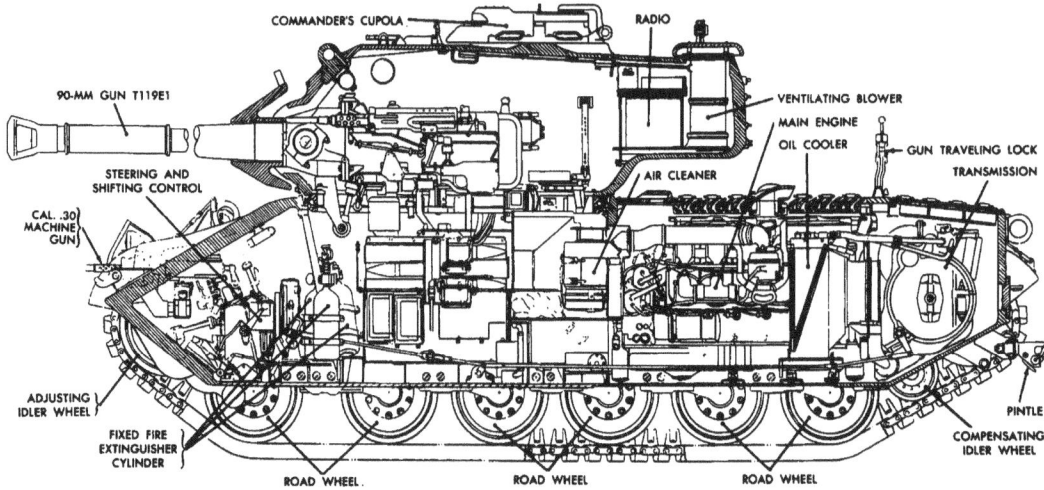

The interior arrangement of the early M47 can be seen in the sectional view above. Like the T42, this tank is armed with a .50 caliber coaxial machine gun. Note the stowage of the 90mm ammunition along the hull side wall.

During the tests at Fort Knox and Aberdeen, numerous deficiencies were detected. As expected, the automotive performance was generally satisfactory and the problems mainly concerned the new turret and fire control system. Fifteen modifications were considered essential before the M47 could be issued to the troops. These included changes in the power traverse and elevation system to eliminate oscillation or lag. A minimum requirement was that the performance should be equal to that obtained in the medium tanks M26 and M46. Most of these problems resulted from the fact that the new system had been designed originally to incorporate the IBM stabilizer. The components had not been tested

when the stabilizer was dropped and immediate production was ordered under the emergency conditions prevailing in the Summer of 1950.

The ballistic box was changed to include appropriate scales for the 90mm gun M3 ammunition, thus allowing the use of these standard rounds in addition to the high pressure ammunition for T119E1 gun. The T35 periscopes were modified to prevent water penetration and some interior turret components were shifted to improve accessibility. The pedestal for the tank commander's .50 caliber machine gun was increased in height and relocated from between the two turret hatches to the left front of the tank commander's cupola. These

A .30 caliber coaxial machine gun is installed in late M47 shown in the drawing below. It also has the taller, relocated mount for the .50 caliber machine gun on the turret roof. The 90mm gun is fitted with the T-shape blast deflector. Compare with the original muzzle brake in the top drawing. However, the cylindrical blast deflector was the most widely used.

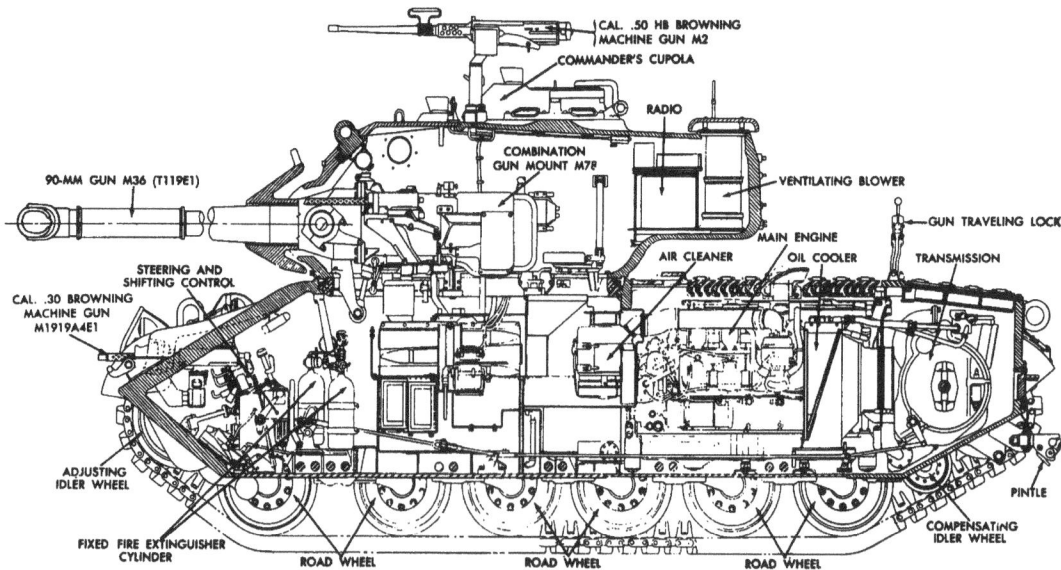

The views above and at the right show the details of the late M47. Except for the T-shape blast deflector which appeared very late in the life of the M47, these photographs illustrate the main features of the majority of the production tanks.

modifications were applied to the tanks already built and were introduced into the production lines at both Detroit Arsenal and American Locomotive.

On 8 April 1952, OTCM 34265 recommended standardization of the M47. This action, which was approved on 22 May, also nicknamed the M47 as the General Patton II. However, OTCM 34658 changed the name to the Patton 47. Later action also standardized the T119E1 gun as the 90mm gun M36. On later models of these guns, the cylindrical blast deflector was replaced by the new T-shaped deflector.

Production of the M47 continued until November 1953 for a total run of 8576 tanks. Detroit Arsenal produced 5481 M47s and the remaining 3095 were built at the American Locomotive Company. By this time, all efforts were being concentrated on the production of the new T48 tank and there was no further need for an interim vehicle.

A—SIDE ACCESS DOOR
B—CENTER ACCESS DOOR
C—BRAKE LINKAGE ACCESS HOLE COVER
D—TRANSMISSION DRAIN ACCESS HOLE COVER
E—OIL COOLER SERVICE HOLE COVER
F—ENGINE MOUNTING BOLT HOLE

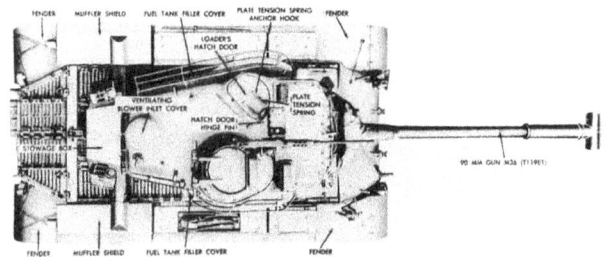

The bottom of the M47 hull can be seen from the front (below) and rear (at right) in these photographs. Note the escape hatch doors for the driver and bow gunner as well as the numerous access hole covers.

G—MAIN ENGINE DRAIN ACCESS HOLE COVER
H—DRAIN VALVE
J—FUEL TANK DRAIN ACCESS HOLE COVER
K—HULL DRAIN HOLE COVER
L—ESCAPE HATCH DOOR
M—MAIN ENGINE OIL FILTER ACCESS HOLE COVER
N—MAIN GENERATOR ACCESS HOLE COVER

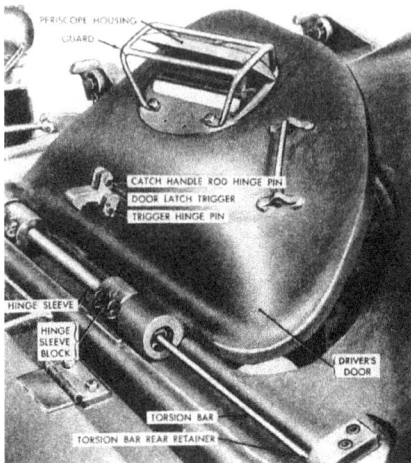

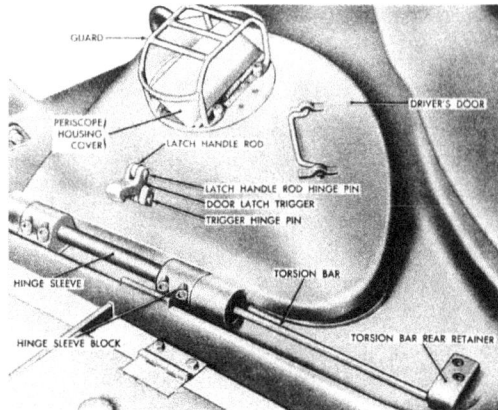

The driver's hatch covers from the early (left) and late (right) production tanks can be compared above. Note the change in the periscope guard design. Below are views of the early (left) and late (right) M47 driving compartment looking through the turret ring with the turret removed.

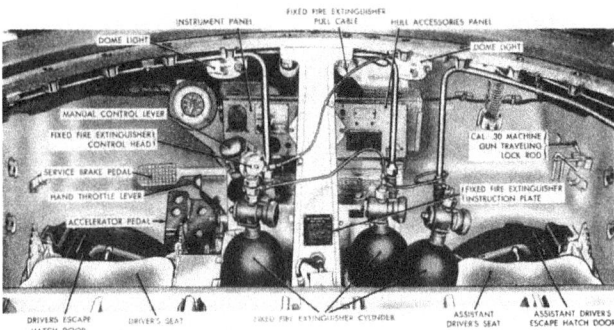

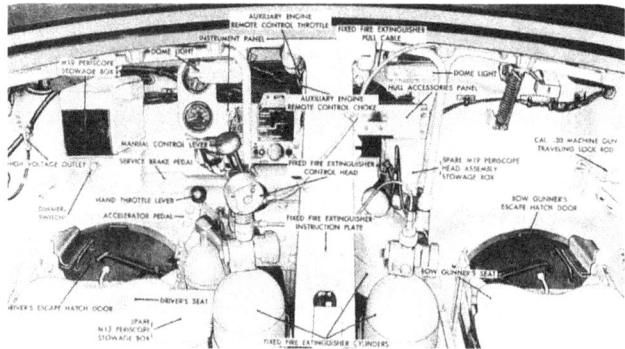

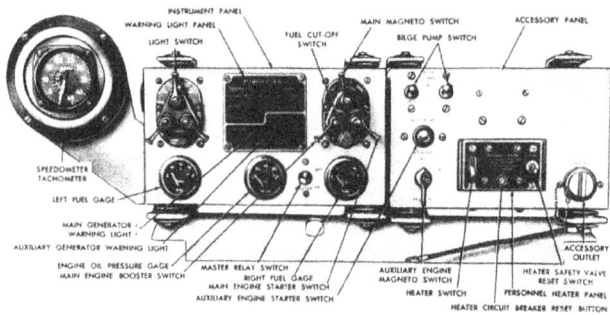

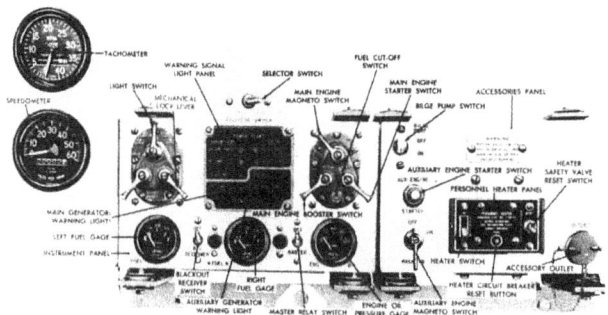

Above are the early (left) and late (right) M47 driver's instrument panels. The later version differed mainly by replacing the combination speedometer-tachometer with two separate instruments and adding controls for the infrared driving lights. Below, the driver's controls for the early (left) and late (right) vehicles are compared. Unlike in the M46, the bow gunner has no controls.

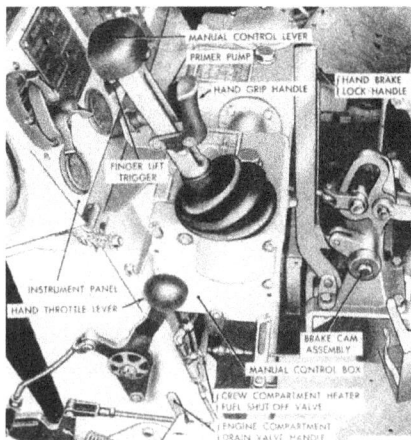

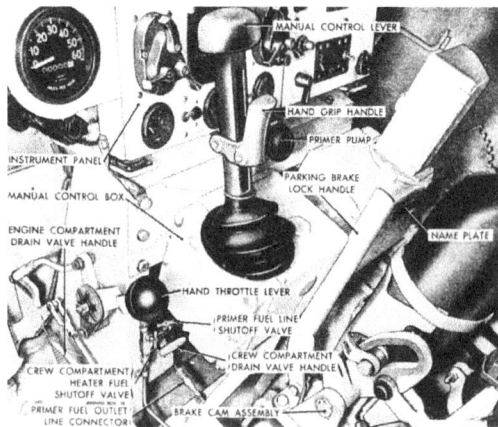

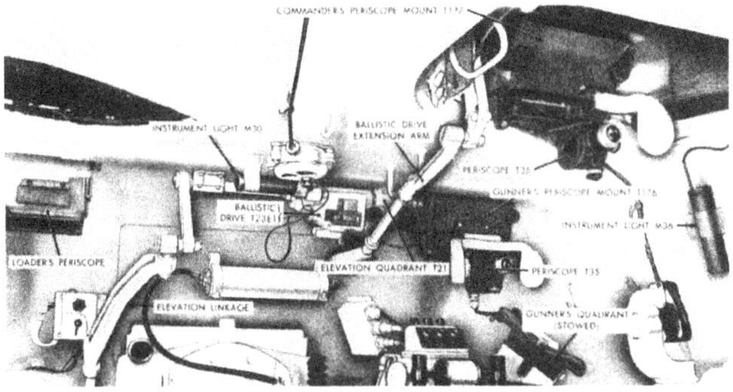

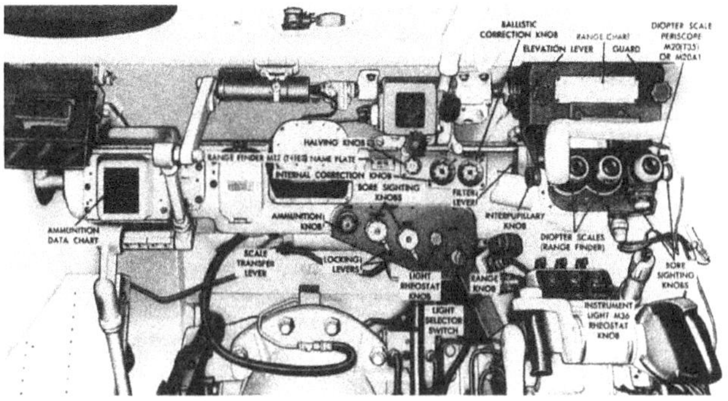

Fire control equipment for the early (left) and late (right) M47s is shown above. The early tank was not equipped with a range finder and relied on the T35 periscope as the primary sighting device. The late tank was fitted with the M12 range finder. No coaxial machine gun is installed in the photograph of the early tank. However, the view at the right shows a .30 caliber coaxial weapon. Below are the gunner's controls for the early (left) and late (right) production tanks.

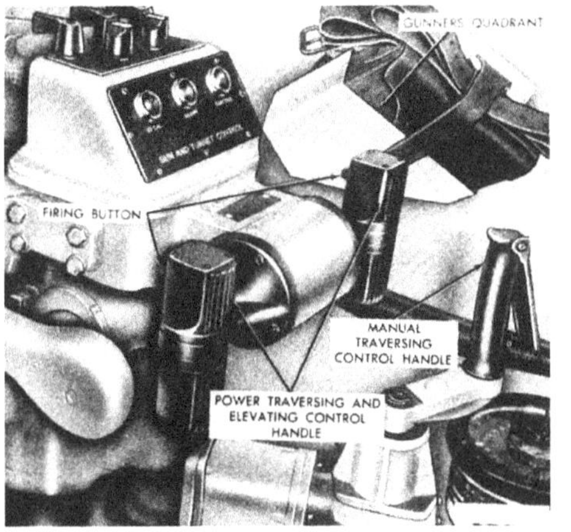

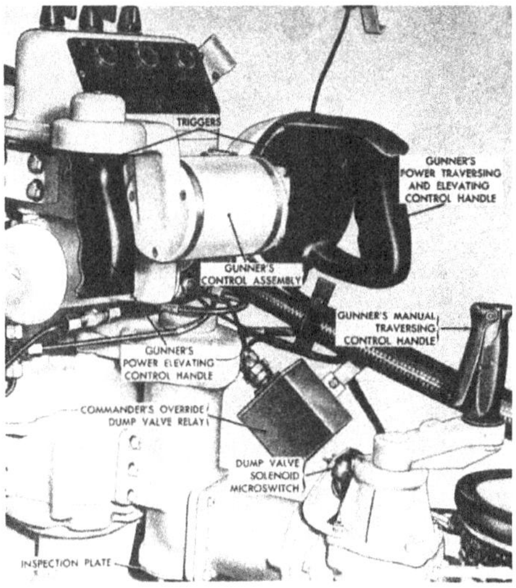

The commander's station in the M47 can be seen directly below and the loader's side of the turret appears below at the right. Both views represent the late production tank. The empty 90mm ammunition ready racks are visible in both photographs.

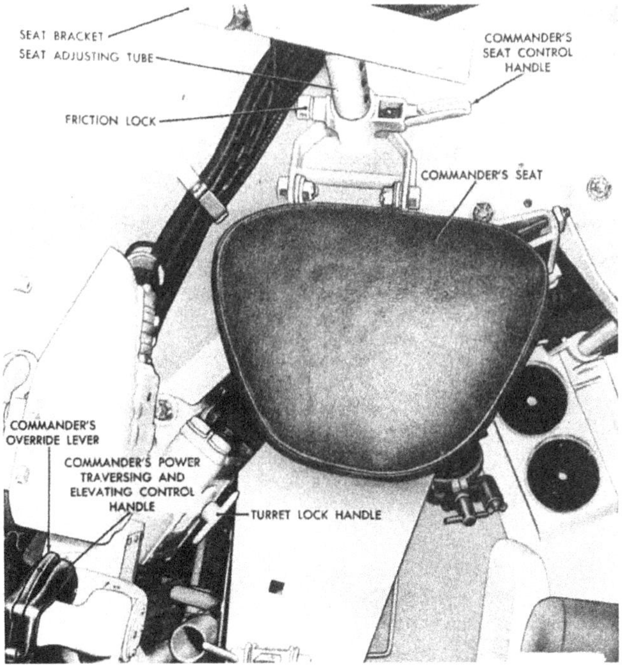

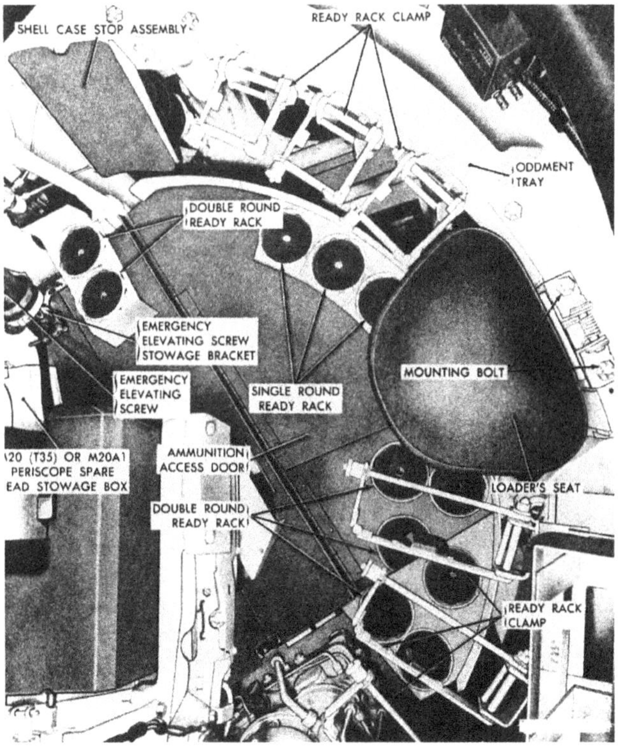

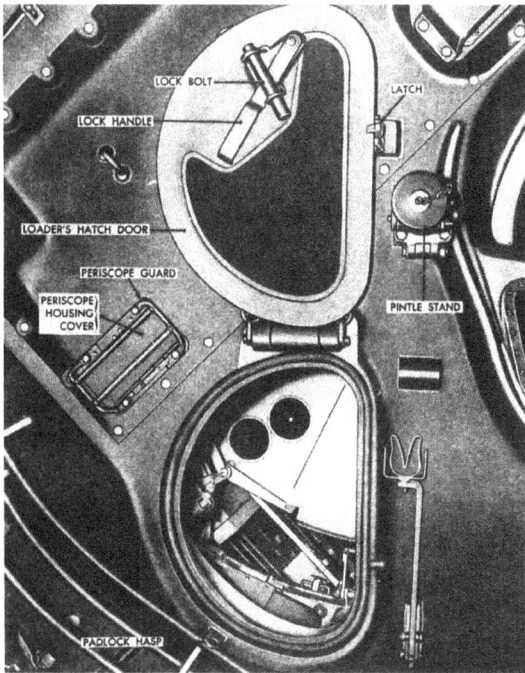

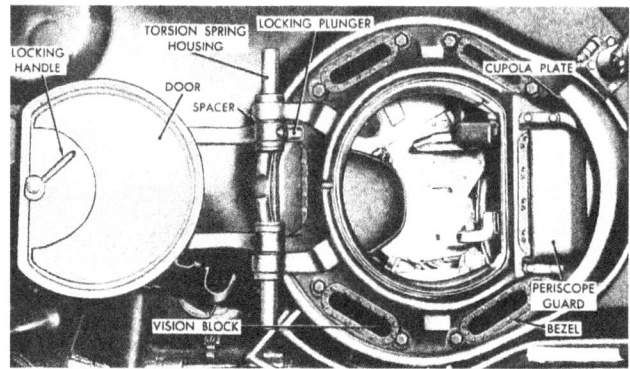

The loader's hatch cover is latched in the open position at the left. The front of the turret is at the upper left in this photograph. This is a late production tank as can be seen by the location of the pedestal for the .50 caliber machine gun. Note the stowage bracket for the .50 caliber machine gun barrel folded flat on the turret roof. Above is the commander's cupola on the late production M47. The front of the turret is at the right in this view.

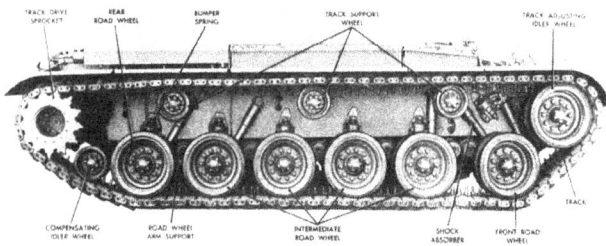

A right side view of the M47 suspension appears above. Unlike the M46, only three track return rollers are fitted. The early nomenclature refers to the compensating idler and the track tension idler as the track adjusting idler wheel and the compensating idler wheel respectively. At the right is an exploded view of the T84E1 double pin rubber track showing the top surface of the track links.

The right-hand front light group (left) and the left-hand front light group (right) can be seen below.

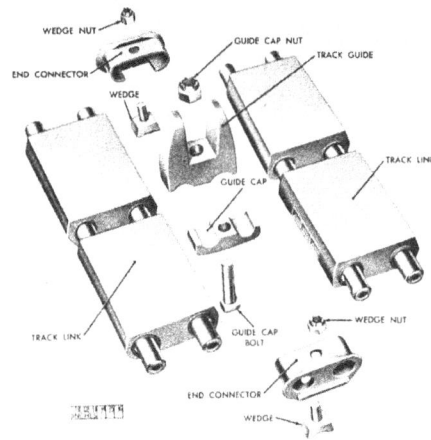

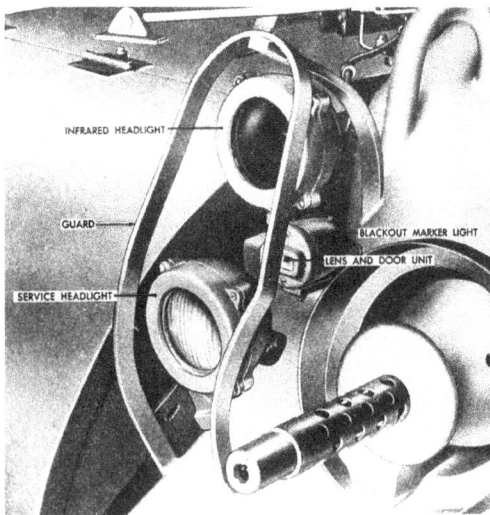

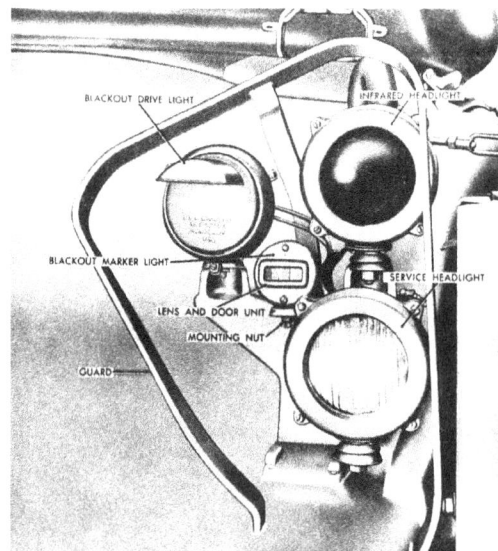

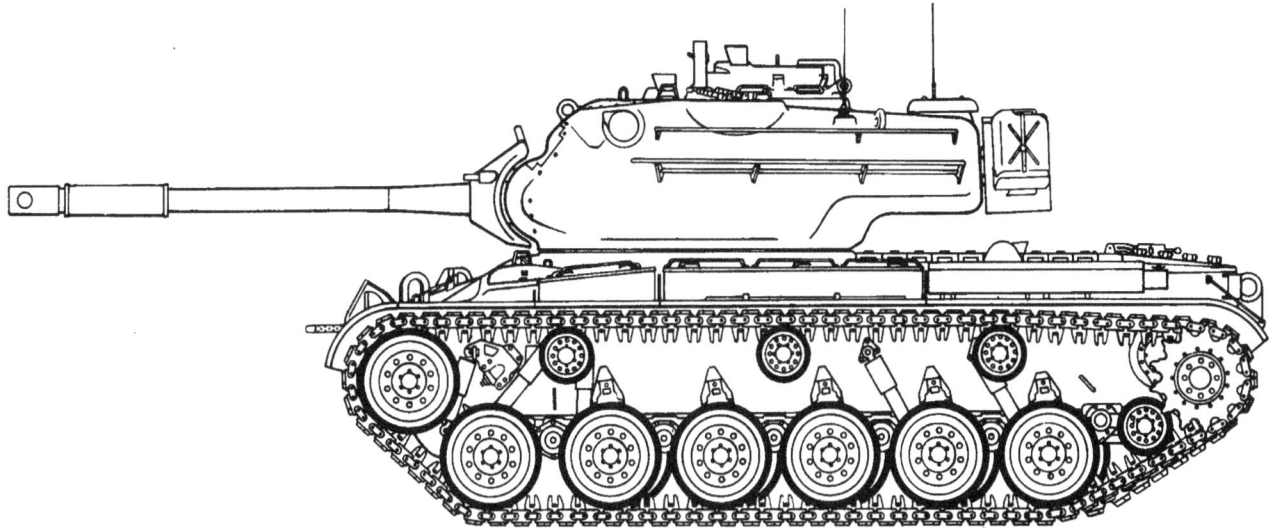

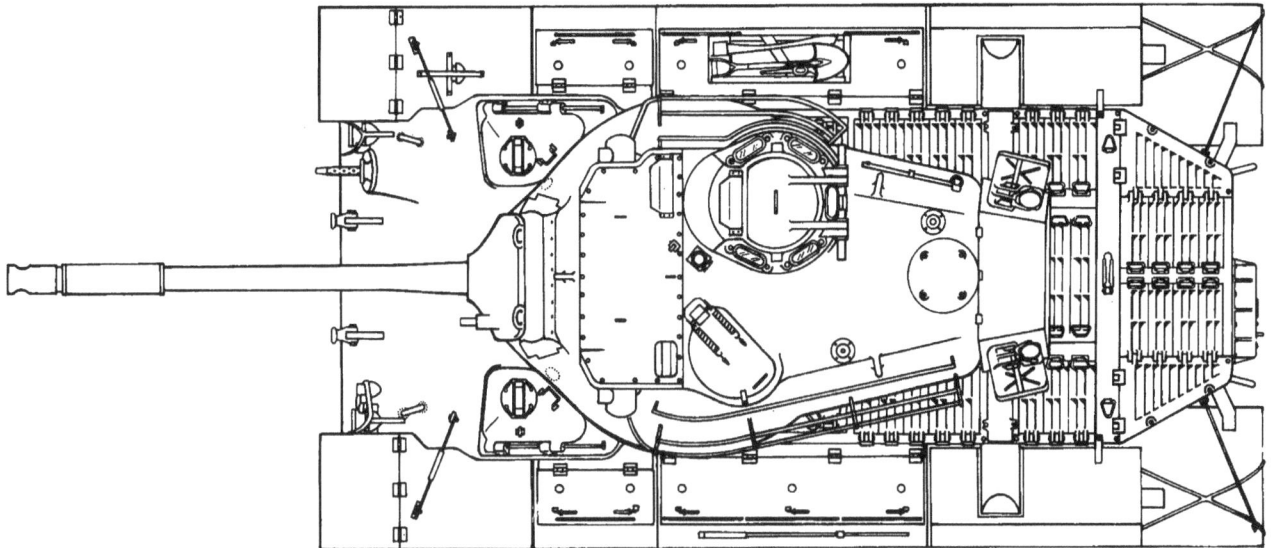

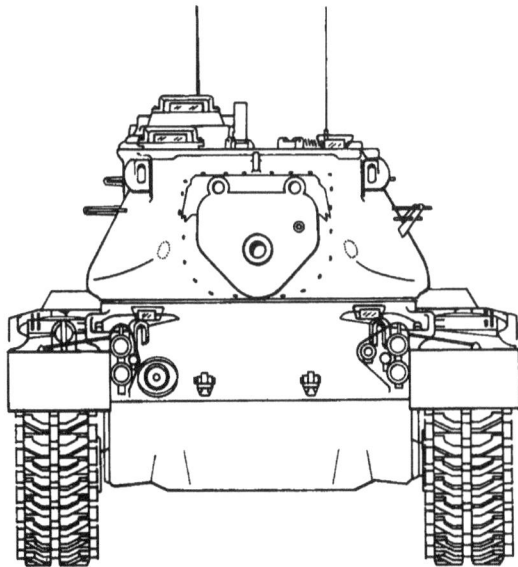

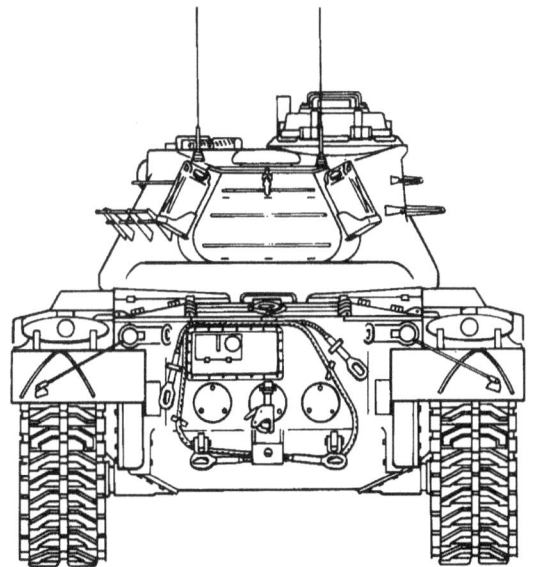

Scale 1:48

90mm Gun Tank M47

Above is a line up of M47s on the firing range at Vilseck, Germany on 4 October 1952. Below left, other M47s from the 66th Tank Battalion, 2nd Armored Division operate near Baumholder, Germany on 9 October 1952. The view below at the right shows an M47 during field testing near Big Delta, Alaska in April 1952. The latter tank is one of the early vehicles and has not been fitted with a range finder.

After completion of the essential modifications, the M47s were available for issue to the troops. At that time, the most dangerous threat appeared to be in Europe and early shipments of the new tanks were rushed to the American units in Germany. A small number also were shipped to Korea for battle testing, although by this time enemy armor had virtually disappeared and tank operations consisted primarily of infantry support. Some of these vehicles were equipped with 18 inch searchlights.

The M47 continued to serve in the U.S. Army until it was gradually replaced by the later tanks of the M48 series. It was then relegated to training duties and was supplied as military aid to numerous friendly nations.

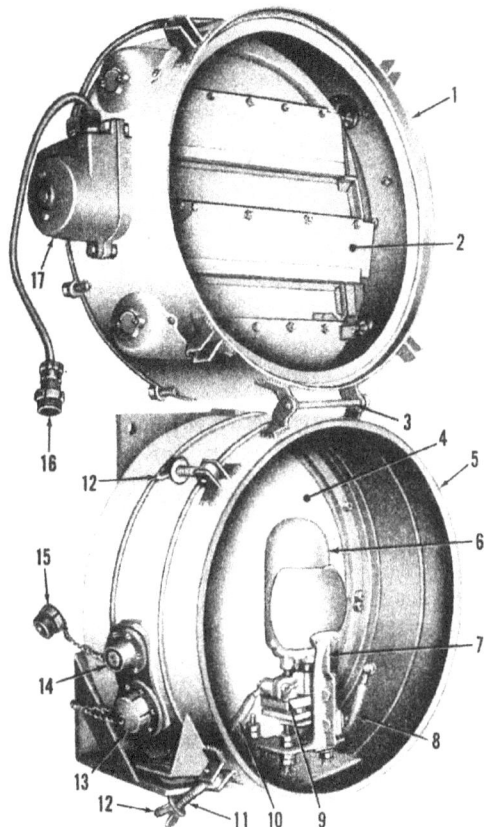

1	Mounting bracket	5	Wingnut
2	Mounting stud	6	Hinge pin
3	Shock mount retaining plate	7	Lamp drum assembly
4	Lamp housing assembly	8	Rubber mount
	9 Relay-to-lamp housing cable		

1	Lamp drum assembly	9	Receptacle thumbscrew
2	Shutter assembly	10	Receptacle-to-lamp lead
3	Hinge pin	11	Eyebolt
4	Main reflector	12	Wingnut
5	Lamp housing assembly	13	Large receptacle
6	Lamp	14	Small receptacle
7	Focusing mechanism assembly	15	Receptacle cap
8	Lamp-to-case ground lead	16	Small cable plug
	17 Solenoid housing		

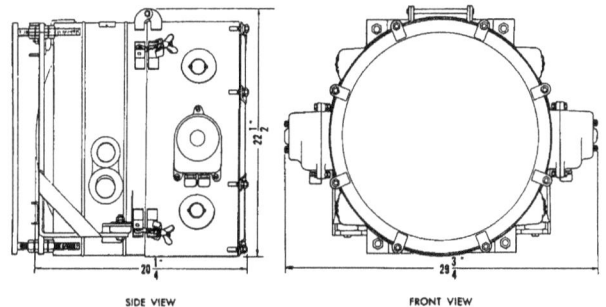

SIDE VIEW   FRONT VIEW

Details of the 18 inch diameter Crouse-Hinds searchlight can be seen above. The view at the top right shows the searchlight mounted on the gun shield of an M47. This searchlight operated at 1000 to 2000 watts depending upon the generator capacity. This is the same searchlight mounted on the M46 and early M48 series tanks. Below, M47s of the 143rd Tank Battalion pause during training near Vilseck, Germany on 26 November 1952.

Details of the rear deck and stowage arrangements can be seen on the M47 above during operations near Vilseck, Germany on 22 November 1952. Below, the driver (left) and the tank commander (right) man their stations on an M47 of A Company, 67th Tank Battalion, 2nd Armored Division at Belsen, Germany on 27 October 1952.

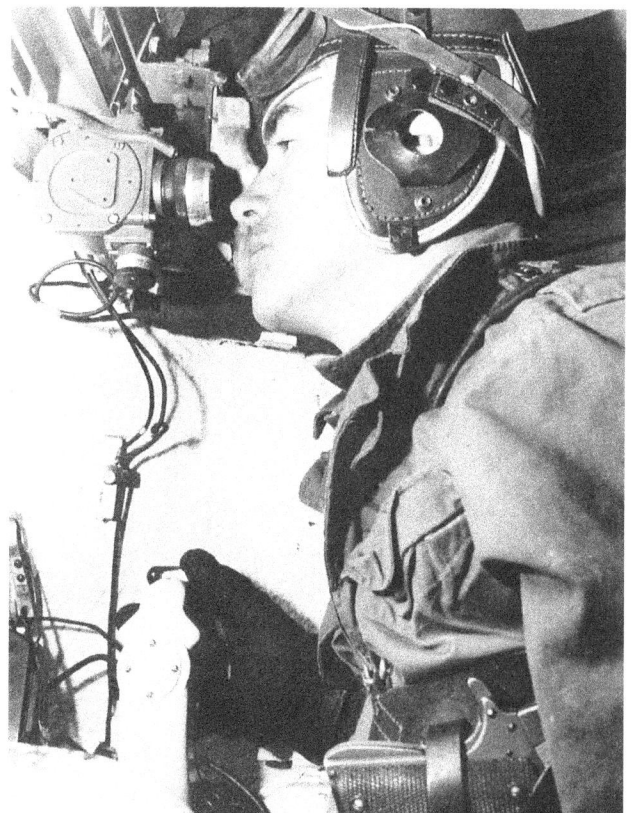

These M47s from the 143rd Tank Battalion operating near Vilseck, Germany have the typical stowage added by troops in the field. Both have the range finder and the tall forward mounted pedestal for the .50 caliber machine gun. Note the blocking timber carried on both vehicles.

Above, the external interphone box is in use on this U.S. Marine Corps M47 during training at Quantico, Virginia on 18 February 1953. Below are two views of another marine M47, also at Quantico.

Above, an M47 crew from the 3rd Platoon, A Company, 89th Tank Battalion displays their equipment for inspection in Korea on 3 November 1953. This battalion was assigned to the 25th Infantry Division. Below (left), the power pack is removed from an M47 at Fort Hood, Texas during field exercises and (right) a king-size wrench is used to adjust the track tension on an M47 from A Company, 40th Tank Battalion at Freidberg, Germany.

A new M47 is being unloaded above at Bremerhaven, Germany. Details of the suspension are clearly visible in this photograph. Below, a variety of stowage arrangements can be seen on these M47s during field exercises in Germany. The tank on the left is from the 29th Tank Battalion of the 2nd Armored Division and the one on the right belongs to B Company, 40th Tank Battalion. Both photographs were dated June 1953.

The M47s below were from the 141st Tank Battalion in Germany. The tanks at the left are loaded on an LCT for crossing the Rhine River as part of the Blackland Forces during the Power Play exercises on 24 October 1953. The photograph at the right was taken at the Grafenwoehr training grounds on 12 May 1954.

Above left, an M47 from C Company, 759th Tank Battalion swings its turret to fire at Aggressor forces during exercise West Wind near Schwabisch Gmeund, Germany on 23 October 1954. Note that the .50 caliber machine gun on the turret has been replaced with a .30 caliber weapon. At the right above, an M47 from B Company, 759th Tank Battalion overwatches advancing infantry during operations near Munich on 25 January 1955.

Above left, an M47 from B Company, 29th Tank Battalion fires its 90mm gun during exercise Cordon Bleu in October 1955. This photograph clearly reveals the contour of the front armor around the bow machine gun. Above right, a U.S. Marine Corps M47 operates with the 9th Marines in training on Iwo Jima during November 1956. Below, the 70th Tank Battalion moves into firing position while on maneuvers near Mount Fuji, Japan on 9 February 1956.

The installation of pancake mufflers required changes in the rear deck of the M47. Above, the modified grill arrangement necessary to clear the new mufflers can be seen at the left and the mufflers themselves, with the rear deck removed, appear at the right. The original fender mufflers are still in place in the left view, although they are no longer in use. The tank used for these tests was serial number 6, registration number 30164213.

The M47s at Aberdeen and Fort Knox were used to evaluate numerous components many of which were applied to later vehicles. Such tests improved the reliability of standard equipment and evaluated new designs such as the pancake mufflers. Use of these mufflers resulted in a somewhat higher noise level, but the red glow and exhaust flames were greatly reduced when compared to the standard fender mounted mufflers. M47, registration number 30164240, was used to evaluate a non-recoil mount for the 90mm gun. The firing tests with the cannon rigidly mounted in the turret indicated that such a mount might be suitable for future tank designs. Another experimental version of the M47 was proposed by the Yuba Manufacturing Company. Their concept study recommended the installation of a steam power plant as the main propulsion unit. Although the original concept study indicated advantages in weight and operating range for such a tank, there were many unknown factors requiring a considerable development effort. The Department of Defense did not feel that the necessary development costs were justified and the proposal was rejected.

Although the stabilizer had been eliminated when the T42 turret was applied to the M47, there was still great interest in fully stabilizing the main armament. The successful application of a stabilizer in the British Centurion tank resulted in the consideration of the same system for the M47. OTCM 33598 authorized the procurement of 20 M47s with the Centurion stabilizer on 1 March 1952. The vehicle was designated as the 90mm gun tank M47E1 and the mount became the 90mm gun mount T171. Other experiments utilized the Minneapolis-Honeywell stabilizer system. Six of these units were authorized for procurement by OTCM 35074 on 23 October 1953. Although tests indicated improved gunnery performance when using a stabilizer, production of the M47 tank was coming to an end and they were not adopted.

The T139 90mm gun from the later M48 tank also was modified for installation in the M47. Designated as the 90mm gun T139E1, it gave the M47 a lighter weight weapon with a quick change tube. However, it was never introduced into production tanks.

Below, the 90mm gun M36 (T119E1) (left) is compared with the 90mm gun T139E1 (right). The bore evacuator and blast deflector have been removed from the standard weapon. The T139E1 gun required a new concentric recoil system for installation in the M47. This was proposed for the M47E2.

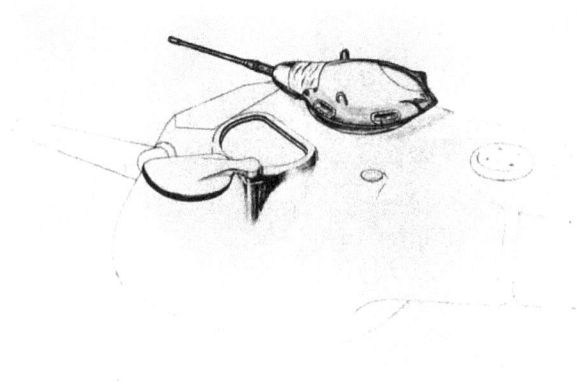

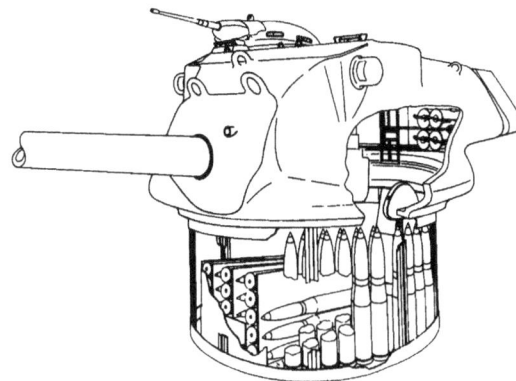

Above (left) is a proposal drawing for the turret of the M47E2 showing a new loader's hatch and the installation of the Aircraft Armaments Model 108 cupola for the tank commander. At the right is a sketch of the proposed turret ammunition stowage in the M47E2. With 23 rounds in the hull, the 44 in the turret brought the total 90mm ammunition stowage to 67.

As the results of the testing program became available, it was obvious that major modifications could greatly improve the operational efficiency of the M47. For example, the limited access to the ammunition stowed beneath the turret basket made it extremely difficult to transfer rounds into the turret. A program was initiated in November 1952 to modify the design with particular attention to improving the fighting compartment layout as well as the gun and turret control mechanisms. Two tanks were authorized for modification and were designated as the 90mm gun tanks M47E2 by OTCM 35000. One major change proposed was the elimination of the bow machine gunner and his weapon and the installation of a 90mm ammunition rack in the right front hull. This arrangement was similar to that in the T42. It was not intended to place the M47E2 in production at that time, but the improved design was to be available if new production was required. It also could be used if it became desirable in the future to rework and upgrade the M47 tank fleet.

The M47E2 mock-up was reviewed at a meeting on 29 June 1954. After some changes, approval was granted at a second meeting on 2 December to start construction of the two pilot tanks. Modification of the first pilot from a standard M47 began the same month and was completed in January 1955. The tank was then shipped to Aberdeen Proving Ground for general automotive, engine cooling, and infrared suppression tests which continued for the rest of the year.

Conversion of the second pilot M47E2 was completed during July 1955 and it remained at the Vehicle Design Agency, ALCO Products, Inc. (formerly American Locomotive Company) to await the results of tests on the first prototype. It was intended that this vehicle would be shipped to Fort Knox for user evaluation after incorporation of any changes indicated by the tests at Aberdeen. The drawings for the modified tank were prepared in kit form so that the various changes could be applied to the standard M47 independent of one another.

Although the M47E2 design was never produced, the widespread use of the M47 by other nations led to a large number of similar modernization programs. These efforts were concentrated in two general areas. The first was to improve the automotive performance of the tank and the second was to increase its firepower. The various rework programs covered one or both of these areas depending upon the requirements of the user and the financial resources available.

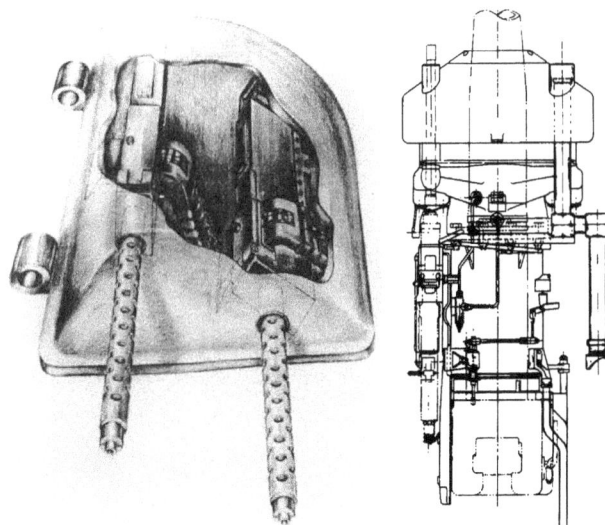

At the near right is a proposed twin .30 caliber machine gun mount for installation in the former bow gunner's hatch on the M47E2. The weapons were to be loaded and fired by the driver. At the far right is a sketch of an offset telescope proposed for the M47E2. Consideration also was given to extending the optical system of the range finder to permit operation by the tank commander.

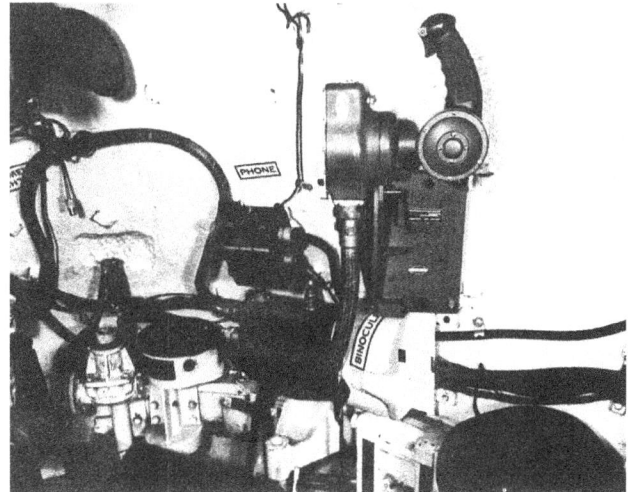

Efforts to improve the gun controls in the M47 resulted in the evaluation of several new systems. These are illustrated on this page with, in each case, the gunner's station appearing in the view at the left and the commander's control handle at the right. Above is the electric or amplidyne gun control system and immediately below is the constant or controlled pressure hydraulic system developed by the Cadillac Gage Company. The British Vickers system appears at the bottom of the page.

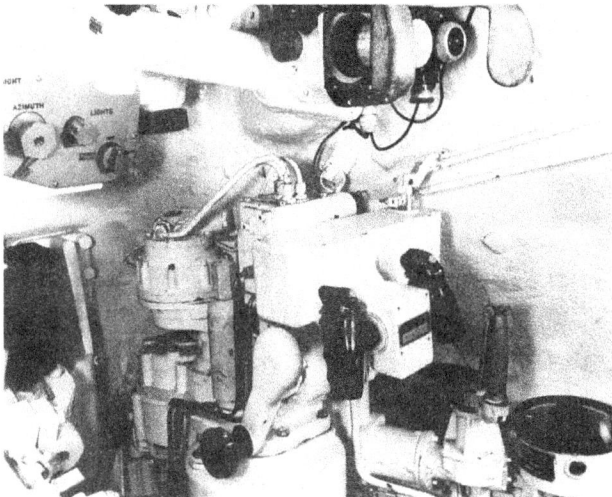

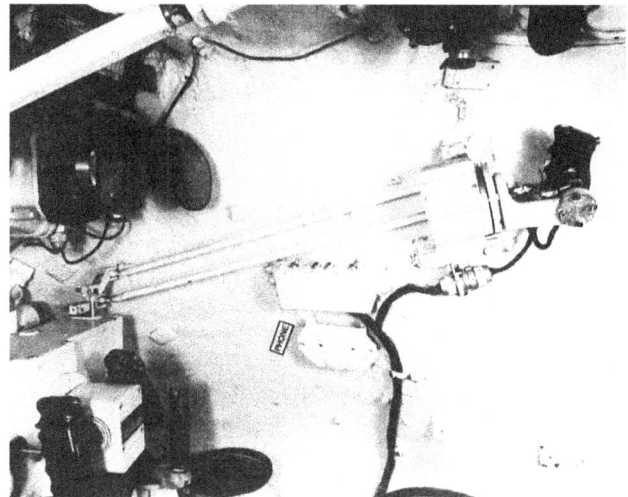

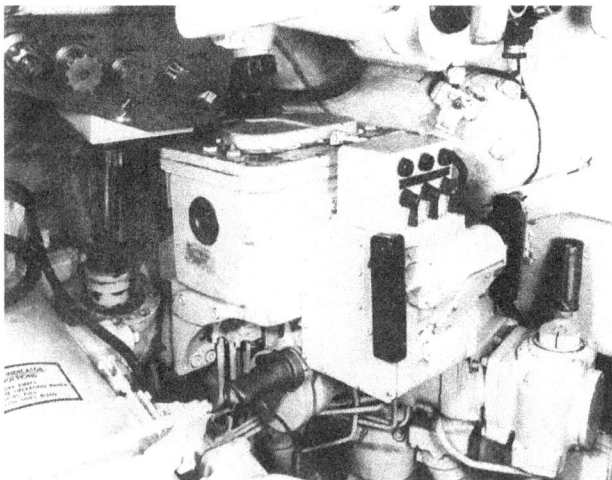

At the right is the M47 converted in France to fire the 105mm non-rotating shaped charge round developed for the AMX-30 tank. The M47 was modified by installing a 105mm barrel, reworking the breech assembly, and replacing the ammunition racks.

The firepower of the M47 could be increased easily by replacing the M36 90mm gun with a more powerful weapon. In 1967, the French 105mm gun, which fired a non-rotating shaped charge projectile, was successfully mounted in the M47 turret. The hollow charge round would penetrate 5.9 inches (150mm) of armor at 60 degrees obliquity. The British 105mm L7 gun or its U.S. equivalent, the 105mm gun M68, could be installed with a minimum of modification. Thus the M47 could be provided with the same armament as some of the most modern battle tanks.

The greatest deficiency in the automotive performance of the M47 was the short range resulting from the high fuel consumption of its gasoline engine. The MB 837 Ea-500 diesel engine manufactured in Germany by Motoren-und-Turbinen Union (MTU) provided one solution to this problem. Used successfully in the Swiss Pz 61 main battle tank, the MTU engine was readily adapted to the M47. Coupled to a slightly modified CD-850-4 transmission, the water-cooled 750 hp diesel consumed fuel at slightly more than half the rate of the AV-1790. This installation was successfully tested in both Italy and Pakistan.

The worldwide interest in upgrading the performance of the M47 led to another design study at the U.S. Army Tank Automotive Command in 1968. This study proposed a modernization program for the M47 utilizing a maximum number of components from the latest M60A1 battle tank. Since these components were readily available in the U.S. Army supply system, the logistical support of such a modified M47 would be greatly simplified. This concept was further refined and the detail design work

At the right is the M47 powered by the MTU diesel engine MB837 Ea-500. Below, the power plant is installed in the M47 (left) with the top deck and grill doors removed. As the power pack is removed (below right), the modified CD-850-4 transmission also can be seen.

A new M47-M appears here after completion at Bowen-McLaughlin-York Company. Note the personnel heater exhaust is now installed in the former bow gunner's hatch and extends toward the right side of the tank. The M60 type rear deck, grill doors, and fender mounted air cleaners also are apparent.

The modified suspension of the M47-M is clearly visible in this side view. The track tension idler has been removed and the last road wheel on each side has been shifted to the rear.

was completed by Bowen-McLaughlin-York Company (BMY) in 1969. Referred to as the M47-M, the modified design replaced the gasoline engine with the new Continental AVDS-1790-2A diesel used in the M60 tank series. The rear of the hull was modified to use the same top deck and grill doors as the M60s. To provide complete interchangeability with the standard M60 engine and hull components, relocation of the last pair of road wheels $3^{13}/_{16}$ inches to the rear was necessary to give adequate clearance between the engine oil pan and the torsion bars. The original cross drive transmission was modified to the latest CD-850-6A standard. The M47's hydraulic shock absorbers were replaced with friction snubbers and the track tension idlers were removed.

Details of the rear hull modification can be seen below at the right. Note that the right-hand access hole cover is removed in this photograph. At the left, the welded steel plug is shown sealing the former port for the bow machine gun.

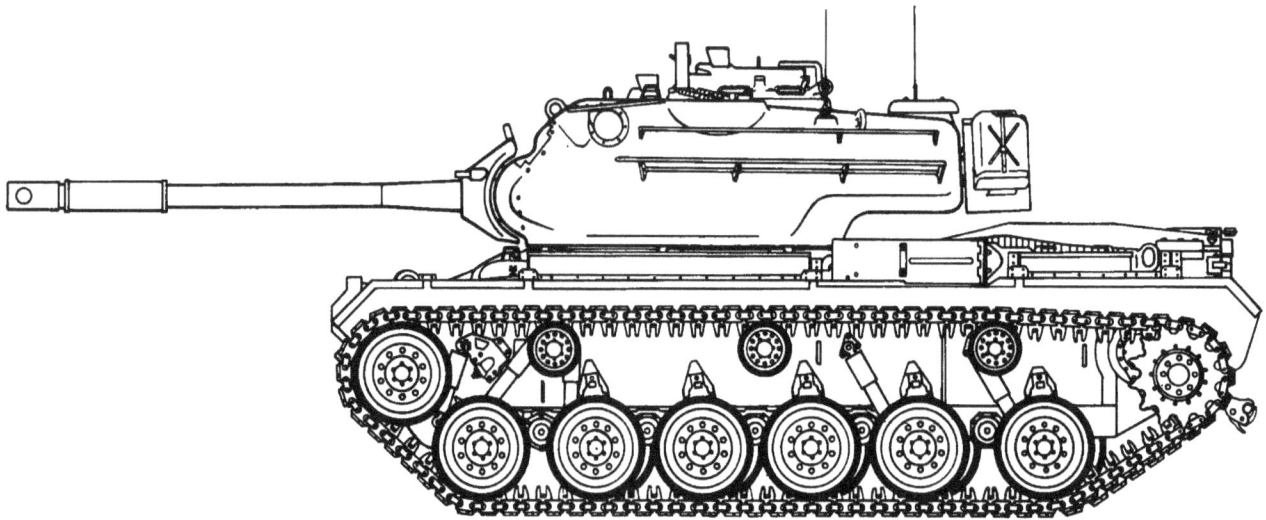

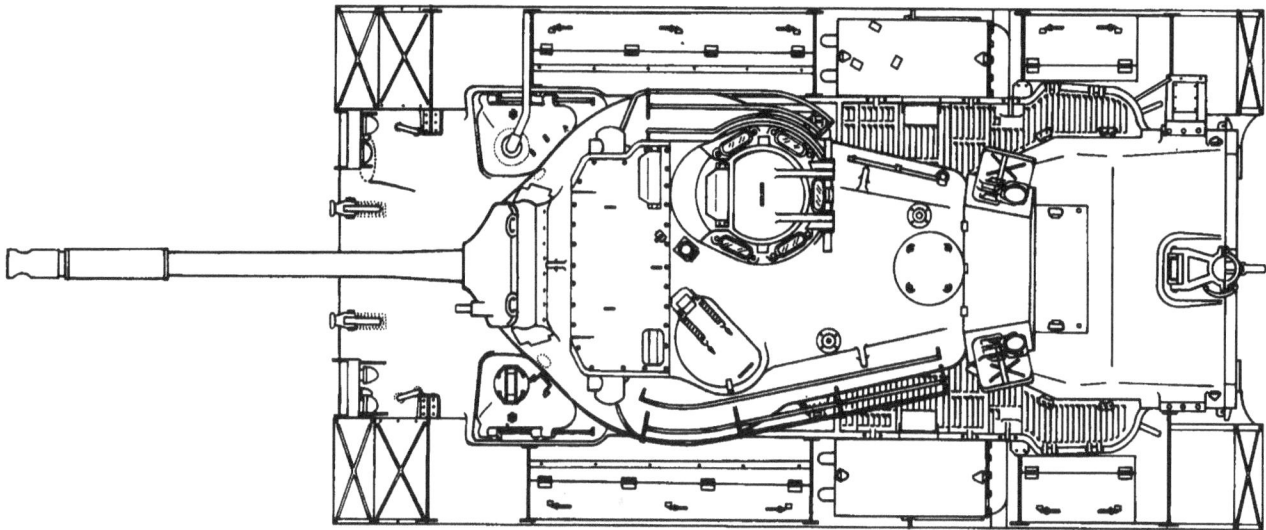

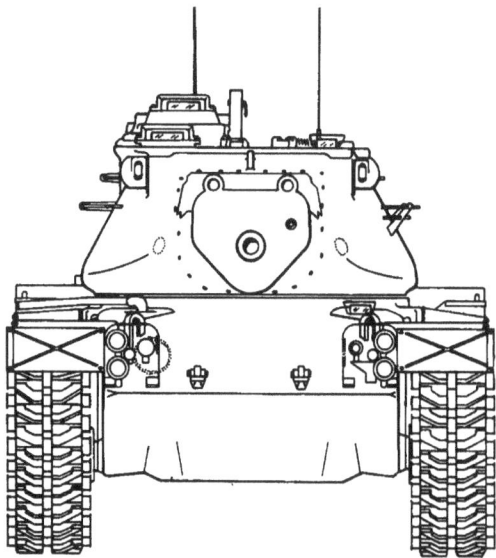

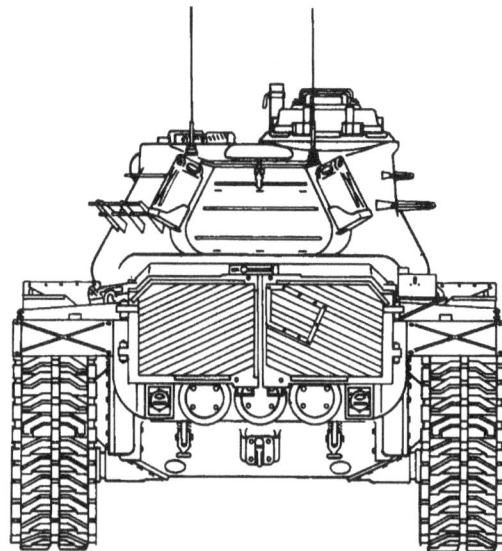

90mm Gun Tank M47-M

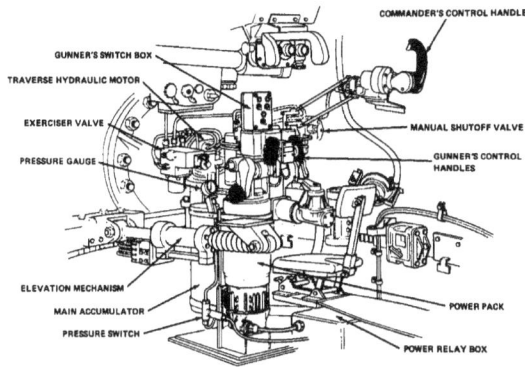

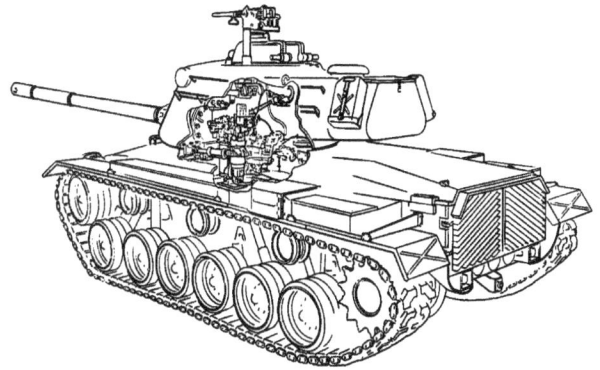

On the M47-M the bow gunner was eliminated and the opening for the bow machine gun was plugged to maintain the full protection of the front armor. The space obtained was used for the installation of a 22 round 90mm ammunition rack. The other ammunition stowage was rearranged greatly improving the access to the rounds. Also, the total number of 90mm rounds carried was increased to 79 compared to the 71 in the original M47. A new Cadillac Gage hydraulic gun control system was installed, identical to that in the latest U.S. tanks.

Between February 1972 and March 1974, BMY converted 400 M47s to the M47-M configuration for Iran and between March 1976 and August 1977, an additional 147 of the tanks were modernized for Pakistan. Both conversion jobs were carried out in Iran. Later, a similar project in Spain modernized over 300 M47s for the Spanish Army. This work was performed by Chrysler España and the modernized tanks were referred to as the M47E. In Israel, Urdan RKM and Israel Military Industries developed a modernization program for Israeli M47 tanks to bring them up to the standard of the latest M60A3s. This work was divided into two phases with phase 1 covering the automotive modifications such as the AVDS-1790-2C diesel engine installation. Phase 2 replaced the 90mm gun with the 105mm gun M68 in a modified M140 mount and installed a new fire control system. The improved tank was designated as the M47RKM. Similar modernization programs covering both armament and automotive components were carried out by Oto Melara in Italy.

Although designed as an interim tank to meet the emergency requirements of the early 1950s, the M47 was basically a sound design. Easily modified, it was capable of modernization to meet rapidly changing battlefield conditions and will continue in service for many years in armies all over the world.

The installation of the Cadillac Gage gun control system in the M47-M can be seen above. details of the system appear at the top left. The drawings below show the ammunition stowage arrangements in the hull and turret.

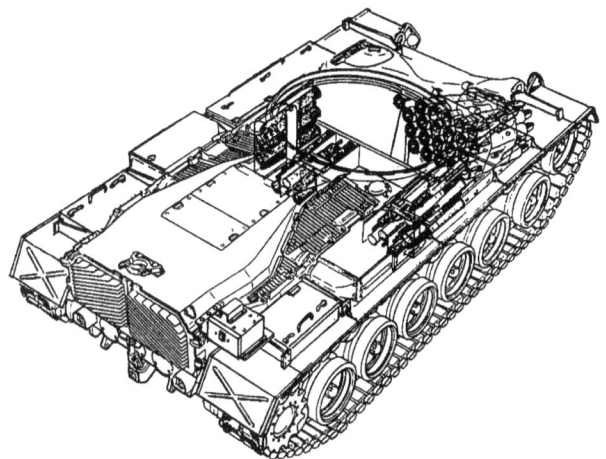

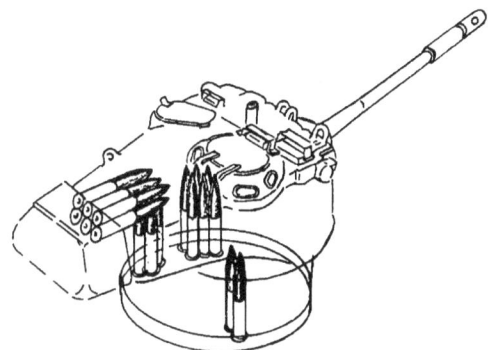

The two views below show the M47E as modified by Chrysler España. The tank still retains the 90mm gun.

This is the M47 modified by Oto Melara in Italy. With the diesel power plant, infrared suppression rear deck, and 105mm armament, this vehicle illustrates how the M47 could be brought up to the standard of a modern main battle tank.

PART II

THE NEW STANDARD MEDIUM TANK

T·48

Above is an artist's concept of the T48 late in the design program. Note the two coaxial machine guns and the track tension idler. Although included in the design, the latter was not installed on the first T48s.

DEVELOPMENT OF THE BASIC TANK

The decision taken during the Summer of 1950 to produce the M47 was clearly recognized as a stopgap solution to the medium tank problem. Introduction of the M47 would allow production to continue after the end of the M46 conversion program filling the interval required to design a modern medium tank acceptable to the Army Field Forces. The situation was in many ways similar to that a decade earlier when the medium tank M3 was rushed into production after the fall of France in June 1940. Using a sponson mounted 75mm gun, the M3 could be produced immediately without waiting for the development of the M4 Sherman with its turret mounted cannon. Although the M3 design was recognized as inadequate, the early availability of new medium tanks in large numbers had a decisive effect on the course of World War II, both in battle and in the training of armored troops. A similar crisis was perceived during the Summer of 1950 and the M47 was intended to fill the gap. Even the unusual practice of assigning the M47 designation immediately without waiting for standardization was similar to that with the earlier tank which was known as the M3 from the beginning without ever having an experimental T number.

Parallel with approval of M47 production, a maximum effort began to design its successor. The new tank was to be powered by the same engine and transmission as the M47. Use of this proven combination would ease the transition between the two vehicles and would help to answer the criticism of the Army Field Forces about the low power of the T42. The proposed design concept was essentially a shorter, lighter version of the experimental heavy tank T43. An outstanding feature of that vehicle was an elliptically shaped cast hull and turret. This shape permitted the maximum armor protection for a given volume with a minimum of weight. Concept studies of a medium tank using the elliptical hull configuration had been prepared by Joseph Williams of the Ordnance Tank Automotive Command as early as 1948. Williams had developed the elliptical hull and turret design for the T43 and also played a major role in the design of the T37/41 light tanks and the T42/M47 turrets.

In addition to the elliptical hull and turret configuration, the new medium tank retained the 85 inch diameter turret ring of the T43, although a new 90mm gun replaced the 120mm weapon on the heavy tank. The large ring diameter allowed the turret walls to slope smoothly down to the hull top, eliminating any shot traps except in the rear or directly under the gun mount. It also provided adequate space for heavier armament in the future.

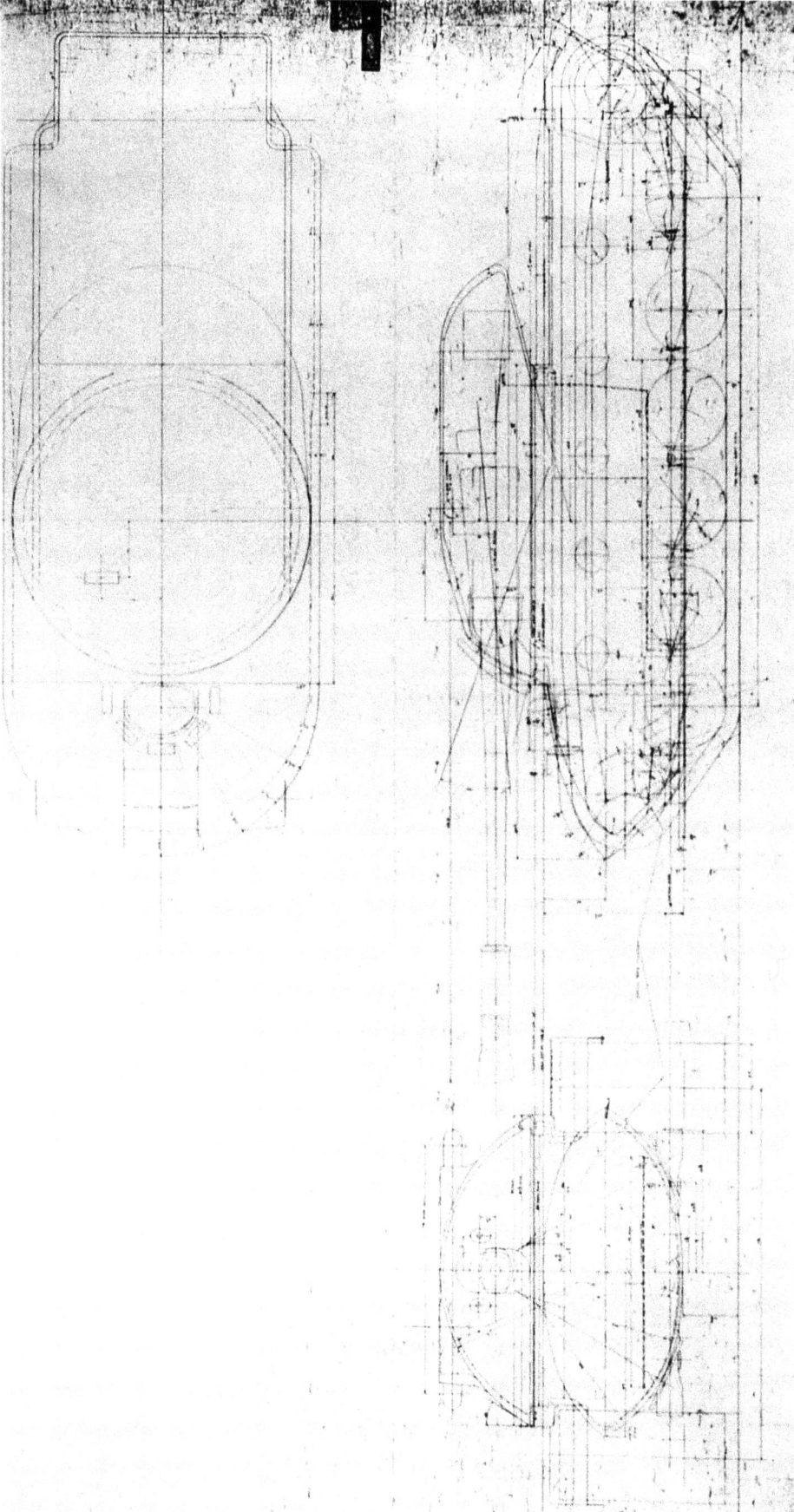

The concept of a tank with a cast elliptical hull and turret was proposed in this drawing by Joseph Williams dated 1948. The drawing was delivered to Chrysler and provided the basis from which the T48 design was developed. The original drawing is in the Patton Museum at Fort Knox, Kentucky.

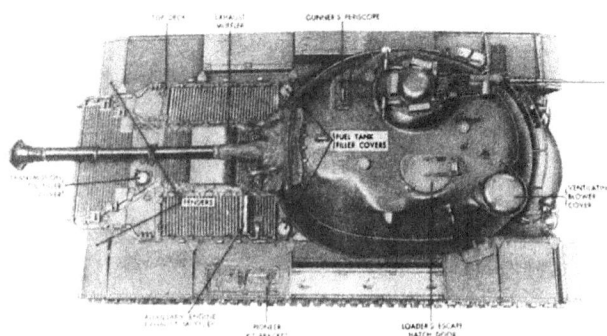

This is the T48 as originally assembled with a muzzle brake on the 90mm gun. It was soon replaced by the cylindrical blast deflector. Note that the main engine exhaust impinges directly onto the gun travel lock. These views were taken from the T48 Technical Manual and appear to have been the subject of some over zealous air brush work. Both the driver's hatch and the personnel heater exhaust have been obliterated in the photograph at the top right.

During October 1950, Ordnance indicated that it intended to give a letter order to Chrysler Corporation to complete the design and start production of the new medium tank. In a meeting at Detroit Arsenal on 8 November, Colonel William Call provided a ⅛ scale drawing of the new tank concept to Mr. Chester C. Utz, Manager of Chrysler's Ordnance Development Department. Chrysler's initial job was to complete the detailed design and build six pilot tanks in the shortest possible time. Five vehicles were for test by the Army and one was for the U.S. Marine Corps.

Work began immediately on layout drawings and the construction of a ½ size clay model of the hull and turret. The model was approved by Ordnance on 2 February 1951 and advance drawings were provided to the foundries for the hull and turret castings. To simplify future production, it was intended to maintain the maximum interchangeability of parts between the new tank and the heavy tank T43. This required considerable redesign of the earlier vehicle, particularly in the suspension system.

The program was already well advanced when Ordnance Committee action officially initiated the project on 27 February 1951. OTCM 33791 outlined the characteristics of the new vehicle and designated it as the 90mm gun tank T48. A maximum weight of 90,000 pounds was specified with a track width of 28 inches to reduce the ground pressure. A four man crew consisted of the tank commander, gunner, and loader in the turret plus the driver located in the center front of the hull. A new lightweight 90mm

gun, identical in ballistic performance with the M47's T119, provided the main armament. Later designated as the 90mm gun T139, the new weapon featured a quick change tube. Attached to the breech ring by interrupted threads, the tube could be rapidly replaced without removing the mount or breech ring from the turret. The weight reduction of the new gun over the earlier T119 was achieved by reducing the recoil surface on the tube and the removal of balancing weight from the breech ring. The new tank's secondary armament included two coaxial machine guns. A .50 caliber weapon was mounted to the left of the cannon with a .30 caliber gun on the right. An additional .50 caliber machine gun was carried in a remote control mount on the tank commander's cupola. Although it could be aimed and fired with the hatch closed, the tank commander had to expose himself to reload the weapon. The turret was designed to incorporate a range finder operated by the tank commander, but it was not expected to be available for the early production tanks.

The chassis of the first pilot T48 was shipped from Chrysler's Ordnance Development Department at Highland Park, Michigan to the company's new Chelsea, Michigan proving ground on 29 January 1952 for a two week shakedown run. It was then returned to Highland Park and fitted with the turret which had been under test in the Chrysler Ordnance Development Laboratory. The tank was returned to Chelsea on 27 February to begin a long series of tests observed by the Army Field Forces representatives from Fort Knox.

These views show T48, serial number 2, at Aberdeen Proving Ground on 12 April 1952. The muzzle brake on the 90mm gun has been replaced by the cylindrical blast deflector adopted for the production tanks. Both coaxial machine guns are installed, but only the long barrel of the .50 caliber weapon on the left side protrudes through the gun shield. The personnel heater exhaust can be seen above on the hull roof just below the .50 caliber machine gun.

As early as July 1951, Chrysler had received approval to build the second of the six T48 pilots at their new plant at Newark, Delaware. This plant had been built specifically for the production of the T48. Although assembled during December 1951, T48 number 2 was held at Newark until modifications resulting from the early tests on pilot number 1 could be incorporated. It was then shipped to Aberdeen Proving Ground arriving on 11 April 1952. Tank number 3 was completed at the Ordnance Development Laboratory and delivered to Fort Knox on 24 May where it was followed by pilot number 4 on 24 July. Further engineering changes were incorporated in tanks 5 and 6 with the former being shipped to Aberdeen in late November. Pilot number 6 was assigned to the U.S. Marine Corps.

Above and below, the T48 can be seen with the gun forward and at the rear in the travel lock. The four periscopes (three in front and one in the hatch cover) in the commander's cupola are clearly visible.

Below, details can be seen of the rear hull (left) and the suspension system (right) on T48 number 2. The track tension idler has been omitted as on the other early T48s.

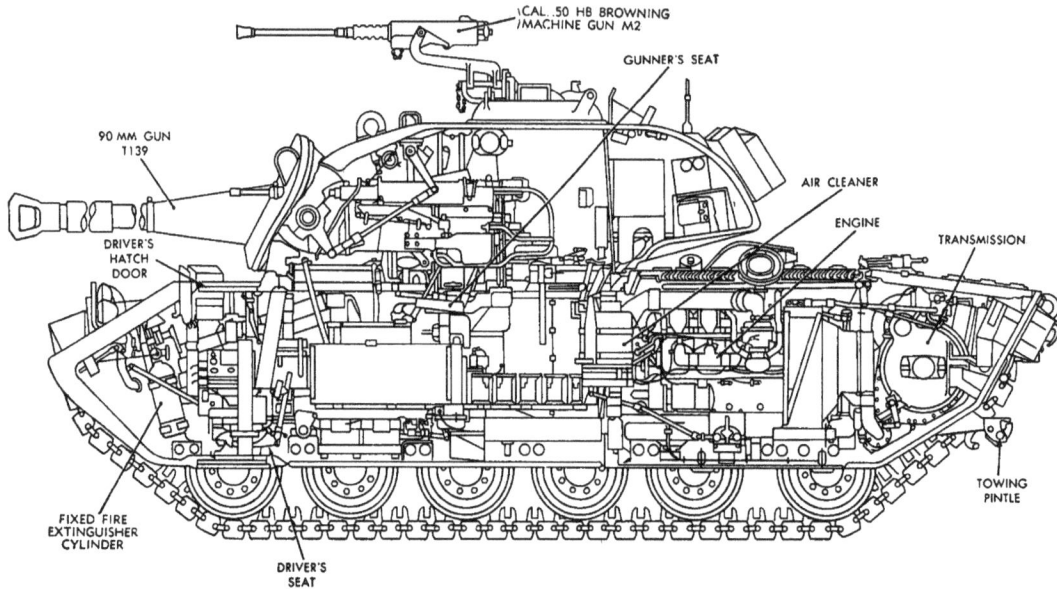

The sectional view above reflects the configuration of the early T48 with a .50 caliber coaxial machine gun on the left side and no track tension idler. The exhaust deflectors have not yet been installed to shield the gun travel lock.

The first series of tests on the pilot tanks continued until the end of 1952 at both Aberdeen and Fort Knox. Although the original design goal was for a maximum weight of 45 tons, the pilot tanks weighed between 49 and 50 tons when combat loaded. However, the critical world situation demanded large numbers of new tanks at the earliest possible date and full production of the T48 began before completion of the test program. Thus, it was vital that corrections for any design defects revealed by the tests be incorporated in production as quickly as possible.

Here is T48, serial number 3, during evaluation at Fort Knox. The .50 caliber machine guns have been fitted with flash hiders on this tank and a new turret bustle stowage rack has been installed. Details of this rack are shown in the view at the bottom right.

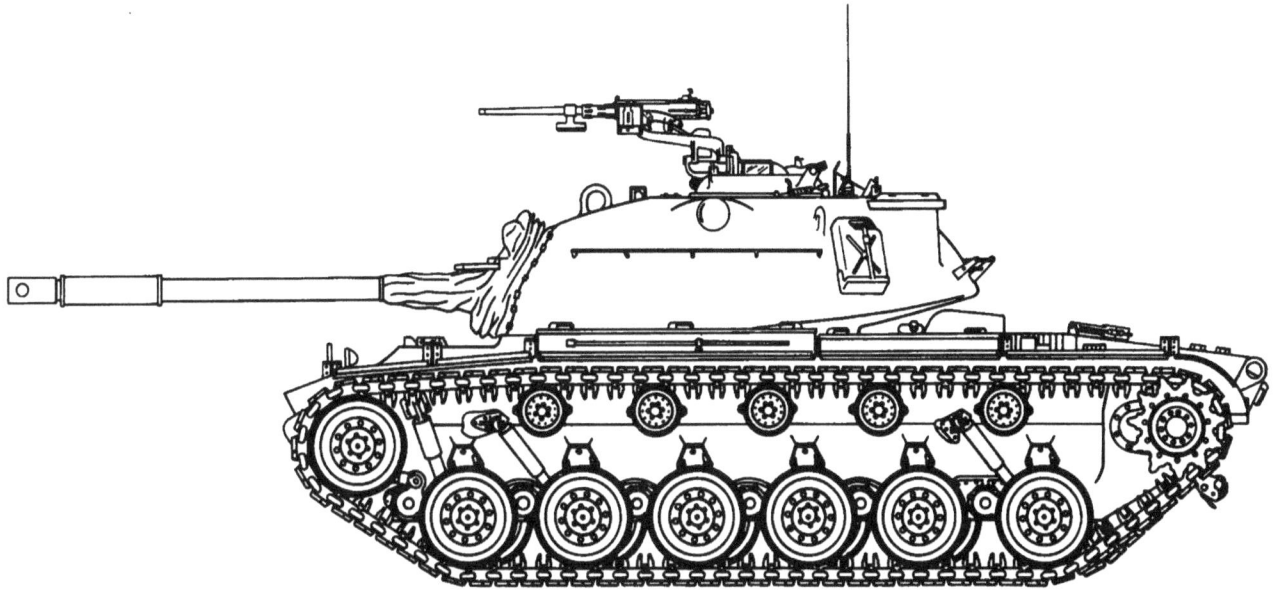

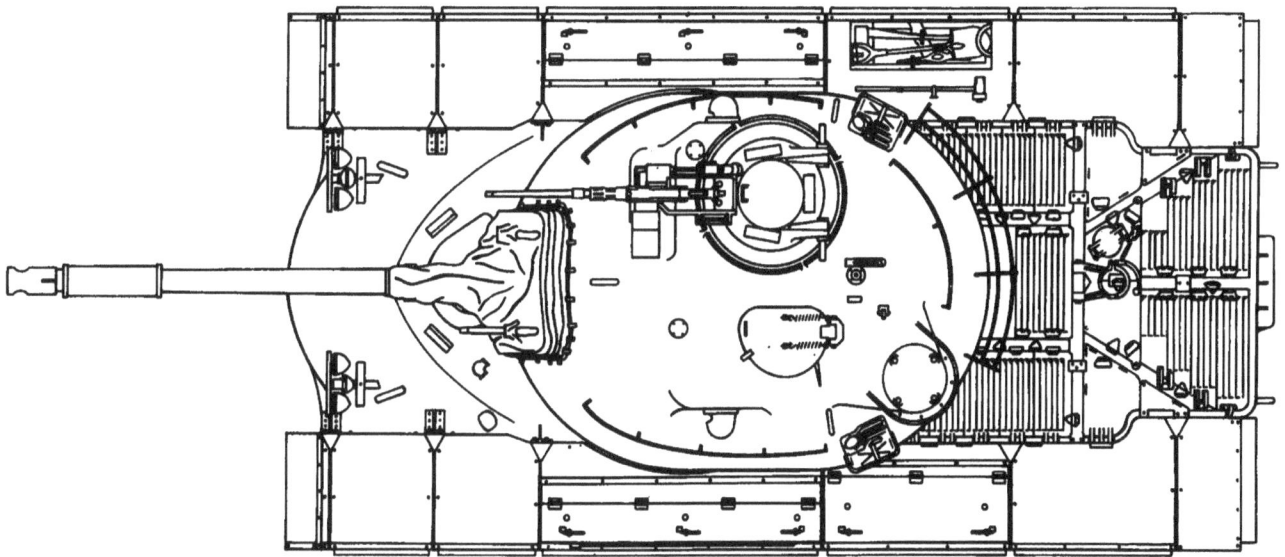

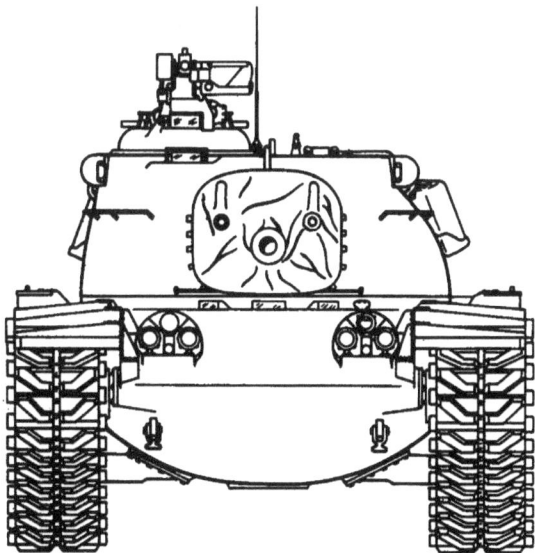

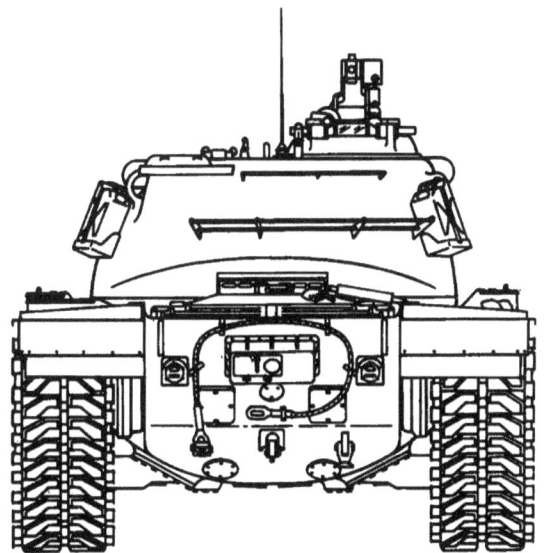

90mm Gun Tank T48

89

Engineers at the Fisher Body Division of General Motors Corporation study a scale model of the T48 during the planning for tank production.

Production of the T48 began at the Chrysler tank plant in Newark, Delaware with the first two tanks delivered in April 1952 followed by six more in May. To meet the need for large numbers of new tanks, Ford Motor Company and the Fisher Body Division of General Motors Corporation had been brought into the program early in 1951. At that time Chrysler was designated as the Vehicle Design Agency and required to furnish drawings and engineering data to the other manufacturers. Production was in full swing by early 1953 with a total of 893 tanks completed by 27 March. At the same time, reports from the pilot tank tests at Aberdeen and Fort Knox indicated the need for numerous modifications before the vehicles would be acceptable for troop use. To resolve the problems resulting from these changes, meetings were held between the three manufacturers, Ordnance, and the Army Field Forces starting in July 1952. This group later became known as the Design Coordinating Committee.

Above, the first two production T48s roll out of the Chrysler Delaware Tank Plant on 11 April 1952. Below, the first T48 completed by General Motors is loaded on a railway car for shipment on 15 May 1952. This tank had been completed on 28 April 1952.

These photographs of the T48 built at Ford Motor Company were dated May 1952. Like the other early T48s, this tank does not have track tension idlers and the exhaust deflector has not been added to shield the gun travel lock.

The fender stowage on this Ford T48 was reversed compared to the tanks built at Chrysler and General Motors. Note that the pioneer kit is on the left fender and the rear stowage box is on the right. This soon was changed to conform to the standard arrangement on the other production tanks. The gunner's sight has not been installed on this vehicle and the opening is sealed.

Above, T48 number 2 climbs the 36 inch step (left) and crosses the test trench (right) during the evaluation program at Aberdeen Proving Ground. Below, the exhaust deflector added to protect the gun travel lock can be seen at the left and the diagram at the right shows the installation of the two personnel heaters in the T48.

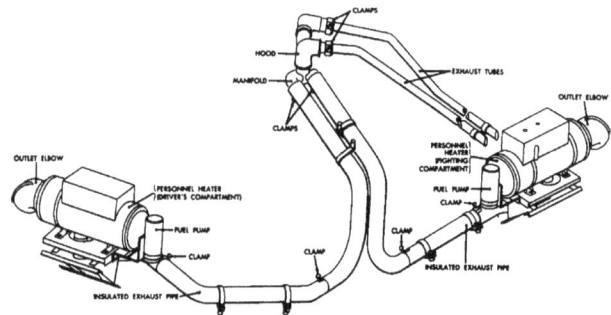

Although the pilot T48s performed remarkably well for an experimental vehicle, the tests at Chrysler, Aberdeen, and Fort Knox revealed the need for numerous important modifications. It was noted very early that the main engine exhaust impinged directly on the travel lock when the gun was locked in the travel position. Heated by the exhaust, the gun lock would freeze if it had been completely screwed down and in any case, it was too hot to touch without asbestos gloves. Deflectors were designed and installed to direct the hot exhaust away from the gun lock.

At Fort Knox, the .50 caliber coaxial machine gun on the left side of the combination mount was replaced by a .30 caliber weapon and a direct sight telescope was installed in place of the .30 caliber gun on the right side. This arrangement was adopted for the production tanks. Two gasoline burning personnel heaters were installed in the T48 with one at the left front of the driver's compartment and the other on the left side hull wall. The two exhausts from these heaters protruded through the top of the hull at the left rear of the driver's hatch. During operation, the exhaust fumes blew into the driver's face and could penetrate into the fighting compartment through the turret ring. This situation was corrected by the installation of extension pipes to carry the fumes to the side of the vehicle. Although the tracks performed well during test, Fort Knox requested the installation of track tension idlers between the sprockets and rear road wheels to minimize track throwing. Minor modifications included an improved ammunition feed for the coaxial machine gun, a better cupola hatch lock, and a guard over the gunner's periscope.

The T48 suffered from the same turret control problems as the early M47 and like on it, the interim electric-hydraulic Oilgear system was installed pending the development of controls which fully met the requirements of the Army Field Forces. The interim equipment was required only to equal the performance obtained on the medium tank M46.

At the left, the T48 poses alongside an M47 during tests at Fort Knox. The difference is obvious between the 28 inch wide tracks of the T48 and those with a 23 inch width on the M47.

The limited number of foundries capable of casting the single piece T48 hull made it highly desirable to develop a hull assembled from smaller castings. Above, a hull fabricated by welding seven sections (left) consisting of castings and rolled plate is compared with the single piece cast hull (right) at Ford. The white lines indicate the weld joints on the seven piece hull.

In January 1953, the Army Field Forces indicated that the T48 would be acceptable for restricted issue to the troops if the direct sight telescope was installed and the other major deficiencies were corrected. Such tanks could then be used for training in the United States, but they were not to be shipped overseas. If the direct sight telescope was not available, a vane sight could be substituted for training purposes. OTCM 34765 standardized the new vehicle as the 90mm gun tank M48 on 2 April 1953. The same action also named the tank the Patton 48 in honor of General George S. Patton, Jr.

At the right is the Ford tank production line on 14 August 1952. Note that the fender stowage now conforms to the standard arrangement. These tanks have the small driver's hatch and the extension pipes have not yet been added to the personnel heater exhausts. Below, newly completed M48s are lined up in the storage yard at General Motors Grand Blanc Tank Arsenal.

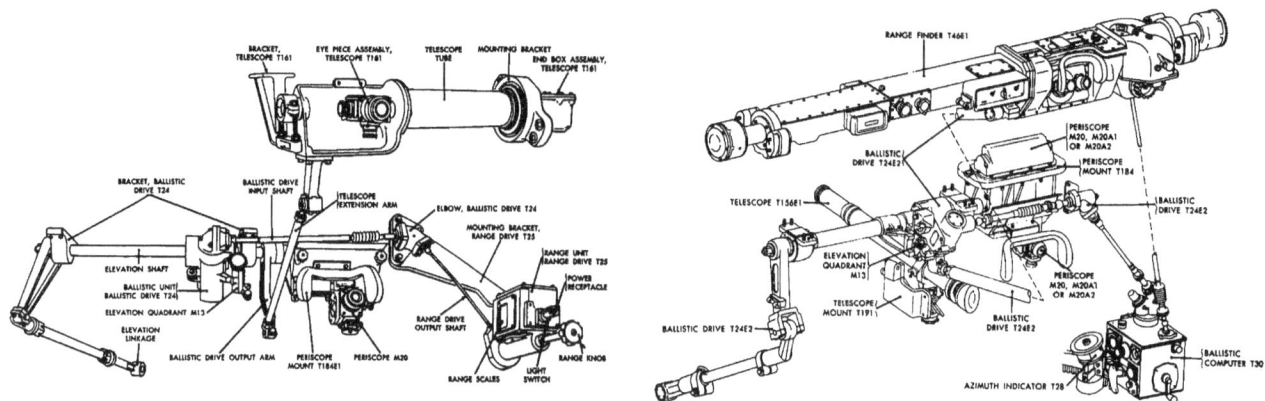

The drawings above compare the phase I (left) and the phase IV (right) fire control systems.

Problems with the T30 mechanical ballistic computer continued to delay the introduction of the complete fire control system. To get the tanks into the hands of the troops as soon as possible, the fire control equipment was installed in four phases as components became available. The first three interim phases were to be upgraded to the phase IV or ultimate system when its development was complete. The relatively simple phase I system was installed on the pilot tanks. On these vehicles, the gunner used the T35 periscopic sight, later standardized as the M20, supported by the mount T184E1 in the turret roof. The tank commander's telescope T161 (formerly elbow sight T157) was installed in the right range finder blister and was used for both sighting and target destination. The T24 ballistic drive provided linkage between the gun mount elevation system, the T35 sight, the T161 telescope, and the range drive T25. The range unit of the latter was used to introduce the estimated range into the system. An elevation quadrant M13(T21) was attached to the right side of the T24 ballistic drive to lay the gun for indirect fire. An azimuth indicator T28 was mounted at the right side of the turret, also for indirect fire.

The phase II and III fire control systems replaced the T161 telescope with the range finder T46E1 for use by the tank commander. The M20 periscopic sight was retained and the T156E1 telescope was added as an alternate sight for the gunner. It was installed on the right side of the gun mount replacing one of the coaxial machine guns. The ballistic drive T24E1 connected the gun elevation mechanism with the M20 sight and the range drive T25. It differed from the T24 in that the output arm was shortened and it was not linked to the range finder. After the tank commander determined the range, the gunner entered it manually into the T25 range drive and then used the periscope M20 to lay the gun. If the periscope was inoperative, the gunner used the direct sight telescope elevating the weapon until the appropriate range line in the reticle appeared on the target.

The phase II and phase III systems were essentially the same, differing only in minor details. For example, the M20 periscope used the T184E1 mount in phase II and either the T184 or T184E1 mount in phase III. The T184 had a machined mounting surface with four tapped holes. This was intended to accommodate a new type range drive which was never put into production.

The phase IV or ultimate system replaced the range drive T25 with the ballistic computer T30. The ballistic drive T24E2 replaced the T24E1. The new drive was heavier in construction and linked the range finder directly to the ballistic computer thus tying the whole system together. The elevation quadrant M13 was mounted on the shaft assembly rather than on the right side of the ballistic drive.

In normal operation, the tank commander announced the type of ammunition to be used which was then set into the computer by the gunner. The commander ranged on the designated target and the range data were transmitted mechanically from the range finder to the computer. The gunner then manually or electrically matched the indices on the computer dial which mechanically transmitted the proper superelevation through the ballistic drive to both the M20 periscopic sight and the range finder. The gunner then elevated the gun tube to place the aiming cross on the target and fired.

Equipment for indirect fire was the same as on the earlier systems and the T156E1 telescope was retained as an alternate sight as in phases II and III.

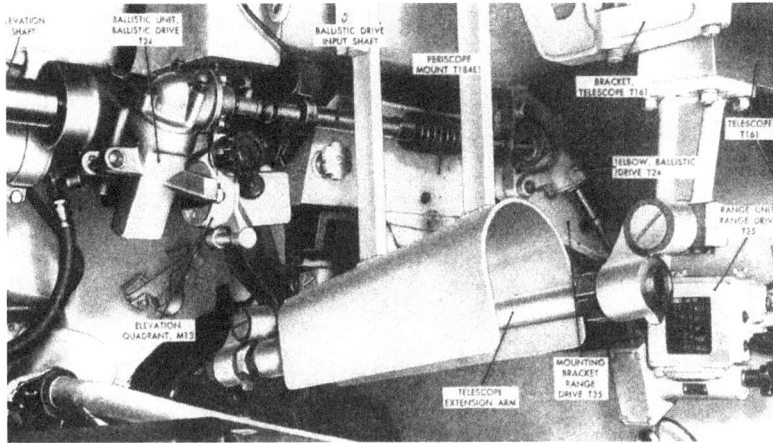

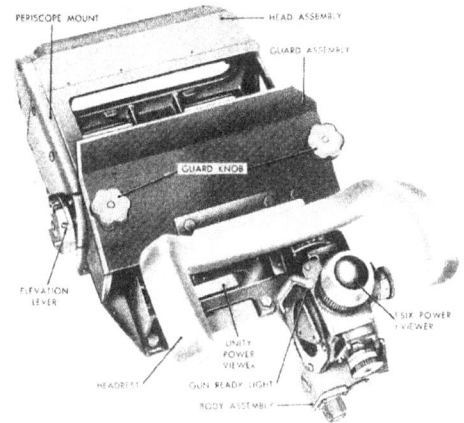

Above is the phase 1 fire control system (left) installed in the T48 and the gunner's M20 periscopic sight (right). Below are left and right side views of the T46E1 stereoscopic range finder.

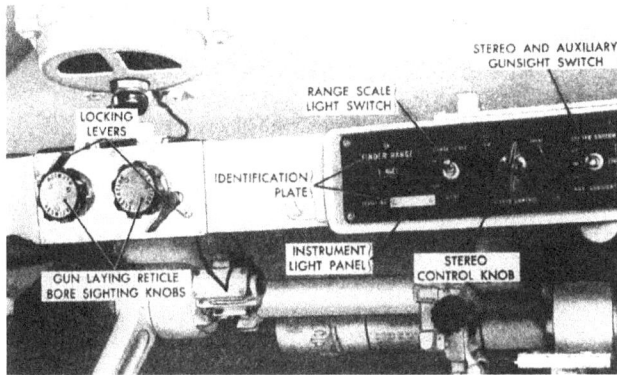

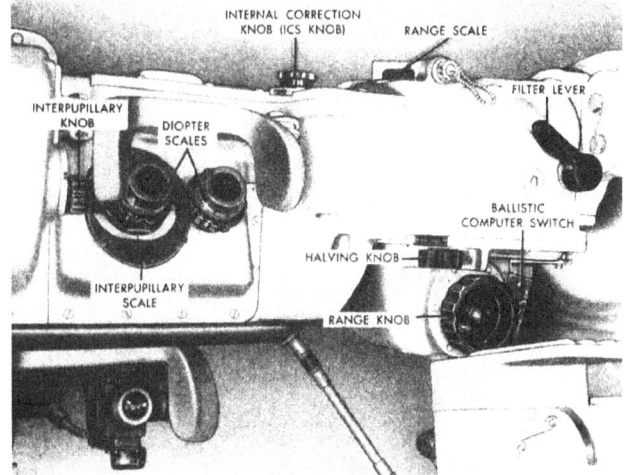

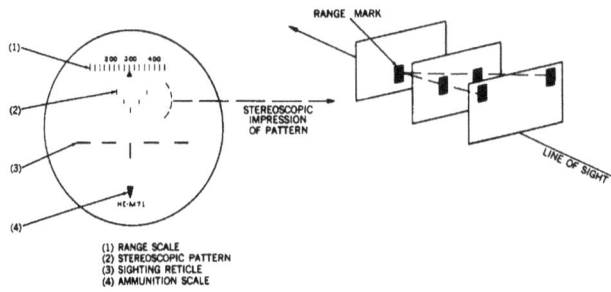

(1) RANGE SCALE
(2) STEREOSCOPIC PATTERN
(3) SIGHTING RETICLE
(4) AMMUNITION SCALE

The range finder reticle pattern and its three dimensional appearance to the gunner is illustrated at the left. To range on a target, the range finder was adjusted until the lower front marks appeared to be on the target. This procedure was often called "flying the geese" referring to the V formation pattern of the reticle. Below, the T30 ballistic computer is at the left and the T156E1 telescope at the right replaces one of the coaxial machine guns.

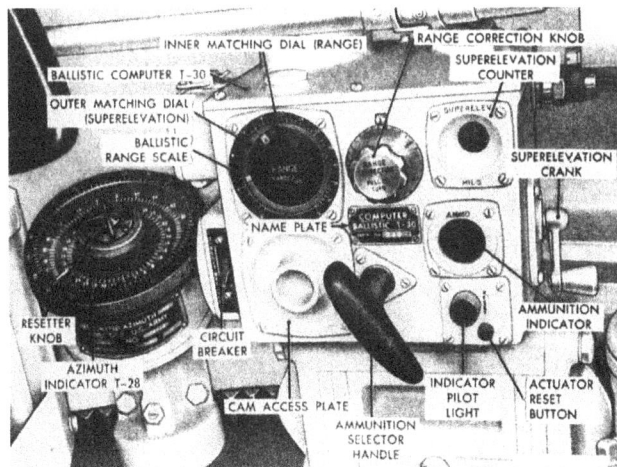

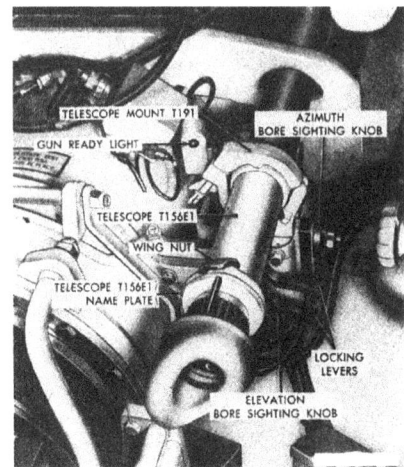

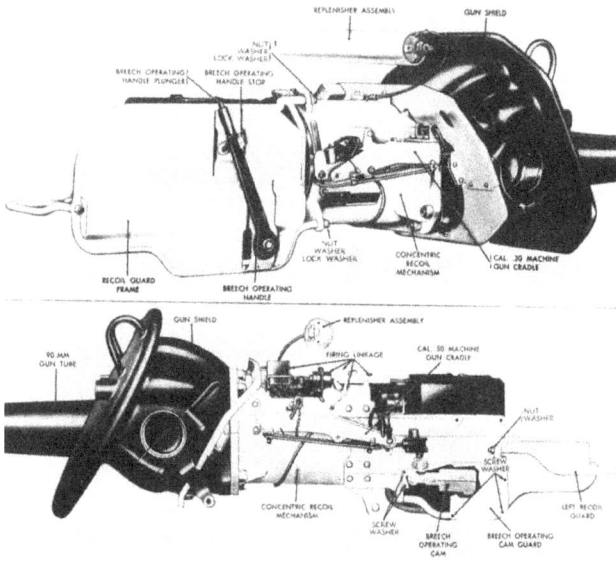

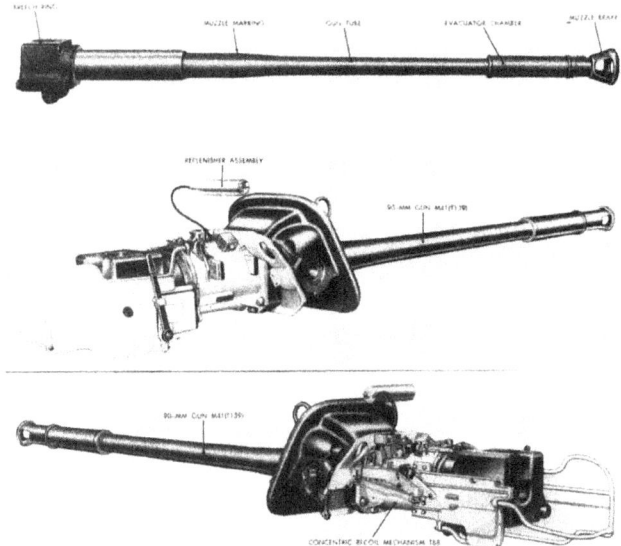

Above, two views of the early T148 gun mount (left) in the original T48 tanks are compared with the later version (right) as modified after tests. Note that the coaxial machine gun cradle has been removed from the right side. At the top right is a view of the 90mm gun M41 (T139) with the original muzzle brake.

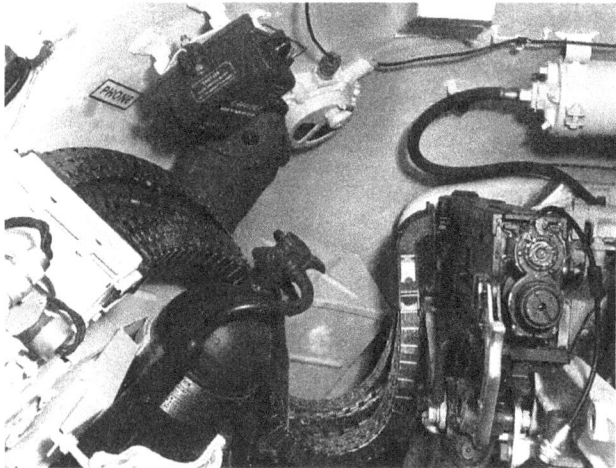

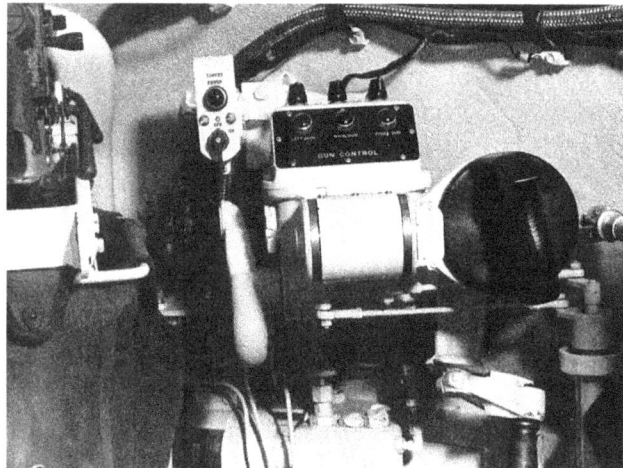

Above at the left is a view of the loader's side of the early gun mount with the coaxial .50 caliber machine gun and its ammunition supply. At the right above is the early gunner's station. Part of the coaxial .30 caliber machine gun appears at the extreme left of the photograph. Below are views of the loader's station (left) and the gunner's controls (right).

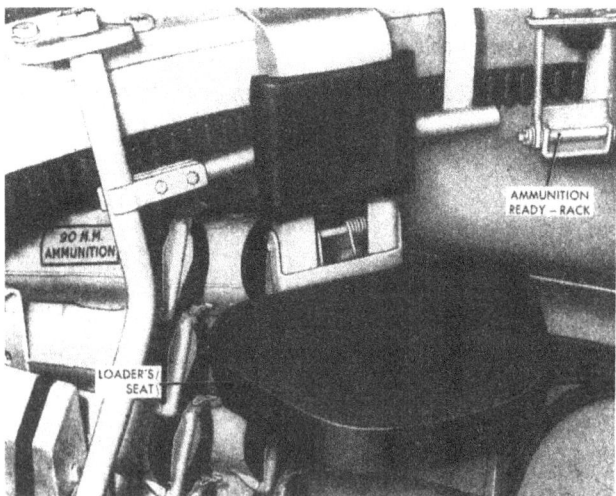

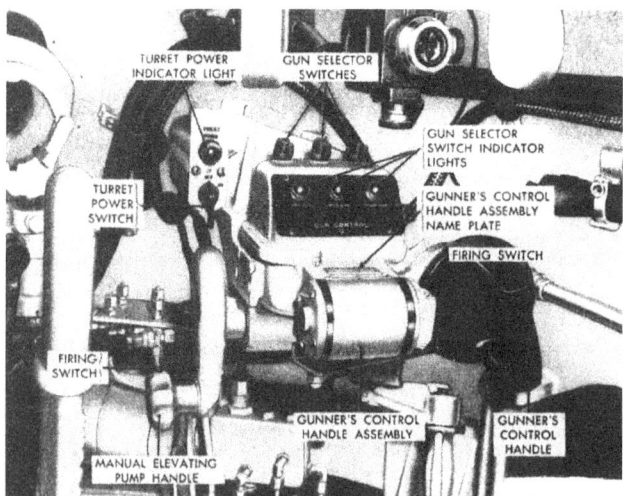

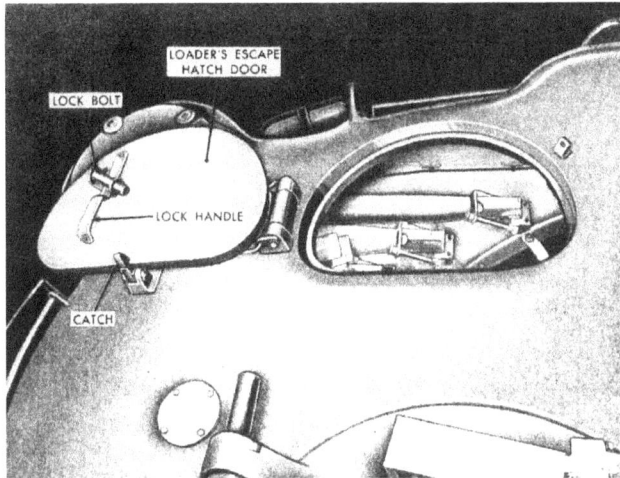

Labels: LOADER'S ESCAPE HATCH DOOR, LOCK BOLT, LOCK HANDLE, CATCH

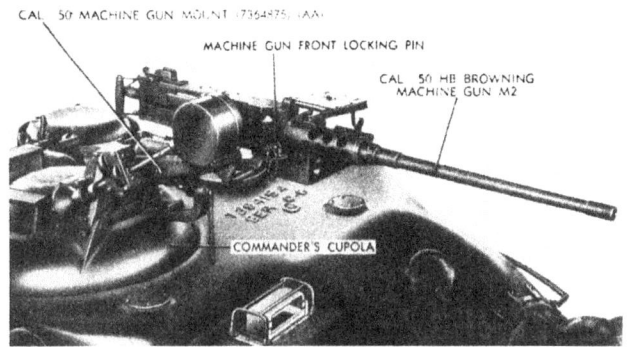

Labels: CAL .50 MACHINE GUN MOUNT (7364875) (AA), MACHINE GUN FRONT LOCKING PIN, CAL .50 HB BROWNING MACHINE GUN M2, COMMANDER'S CUPOLA

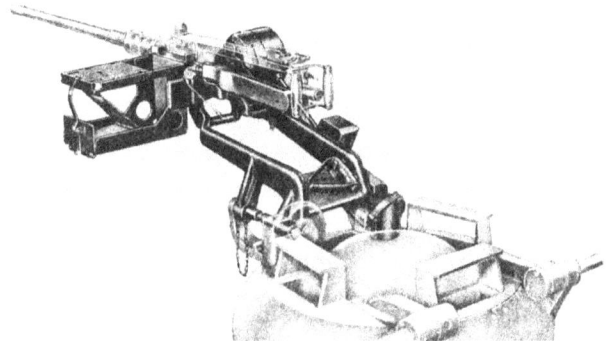

Directly above is the loader's hatch and at the right details can be seen of the commander's cupola and .50 caliber machine gun mount. Below, from left to right, are views of the gunner's seat, the bottom of the commander's cupola, and the commander's seat in the folded position.

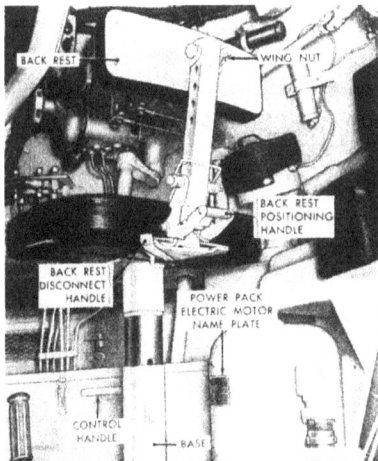

Labels: BACK REST, WING NUT, BACK REST POSITIONING HANDLE, BACK REST DISCONNECT HANDLE, POWER PACK ELECTRIC MOTOR NAME PLATE, CONTROL HANDLE, BASE

Labels: HATCH LOCKING HANDLE, CRASH PAD

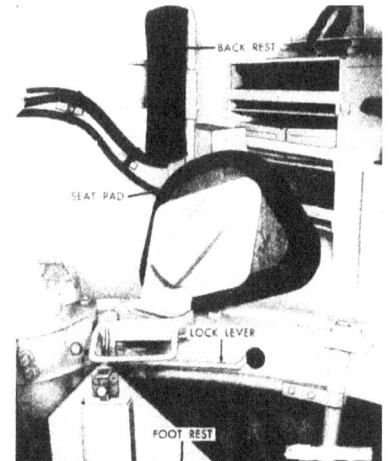

Labels: BACK REST, SEAT PAD, LOCK LEVER, FOOT REST

Above is the AN/GRC-3 radio installed behind the tank commander in the turret bustle. At the right is the commander's gun control handle.

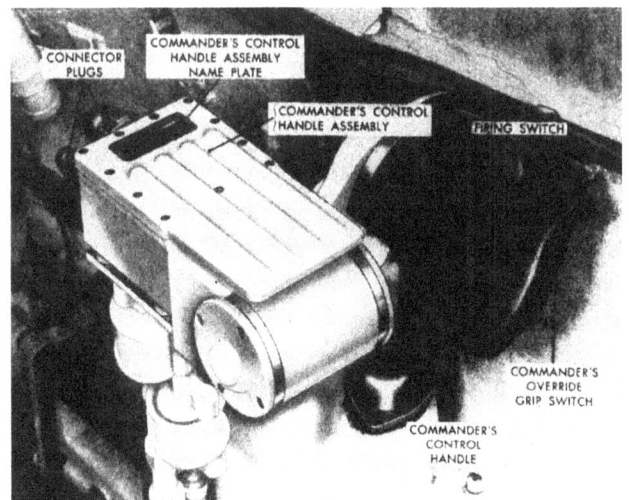

Labels: CONNECTOR PLUGS, COMMANDER'S CONTROL HANDLE ASSEMBLY NAME PLATE, COMMANDER'S CONTROL HANDLE ASSEMBLY, FIRING SWITCH, COMMANDER'S OVERRIDE GRIP SWITCH, COMMANDER'S CONTROL HANDLE

Above is a comparison of the small early design driver's hatch (left) with the later large hatch (right). On the latter, the hatch cover is raised to clear the fixed periscopes. Note the supports for the hatch cover in the open position.

During the tests of the pilot tanks, it was noted that the small size of the driver's hatch resulted in an uncomfortable position for the driver when operating the tank with the hatch open. It also was extremely difficult to dismount if the gun was in the travel lock position. This hatch was redesigned and enlarged with the new version applied to the production vehicles as early as possible. When opened, the new hatch cover lifted prior to rotation thus clearing the periscope heads which remained fixed in position. The original small hatch design required the periscopes to drop down automatically before the hatch cover opened. They then had to be pushed up manually into the operating position after the hatch was closed. The new large hatch cover, as well as later production of the small early model, was fitted with a mount for the T41 infrared periscope.

The T25 and M7 (T36) driver's periscopes can be seen directly below. The T25 was used with the small hatch tanks and dropped down when the hatch was opened. The M7 (T36) was a fixed periscope used with the large hatch tanks. The two were not interchangeable. At the right is the driver's M24 (T24) infrared periscope and the view below shows it installed along with the M7 (T36) fixed periscopes. The M24 had to be removed before opening the hatch.

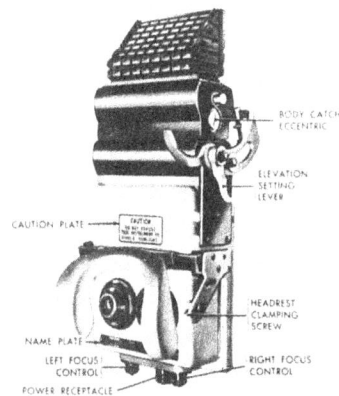

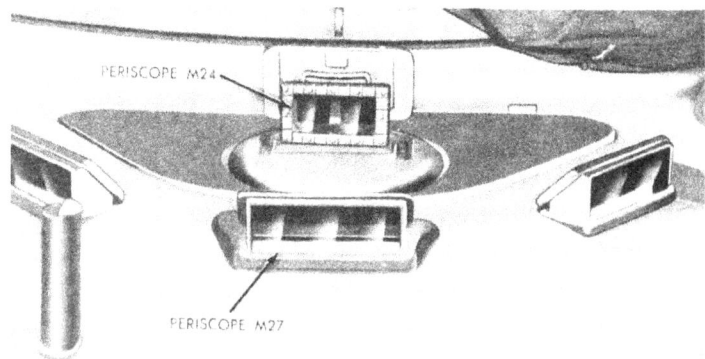

The driver's controls in the M48 differed from those in the earlier M46 and M47 tanks. The wobble stick control was replaced by a steering wheel. When the cross drive transmission was placed in the high or low range, the tank could be steered to the left or right by turning the wheel in the appropriate direction. With the transmission in neutral, a full turn of the wheel would cause the tank to pivot in place with the tracks moving in opposite directions.

DRIVER'S PERISCOPE M27

PERISCOPE M24

PERISCOPE M27

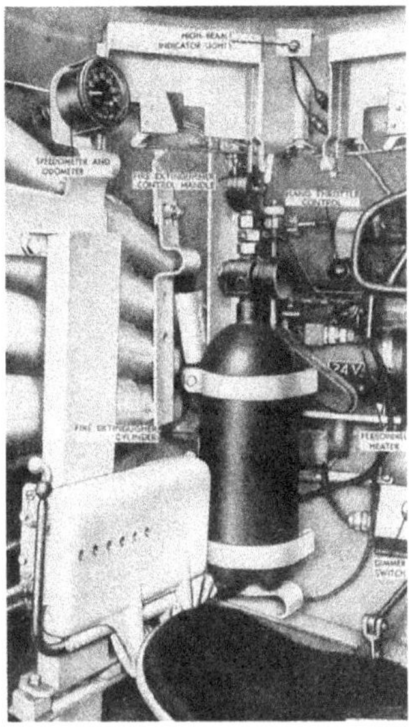

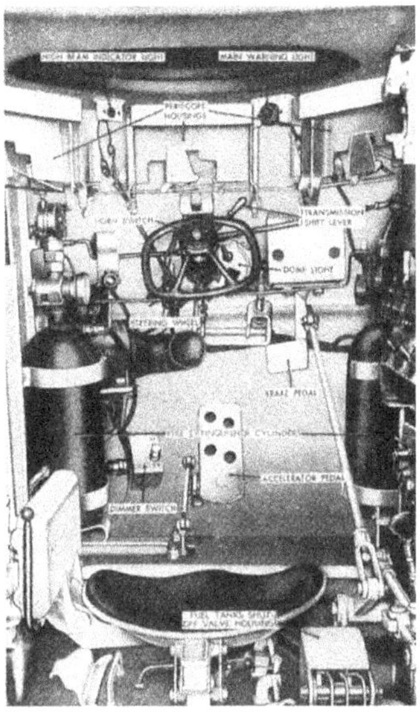

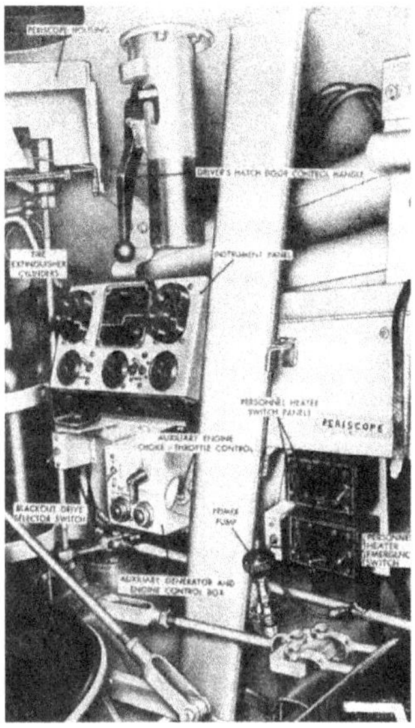

The left, center, and right sides of the driver's compartment in the early small hatch tanks (above) are compared with the same areas in the later large hatch vehicles (below) in these photographs. Note the linkage between the periscope mounts and the hatch control in the top views. This was used to drop the periscopes when the hatch was opened. Also, in the later tanks, the personnel heater outlet was directed away from the driver to prevent overheating. The back to the driver's seat is removed in these photographs.

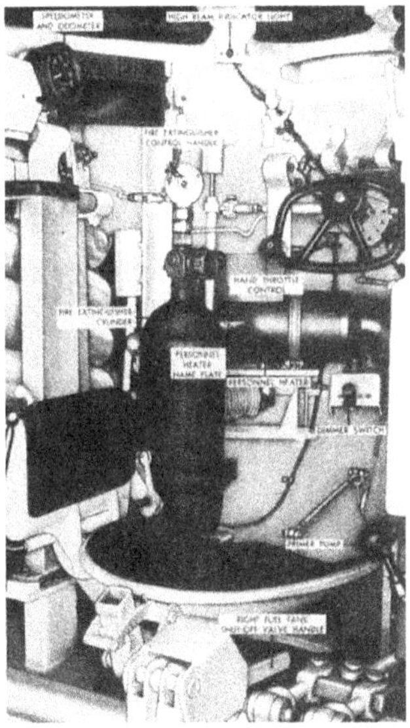

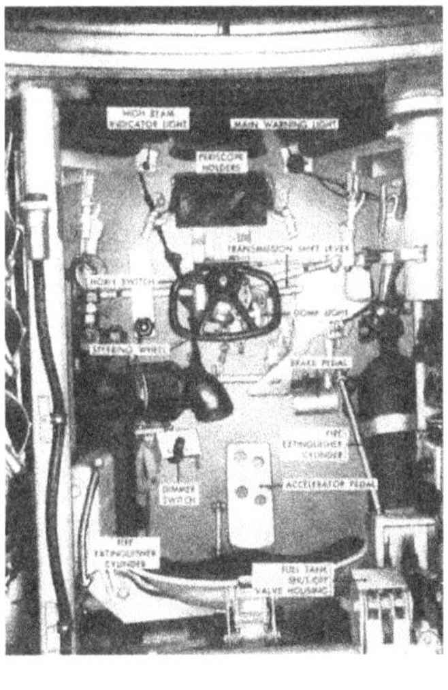

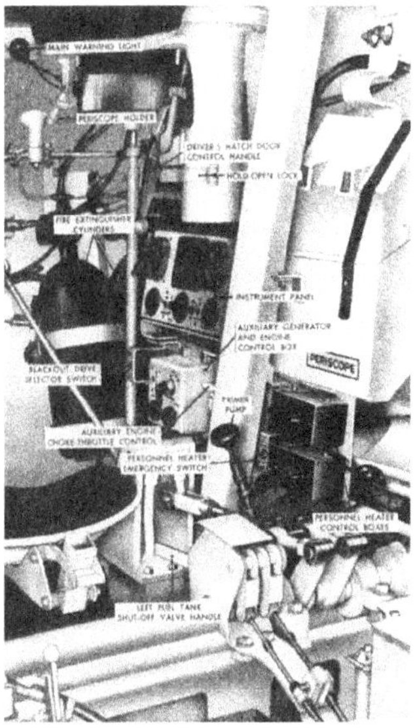

Below, the driver's instrument panel can be seen at the left and his escape hatch in the hull floor appears at the right. His seat had to be tipped sideways into the dumped position to permit use of the hatch.

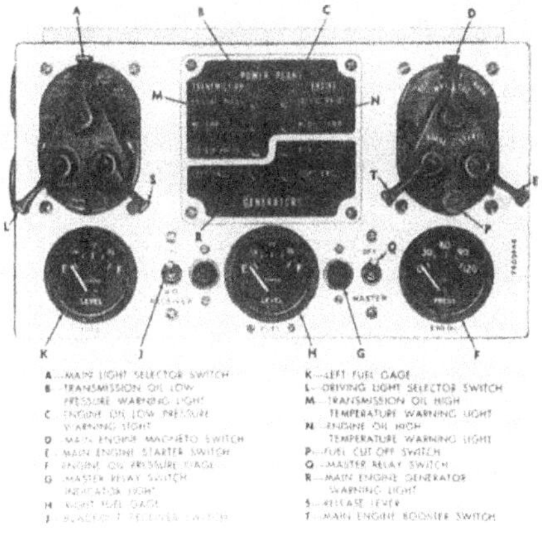

A—MAIN LIGHT SELECTOR SWITCH
B—TRANSMISSION OIL LOW PRESSURE WARNING LIGHT
C—ENGINE OIL LOW PRESSURE WARNING LIGHT
O—MAIN ENGINE MAGNETO SWITCH
E—MAIN ENGINE STARTER SWITCH
F—ENGINE OIL PRESSURE GAGE
G—MASTER RELAY SWITCH INDICATOR LIGHT
H—RIGHT FUEL GAGE
J—BLACKOUT DRIVING LIGHT

K—LEFT FUEL GAGE
L—DRIVING LIGHT SELECTOR SWITCH
M—TRANSMISSION OIL HIGH TEMPERATURE WARNING LIGHT
N—ENGINE OIL HIGH TEMPERATURE WARNING LIGHT
P—FUEL CUT-OFF SWITCH
Q—MASTER RELAY SWITCH
R—MAIN ENGINE GENERATOR WARNING LIGHT
S—RELEASE LEVER
T—MAIN ENGINE BOOSTER SWITCH

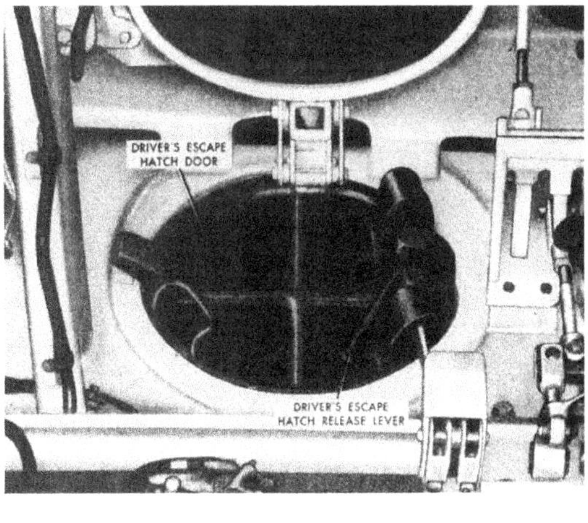

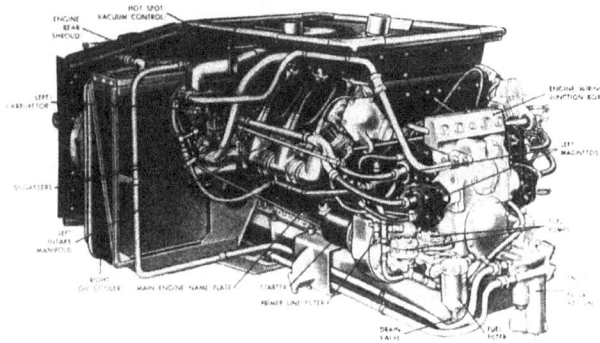

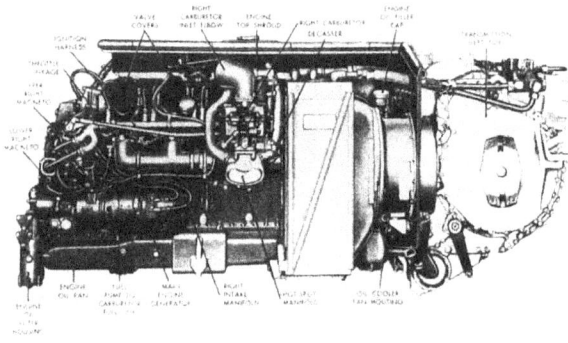

Above at the left and below are photographs of the AV-1790-7B engine with the new 300 ampere generator. At the top right is the complete power pack with the CD-850-4 transmission.

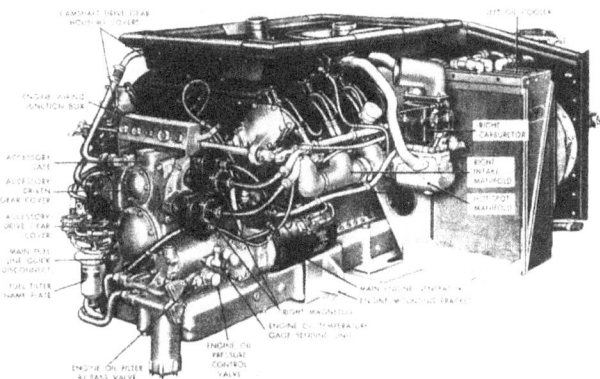

The first production tanks were powered by the Continental AV-1790-5B engine then being installed in the M46A1 and the M47. In November 1952, the AV-1790-7 entered production. This engine was fitted with the same cylinders as the AOS-895-3 which powered the light tank M41. The new engine provided interchangeability between the two vehicles which would simplify the spare parts problem. A later modification appearing in August 1954 was the AV-1790-7B which had a 300 ampere generator to meet the higher current requirements of the new tanks. The AV-1790-7C was introduced in November 1954 with twin fuel filters attached to each oil cooler to facilitate service. All of the −7 series retained the same arrangement of oil coolers and outrigger fans as on the earlier AV-1790 engines.

At the left is the rear deck of the production M48 with exhaust deflectors to protect the gun travel lock. On later vehicles, the exhaust muffler for the auxiliary generator engine was shifted from the deck grill to the right fender. The pioneer kit also was moved to the left rear fender. Below, the power pack can be seen installed with the rear deck removed.

A - FUEL TANK FILLER COVER
B - ENGINE ACCESS SIDE GRILLE DOOR
C - AUXILIARY ENGINE EXHAUST MUFFLER
D - LIFTING EYE
E - TRANSMISSION ACCESS SIDE DOOR
F - TRANSMISSION ACCESS SIDE PLATE
G - TRANSMISSION ACCESS GRILLE DOOR
H - CROSS BEAM
J - ENGINE EXHAUST
K - ENGINE ACCESS CENTER GRILLE DOOR
L - FRONT SUPPORT BEAM
M - LOCKING PLATE
N - TRANSMISSION OIL FILLER COVER
P - TRANSMISSION ACCESS CENTER PLATE
Q - REAR TRAVERSING SUPPORT BEAM
R - REAR TRAVERSING LOCK

A—CARBURETOR INTAKE DUCT
B—FIRE EXTINGUISHER RIGHT DISCHARGE NOZZLE
C—TRANSMISSION MOUNTING NUT AND WASHER

100

The original T48 suspension (above) can be compared with the later production version (below). Note the addition of the track tension idler. At the right is a view of the hull bottom from the rear.

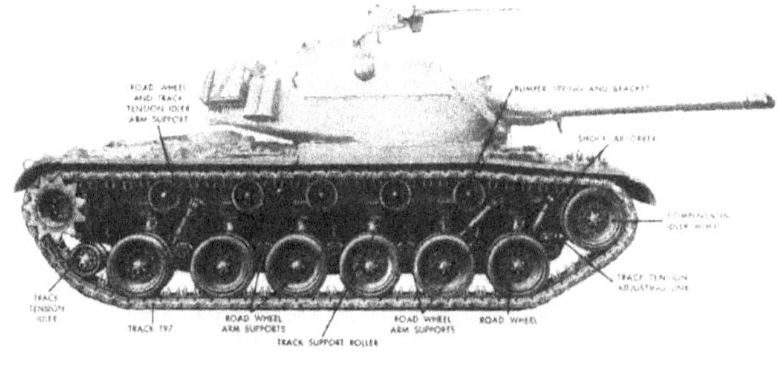

A — TRANSMISSION LEFT ACCESS PLATE
B — TRANSMISSION CENTER ACCESS PLATE
C — TRANSMISSION RIGHT ACCESS PLATE
D — AUXILIARY ENGINE DRAIN HOLE COVER
E — MAIN ENGINE DRAIN HOLE COVER
F — FUEL TANK DRAIN HOLE COVERS
G — FUEL TANK HATCH DOOR
H — MAIN ENGINE OIL FILTER ACCESS HOLE COVER
J — BRAKE ROD ACCESS PLATES
K — TRANSMISSION DRAIN HOLE COVERS

Below, details of the headlight groups can be seen. The right-hand group appears at the left and the left-hand group is shown at the right.

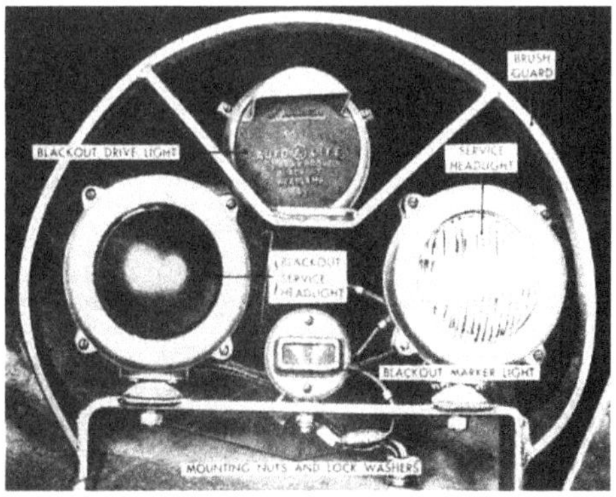

The taillight installations are shown below with the left-hand light at the left and the right-hand light at the right.

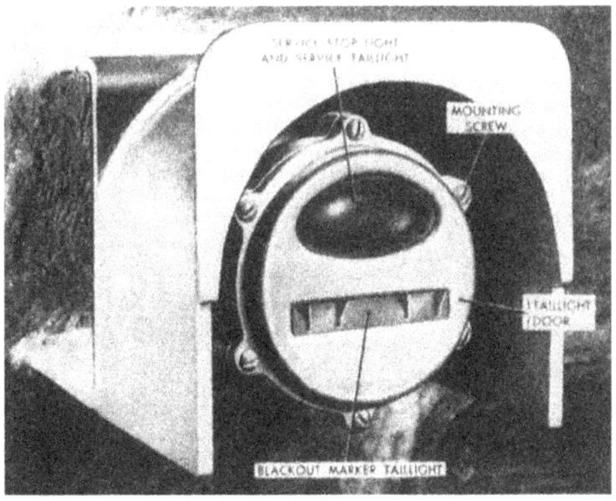

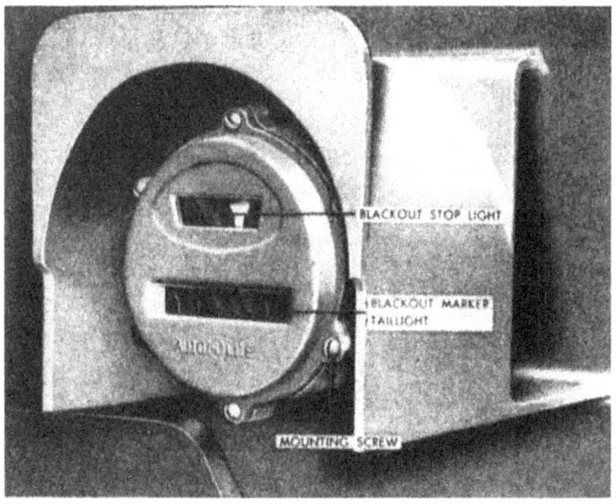

101

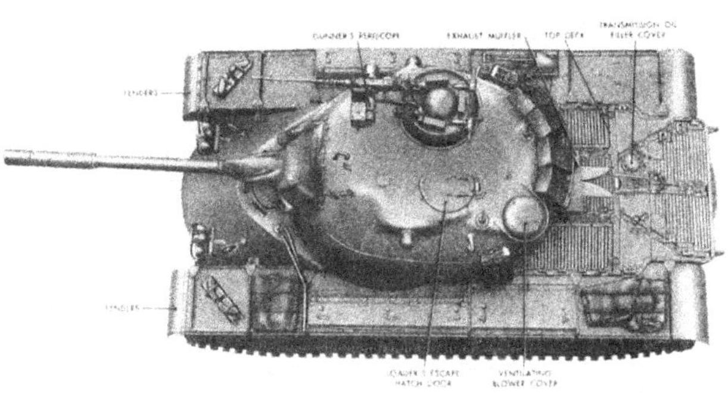

The production M48 incorporating many changes resulting from the test program is shown here. The most obvious new features are the track tension idlers and the main engine exhaust deflectors. The .50 caliber coaxial machine gun also has been replaced by a .30 caliber weapon which does not extend beyond the gun shield.

Production tanks at all three manufacturers were fitted with the Chrysler designed commander's cupola. It consisted of a flat low silhouette cupola with all round vision provided by four periscopes. The .50 caliber machine gun externally mounted on the cupola top could be aimed and fired by remote control with the turret closed. As mentioned earlier, the tank commander had to open the hatch and expose himself to reload the weapon.

Test reports on a new commander's cupola developed by Aircraft Armaments, Incorporated, indicated that it was superior to the earlier model. Designated as the Aircraft Armaments model 30, it was approved in modified form for use on production tanks in August 1953. The .50 caliber machine gun was mounted inside the new cupola and it could be loaded and fired without opening the hatch. Five vision blocks provided a view on all sides of the tank and a T42(M28) periscopic sight in the cupola roof was used to aim the .50 caliber machine gun. The original model 30 cupola used a drum magazine mounted on a platform at the bottom of the fighting compartment. With a capacity of approximately 600 rounds, it fed the .50 caliber machine gun through a flexible chute extending up into the cupola. Tests showed that the chute arrangement interfered with the tank commander, particularly when using the range finder. The drum magazine and feed chute were then replaced by a 100 round ammunition box attached to the cupola mounting ring next to the machine gun. The rounds passed through a short

flexible chute inside the front of the cupola to the gun. The empty shell cases and belt links were ejected from the right side of the cupola. The modified model 30 was adopted as the .50 caliber machine gun mount M1 and later referred to as the M1 cupola. However, the limited interior space and the problem of providing an adequate ammunition supply to the gun was to plague this design throughout its service and eventually resulted in its replacement.

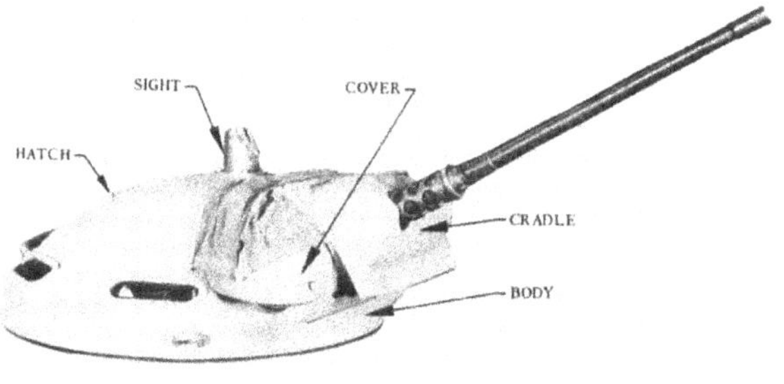

The original Aircraft Armaments model 30 cupola can be seen at the right. In the lower view, it has been installed on T48 serial number 49 at Aberdeen Proving Ground.

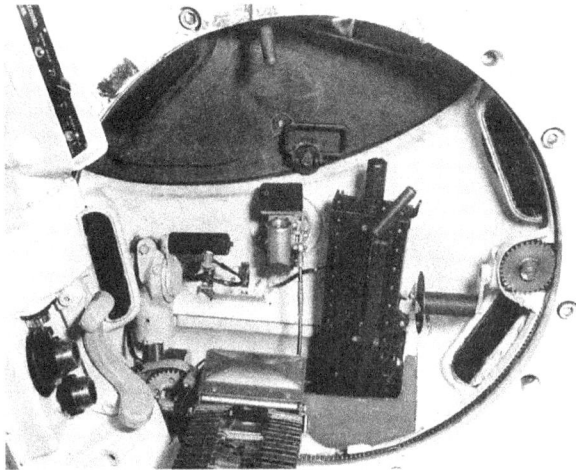

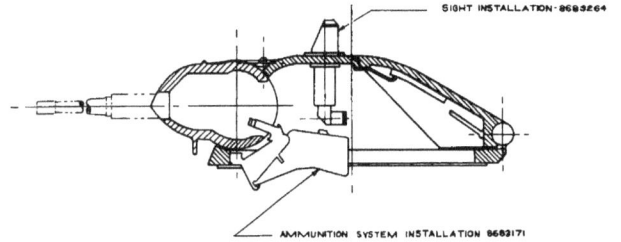

At the left is the interior of the model 30 cupola showing the controls (top) and the original ammunition feed system (bottom) with the drum magazine. Above is a sectional drawing of the design finally adopted with the 100 round ammunition box.

nomenclature on 25 October 1954. The tanks with the small driver's hatch and the Chrysler cupola were designated as the 90mm gun tank M48. This group included 120 early tanks with ballistically deficient hull castings. They were identified as the 90mm gun tank M48C and were to be marked by welding on the upper glacis plate "non-ballistic training tank only". The Chrysler cupola on the large hatch tanks was replaced by the M1 cupola and all vehicles so equipped were then designated as the 90mm gun tank M48A1. Both the M48 and M48A1 were classified as standard types, but the M48A1 was approved for overseas shipment on a preferential basis.

The introduction of these many modifications into production resulted in a wide variety of M48 tanks. For example, the Chrysler commander's cupola was fitted on all of the tanks with the small driver's hatch as well as 1805 of the large hatch vehicles. The latest tanks had the M1 cupola. To clarify the situation, OTCM 35619 revised the

The weld metal marking on the front of an M48C appears above at the right. Note that it does not precisely conform to the wording specified. Below is an M48A1 with the production version of the commander's cupola.

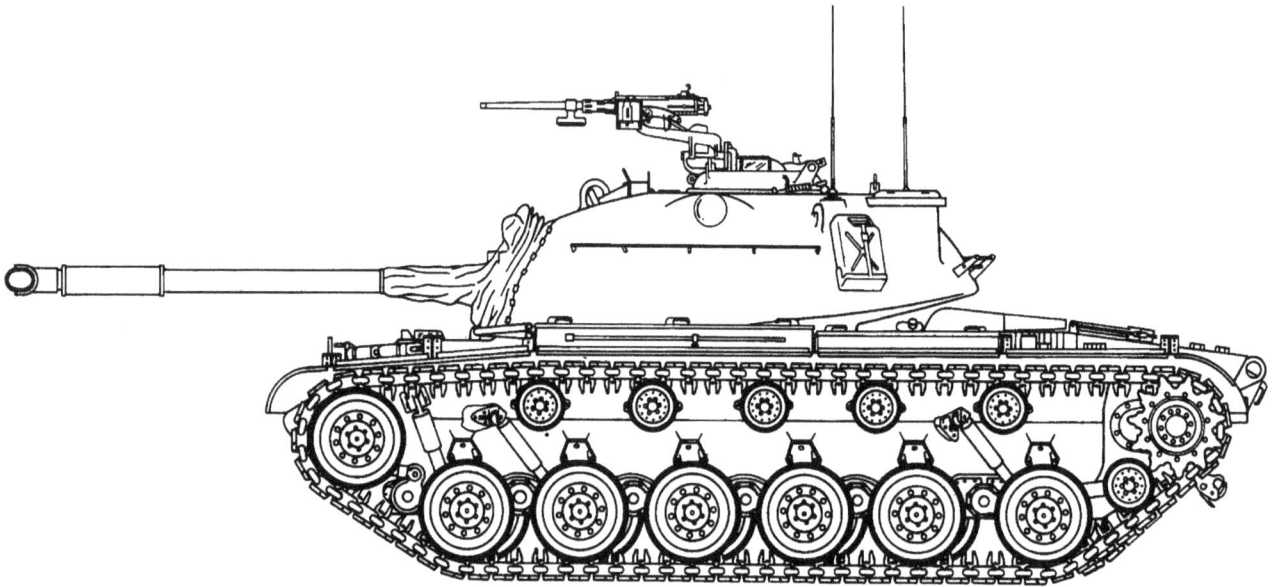

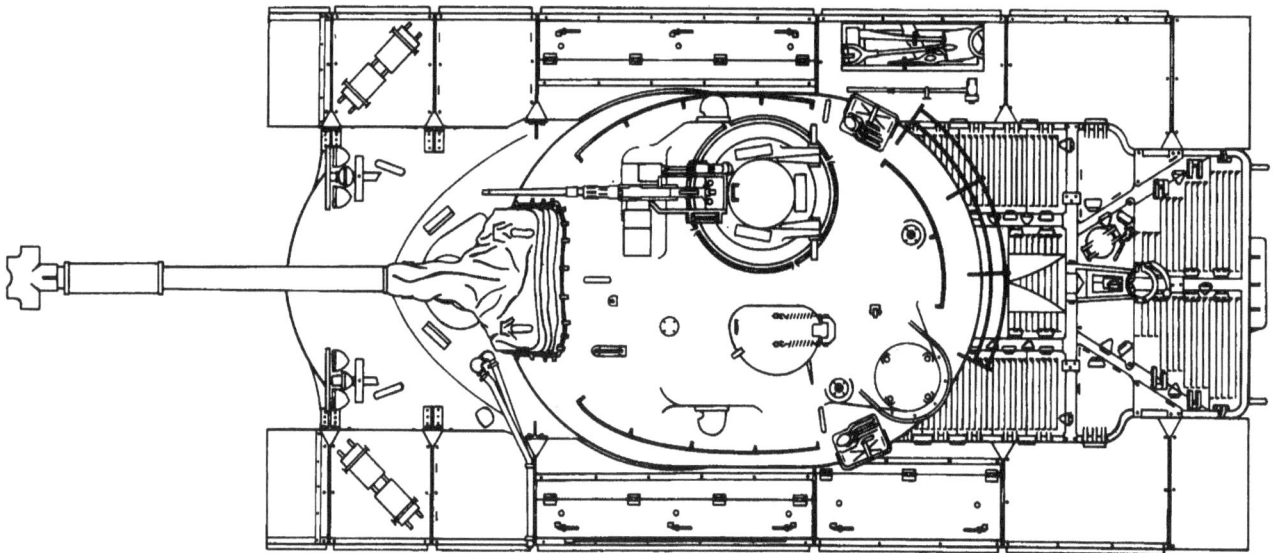

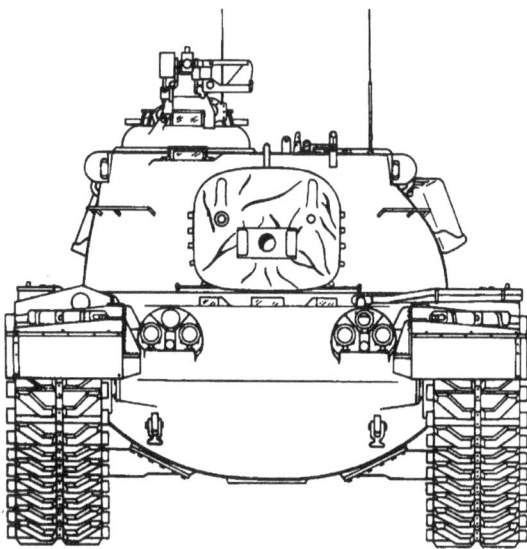

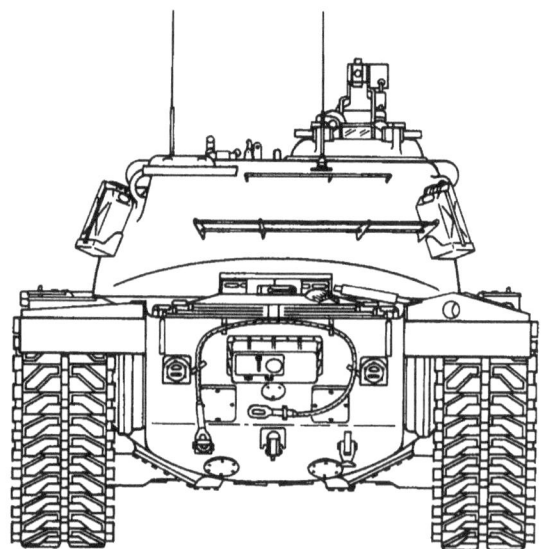

90mm Gun Tank M48

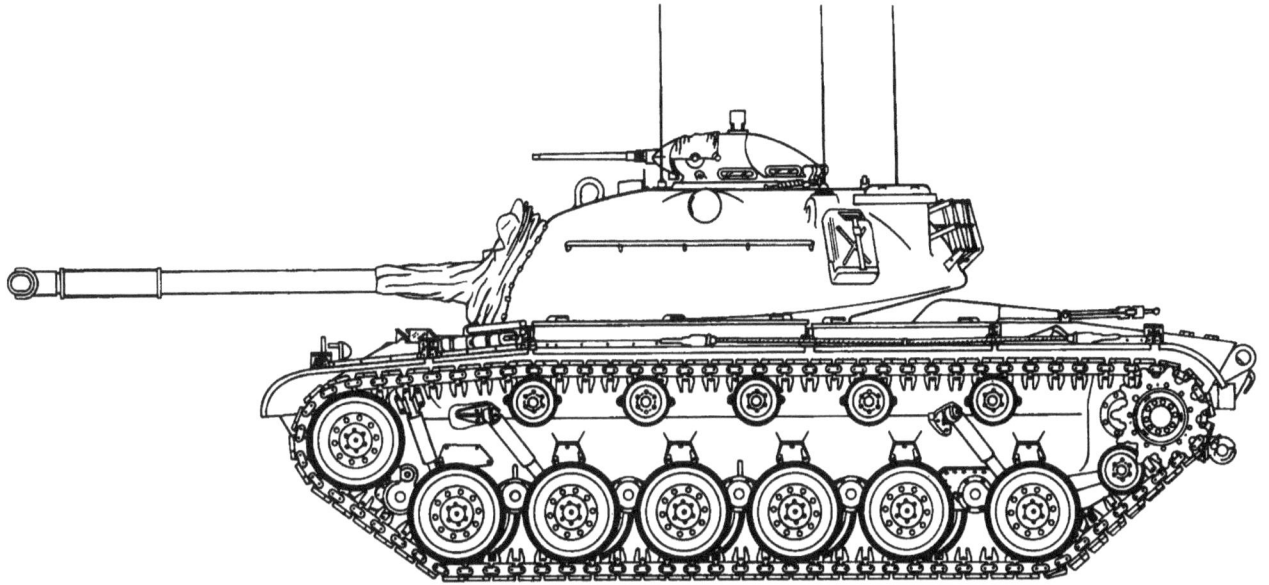

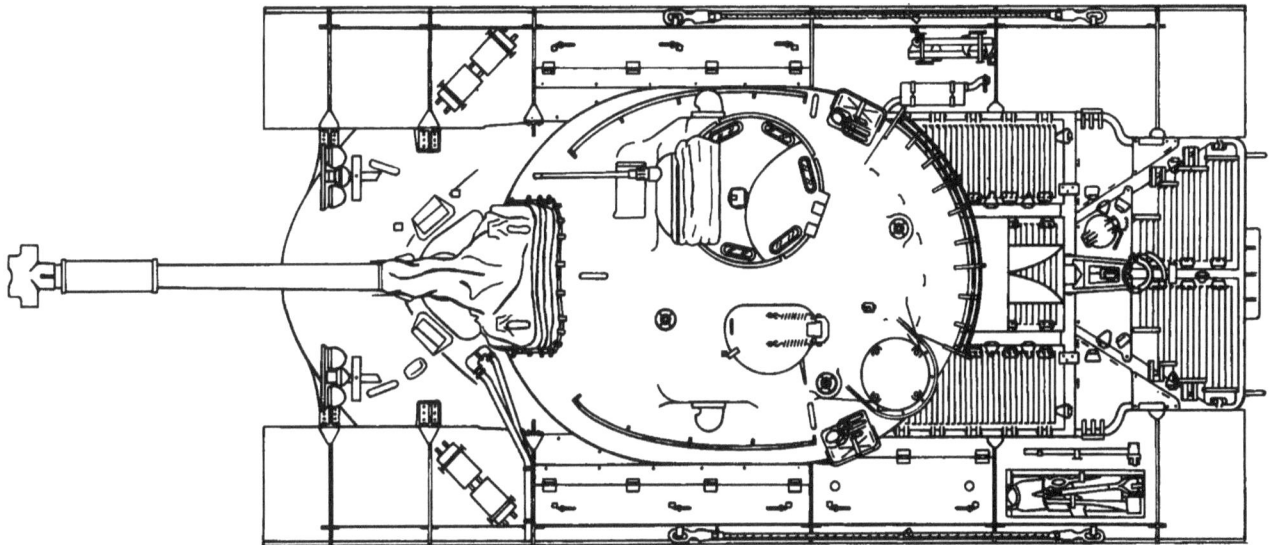

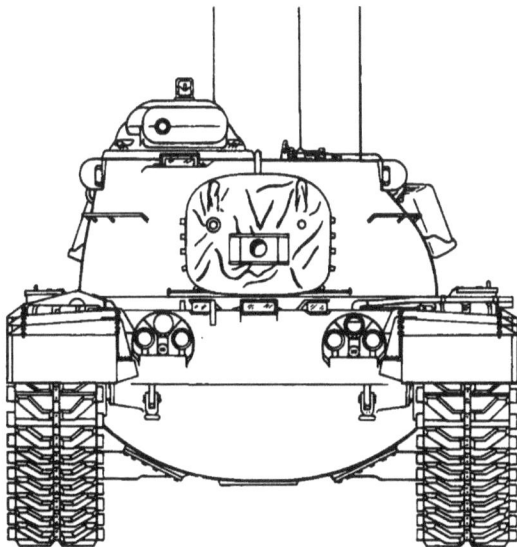

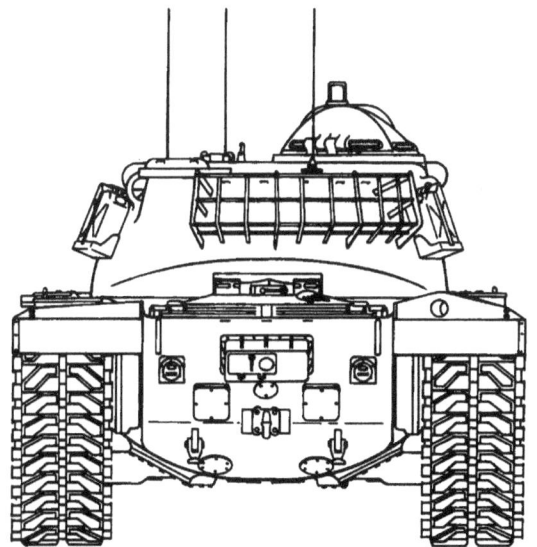

Scale 1:48

90mm Gun Tank M48A1

After the .50 caliber coaxial machine gun was eliminated, other methods were sought to increase the secondary firepower of the M48. One experiment (above) mounted a .30 caliber machine gun on the turret roof in front of the loader's hatch.

Another experimental feature that was not adopted for production was the cover for the range finder port illustrated at the right. It protected the optical system from dust and could be opened and closed from inside the turret. Such a device appeared much later with the laser range finder on the M60A3.

Above are two additional views of the Aircraft Armaments model 30 cupola as installed on T48 number 49 at Aberdeen Proving Ground. Compare with the production cupola (below right). Note the addition of the guard for the M28 periscopic sight on the cupola roof. At the left below is a late M48A1 during arctic tests at Fort Greely, Alaska. Note the protective cover for the driver and the T-shape blast deflector on the cannon.

Above, M48A1s fitted with jettisonable fuel drums cross the Main River near Gross Auheim, Germany during exercises with the 8th Infantry Division.

Although many of the deficiencies revealed in the T48 testing program could be easily corrected, a more serious problem was presented by the rapid rate of fuel consumption and the resulting short cruising range. An interim solution for troops in the

Above is the AVI-1790-8 engine with the oil coolers mounted around the top. Without the outrigger type coolers, it is much more compact than the AV-1790-7B shown on page 100. Below is the XT-1400 transmission used in the T48E1.

field was provided by the use of jettisonable fuel tanks. Four 55 gallon fuel drums were mounted on a rack at the rear of the hull and connected to the tank's fuel system. Use of these unprotected drums extended the range of the vehicle to approximately 135 miles and they could be jettisoned from inside the tank in an emergency. However, the development of an improved version of the Continental engine and a new transmission appeared to offer a much better solution to the problem.

The new engine was the Continental AVI-1790-8 which differed from its predecessors in several important features. The outrigger oil coolers with their separate fans on the earlier engines were replaced by new unit coolers mounted around the top of the engine itself. The main engine fans drew the air through these coolers. Elimination of the bulky outriggers provided additional space in the engine compartment. Another feature of the new engine, as indicated by its AVI designation, was the use of a Simonds SU fuel injection system replacing the two carburetors on the earlier models. This equipment injected a metered amount of fuel directly into the intake manifold adjacent to the intake valve. This arrangement provided greater efficiency and reliability.

The new XT-1400 transmission was a more efficient design than the CD-850. Similar to the smaller XT-500 described earlier, it was designed for heavier tracked vehicles and included regenerative steering. The XT-1400 featured a triple differential steering system with controlled planetary gear trains at each drive sprocket replacing the clutch-brake steering of the XT-500. With three forward gears and a throttle velocity controlled lockup clutch, the XT-1400 provided more effective middle range performance and better fuel economy than the cross drive transmission. The XT-1400 also was less expensive to manufacture with 35 per cent fewer parts than the CD-850.

Above are two views of T48E1 pilot number 3 during air flow tests to develop a new infrared suppression rear deck and exhaust system. The view at the left shows the AVI-1790-8 fitted with pancake mufflers. The top deck is removed in this photograph. At the right is the exhaust grill at the rear of the tank.

The problem of increased range for the M48 was discussed by the Design Coordinating Committee at Chrysler during May 1953 and in June a project was started to adapt the new engine and transmission to the tank. It was recommended that three M48s be converted to component test vehicles and a crash program began at Detroit Arsenal in October with the first tank completed on 23 November.

The additional space provided in the engine compartment by the elimination of the bulky outrigger fans allowed the fuel capacity to be increased from 200 to 380 gallons. However, installation of the XT-1400 transmission required the modification of the standard M48 hull with some decrease in ground clearance. The planetary final drives on the modified tank were fitted with new 13 tooth sprockets.

Shipped to Fort Knox for test, the Army Field Forces found only a marginal decrease in fuel consumption, although the range was greatly increased because of the larger capacity. Fort Knox objected to the reduction in ground clearance pointing out that this decreased the cross-country mobility.

Although the tank was referred to as the M48E1 during the development and test program, OTCM 35154 officially designated it as the 90mm gun tank T48E1 on 28 January 1954. The project had succeeded in its objective of extending the cruising range, but the cost of the necessary production changes to incorporate the new transmission was considered excessive. Also, if the earlier production tanks were not modified to the new standard, logistical support would be required for vehicles with two different transmissions. After the unfavorable report from Fort Knox, further work was stopped and efforts turned to the evaluation of the AVI-1970-8 engine with the CD-850 transmission.

Parallel with the development of the T48E1, Detroit Arsenal initiated studies to increase the range of the M48s already produced by installing the new AVI-1790-8 engine with a modified version of the CD-850-4B transmission. As on the T48E1, the compact AVI-1790-8 would permit much larger fuel tanks in the engine compartment. A modified rear deck was required to provide sufficient air flow for cooling the new engine and a meeting at Chrysler on 12 March 1954 concluded that the new design also should reduce the infrared radiation making the tank more difficult to detect. Procurement of at least two prototypes was authorized by OTCM 35539 which also assigned the designation 90mm gun tank T48E2 to the new vehicle. The number of pilot tanks authorized was increased to four in December 1954.

At the right is the artist's concept of the first proposed design for an infrared suppression rear deck and exhaust system. At this stage, it was planned for the T48E1, but the results of the study were applied to the T48E2.

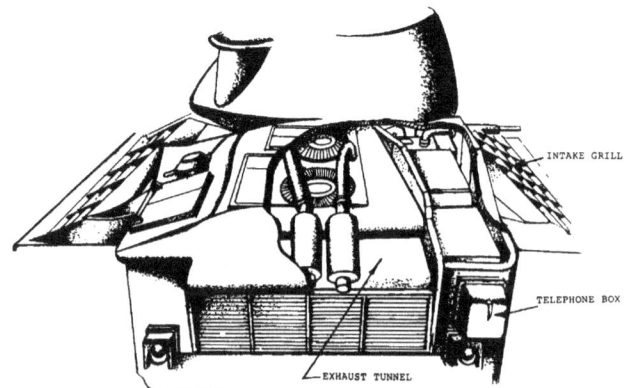

INTAKE GRILL

TELEPHONE BOX

EXHAUST TUNNEL

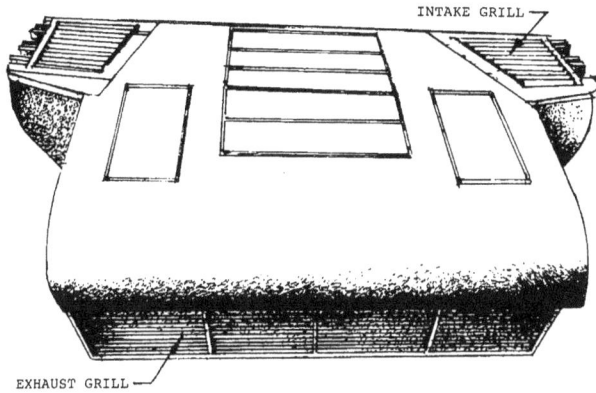

INTAKE GRILL

EXHAUST GRILL

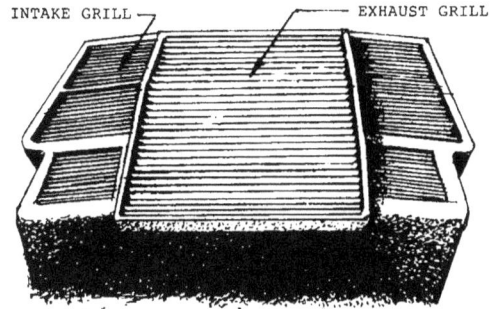

INTAKE GRILL — EXHAUST GRILL

The second rear hull design concept for the new tank appears at the top left and the completely louvered top deck of the third proposal is at the right. The fifth design (at the left) had an extremely high rear silhouette.

Several designs were studied to determine the optimum configuration for the engine compartment and rear deck. The first involved a complete redesign of the hull body. It provided air intake grills along each side of the hull above the fenders with a solid deck over the engine. The cooling air and exhaust from the mufflers discharged into a tunnel below this deck and passed out through vertical louvered grills at the rear of the hull. Although the design provided sufficient air inlet and exhaust area, the intakes were vulnerable to plugging from mud thrown up by the tracks.

A second approach retained the exhaust tunnel and rear grills, but changed the air intake to horizontal grills. However, further study showed that the air inlet area was inadequate.

A completely louvered top deck design increased both the inlet and exhaust areas, but the elimination of the exhaust tunnel resulted in excessive infrared radiation.

The fourth concept also must have been unsatisfactory, but no details have survived. However, the fifth design returned to the insulated exhaust tunnel with horizontal air intake grills and louvered exhaust doors in the rear. To increase the exhaust area, the grill doors were extended upward and outward. This resulted in a high rear silhouette which was considered unsatisfactory.

The sixth and seventh proposed layouts provided the basis for the final hull design. These arrangements used horizontal air intake grills along each side of the deck. The solid center area covered an insulated exhaust tunnel which discharged through rear grill doors hinged at each side of the hull. These louvered grill doors provided an easy means to identify the T48E2 from the later production tanks. As originally installed on the T48E2, the louvers, extended across each door in the vertical

At the left is the seventh proposal which evolved into the design used on the pilot tank. One of the T48E2 pilots is shown below.

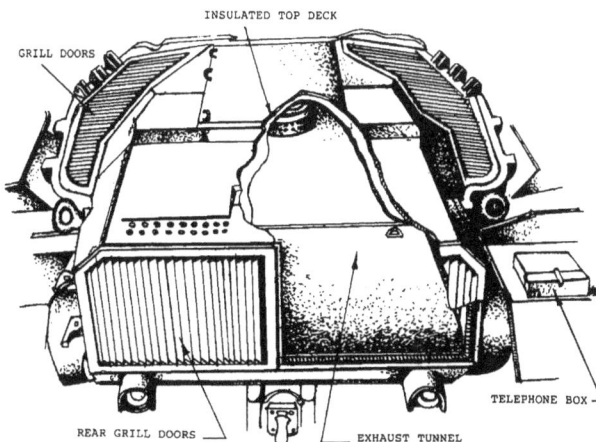

INSULATED TOP DECK

GRILL DOORS

REAR GRILL DOORS — EXHAUST TUNNEL

TELEPHONE BOX

The views on this page show the first pilot **T48E2** as received at Aberdeen Proving Ground. These photographs were dated 23 June 1955. The single exhaust pipe for the new personnel heater can be clearly seen. The cupola machine gun has not been mounted in this test vehicle. Note that the new tank has only three track return rollers on each side.

Below, the new headlight groups are visible in the front view (left) and the discoloration on the rear vertical grill doors (right) reveals the location of the main engine exhaust.

Above is another T48E2 pilot (left) during test operations at Fort Knox. The vertical grill doors have been replaced by the sloped design, but no provision has been made for attaching the deep water fording exhaust stack. At the right above, the power pack is removed from the tank. The AVI-1790-8 engine is fitted with the original pancake mufflers in this photograph. The pancake and Hollywood mufflers are compared in the view below.

direction deflecting the exhaust gases toward the center. During the test program, new doors were installed with the louvers angled at 22½ degrees from the vertical. These angled louvers deflected the exhaust gases upward and outward reducing the dust disturbance at the rear of the tank. This design was released for production after the louvers on the right hand door incorporated a square mounting pad for the deep water fording exhaust pipe.

The pilot T48E2s originally used pancake mufflers in the tunnel over the engine cooling fans, but during the test program they were replaced by two oval Hollywood mufflers exhausting directly into the air tunnel over the engine shroud. This arrangement was adopted for the production tanks.

With more space available in the engine compartment, the fuel tanks were redesigned. Four tanks were installed, two on each side of the power plant, raising the total fuel capacity to 335 gallons. This increased the cruising range to about 160 miles. The air cleaners were relocated into the fighting compartment within recesses in the engine compartment bulkhead. This permitted easy access for service and eliminated the expensive bulkhead doors in the previous design.

Below is a sketch of the new engine compartment bulkhead from the rear. The enclosures for the air cleaners can be seen at each side.

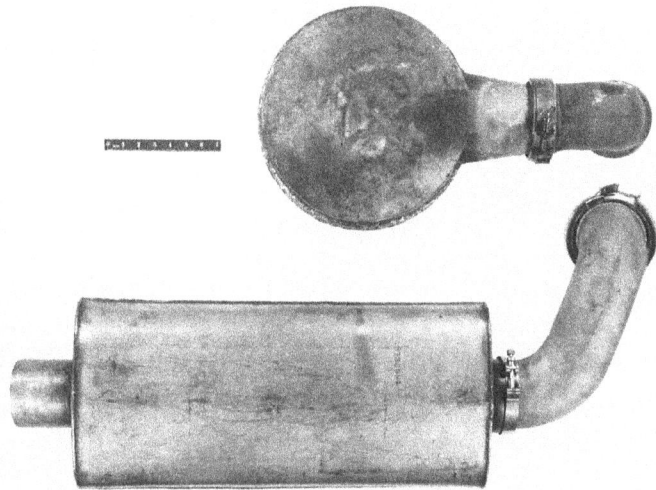

The interior of the rear hull is shown below with the power pack removed. The new fuel tanks can be seen on each side of the compartment utilizing space formerly occupied by the outrigger oil coolers on the earlier engines.

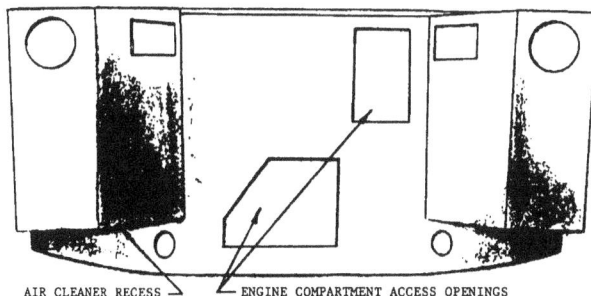
AIR CLEANER RECESS ─   └─ ENGINE COMPARTMENT ACCESS OPENINGS

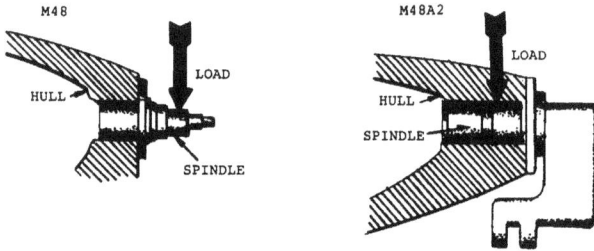

Above, the old (left) and new (right) spindle bearing mounts are compared. The hull modification of the T48E2 to the new configuration is shown at the right.

Other modifications reflected the experience gained from the service tests and troop use of the M48 and M48A1. Cross-country operation had resulted in numerous failures in the suspension system. The compensating idler spindles were particularly vulnerable to damage. To provide a more rugged suspension, several changes were made, including a new spindle and hull mount design. The hull casting was modified to have a larger boss with a greater bearing area for the spindle. A modification kit also was developed which could be applied to the earlier tanks by welding adapters to the hull to carry the new spindle bearings. Double bump springs were installed for the front road wheel arms and a new internal screw type track adjusting link was fitted between the compensating idlers and the front road wheel arms. The second and fourth track return rollers on the earlier suspension were eliminated leaving three rollers on each side.

The drawing below shows the layout of the new T48E2 suspension with three track return rollers and double bump springs for the front road wheel arm. The forward snubbers and part of the double bump spring can be seen in the photograph at the right.

Friction type snubbers replaced the hydraulic shock absorbers on the two front and the rear road wheels. Similar in outward appearance to the hydraulic shock absorbers, the snubbers absorbed energy by the friction between a brake lining material and the inside steel surface of the tubular assembly. Tests showed that the snubbers were more reliable with a longer life than the hydraulic shock absorbers. However, the linear response of the friction devices resulted in a harsher ride, particularly at high speeds. This latter characteristic made the snubbers unsatisfactory for use on the higher performance light tanks.

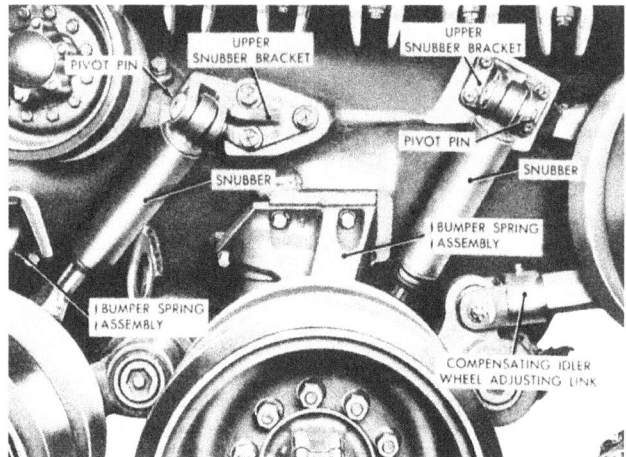

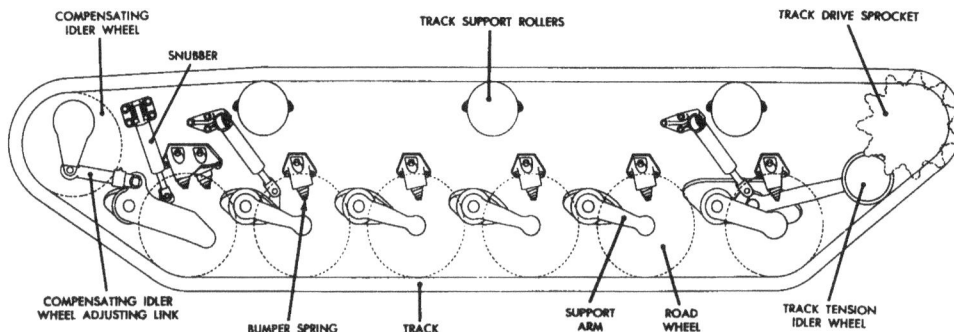

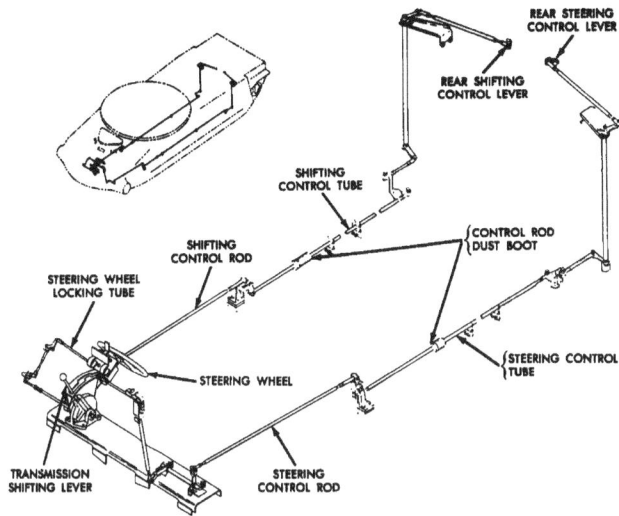

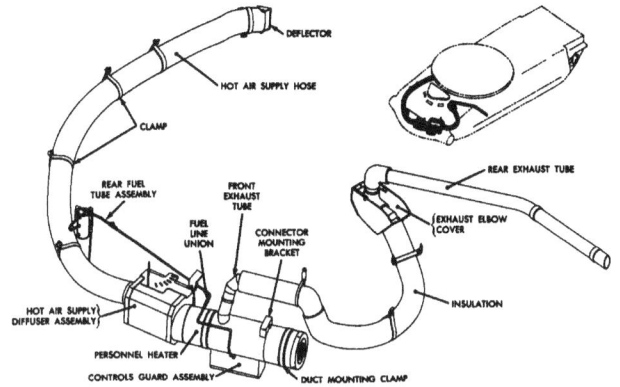

The steering linkage on the T48E2 appears at the left. The sketch above shows the installation of the new single personnel heater replacing the two heaters in the M48 and M48A1.

The new arrangement of the engine compartment required the redesign of the control linkage to the rear of the vehicle. At the same time, several improvements were incorporated. A new 14 inch diameter steering wheel, mount, and steering head reduced driver fatigue. The transmission shift controls were removed from the steering head and mounted on the floor giving the driver an easy visual check as well as a feel of the shift position. A larger capacity gasoline burning personnel heater was installed in the left front of the driver's compartment replacing the two in the earlier tanks. A single exhaust pipe extended through the hull roof at the left of the driver's hatch to the left side of the tank.

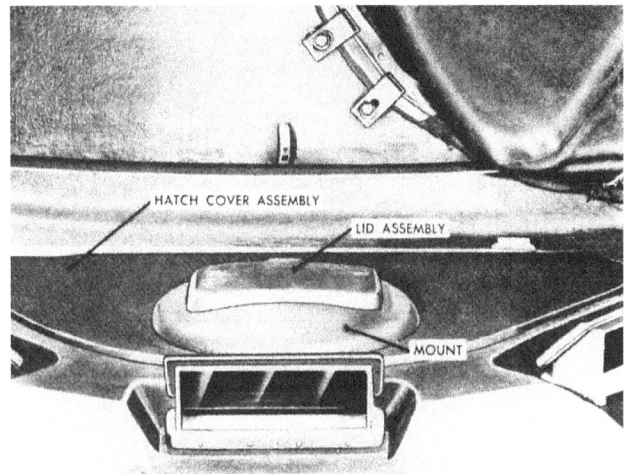

Above is the driver's hatch surrounded by three fixed periscopes and fitted for the installation of the M24 infrared periscope. Below, the driver's seat appears at the left in the dumped position required for the use of the floor escape hatch. An additional view of the driver's controls is at the bottom right.

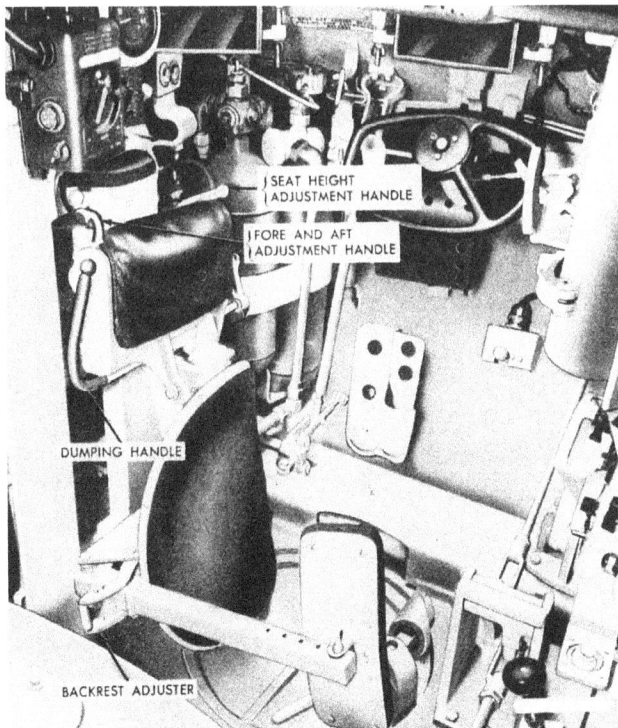

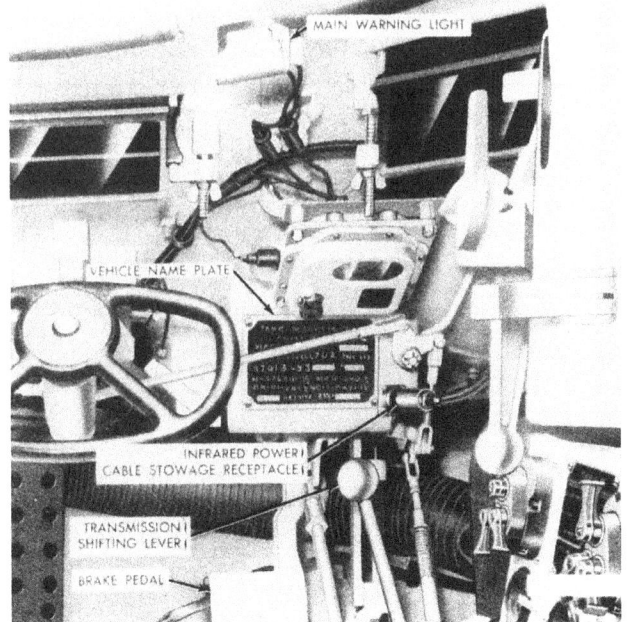

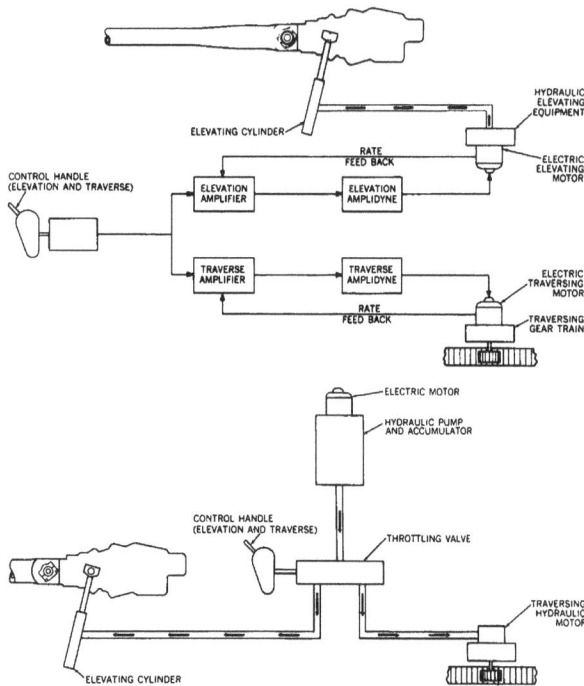

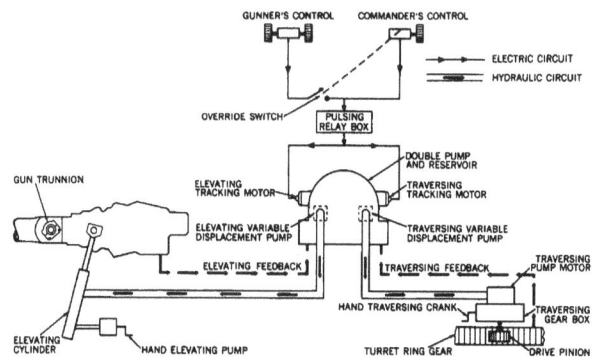

Simplified block diagrams illustrating the operation of the electric amplidyne (top left), electric-hydraulic (above), and the constant pressure hydraulic (left) gun control systems are shown here. Below is a sketch of the constant pressure hydraulic (Cadillac Gage) system selected for the production tanks.

Parallel with the work on the T48E2, two types of new turret control systems were under development. The first was an electric amplidyne system designed by Minneapolis Honeywell which was installed for test purposes in five experimental tanks. The second was a constant pressure hydraulic system developed by the Cadillac Gage Company. It also was installed in five tanks for evaluation. After exhaustive tests at Fort Knox as well as at the Yuma and Aberdeen Proving Grounds, the Cadillac Gage system was selected for production.

Both the electric amplidyne and the constant pressure hydraulic equipment provided precise control in elevation and traverse which was superior to that obtained with the more complex electric-hydraulic (Oilgear) system already in use. Both new control systems were considered preferable in terms of reliability and durability to the older equipment. Space requirements were roughly equivalent for the hydraulic and electric apparatus in light tanks. However, with heavier turrets, the Cadillac Gage controls had a definite advantage when the small size of the hydraulic traverse motor was compared to the larger electric motor required. Also, much less heat was generated by the hydraulic components.

After completion of the T48E2, Chrysler was directed to design a new vehicle combining a turret using the Cadillac Gage controls with all of the hull and engine features of the T48E2. For a time, this vehicle was unofficially referred to as the T48E3. However on 6 October 1955, it was standardized by Ordnance Committee action as the 90mm gun tank M48A2 and in December it was approved for production.

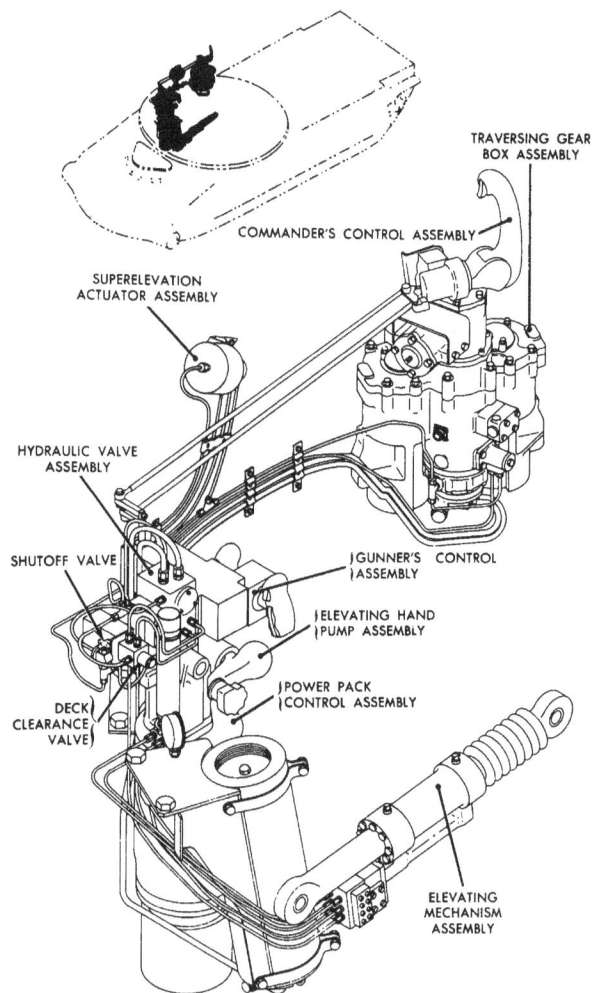

These photographs show the fully equipped M48A2 as released for production. The T-shape blast deflector was standard for the M41 cannon at this time.

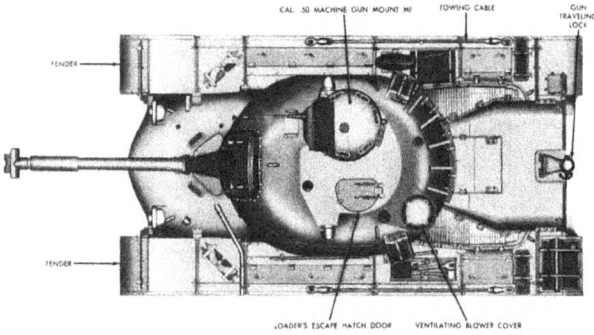

The major points of identification for the M48A2 are visible above and at the left. These include the infrared suppression top deck and rear grill doors, the three track return or support rollers, and the single personnel heater exhaust pipe extending to the left side of the tank.

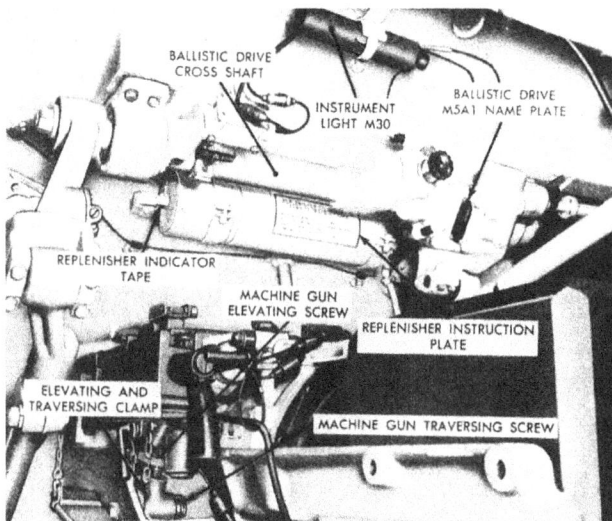

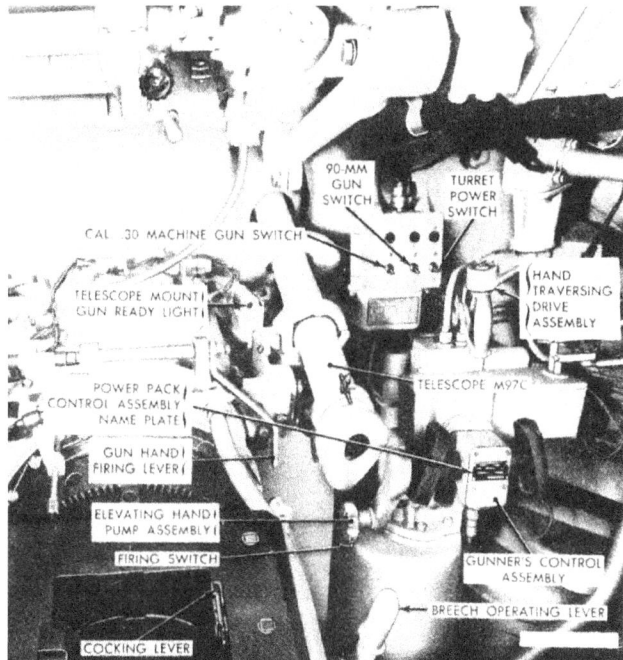

Above, the coaxial machine gun can be seen at the left on the loader's side of the turret and the gunner's controls are at the right. Below is the radio set (left) installed behind the tank commander and the ventilating blower (right) at the left rear of the turret.

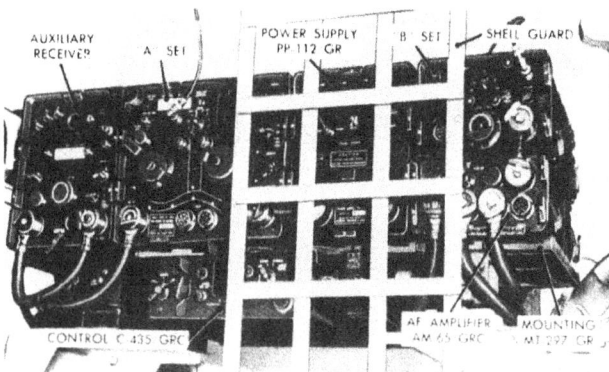

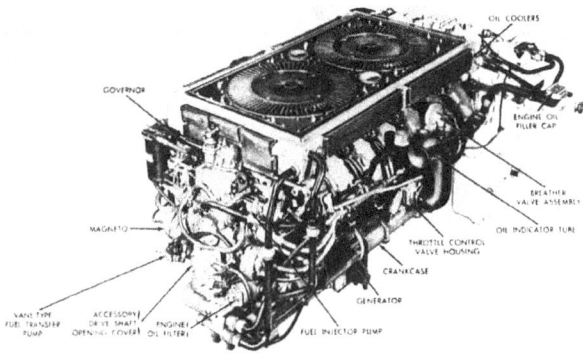

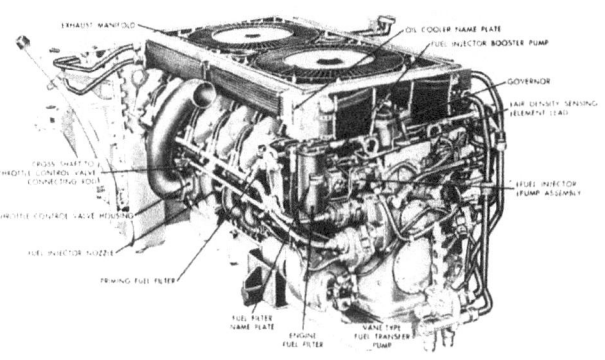

Details of the production AVI-1790-8 engine and CD-850-5 transmission can be seen above. Below at the right, another view shows the Hollywood mufflers installed. The letters in the photograph refer to a parts list. At the left below is the right headlight group and horn assembly. Each headlight group consisted of a service headlight, an infrared headlight, a blackout drive lamp, and a blackout marker lamp. The two headlight groups were identical.

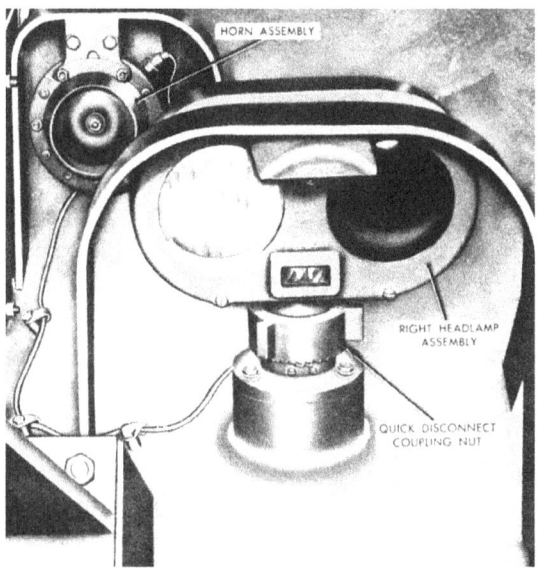

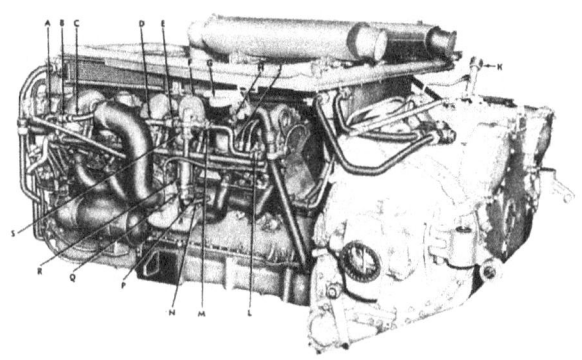

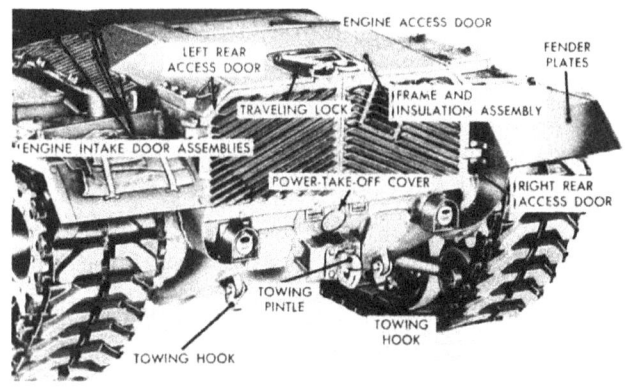

Details of the rear grill doors and other components on the rear of the tank are visible in the photograph at the right. Below at the right, the grill doors are open revealing the engine shroud and the end of one of the Hollywood mufflers. The small muffler attached to the deck is for the auxiliary generator engine. Directly below is a view of the auxiliary generator and engine. This unit is essentially the same as that installed in the earlier tanks.

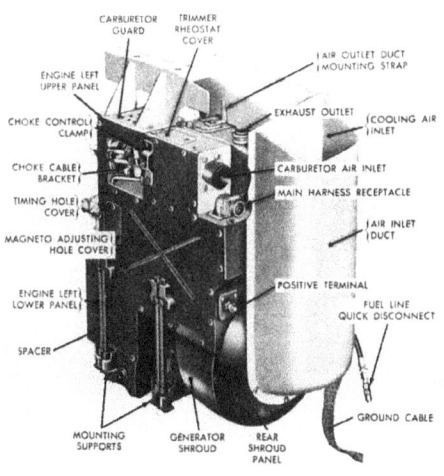

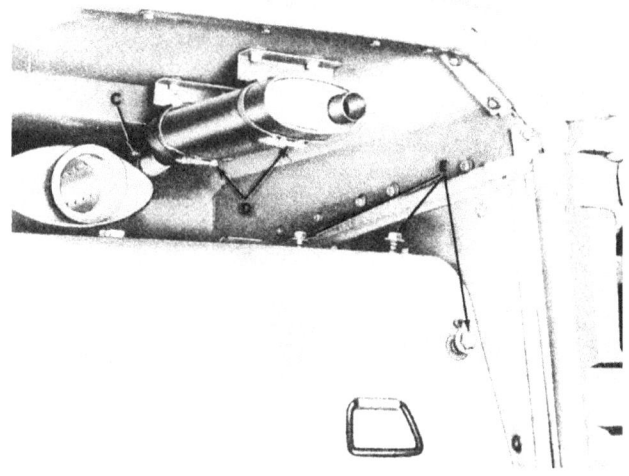

Above is a new M48A2 at Chrysler on 16 January 1957. This is the same tank shown on page 115. The square mounting pad for the deep water fording exhaust stack is visible on the right-hand grill door. Below is a production M48A2 after its arrival at Fort Knox in December 1956.

Despite the increased fuel capacity, the range of the M48A2 still was considered inadequate. As a result, a jettisonable fuel tank system was designed similar to that for the earlier tanks. A view of this arrangement during tests at the Yuma Proving Ground is shown at the right.

The M48A2 was built at both ALCO Products, Inc. (formerly American Locomotive Company) in Schenectady, New York and at Chrysler's Delaware plant with a total production of 2328 tanks. Further tests and troop use resulted in the introduction of additional modifications to the new tank. An M17 (M13A1E1) coincidence type range finder replaced the stereoscopic M13A1. The ballistic drive M5A1 was superseded by the M5A2 which featured temperature compensated linkage. The new fire control equipment utilized the metric system replacing yards with meters. A larger capacity bore evacuator was installed on the 90mm gun and the track tension idlers were eliminated between the rear road wheels and the sprockets. With all of these changes, it was originally intended to redesignate the vehicle as the M48A3. However, the final version of OTCM 37025 on 14 January 1959 assigned the designation 90mm gun tank M48A2C to the improved vehicles. The M48A2C modifications were applied to 1344 M48A2 tanks already produced.

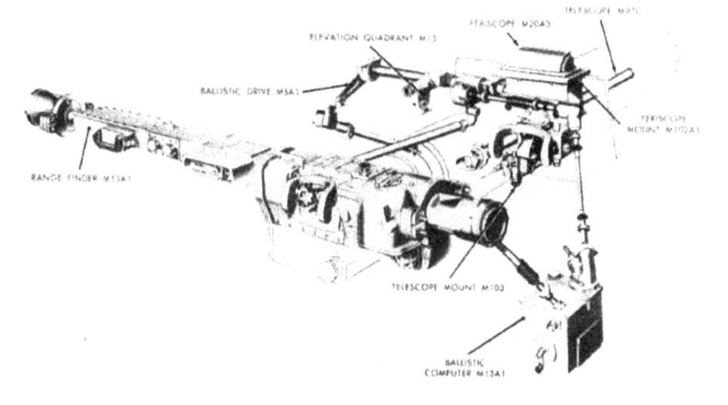

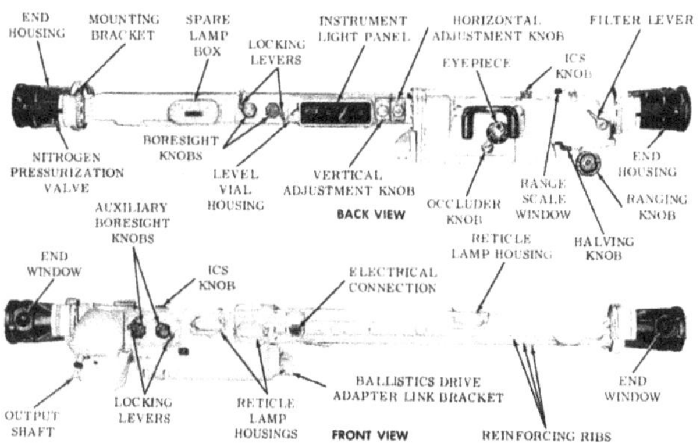

The sighting and fire control system for the M48A2 appears above at the right. New components which were introduced with the M48A2C included the M17C coincidence type range finder (directly at right), the M5A2 ballistic drive (below left), and the M13A1C ballistic computer (below center). The sketch at the right below illustrates the method of ranging with the M17C by bringing the images of the target into coincidence. Compare with the stereoscopic system on page 95.

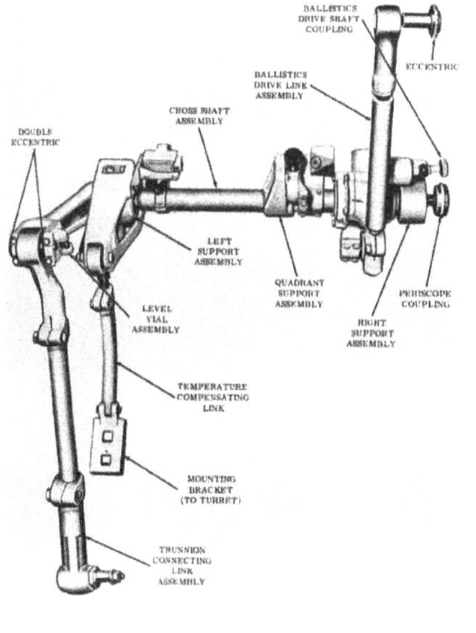

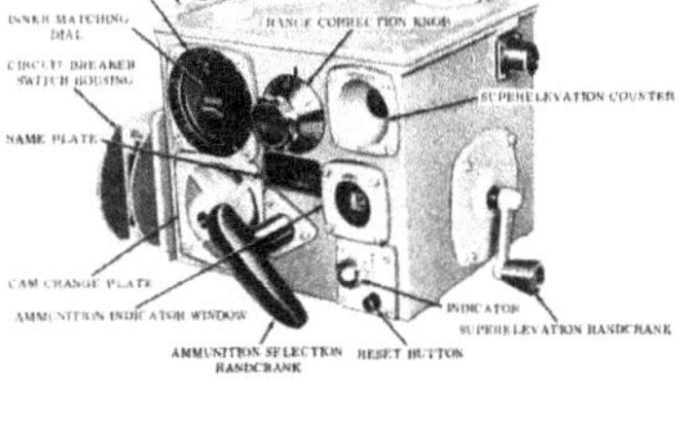

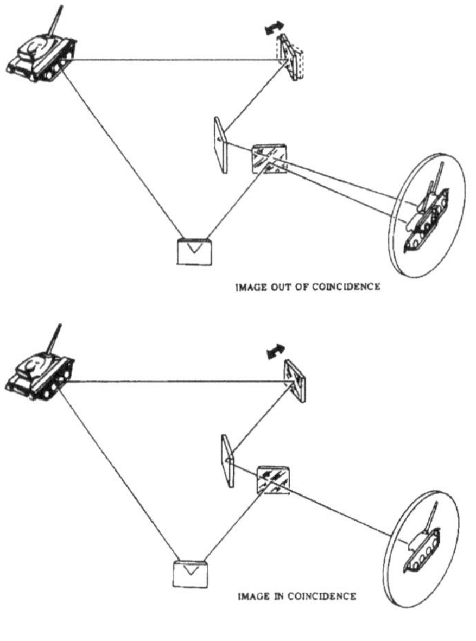

CAL. .50 HB BROWNING MACHINE GUN M2   COMBINATION GUN MOUNT

90-MM GUN M41

COMMANDER'S PLATFORM

DRIVER'S SEAT

CAL. .30 BROWNING MACHINE GUN M1919A4E1 or M37   GUNNER'S SEAT

AIR CLEANER

ENGINE

TRANSMISSION

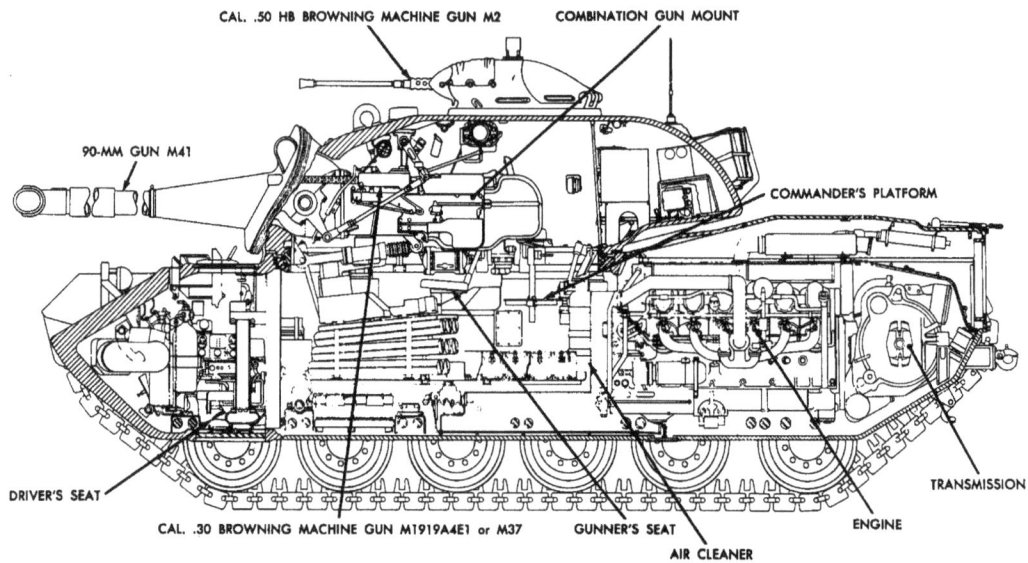

Above is a sectional view of the M48A2C. The arrangement of the exhaust mufflers in the tunnel under the insulated top deck can be seen here.

The photographs above show an M48A2C after modification at Chrysler. Note the absence of the track tension idler. The headlight groups have been removed on this tank. Below is a new fully equipped M48A2C. Even the fixed periscopes are in place around the driver's hatch.

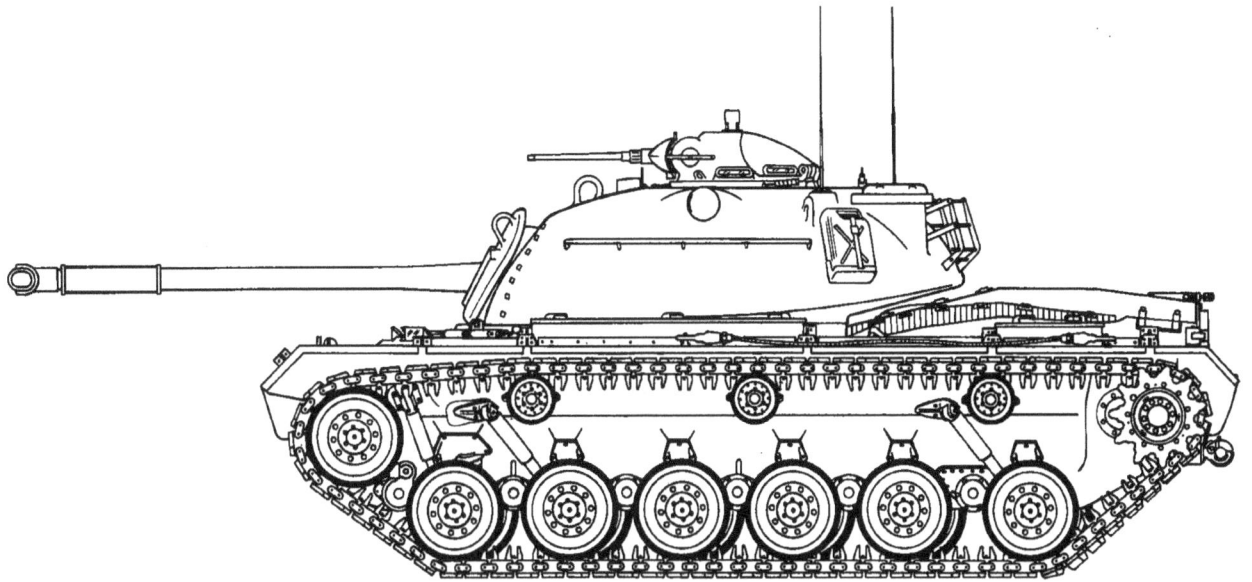

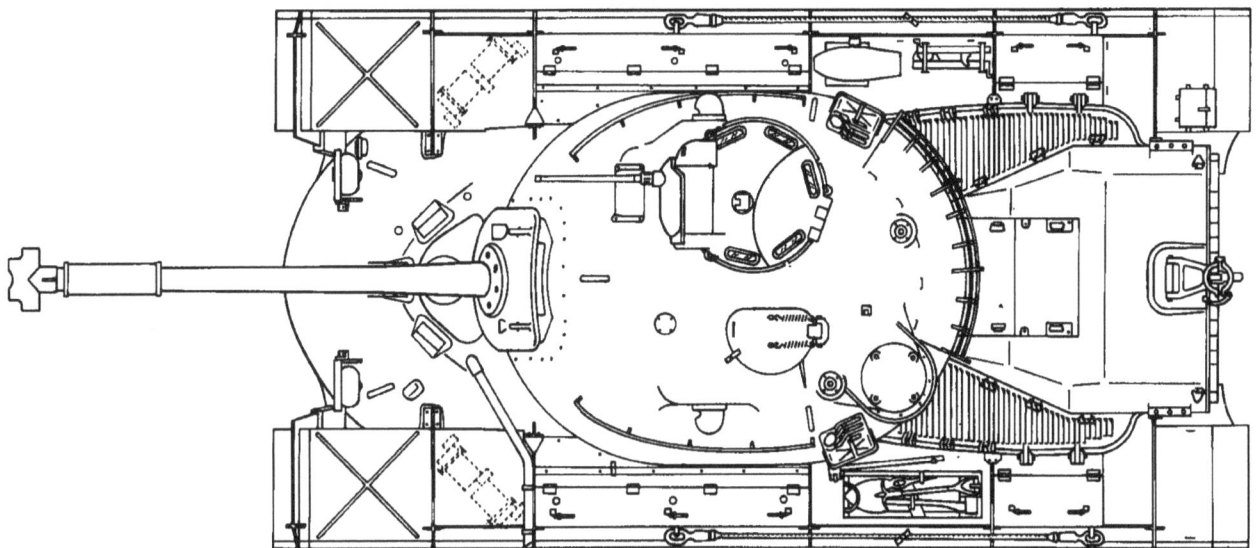

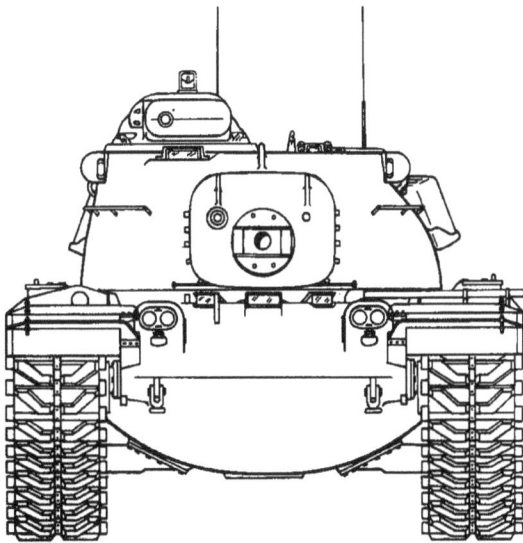

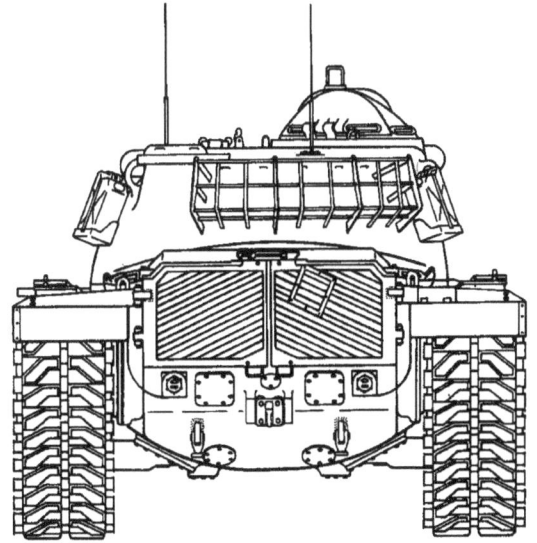

Scale 1:48

90mm Gun Tank M48A2C

A single M48A2C was rearmed with the 105mm M68 gun by Chrysler in early 1966. It was intended to serve as a prototype for modifications suitable for use in the middle east. Other features included 105mm ammunition racks, elimination of the personnel heater, and installation of the xenon searchlight. A plug can be seen welded in the hole for the personnel heater exhaust pipe. The views below at the right show the searchlight mounted in the operational and stowed positions. Note the interrupters installed to prevent the cupola machine gun from damaging the searchlight in either location.

The models above compare the standard M48 cast hull (left) with a new hull (right) designed to use siliceous cored armor. The latter model is sectioned and it is opened below at the right to show the internal arrangement of the fused silica cores.

EXPERIMENTAL MODIFICATIONS

The report of the Stilwell Board as well as the Army Equipment Development Guide recommended the development of special armor to defeat shaped charge as well as kinetic energy projectiles. Research on this problem indicated that glass or ceramic materials were particularly effective against the HEAT (High Explosive AntiTank) or hollow charge round. The combination of such materials with steel armor offered a solution to the problem of providing protection against the HEAT projectiles within practical weight limitations. It also offered protection against the effect of the HESH (High Explosive Squash Head) ammunition. Developed in the United Kingdom, it was referred to in the U.S. Army as the HEP (High Explosive Plastic) round.

Two methods of application were considered for the new special armor. One approach cast the steel armor around a fused silica core. Appropriately named siliceous cored armor, this material was considered for several design concepts and it was proposed for future use on the M48 series. However, production of these tanks was complete before any action could be taken.

The protection provided by homogeneous steel armor and siliceous cored armor is compared at the right for kinetic energy projectiles (AP) and shaped or hollow charge rounds (HEAT). Note the superiority of the siliceous cored armor with the latter.

SOLID STEEL ARMOR vs. SILICEOUS CORED ARMOR

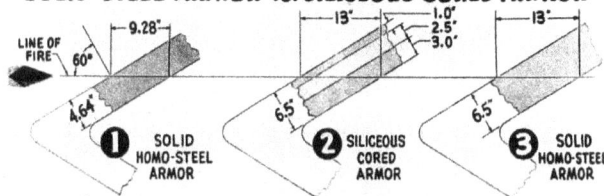

UNDER THE CONDITIONS SKETCHED ABOVE, THE SILICEOUS CORED ARMOR (2) IS EQUAL IN WEIGHT TO THE SOLID HOMOGENEOUS STEEL ARMOR (1) AND GIVES EQUIVALENT PROTECTION AGAINST KINETIC ENERGY ARMOR PIERCING AMMUNITION. HOWEVER, (2) PROVIDES GREATER PROTECTION AGAINST SHAPED CHARGE ROUNDS THAN THE 40% HEAVIER SOLID HOMOGENEOUS STEEL ARMOR (3).

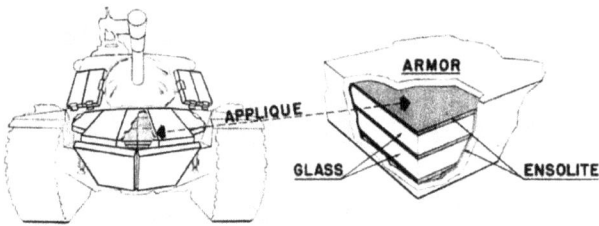

The sketch above illustrates the use of applique armor on the standard tank. Such a kit installed in the field weighed about 3000 pounds for the hull alone, but it provided about the same protection as the siliceous cored armor. It was estimated that about 20 minutes would be required for installation on a prepared vehicle.

The second method of utilizing the new protective material combination made use of applique armor which could be attached when required to reinforce critical areas of the tank. Consisting of glass slabs encased in steel, such armor could be installed in a very short time at battalion level and easily removed when not required. Examples of the applique armor were tested at Fort Knox, but it was not adopted for general use.

The crowded space inside the M1 commander's cupola continued to cause problems. The 100 round ammunition box was replaced by a smaller 50 round box, but this required frequent reloading of the machine gun and did nothing to improve the limited head room for the commander.

The model above shows the installation of an applique armor kit on the M48 hull. The fittings required to attach the kit to the hull armor can be seen in the view at the right. Below, an M48A2 has been fitted with a modified version of the applique armor kit. However, it appears that the three segments have been tack welded together which might make it a bit cumbersome to remove.

CROSS SECTION

The model of the M47 above is fitted with protective grills intended to defeat shaped charge (HEAT) or squash head (HESH or HEP) projectiles. Tests at Aberdeen indicated that it was effective against the 3.5 inch antitank rocket at 60 degrees obliquity.

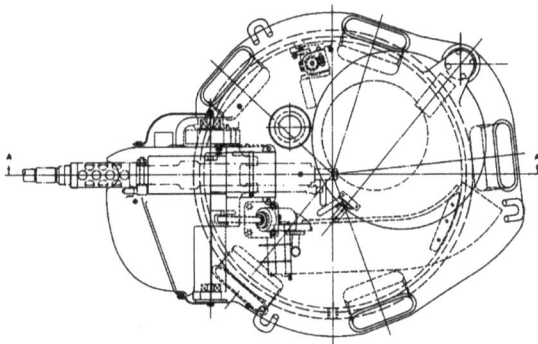

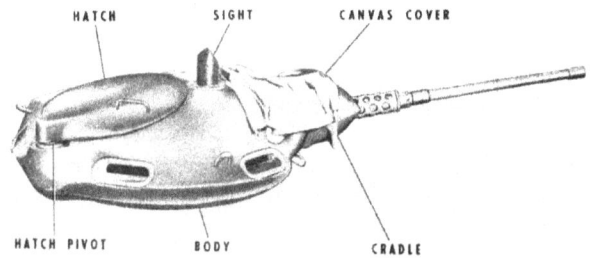

SECTION A–A

The Aircraft Armaments model 108 cupola is shown in the photograph above and in the drawings at the left. In the top view of the latter, the 200 round ammunition box can be seen extending back along the left side of the cupola.

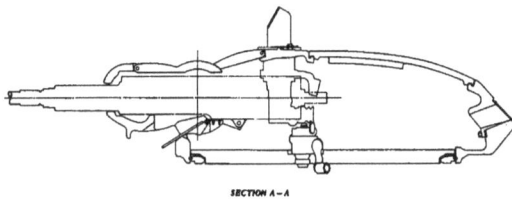

A new experimental cupola was installed on an M48A2C for evaluation. Designated by Aircraft Armaments, Incorporated as the model 108, it fitted the same 29.75 inch diameter opening in the turret roof as the M1. The hatch on the new cupola was pivoted rather than hinged as on the M1. The body of the cupola extended outward on all sides providing greater interior volume, although this resulted in reentrant angles in the armor between the cupola and the turret. The M2 .50 caliber machine gun was fed from a 200 round ammunition box and the empty

cases and belt links were ejected from the cupola. The armor varied from 1 inch in the front to ¾ inches in the rear for a total cupola weight with gun and ammunition of 1287 pounds. Problems with the ammunition feed resulted in the termination of the test program on the M48A2C without a report being submitted.

At the right, a model 108 cupola is installed on an M48A2C during test operations.

125

The design of an automatic loader for the 90mm T139 gun in the T48 tank was the subject of a study contract at the Rheem Manufacturing Company. This project ran from May through December 1953 and was a paper study only. No pilot models were built.

The proposed automatic loader was based on a design prepared by Rheem for use with the 76mm gun in the light tank T41E1. A low (18½ inches high) rectangular magazine, open on all four sides, was located below the gun mount containing 22 rounds of 90mm ammunition. It was reloaded through a door in the top surface. An hydraulic indexing cylinder moved the rounds into position for transfer to the hoist. Any one of three types of ammunition could be selected. The hoist was supported by a vertical aluminum casting and followed the gun during changes in elevation. The carrier rammer lifted the round to the center line of the gun and rammed it into the breech. After firing, it caught the ejected case and returned it to the magazine. The design rate of fire was 22 rounds/minute. The cannon was rotated 90 degrees in the mount so that the breechblock moved in the horizontal direction. The open part of the breech ring then faced the loader permitting easy hand loading, if required. Five additional ready rounds were stowed on the platform floor and the normal hull and bustle stowage remained the same.

The Army Equipment Development Guide of December 1950 recommended the development of a 105mm tank gun. The future need for such a weapon was considered during the design of the T48 and its 85 inch turret ring provided ample room for a larger cannon. OTCM 33842 officially initiated a development project to improve the armament on 6 July 1951 designating two new vehicles. These were the 105mm gun tank T54 when fitted with a conventional

turret and the 105mm gun tank T54E1 when an oscillating turret was installed. Both vehicles used the T48 chassis.

The new cannon was designated as the 105mm gun T140 and it was a lighter weight version of the 105mm gun T5E2, although fixed rounds replaced the separated ammunition of the earlier weapon. A muzzle velocity of 3500 feet/second was attained with a 35 pound armor piercing projectile. When modified for installation in the oscillating turret, the designation was changed to the 105mm gun T140E2. Both the T140 and the T140E2 guns were inverted for use with the automatic loaders specified for both vehicles. In this position, the breechblock moved up to open and down to close. Four separate recoil cylinders replaced the concentric recoil mechanism used with the 90mm gun in the T48.

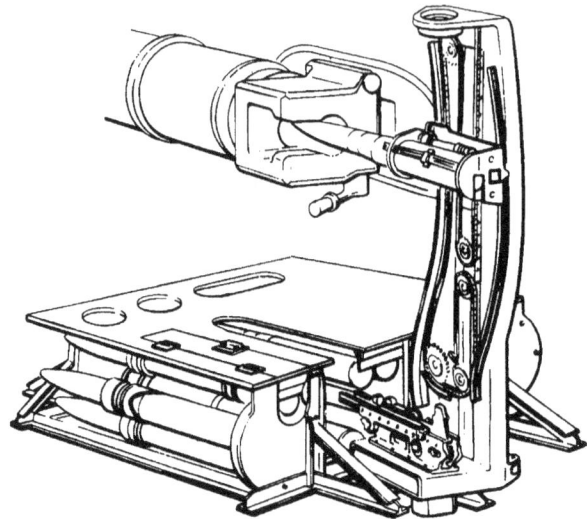

Although no usable photographs of the T54 were located prior to publication of this volume, the model above was constructed from the original drawings prepared at Rheem Manufacturing Company. Ron Johnson started the project which was completed by Jacques Littlefield.

Rheem Manufacturing Company received a contract to design and build two T54 pilot tanks. A similar contract was let to the United Shoe Machinery Corporation to produce two T54E1 vehicles. M48 chassis were diverted from production to be modified for the installation of the new turrets.

The T54 turret was assembled using a large single piece homogeneous armor casting with a highly sloped front and a large bustle at the rear to balance the combination gun mount T156. The armor was equivalent to 4¼ inches at 50 degrees from the vertical in the front tapering down to 2¾ inches at 25 degrees on the sides. The small tapered main gun shield did not cover the port for the .30 caliber coaxial machine gun on the left side of the mount. The turret crew was located in the usual arrangement with the gunner in front of the tank commander on the right and the loader to the left of the gun. A .50 caliber machine gun was installed in an Aircraft Armaments model 108 cupola for the commander on the right side of the turret roof. A flat hatch was provided over the loader's position. Both the cupola and loader's hatch were fitted in a large hinged section of the turret roof which could be opened for rapid access or to permit the installation of large turret components. The fire control system consisted of the T46E3 range finder used by the tank commander connected to the T32 ballistic computer and the gunner's M16E1 periscopic sight. Although the original design did not include a direct sight telescope, later specifications called for a T174 telescopic sight for the gunner.

The automatic loader was fed from a magazine holding 14 105mm rounds. It was possible to select from three types of ammunition. The T54 design called for a total 105mm ammunition stowage of 47 rounds with eight in the turret bustle, five in the basket, and 20 in the hull in addition to the automatic loader magazine. The magazine was loaded through a hole in the back plate. Like the design study for the T139 gun, the automatic loading equipment consisted of two main assemblies, the magazine and the hoist. The latter was a vertically mounted aluminum casting fitted with a carrier rammer and a round transfer mechanism. It served to convey the rounds from the magazine into the gun and to return the empty cases to the magazine after firing. A large hydraulic cylinder in the hoist provided the force to transfer the round. and to ram it into the cannon. An arm connected the gun to a control chain on the hoist. When the gun was elevated or depressed, the chain caused the carrier to follow the motion maintaining its alignment with the gun tube. The control panel for this equipment was located at the gunner's station, but it could be reached by the tank commander in an emergency.

Despite the larger gun, the T54 turret had a smaller frontal target area than that on the standard M48 tank. The large hinged access cover in the turret roof was power operated providing a quick escape route in an emergency. The radio and a ventilation blower were installed in the turret bustle in addition to the ammunition stowage.

When the T54 turret was installed on the M48 chassis at Rheem, a new travel lock was fitted for the 105mm gun. However, the travel lock for the 90mm gun was not removed from the test vehicle. Also, the track tension idlers were retained, although their removal was included in the design specification.

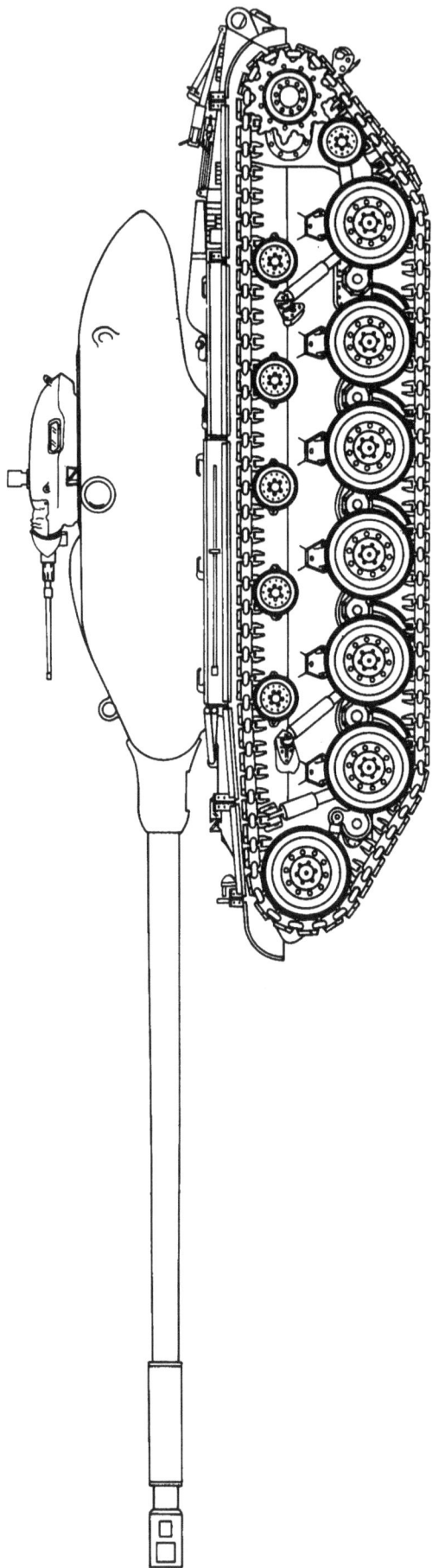

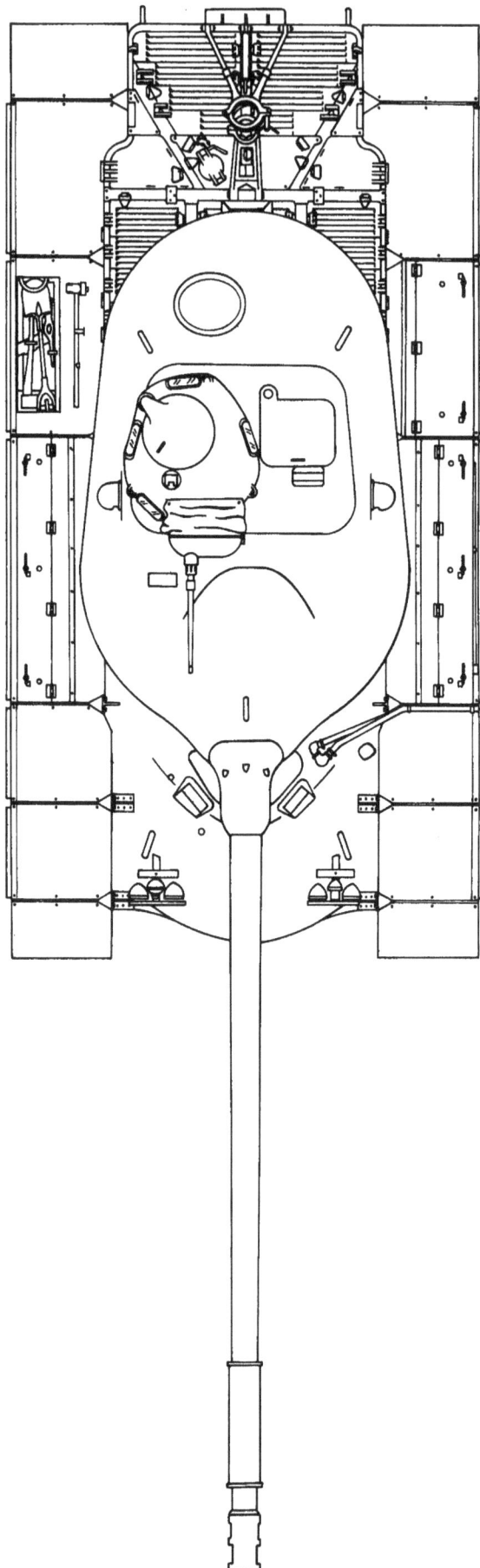

Scale 1:48

105mm Gun Tank T54

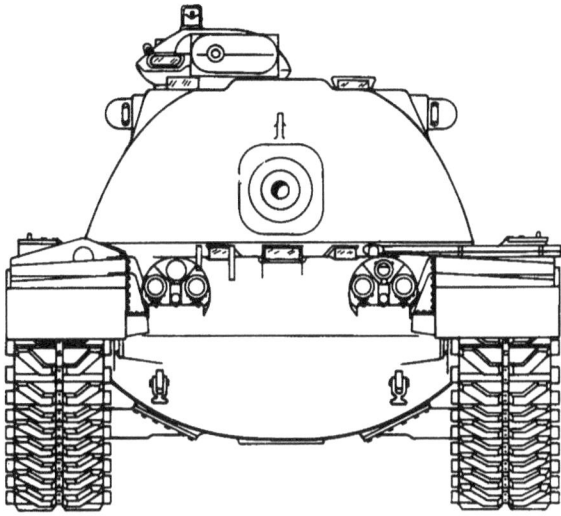

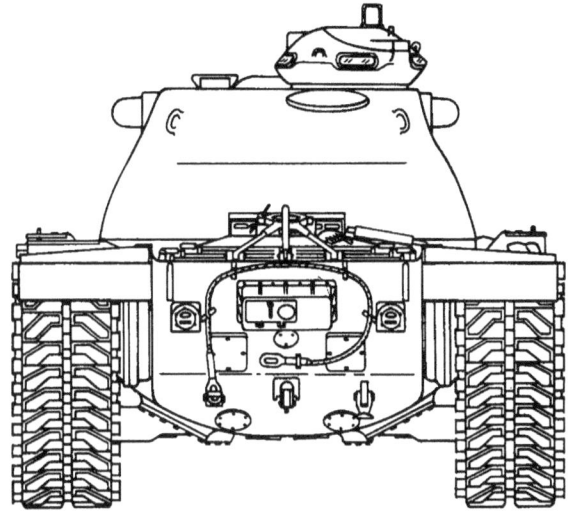

Below are three additional views of the T54 model. It is intended to reflect the proposed production tank without track tension idlers and with the travel lock only for the 105mm gun. The 4-view drawing retains both the track tension idlers and the 90mm gun travel lock as did the prototype tank.

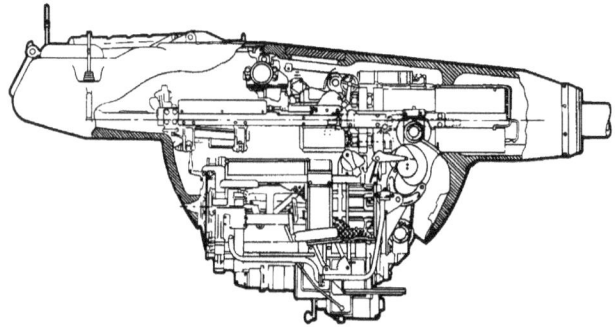

The model above dates from the development period and illustrates the original concept of the T54E1. At the top right is a sectional view of the oscillating turret showing the automatic loader. The 9 round magazine for the loader can be seen at the right. At the bottom of the page is the T54E1 after final assembly.

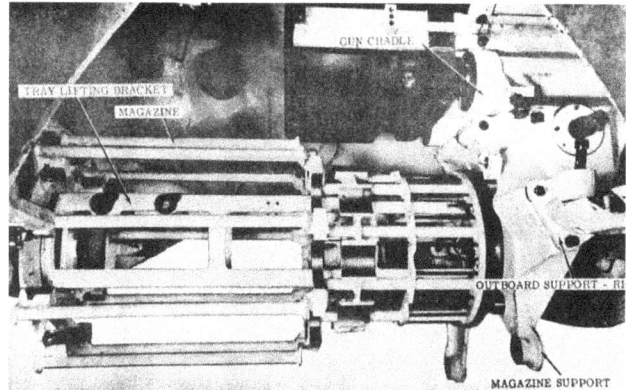

At United Shoe Machinery, work on the T54E1 turret paralleled that on the T54 at Rheem. The oscillating turret consisted of a cast turret body mounted on the trunnions of the ring casting which was bolted to the bearing race assembly of the M48 tank. The turret casting varied in thickness from 5 inches at 60 degrees from the vertical in front to 2 inches at 30 degrees in the rear. The sides were equivalent to 2½ inches at 30 degrees. The turret basket was attached to the trunnion ring casting. Seats for the tank commander and the gunner were installed on the right side and moved with the turret in both azimuth and elevation. The loader's seat on the left moved only in azimuth. The 105mm gun T140E2 was installed in the T157 mount with a .30 caliber coaxial machine gun to the left of the cannon. An interim design cupola for the tank commander was located on the right side of the turret roof. It essentially consisted of the standard World War II type vision cupola modified just above the six vision blocks to include a rotating section. The rotating part incorporated a hatch with a periscope and a mount for a .50 caliber machine gun. A large

The single remaining example of the T54E1 is displayed above by the Ordnance Museum at Aberdeen Proving Ground. However, the sign and registration number on this vehicle are incorrect. The registration number should be 30172551.

section of the left turret roof was hinged along the left side and could be opened for easy access during maintenance or ammunition loading. A small flat hatch for the loader was installed in the hinged section for normal use. The primary fire control system included the tank commander's T50 range finder, the T34 ballistic computer, and the gunner's M20(T35) periscopic sight. A T170 or T170E1 telescopic sight was provided for the gunner as a backup system.

A major reason for the oscillating turret was to simplify the design of the automatic loader. Since the cannon moved only in recoil relative to the turret, the position of the loader could be fixed. In the T54E1 turret, the automatic loader consisted of a nine round rotary magazine and a main drive assembly which operated the loading tray and ramming mechanism. The tray lifted the selected round from the top of the magazine until it was aligned with the cannon bore. The rammer then propelled the round into the gun at a velocity of 15 feet/second. After firing, the case was ejected along a continuous path provided by the ejection chutes and

bustle groove to the port in the back of the turret. This port was opened automatically by the gun recoil. Any one of three types of ammunition could be selected and the maximum design firing rate was 35 rounds/minute. In addition to the nine rounds in the magazine, six were stowed in the turret bustle, two in the basket, and 17 in the hull for a total of 34. Later specifications increased the hull stowage to 19 raising the total number of 105mm rounds to 36. The T54E1 also differed from the T54 by being equipped with the electric amplidyne instead of the constant pressure hydraulic turret control system.

Work on both the T54 and T54E1 began in earnest during 1952, but progress was slow because of design problems and difficulty in obtaining much of the required government furnished equipment. The latter problem reflected the demands of higher priority projects. Because of the delays, the program was overtaken by later developments of more powerful guns and lighter more mobile tank chassis such as the T95 series.

The design of the interim commander's cupola is illustrated at the left. This cupola also can be seen in the view below of the T54E1 at Aberdeen.

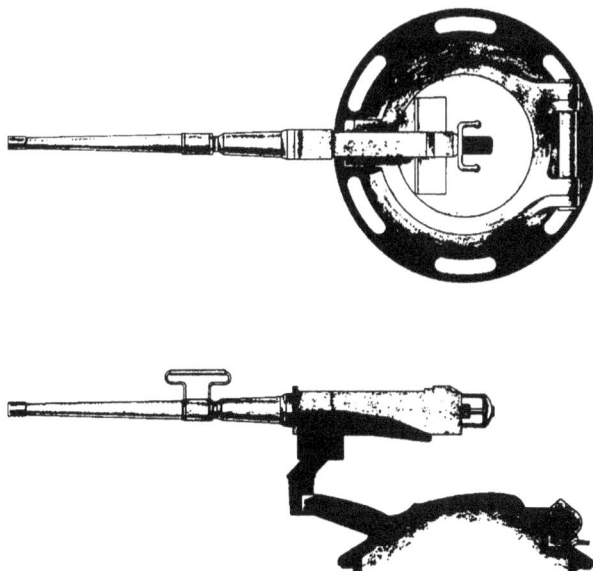

131

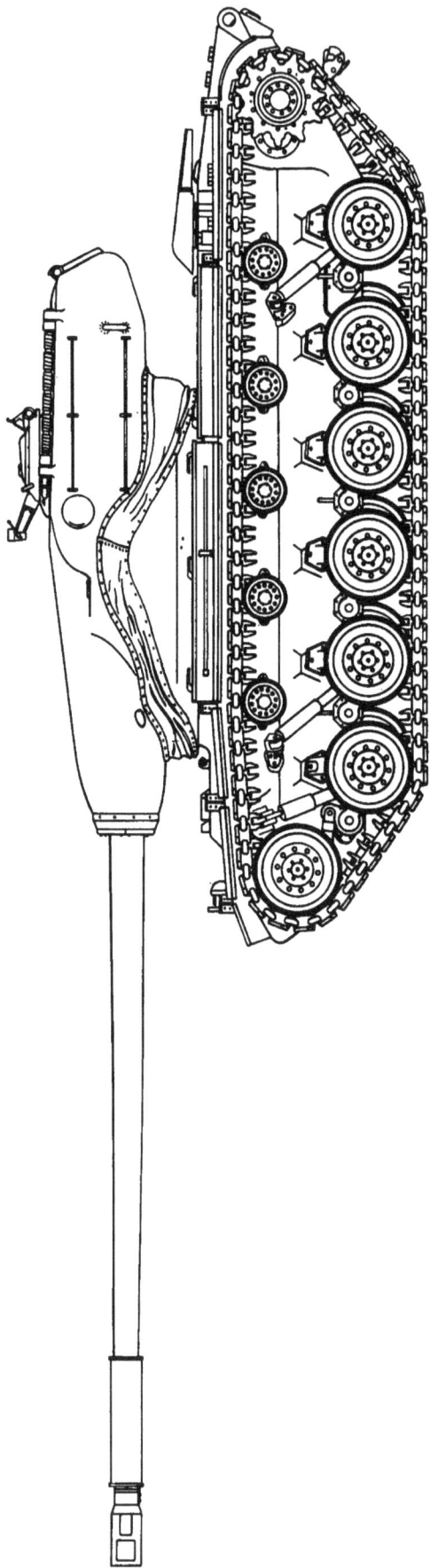

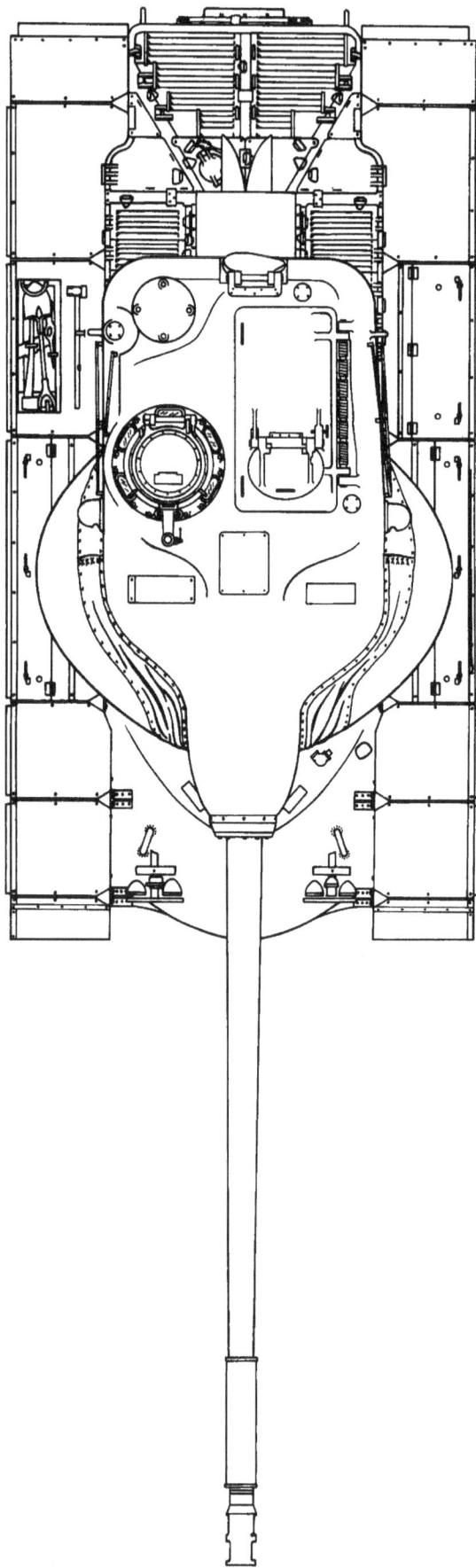

Scale 1:48

105mm Gun Tank T54E1

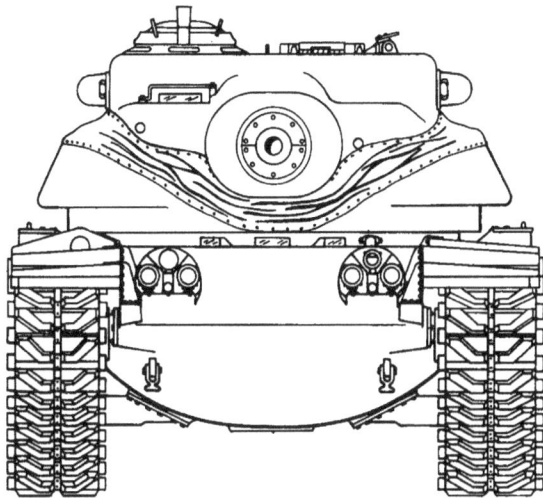

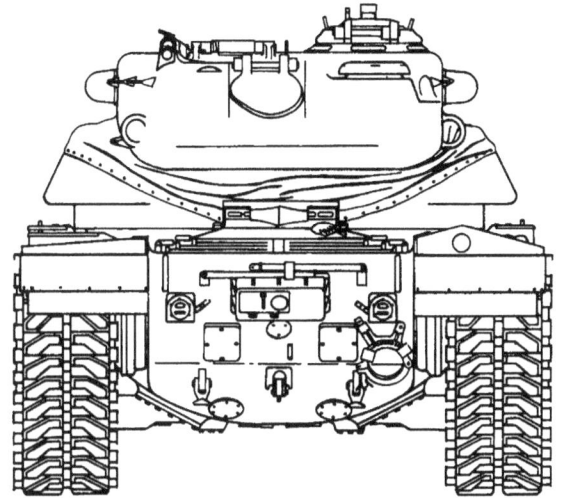

Additional views of the T54E1 at the Ordnance Museum are shown here. Note that the engine exhaust is shifted to the rear to provide clearance for the bustle of the oscillating turret at maximum elevation. The T54E1 turret was installed on an M48 hull with the small driver's hatch. However, the track tension idler has been removed from the suspension and the extension pipes are missing from the personnel heater exhausts. The travel lock for the 105mm gun can be seen above.

Above is the T54E2 as originally assembled with the flat circular door covering the direct sight telescope port. Note that a protective shield has been provided for the periscopic sight in the model 108 commander's cupola.

In late 1952, Detroit Arsenal began a study of a third type of 105mm gun turret. The new approach utilized a conventional turret similar to that on the T54, but eliminated the automatic loader to simplify the design and to minimize the weight and cost. On 18 May 1953, OTCM 34795 designated the new model as the 105mm gun tank T54E2 and authorized the manufacture of two pilot vehicles. This was later increased to three with the Cadillac Gage hydraulic turret control system installed in the first two tanks and the electric amplidyne system in the third pilot.

Like the other tanks in the series, the cast turret on the T54E2 had a narrow front with a small gun shield and a large bustle to balance the long 105mm gun. The armor ranged from 4 inches at 60 degrees from the vertical in front to 2½ inches at 0 degrees in the rear. An Aircraft Armaments model 108 cupola was installed on the right side of the turret roof for the tank commander and a small flat

The 105mm gun T140E3 is shown here. Without an automatic loader, the weapon was installed in the conventional manner with the open side of the breech ring on top.

hatch was provided for the loader. The main weapon, now designated as the 105mm gun T140E3, was mounted upright with the breechblock moving down to open. The new T174 gun mount occupied less space and was lighter than previous mounts. It consisted of a concentric recoil mechanism with a separate counter recoil cylinder underneath the gun. The normal recoil distance was only 10 inches compared to 12 inches in the T54 and 14 inches for the 90mm gun in the M48 thus saving additional space in the turret. A .30 caliber coaxial machine gun was fitted in the left side of the mount.

The T54E2 shown here is based on the M48A1 chassis with the large driver's hatch and the track tension idlers have been removed. The considerable overhang with the long barreled 105mm gun even in the travel position is obvious in the side views.

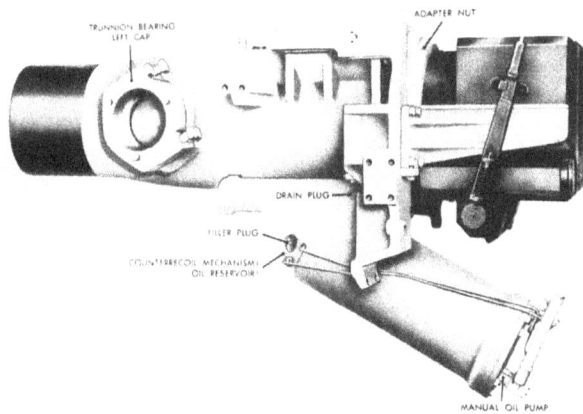

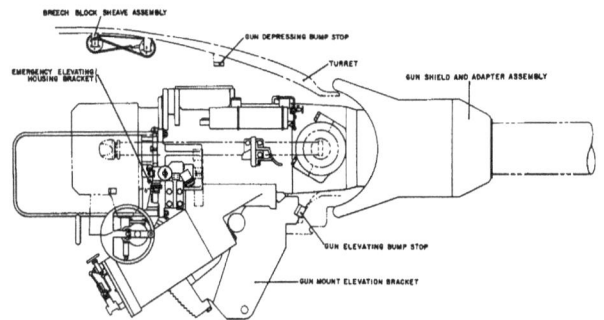

Above at the left is a view of the 105mm gun breech and the cradle assembly of the T174 gun mount. A sectional view of the concentric recoil system and counterrecoil mechanism appears below. At the right are right side (upper) and top (lower) view drawings of the T174 mount. The coaxial .30 caliber machine gun can be seen in the latter.

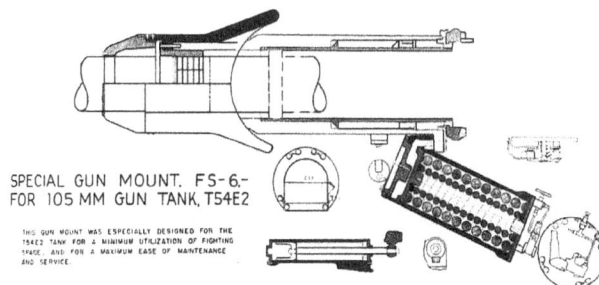

SPECIAL GUN MOUNT, FS-6,- FOR 105 MM GUN TANK, T54E2

THIS GUN MOUNT WAS ESPECIALLY DESIGNED FOR THE T54E2 TANK FOR A MINIMUM UTILIZATION OF FIGHTING SPACE, AND FOR A MAXIMUM EASE OF MAINTENANCE AND SERVICE.

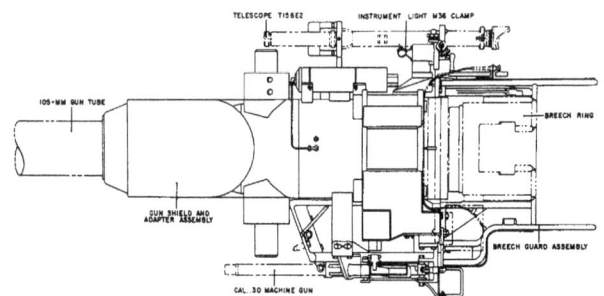

135

Details of the front and rear of the T54E2 are shown above. It appears that the extension pipes were not installed on the personnel heater exhausts when this pilot tank was assembled. Below at the left is a view of the turret casting prior to the installation of the gun mount, cupola, and other fittings.

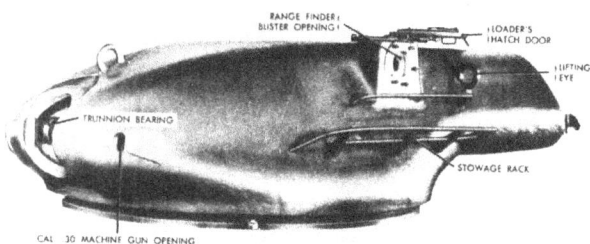

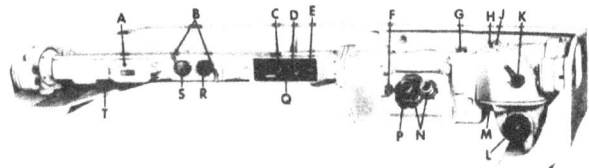

A new feature of the T54E2 was the shock mount arrangement to isolate the primary fire control equipment from the turret armor. The shock mount was bolted to the race ring flange and did not contact any other part of the turret casting. The tank commander's T46E3 range finder, the mount for the gunner's M16E1 periscopic sight, the ballistic computer T32, and the ballistic drive T37 were all installed on the shock mount. This arrangement reduced the transmission of shock to the fire control instruments from any hit on the turret unless it was struck near the race ring. It also eliminated errors resulting from thermal expansion of the turret casting. In addition to the periscopic sight, a T156E2 telescope was installed on the right side of the gun mount. It was protected when not in use by an armored telescope door operated from inside the turret.

A—SPARE LAMP BOX
B—LOCKING LEVER
C—RANGE SCALE LIGHT SWITCH
D—STEREO CONTROL KNOB
E—STEREO AND AUXILIARY GUNSIGHT SWITCH
F—INTERPUPILLARY KNOB
G—INTERNAL CORRECTION KNOB (ICS KNOB)
H—RANGE SCALE
J—RETICLE LAMP RECEPTACLE
K—FILTER KNOB
L—RANGE KNOB
M—HALVING KNOB
N—DIOPTER SCALE
P—INTERPUPILLARY SCALE
Q—INSTRUMENT LIGHT PANEL
R—GUN LAYING RETICLE ELEVATION BORE SIGHTING KNOB
S—GUN LAYING RETICLE AZIMUTH BORE SIGHTING KNOB
T—RANGE FINDER HOUSING SHOCK MOUNT

The T46E3 range finder can be seen above with the shock mount installed. The ends of this mount were supported by the shock mount brackets below which, in turn, were attached to the race ring flange.

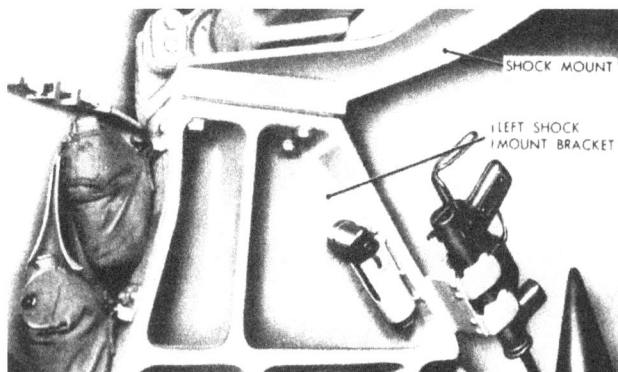

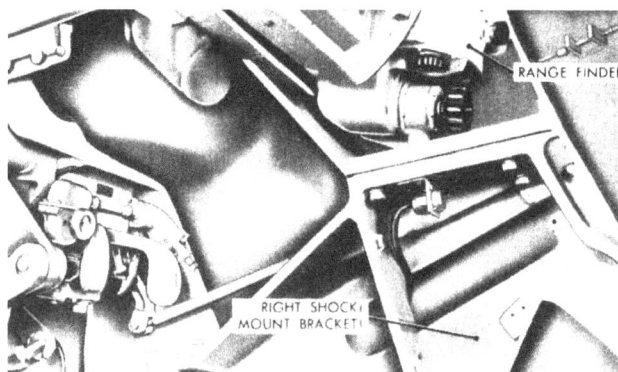

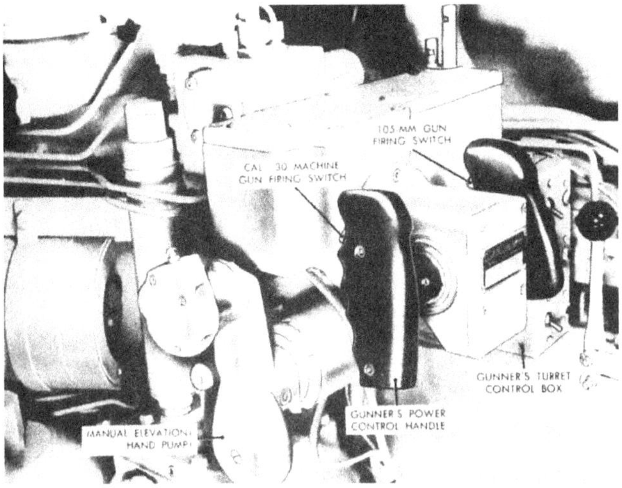

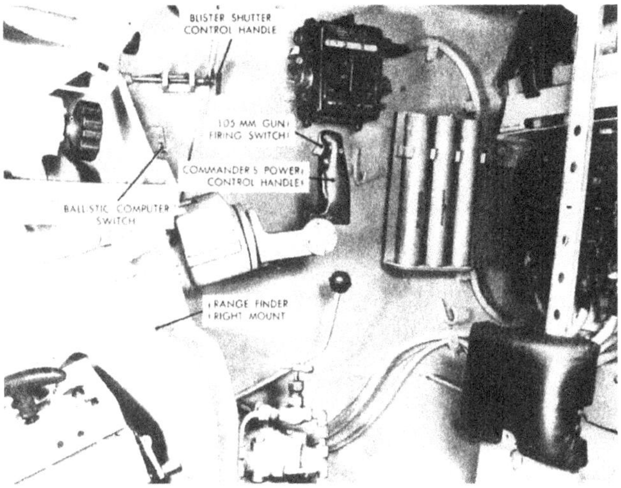

Above are the Cadillac Gage turret controls for the gunner (left) and the tank commander (right). Below are the hydraulically adjustable seats for the gunner (left) and the tank commander (right).

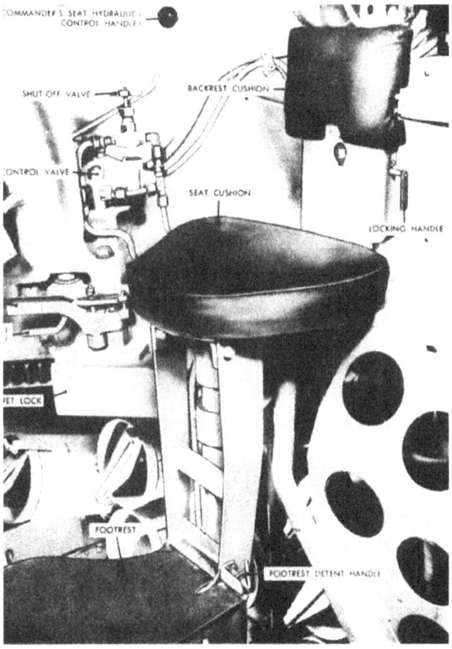

Below at the left, the gunner's M16E1 periscopic sight is installed in the shock mount. Details of this periscope can be seen in the two views below at the right.

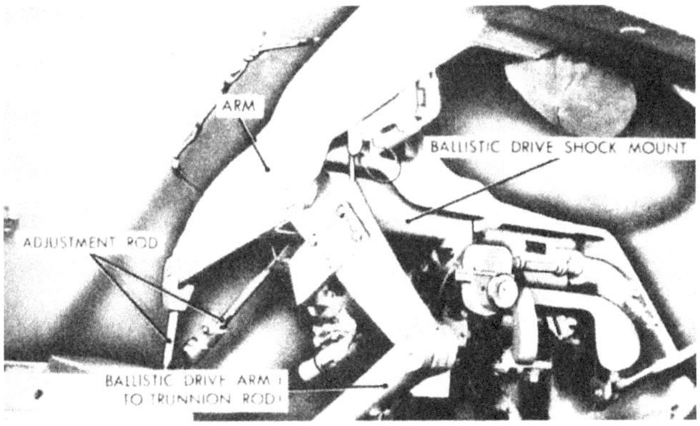

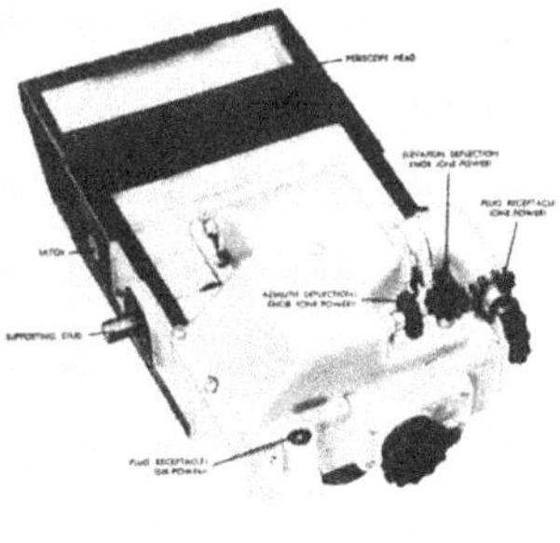

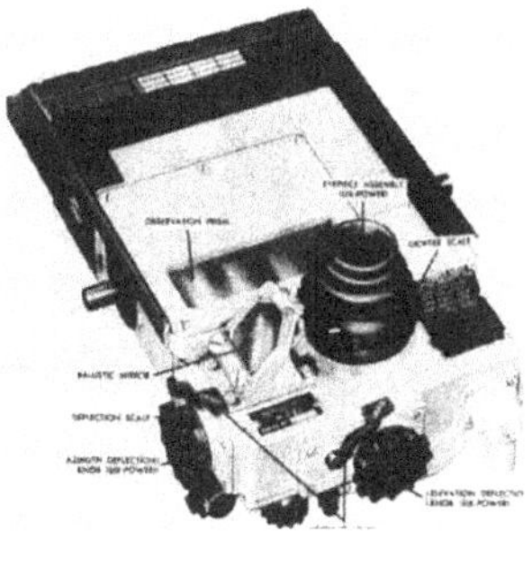

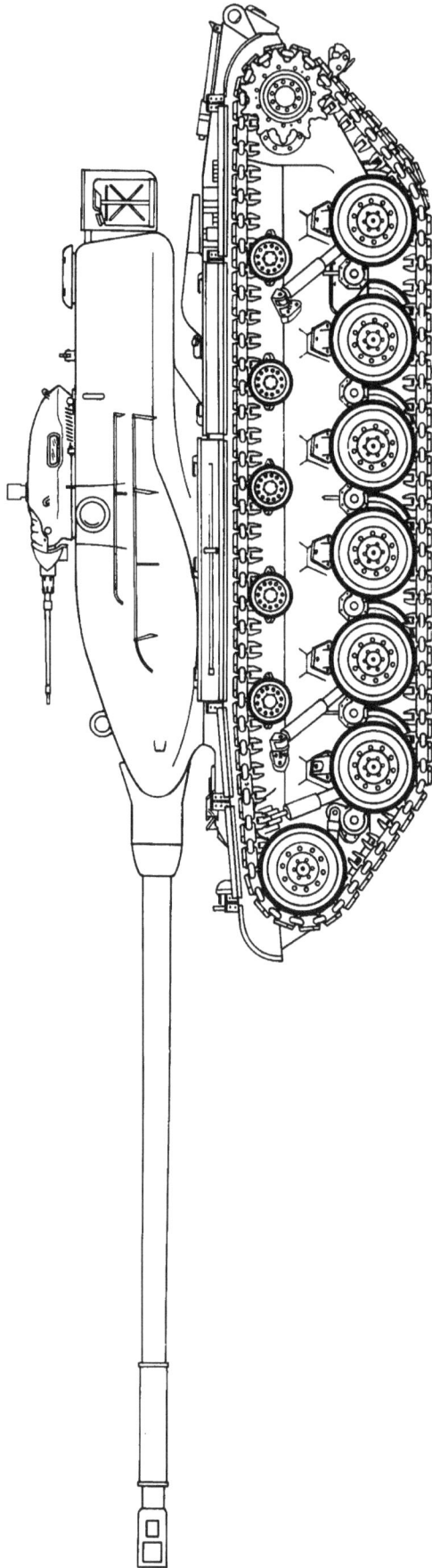

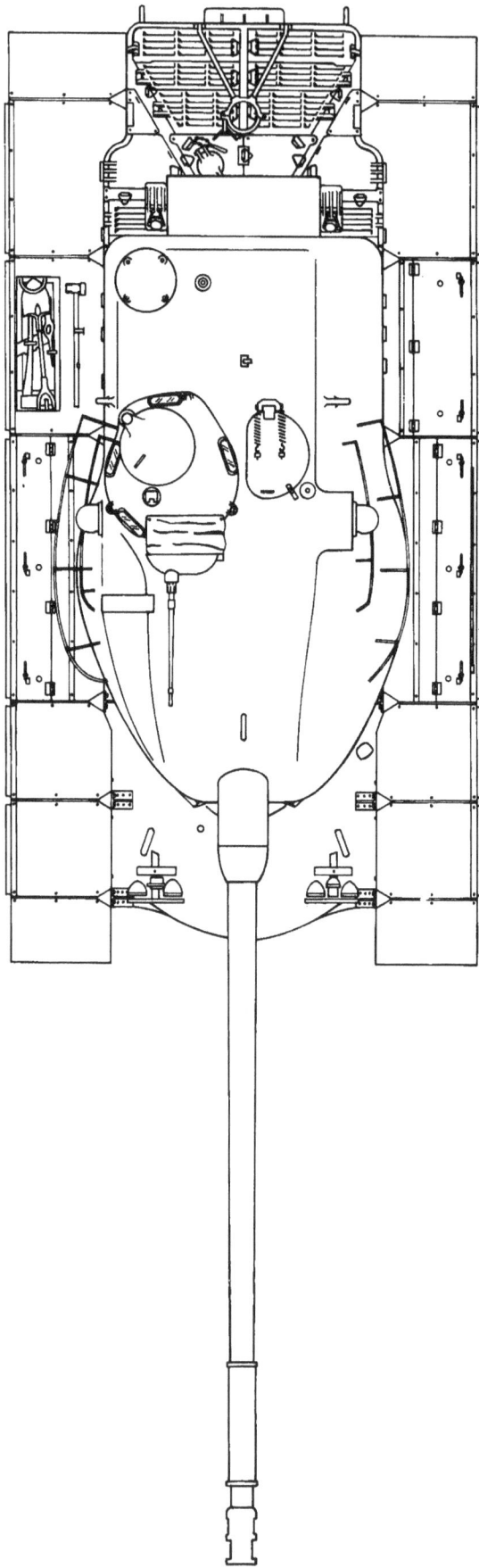

105mm Gun Tank T54E2

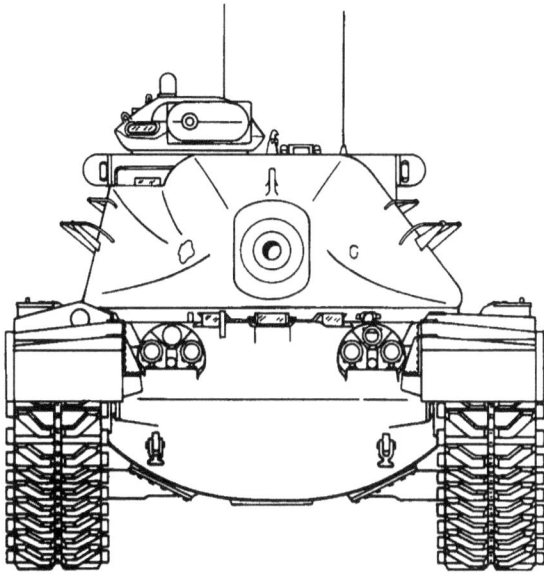

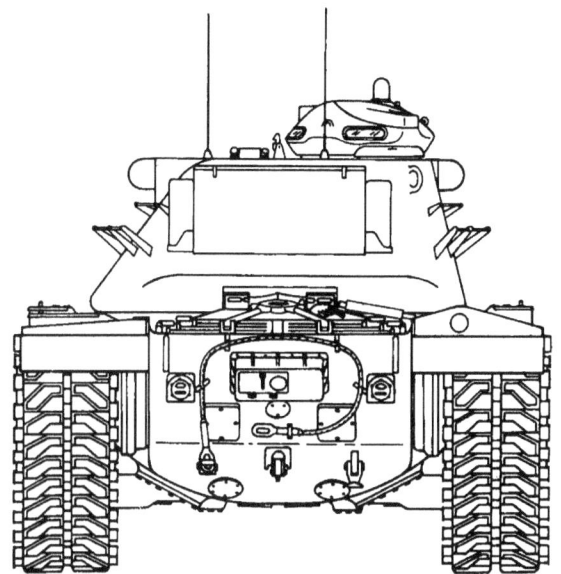

The T54E2 turret was installed on the standard Patton tank hull with a few modifications. The hull stowage was changed to accommodate 19 105mm rounds in two racks, one on each side of the driver. Ten rounds were carried in the left rack and nine on the right. Eleven rounds in the loader's ready rack and nine in the turret bustle brought the total 105mm ammunition stowage to 39 rounds, although some specifications indicate a total of 40. Other

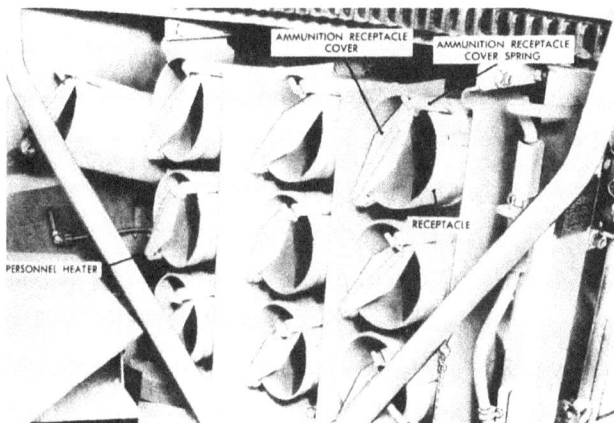

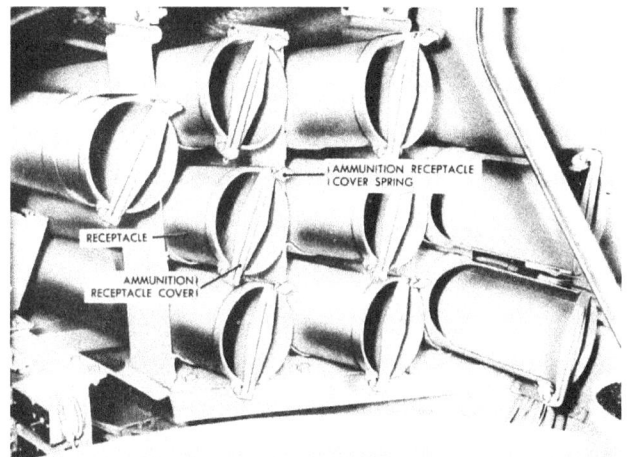

Above, the 105mm hull ammunition racks are installed to the left and right of the driver's position. The 11 round turret ready rack appears below at the left. One round is shown stowed and the breech ring of the 105mm cannon is visible in the edge of the photograph. The turret bustle ready rack is below at the right. Part of the radio appears at the left of this photograph.

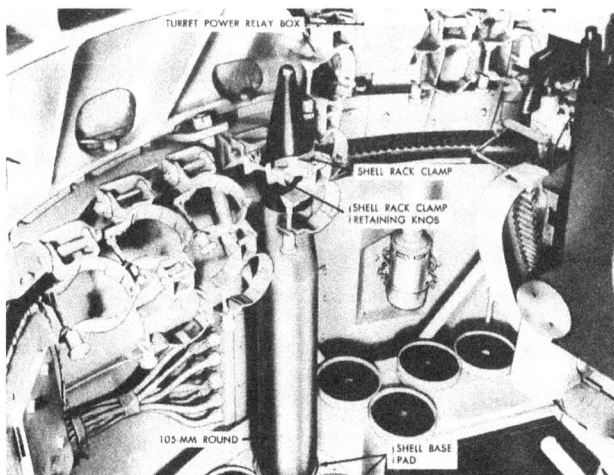

139

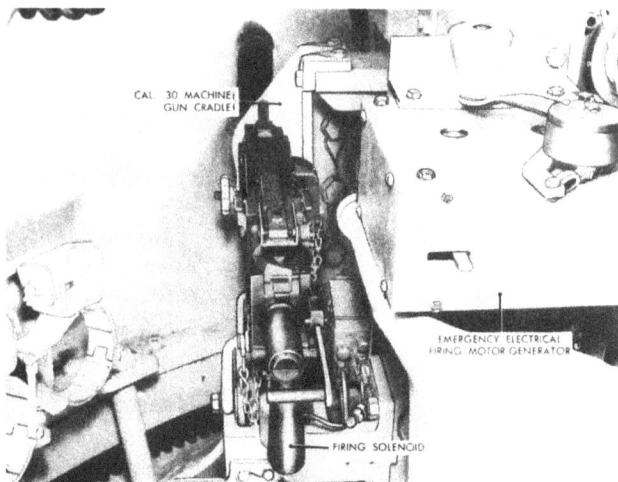

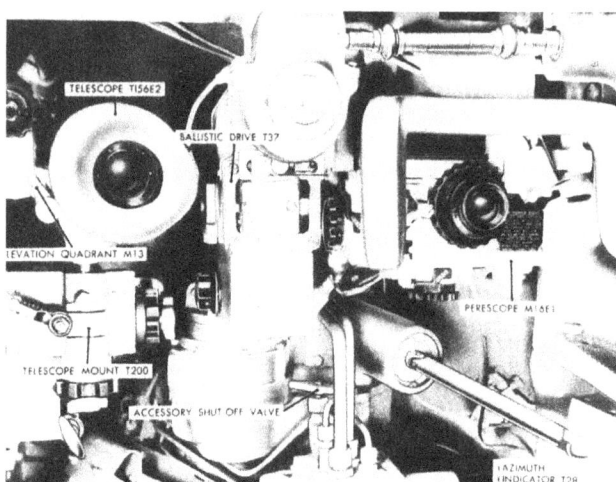

The .30 caliber coaxial machine gun is shown above at the left. At the right, the gunner's sighting and fire control equipment can be seen. Below is the loader's seat attached to the turret ring. Unlike the seats for the gunner and tank commander, it was not hydraulically adjustable.

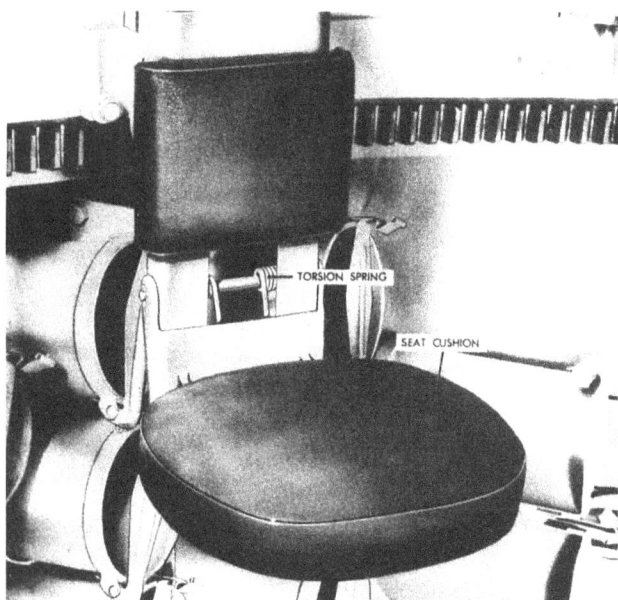

modifications included a new travel lock for the 105mm gun and the relocation of the two personnel heaters, the fixed fire extinguishers, and the control rods to accommodate the new ammunition racks. Modified torsion bar anchors were installed which repositioned the bars three degrees increasing the ground clearance to 17½ inches. Spacer plates were welded to the mounting surfaces of the bumper spring brackets. This lowered the bumper springs to maintain the normal travel between the springs and the repositioned road wheel arms. The shock absorber brackets also were modified to compensate for lowering the road wheel arm assembly.

One pilot tank T54E2 was shipped to Fort Knox for preliminary tests. After 2000 miles of operation, it was returned to Detroit Arsenal. Pilot number 3, with the electric amplidyne turret control system, was sent to Aberdeen Proving Ground for

The view of the suspension at the left shows some of the modifications required for the T54E2. Below is the travel lock for the 105mm gun.

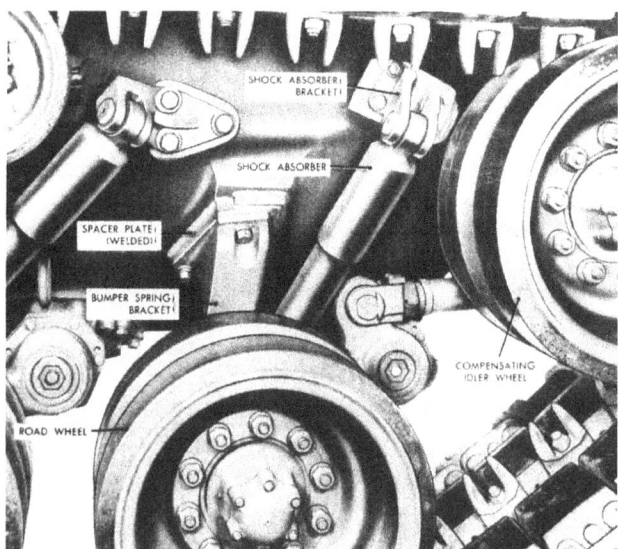

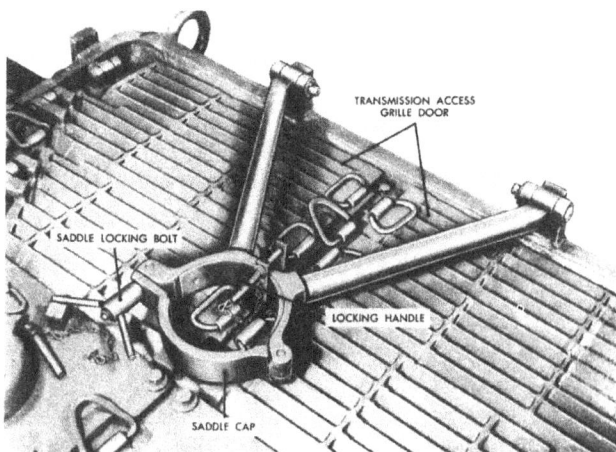

These photographs show the pilot T54E2 during tests at Fort Knox. Note the new configuration of the direct sight telescope port cover. In the front view below, this cover is in the open position. Extension pipes have now been added to the personnel heater exhausts.

fire control shock tests. As mentioned earlier, interest had now shifted to later developments, particularly the T95 tank series, and it was proposed that the T54E2 turret be evaluated on that tank. The turrets from pilots 1 and 2 with the Cadillac Gage turret control system were removed and modified for installation on the T95 chassis. One T54E2 turret also was installed for test on one of the T48E2 pilots.

Development of the T54 series was terminated by Ordnance Committee action in January 1957 and OTCM 36741, dated 12 February 1958, directed that both T54 and one T54E1 turret be scrapped. The hulls were turned in as serviceable items. The remaining T54E1 tank was shipped to the Ordnance Museum at Aberdeen. As stated previously, the T54E2 turrets were assigned to the T95 program. Their hulls also were turned in to supply.

Above and below are additional views of the T54E2 during the Fort Knox test program. Again, the extreme overhang of the 105mm gun is obvious in both the forward and rear positions. The top view below shows the details of the model 108 commander's cupola.

The T54E2 turret also was mounted on the hull of one of the T48E2 pilots (below). This tank, registration number 9A 4161, served as a test bed for several experimental programs.

During the research for this volume, no photographs were located showing the 120mm gun tank T77. The photograph above shows a model that was constructed, as in the case of the T54, to illustrate this vehicle.

As part of the heavy gun tank program, Rheem Manufacturing Company received a contract in 1952 to design and manufacture two pilot tanks mounting the 120mm gun T179 in an oscillating turret. Designated as the 120mm gun tank T57, the turret was installed on the hull of the heavy tank T43E1. The 120mm gun was rigidly mounted in the turret without a recoil mechanism and it was provided with an automatic loader. As the development work progressed, it appeared to be feasible to mount a version of the same turret with lighter armor on the chassis of the T48 tank. Such a combination offered the possibility of a heavy gun tank much lighter than any built to date. In May 1953, a project was opened to develop such a vehicle and it was designated as the 120mm gun tank T77. A contract was let to Rheem for the manufacture of two pilot tanks.

The T77 design was similar to the other oscillating turrets, consisting of a long nosed cast turret body with a large bustle. The turret body was pivoted on a cast trunnion ring supported by the bearing race on the M48 hull. The armor thickness ranged from 5 inches at 50 degrees from the vertical in front to the equivalent of 2 inches at 40 degrees in the rear. A large round cornered square section of the turret roof was hinged along its front edge and could be opened by an hydraulic actuator providing easy access to the interior. This access cover contained the tank commander's hatch on the right side surrounded by six T36 periscopes. A .50 caliber machine gun was pedestal mounted on a rotating ring surrounding the commander's hatch. A small hatch

for the loader was located in the left side of the access cover. The turret was powered in elevation and traverse by the Cadillac Gage hydraulic control system.

The 120mm gun T179 was rigidly mounted in the turret by means of an adapter screwed onto the forward end of the breech ring. This adapter was keyed to the turret to prevent rotation from the torque resulting from a projectile traveling down the rifled tube. The gun was mounted with the breechblock sliding in the horizontal direction. Since the weapon was rigidly mounted, the motion during recoil and counter recoil could not be used to open the breech after firing. However, sensors were provided to detect the firing shock and to actuate the hydraulically operated breech mechanism. The automatic loader, consisting of an eight round rotary drum type magazine and a rammer, was located below and behind the 120mm gun. The loader was powered by an hydraulic system and was intended to handle one piece fixed ammunition. However, an optional design was prepared to use two piece separated rounds. Any one of three types of ammunition could be selected. In addition to the loader magazine, two 120mm rounds were carried in the turret bustle and eight in the hull bringing the total ammunition stowage to only 18. A .30 caliber coaxial machine gun was installed to the left of the cannon in the T169 mount.

Taking advantage of the fixed relationship between the turret and gun, the fire control components were attached directly to the turret casting

143

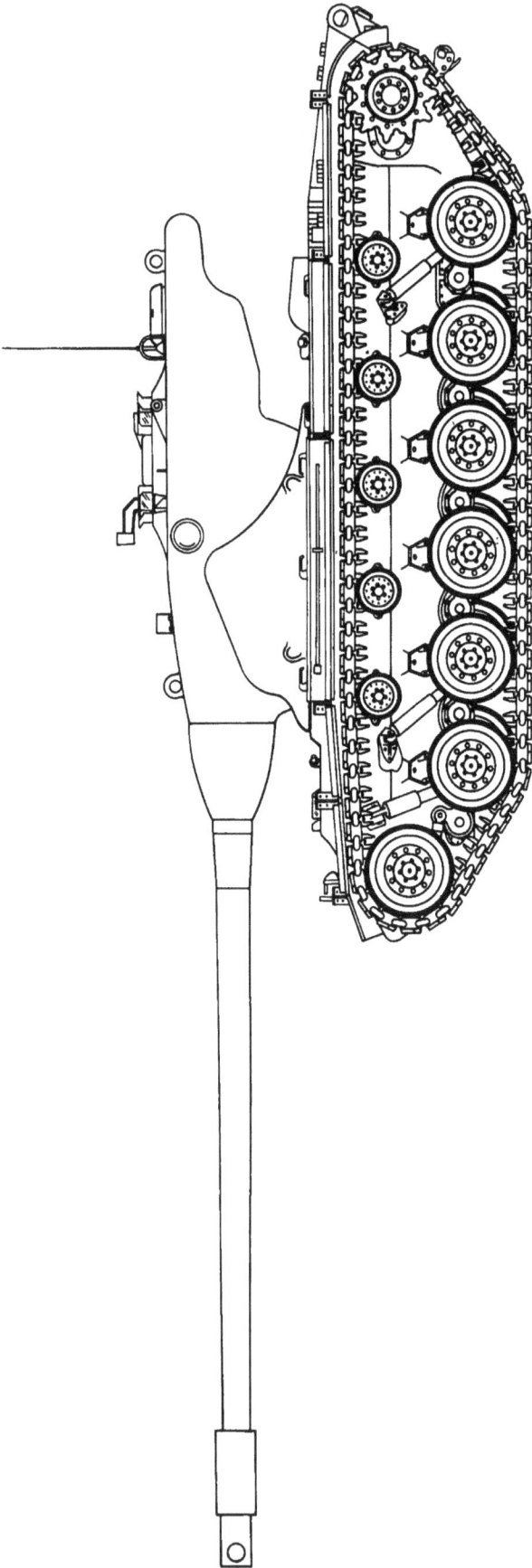

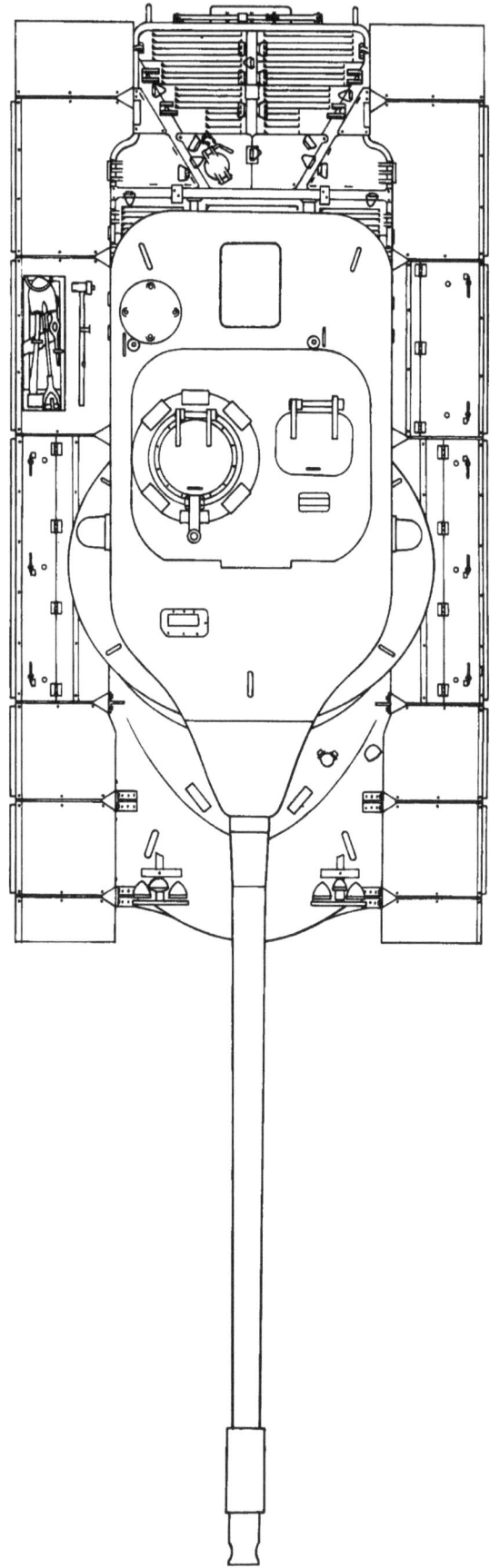

120mm Gun Tank T77

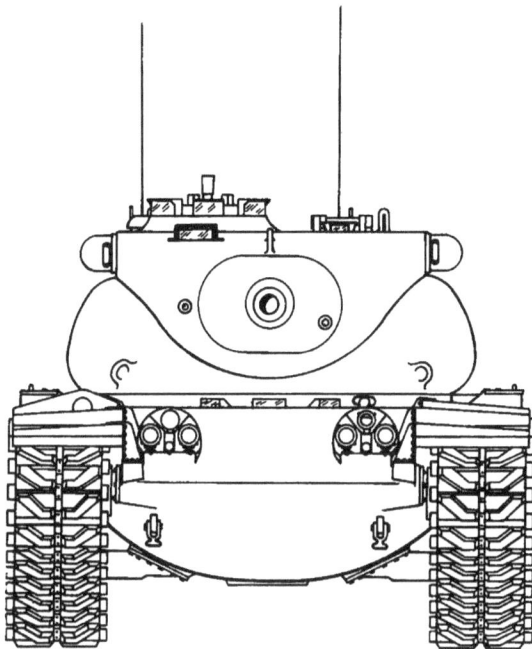

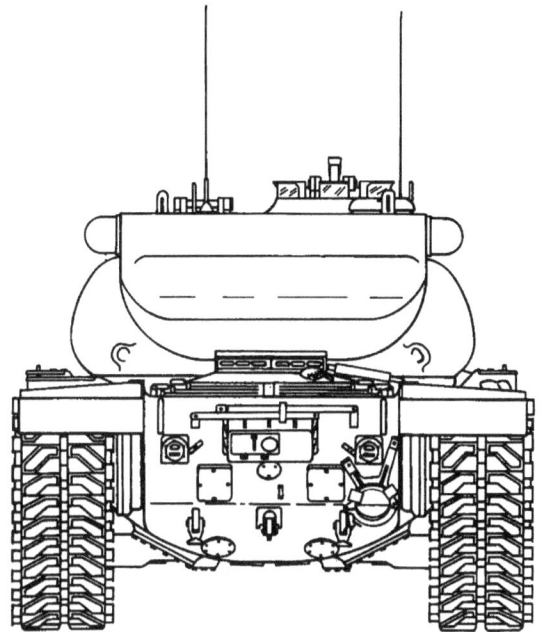

eliminating any linkage between them and the gun. A T50 range finder was provided for the tank commander. This, along with the range drive T33E2, the gunner's M16E1 periscopic sight, and the T30 ballistic computer, comprised the primary fire control system. Secondary fire control was provided by the T173 telescopic sight for the gunner.

Like the T54 series, the T77 suffered from a long slow development cycle and changing requirements of the user. The project was cancelled in 1957 and OTCM 36741 scrapped both T77 turrets on 12 February 1958 before the equipment was ready for testing.

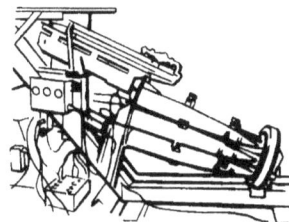

The sketches at the right depict the operating sequence of the automatic loader in the T77. Below are side and transverse section drawings of the T77 turret showing the interior arrangement.

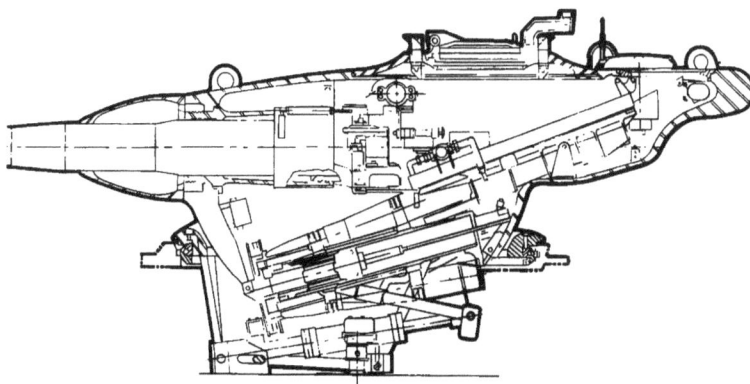

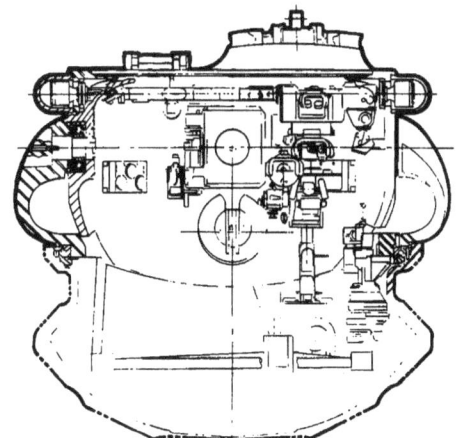

1. Round selected & withdrawn from magazine.

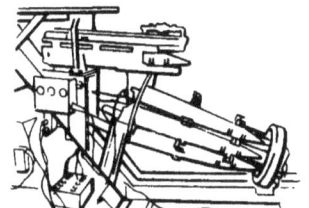

2. Round Aligned with Breech.

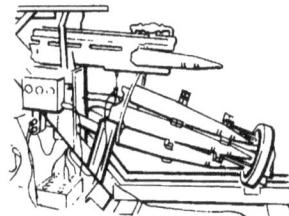

3. Round Being Rammed.

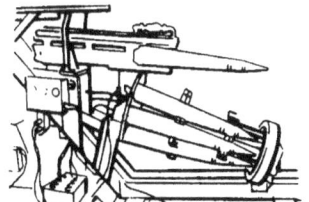

4. End of ramming stroke round continues into Breech.

These views show additional details of the T77 model. Like the model of the T54, it was constructed from the original Rheem drawings by Ron Johnson and Jacques Littlefield.

PART III

THE ADVENT OF THE MAIN BATTLE TANK

Intended to replace the M48 series, the 90mm gun tank T95 was smaller, lighter, and had a new more powerful main weapon. The extremely long thin barrel of the smooth bore 90mm gun T208 can be seen in the view above of the first pilot tank.

THE UNIVERSAL OR ALL PURPOSE TANK

As discussed in an earlier section, the limitations of the M48 series tanks were readily apparent even before they reached the troops. Parallel with the program to improve their performance, several new projects studied possible replacements. Prior to its cancellation, the T87 was proposed as a lighter, more efficient medium gun tank. After the Questionmark III conference in June 1954, a new program was launched to develop a lighter, more powerfully armed vehicle. Designated as the 90mm gun tank T95, it incorporated numerous innovative features, many of which were highly experimental. These included a hypervelocity smooth bore cannon rigidly mounted in the turret as well as a new power train and suspension system. All of these items required considerable development and progress was slow for the overall program.

By early 1957, it was obvious that a complete reassessment of the tank situation was in order. On 14 January, the Joint Coordinating Committee on Ordnance sent a request to the Assistant Secretary of the Army for Research and Development to establish a panel to review the status of tank development. At this same time, a congressional investigation began of the defense program, although as far as tank development was concerned, it was limited to the question of why the United States did not have a light gun amphibious tank comparable to the Soviet PT76. However, the inquiry emphasized the need for a comprehensive review of the entire tank development situation. In February, the Army Chief of Staff, General Maxwell D. Taylor, established the Ad Hoc Group on Armament for Future Tanks or Similar Combat Vehicles (ARCOVE). This group was to study tank armament requirements for the time period after 1965 giving consideration to the effects of atomic weapons.

ARCOVE completed its review and submitted its recommendations in May 1957, although a formal report was not published until January 1958. The group considered that a maximum effort should be made to equip tanks by 1965 with a guided missile weapon system using line of sight command guidance. To provide funds for such a system, it was suggested that work on conventional weapons such as hypervelocity guns and penetrators be sharply curtailed. However, further research was recommended on chemical energy warheads, target detection, armor, and personnel protection.

In August 1957, General Taylor approved a new tank development program which incorporated many of the ARCOVE recommendations. Although the light, medium, and heavy gun tanks were retained for the present, the future fleet was to consist of two tank type vehicles. These were the airborne reconnaissance/airborne assault vehicle, which was essentially a heavily armed amphibious light tank, and a main battle tank. The latter was a universal or all purpose fighting vehicle capable of performing the functions of both the medium and

At the left is the AVDS-1790-P diesel engine, serial number 1. At the right, T48E2 pilot number 4 is being prepared for the installation of this engine. The complete power pack can be seen at the rear of the tank. This was the vehicle referred to as the Yuma Test Rig.

heavy gun tanks. It was to have firepower and protection sufficient for the assault role as well as adequate mobility to perform as a medium tank.

Until the new main battle tank was developed, the M48A2 was to continue in production including any modifications necessary to improve its performance. At the time this program was approved, the Army Staff envisaged the future main battle tank as one of the T95 series armed with a hypervelocity smooth bore cannon and fitted with a compression ignition engine to improve its cruising range.

The development of the compression ignition engine had gained momentum in June 1956 when the Army removed many of the restrictions on the use of middle distillate fuels. The new ruling did not permit Ordnance to use high grade diesel fuel in combat vehicles, but jet engine fuel (JP-4) could be utilized. This final restriction was removed in a review of the Department of the Army petroleum policy in June 1958. The use of diesel fuel in any family of vehicles was then permitted, if it contributed significantly to better fuel economy.

The development of a diesel version of the AV-1790 tank engine had been initiated by OTCM 35221 on 10 February 1954. Designed by Continental Motors to use a large part of the existing AV-1790 production facilities, it was tested at Yuma, Arizona during the Summer of 1957. Designated as the AVDS-1790-P, the new supercharged engine performed satisfactorily on both JP-4 and diesel fuel. AVDS-1790-P serial number 1 was installed in the modified T48E2, registration number 9A4161. Referred to as the Yuma Test Rig, the tank with the new diesel showed better than a 60 percent improvement in fuel economy over the gasoline engine. By October 1957, the first AVDS-1790-P had operated for 2000 miles and a second engine was under test.

It was originally intended that one of the two AVDS-1790-P engines would be rebuilt after tests at

Yuma and would then be shipped to the Armor Board for evaluation in an M48A2. However, a conference in February 1958 decided that an improved version, designated as the AVDS-1790-1, would be provided for the user tests. The new engine differed from the AVDS-1790-P in several respects. A single 12-cylinder fuel pump replaced the two 6-cylinder pumps. The new engine also had a speed sensitive fuel injection timing device. The lubrication system was revised to allow for a 30 degree tilt and one disposable oil filter cartridge and one cleanable oil strainer were provided. The forward fan assembly was moved closer to the flywheel end to improve the cooling air flow and the transmission oil cooler capacity was increased. With a compression ratio of 14.5:1, this diesel version of the 1790 developed 700 horsepower at 2400 rpm.

To match the characteristics of the new engine, the standard cross drive transmission was modified to decrease the input gear ratio and to incorporate the converter from the XT-1400 transmission. With these changes, it was redesignated as the CD-850-6. The new power train was installed in two M48A2s, registration numbers 9B1287 and 9B1288. The first tank was delivered to the Yuma Test Station on 7 July 1958 and the second arrived at Fort Knox on 20 August. Both vehicles were designated as the 90mm gun tank M48A2E1. A third M48A2E1 was powered by the AVDS-1790-2. This was an improved version of the AVDS-1790-1 and it included a thermostatically controlled engine cooling fan clutch, an integral crankcase breather system, a redesigned flywheel, improved turbocharger impellers, revised throttle linkage, and an increase in the compression ratio from 14.5:1 to 16:1. This vehicle was subjected to cold weather tests at Fort Churchill in Canada.

Because of the low diesel engine fuel consumption, the auxiliary generator with its separate power plant was eliminated from these tanks. The fuel used by the new V-12 diesel at idle was less than that

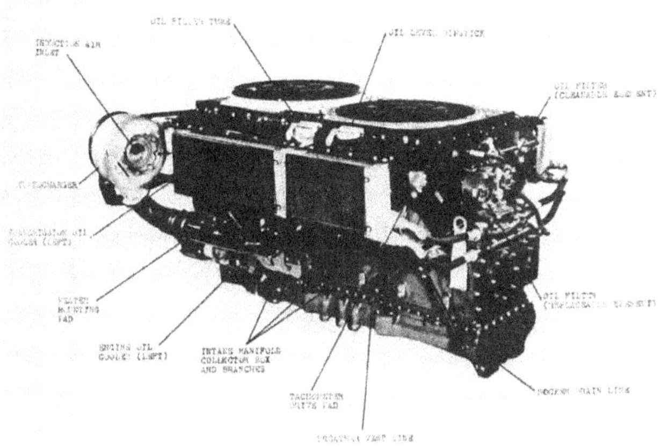

The improved version of the new diesel engine, the AVDS-1790-1, appears above. Below are two views of M48A2E1 number 1 after installation of the new power plant. The latter two photographs were dated 28 August 1958.

required by the A-41-2 auxiliary generator engine at full power and both could provide the same electrical output under these conditions. Thus the main engine could be used to maintain the charge in the batteries and to perform the other duties formerly the task of the auxiliary generator. Also, since the engine exhaust passed through the turbochargers, additional mufflers were not required.

The diesel engine was provided with two fender mounted dry type air cleaners with a combined

capacity of 1800 cubic feet of air per minute. Installation of the new air cleaners required changes in the rear deck grills and the relocation of the tow cables and fuel transfer pump stowage from the fenders to the side of the turret. The fuel tanks were redesigned and the tests at Fort Knox indicated an average usable fuel capacity of 314 gallons compared to 304 gallons for the gasoline powered M48A2. At Fort Knox, the cross-country fuel consumption of the M48A2E1s averaged 2.86 gallons/mile compared

At the left is another view of the first M48A2E1. The opening for the commander's cupola was covered by a flat plate on this test vehicle. Below is M48A2E1 number 2 during the tests at Fort Knox. Note the steel plates added to the turret to simulate the weight of the proposed XM60.

to 3.89 gallons/mile for the M48A2 under the same conditions.

A conference in early February 1958 at the Ordnance Tank Automotive Command (OTAC) had discussed methods of improving both the cruising range and the firepower of the M48A2. This was in line with tank policy approved by the Chief of Staff in August 1957 and reflected in the February 1958 edition of the Five Year Materiel Program. This program called for the yearly production of 900 M48A2 tanks through the fiscal year 1961. After that time, the M48A2 would be replaced by one of the T95 series. The conference at OTAC agreed to expedite the development of the AVDS-1790 engine so that it could be introduced into the M48A2 production in July 1959. The tests of the M48A2E1 were in support of this program. In regard to the improved armament, no decision was made at the OTAC conference pending further investigation of various weapons under development.

The future course of the tank program was the subject of widespread debate within the Army and with the Bureau of the Budget (BOB) from March through May 1958. In the opinion of BOB, the Army was not progressing rapidly enough in its tank modernization program. The Bureau recommended that the Army explore all means of immediately replacing the M48A2 with a better tank. Convinced that BOB would not approve procurement of the M48A2 after fiscal year 1959, the Deputy Chief of Staff, Logistics (DCSLOG) directed that Ordnance prepare a plan for an acceptable tank. Because of the limited time available, DCSLOG proposed a tank based on the M48A2 incorporating more powerful armament and the AVDS-1790 engine. At this time, DCSLOG suggested that the new tank be designated as the M51 rather than as the M48A3, no doubt to emphasize the improvement over the M48A2.

An alternate to an improved M48A2 was to speed up the introduction of a new tank of the T95 series. However, many components of the T95 were still highly experimental and the development of a suitable compression ignition engine for this lighter vehicle was not as far advanced as the AVDS-1790. Also, by May 1958, an influential group of senior officers had concluded that the T95 represented only marginal advances over the M48A2. They believed

that the two most important improvements, greater firepower and fuel economy, could be achieved at far less cost by installing a more powerful gun and a compression ignition engine in the M48A2.

The BOB decision which finally came on 1 May 1958 to prohibit procurement of the M48A2 after fiscal year 1959 was fully expected by the Army planners and confirmed their preference for developing an interim main battle tank. Such a development would be a low risk program based on proven components which could be rapidly introduced without any break in production when the M48A2 was phased out. For the future, the objective would be the development of an ultimate main battle tank armed with a guided missile system and incorporating radiological protection.

OTAC was already at work when the guidelines arrived for the Ordnance proposal to meet the interim tank requirement. A preliminary program was formulated by 4 June for the evaluation of what was now referred to as the XM60 weapon system. This reflected the new system of using an X prefix to designate vehicles prior to standardization replacing the assignment of T numbers such as the T95. Under this new arrangement, the item retained its original number when it was standardized and dropped the X prefix.

After review of the status of major components, the planners settled on the M48A2 chassis and the AVDS-1790 engine as the basis for development. To determine the durability and operating characteristics of the power train and suspension under the increased weight of the proposed tank, the second M48A2E1 at Fort Knox was loaded to a total weight of 56 tons by adding slabs of armor plate on the turret. Components which failed during these tests were noted and corrective action was specified for the production vehicles.

Siliceous cored armor was proposed for both the hull and turret of the new tank depending on the outcome of further tests and production availability. The main gun was to be selected after comparative firing tests scheduled in October at Aberdeen. High among the weapons to be tested was the British X15E8. This gun, 105mm in caliber, was listed in some documents as the Ex 20 pounder, referring to the weapon from which it was developed. The

The outline drawings of the two 105mm guns below compare the British X15E8 (left) with the U.S. T254 (right). The tubes were not interchangeable on these weapons.

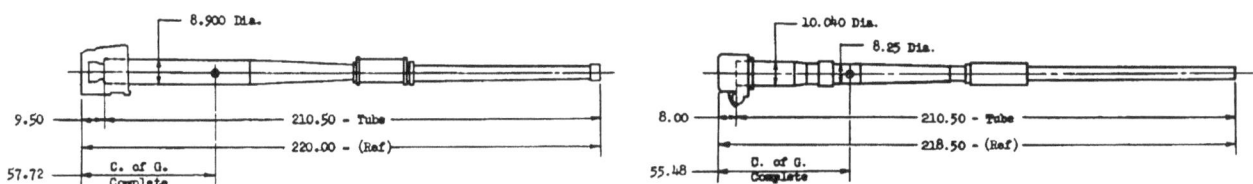

The original 105mm gun T254E2 is shown above. Although its tube was interchangeable with that on the British X15E8, it retained a concentric bore evacuator. The vertical sliding breechblock on this weapon also can be seen in the view below.

X15E8 had shown exceptional penetration performance using armor piercing discarding sabot (APDS) ammunition, despite its relatively small overall size and low weight. An American version, designated as the 105mm gun T254, fired the same ammunition and was intended for installation in the T95E5 tank. This vehicle used the M48A2 type turret. A further development was the 105mm gun T254E2 which used a tube interchangeable with that of the British weapon. However, it had a concentric bore evacuator unlike the eccentric type on the X15E8. The latter permitted greater depression when aimed to the rear over the engine deck. The British gun used a horizontal sliding breechblock while that on the American version moved in the vertical direction. Two guns with chrome plated bores were designated as the T254E3.

The armament preferred for the new tank by the Ordnance Department was the 120mm gun

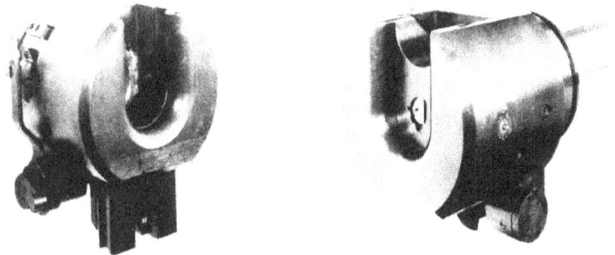

T123E6 originally destined for use in the T95E6. This was a lightweight version of the 120mm gun M58(T123E1) installed in the heavy gun tank M103. The T123E6 weighed 4244 pounds compared to 6370 pounds for the M58. Firing the same ammunition, its performance was essentially the same as the heavier weapon. Ordnance considered it preferable

The Ordnance Department's choice of armament for the new tank is illustrated below. The 120mm gun T123E6 was a very powerful weapon, but its heavy two piece ammunition resulted in a slower rate of fire.

GENERAL VIEW

MUZZLE COUNTERBORE

Above are concepts of the XM60 armed with the 120mm gun T123E6 (left) and the 105mm gun T254E2 (right). The 120mm gun is mounted in the T95E6 type turret with the new vision cupola. The 105mm gun uses the M48A2 turret.

because its ammunition development was further advanced. However, a request in June for authority to develop the XM60 with the 120mm gun was rejected by DCSLOG who stated that the main gun would be selected after review of the October firing tests at Aberdeen. The final plan for these tests also included the smooth bore 90mm gun T208E9 along with the 90mm gun M41 and the 120mm gun M58 for comparison purposes. Since the main gun had not yet been specified, two XM60 weapon system plans were submitted in September by OTAC. The first concept was armed with the 120mm gun T123E6 in the long nosed T95E6 turret as preferred by Ordnance and the second carried the 105mm gun T254E2 in the M48A2(T95E5) type turret.

The comparative tests at Aberdeen evaluated the penetration performance, lethality of a hit, accuracy, and rate of fire for the six weapons and ammunition shown in the table. Combining these factors, the British X15E8 was ranked first, except for the 120mm gun M58 which was not under consideration. One factor against the T123E6 gun was the slow rate of fire with its two piece separated ammunition. This problem was less severe for the M58, since in the M103 it was served by two loaders.

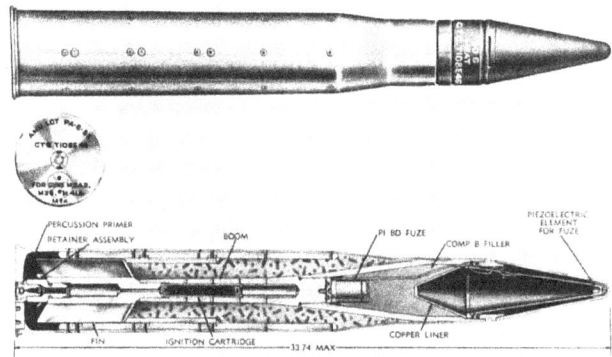

The 90mm gun M41 was included in the comparative tests because of the good performance expected from the new T300 HEAT ammunition. Developed for the M36 and M41 guns, it had a muzzle velocity of 4000 feet/second compared to 2800 feet/second for the T108 round (above) that it replaced. The view of the complete round below shows the T300E59 which was standardized as the M431.

Below are concepts of the XM60 fitted with the 90mm T208 smooth bore gun in the T95E1 turret (left) and the standard 90mm gun M41 in the M48A2 turret (right). The T95E1 turret is shown with the new vision cupola.

The actual vehicles used in the comparative firing tests can be seen in these photographs. Above are the M103 armed with the standard 120mm gun M58 (left) and the M48A2 fitted with the 120mm gun T123E6 in the T54E2 turret (right). Below are the M48A2 with the 90mm gun M41 (left) and the M48A2 armed with the 90mm gun T208 in the T95E1 turret (right).

Based on these tests, the 105mm gun T254E2 was selected for the XM60 and the British tube was specified until tubes of American manufacture with comparable accuracy could be obtained. This was the weapon standardized as the 105mm gun M68 using the eccentric bore evacuator.

Although the weapon tests settled the question of the main gun, the decision on several other major components had not been made as late as December 1958. A long nosed T95E7 type turret was preferred because of its better ballistic protection, but such a turret would not be available for production in 1960. The T95E7 design was the same as the T95E6 turret, but it was fitted with the 105mm gun T254E2. The M48A2 type turret was finally specified for the initial production and a long nosed turret similar to the T95E7 subsequently appeared on the M60E1.

Below, the M48A2 is fitted with the 105mm gun T254. This same vehicle was used later to mount the British X15E8 gun. The horizontal sliding breechblock of the latter weapon can be seen at the right through the loader's hatch.

COMPARATIVE TESTS OF TANK WEAPONS

Gun	Ammunition	Turret	Chassis
90mm M41	HEAT T300E53, HE	M48A2	M48A2
90mm T208E9	APFSDS T320E82	T95E1	M48A2
#105mm T254	APDS	*M48A2 (mod.)	*M48A2
105mm X15E8	APDS	*M48A2 (mod.)	*M48A2
120mm T123E6	HEAT, AP, HE	T54E2	M48A2
120mm M58	HEAT, AP, HE	M103A1	M103A1

The first T254E2 was not available in time for the test.

* The same tank was used for the T254 and X15E8 guns. The T254 was removed after firing 35 rounds and replaced by the X15E8 in the same mount.

155

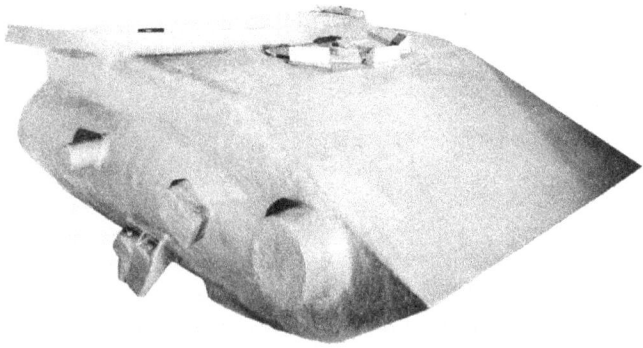
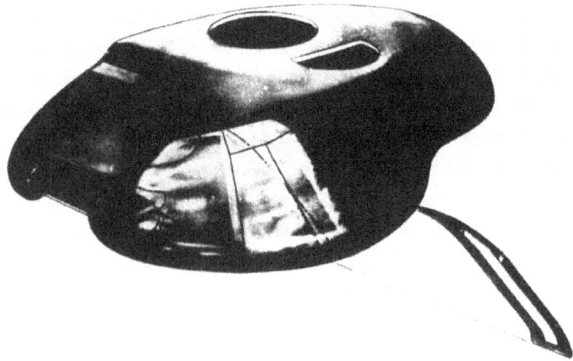

Above is the mock-up of the new lightweight hull (left) proposed for the XM60. The flat contours simplified the use of siliceous cored armor. The design of the siliceous cored armor turret can be seen at the right.

Originally, it was intended to provide siliceous cored armor on both the turret and hull. The T95E7 type turret and the flat square front of the glacis plate on the new hull casting simplified the application of siliceous cored armor. However, cost and the limited availability of production facilities resulted in the cancellation of this feature and all M60 series tanks were protected by conventional homogeneous steel armor.

The original mock-up of the T95E6 type turret for the XM60 was fitted with a new full vision cupola at the commander's station. Seven tiltable head periscopes were arranged to give a 360 degree field of view with overlapping vision between adjacent periscopes. A dual power 7x50 binocular sight or an M19 infrared periscope could be attached to the front periscope mount. A special feature of this

design was that the upper three quarters of the cupola body could be raised any distance up to 3½ inches above the base providing an open 360 degree vision slot. This gave the tank commander overhead protection from air bursts or weather while retaining clear direct vision. Access was through a hinged equilibrated hatch cover in the roof and a .50 caliber machine gun was pedestal mounted on the forward part of the upper cupola structure. The weapon could be aimed and fired with the turret closed. This turret mock-up also featured an 11 inch diameter hydraulically operated port in the left turret wall to permit the loader to eject empty cartridge cases from the turret.

Below are views of the XM60 turret mock-up armed with the 120mm gun T123E6 and fitted with the proposed cupola. At the right, the cupola is shown in the open position with 360 degree direct vision. In the lower right view, the two ports were for the .50 caliber ranging machine gun (upper) and the .30 caliber coaxial machine gun (lower).

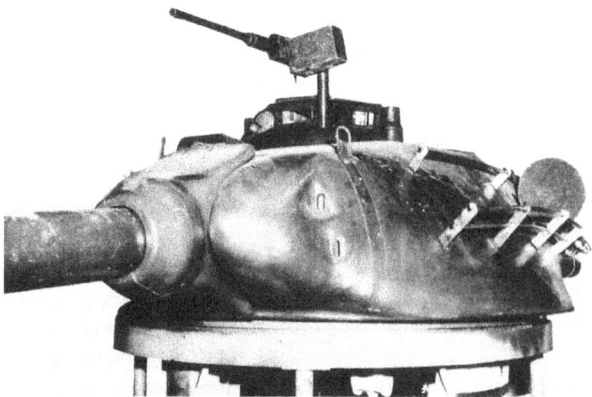

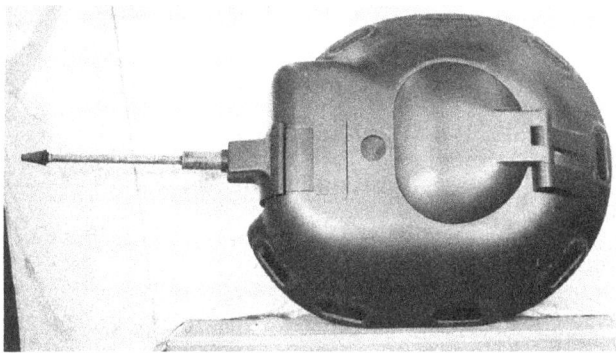

The mock-up of the T9 commander's cupola appears above at the left. The contour was modified around the two left front vision blocks before the design was adopted. At the top right is a concept drawing of the XM60 at a later stage. Here the tank is armed with the 105mm gun T254E2 fitted with the British tube having the eccentric bore evacuator. Note that the interim cupola is installed as on the T54E1 (see page 131).

With the selection of the M48A2 type turret for the final version of the XM60, the full vision cupola design was dropped. However, the problems with the M1 on the M48 tank series had resulted in the design of the new T9 cupola and it was selected for the new tank. With the mounting ring diameter increased to 34 inches and designed to carry the new short receiver T175 .50 caliber machine gun, it was much roomier and easier to use. However, the new cupola was not expected to be available for the first production tanks.

A new short receiver coaxial machine gun designed especially for tank use also was provided. This was the 7.62mm M73(T197E2) which replaced the .30 caliber M37. The original concept of the XM60 using the T95E6 turret envisaged a simplified fire control system with a .50 caliber ranging machine gun, but this was dropped from consideration at an early stage. Initial production tanks were to be fitted with a modified M48A2 fire control system with later vehicles to have an advanced type with a xenon searchlight and infrared night vision. Following a briefing on 11 December 1958, General Maxwell D. Taylor ordered the XM60 into production because of the improvements it offered in firepower, protection, and cruising range.

A contract had been awarded to Chrysler Corporation in September 1958 for the advanced production engineering (APE) of the XM60. The vehicle was standardized as the 105mm gun full tracked combat tank M60 by OTCM 37002 on 16 March 1959. It was proposed in April to revise the nomenclature to the 105mm gun main battle tank M60, but this change was rejected because of a conflict with Federal Cataloging Agency policy. The scope of the Chrysler contract included the fabrication of four pilot tanks. Pilot number 1, registration number 9B3057, was completed at Chrysler Defense Engineering and transferred to Detroit Arsenal on 3 July 1959 for shipment to Aberdeen Proving Ground. Number 2 was completed on 4 August and subsequently used to prepare publications. M60, pilot number 3, was completed and transferred to Detroit Arsenal on 2 September. It was then shipped to Fort Knox for user tests. The fourth pilot tank was completed and turned over to the Arsenal on 26 October.

M60 pilot number 1 can be seen at the right and below on 2 July 1959 after completion at Chrysler. The commander's cupola was removed before it was shipped to Aberdeen Proving Ground.

Above is a group of early M60 tanks at Aberdeen Proving Ground. Note that none of these are fitted with cupolas.

M60 number 1 arrived at Aberdeen on 24 July without the commander's cupola. A T6 cupola was removed from a T95 tank and installed with an adapter ring for the early tests. To permit simultaneous automotive and fighting compartment tests, the M60 turret was removed on 11 September and installed using an adapter ring on an M48A2 hull. The T6 cupola was removed at this time. The first M19(modified T9) cupola arrived on 27 October. It was then immediately installed and the tests continued. For the automotive tests, an up-weighted M48A2 turret was fitted to the M60 chassis.

The M60 pilot tanks were built with a cast hull having an elliptical cross section and a rear similar to the M48A2. However, the front was changed and the elliptical nose of the M48 series was replaced by a wedge shaped front end extending straight across the hull. The flat straight edged surfaces of the new casting would have permitted easier incorporation of the siliceous cored armor originally intended for the M60. Like the earlier tanks, the hull could be cast in a single piece or assembled by welding smaller castings. The hull weighed slightly over 12 tons after machining and without the floor plate or other weldments. The single piece cast turret was similar to that on the M48A2, but it was modified to accept the larger diameter commander's cupola and the M116 mount for the 105mm gun.

The production version of the Continental diesel, the AVDS-1790-2, developed 750 gross horsepower at 2400 rpm and it was installed with the CD-850-6 transmission. As on the M48A2E1, the economical low speed operation of the diesel engine eliminated the requirement for an auxiliary generator providing additional space in the M60 hull. A new personnel heater burning diesel fuel replaced the earlier gasoline model. The exhaust pipe from the new heater passed through the hull roof on the right side of the driver's hatch and extended to that side of the tank.

The torsion bar suspension was similar in appearance to that on the M48A2C, but it was fitted with forged aluminum alloy road wheels, each of which weighed about 65 pounds less than the equivalent steel wheel. Development of the lightweight road wheels started in 1957 and they were originally intended for the M48 series tanks. They were higher in cost and required the installation of a steel ring around the inside surface to prevent excessive wear from the track guides. The road wheels ran on the same 28 inch wide T97E2 tracks as on the M48A2. However, on the original M60, the suspension was free without any hydraulic shock absorbers or friction snubbers. Bumper springs were installed to limit the travel of the front and rear road wheel arms.

The M60 at the left illustrates the original suspension configuration without snubbers or shock absorbers. The view at the right shows snubbers installed on the front and rear road wheel arms after the tests at Aberdeen and Fort Knox.

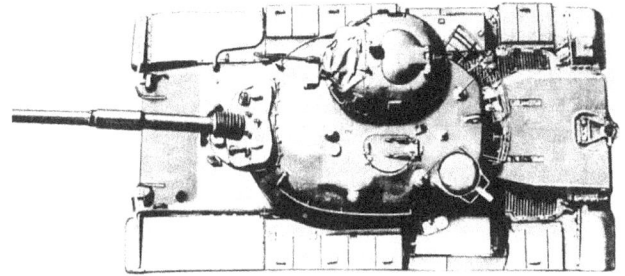

Details of the standardized M60 can be seen above. Note that the exhaust pipe from the new personnel heater extends to the right side of the hull. The M29 cupola appears below. The lower right photograph shows the external mount for the .50 caliber machine gun M2 prior to the availability of the M85 machine gun.

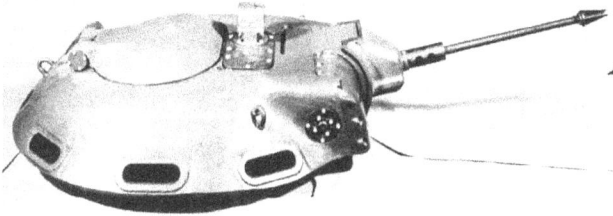

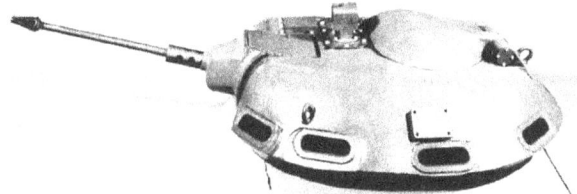

The .50 caliber machine gun T175E2 was standardized as the M85. Installed in the M19 cupola, it was fed from a 180 round ammunition box mounted on the inside of the left rear cupola wall. A chute at the right side of the gun ejected the empty cartridge cases and links. An M28C periscopic sight in the cupola roof was coupled to the machine gun. However, development problems delayed the production of the M85 machine gun and it was directed that the first 300 tanks be armed with the .50 caliber machine gun M2. Since this weapon would not fit inside the new cupola, a pad was welded to the outside of the left cupola wall for a pedestal mount. The M2 guns were to be replaced with M85s as soon as they became available. Production of the new cupola also lagged behind that of the tanks and the first 45 M60s were delivered without cupolas which were installed later in the field.

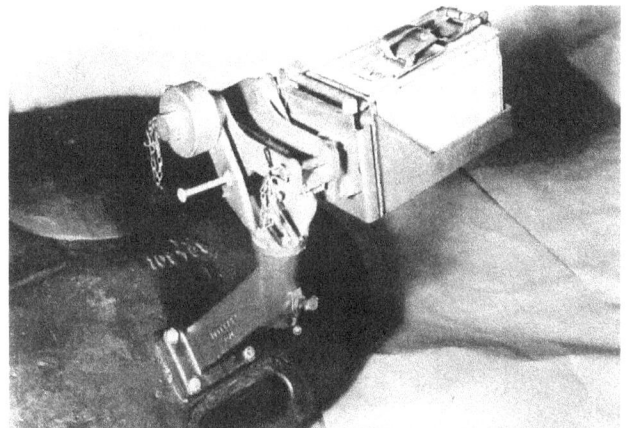

Below new M60 tanks are lined up in the 1st Regiment motor park at Fort Knox on 14 October 1960. These vehicles are fully equipped including the M19 commander's cupola, but they have not received the M85 machine guns.

Labels (top left diagram): ELEVATION QUADRANT M13A1 · BALLISTIC DRIVE M10 · RANGE FINDER M17C · TELESCOPE M105C · TELESCOPE MOUNT M114 · PERISCOPE M31 · PERISCOPE MOUNT M115 · INFINITY SIGHT M44C · GUNNER'S SWITCH BOX · COMMANDER'S CONTROL HANDLE · GUNNER'S QUADRANT M1A1 · GUNNER'S POWER CONTROL HANDLE · AZIMUTH INDICATOR M28A1 · BALLISTICS COMPUTER M13A1D · TRAVERSE GEAR BOX

Labels (top right photo): GUN FIRING TRIGGER SWITCH · PERISCOPE MOUNT M104A2 (FIG. 75) · CUPOLA HATCH DOOR (FIG. 76) · ELEVATING SCREW JACK (FIG. 76) · PERISCOPE SIGHT M28C (FIG. 75) · DOME LIGHT—M60 (FIG. 77) · CUPOLA ACCESS DOORS AND LATCHES (FIG. 77) · GUN FIRING SAFETY SWITCH (FIG. 76) · COMMANDER'S CONTROL HANDLE · AZIMUTH LOCK AND INTERLOCK (FIG. 78) · CAL. .50 MACHINEGUN (FIG. 78) · INSTRUMENT LIGHT M50 (FIG. 76) · AZIMUTH GEAR BOX (FIG. 76)

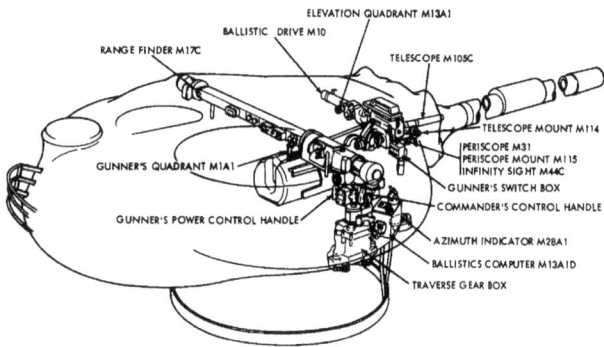

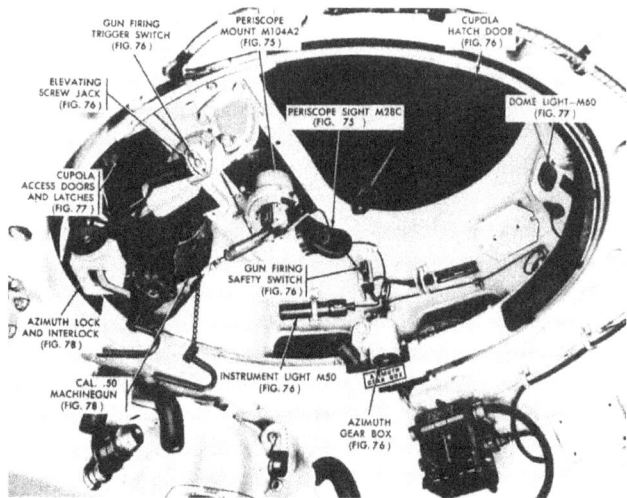

The M60 fire control equipment is illustrated above at the left. Using the M17C range finder, it was basically the same system as in the M48A2C. At the right above is the interior of the commander's M19 cupola. Below, the left front of the turret (left) and the tank commander's controls (right) can be seen.

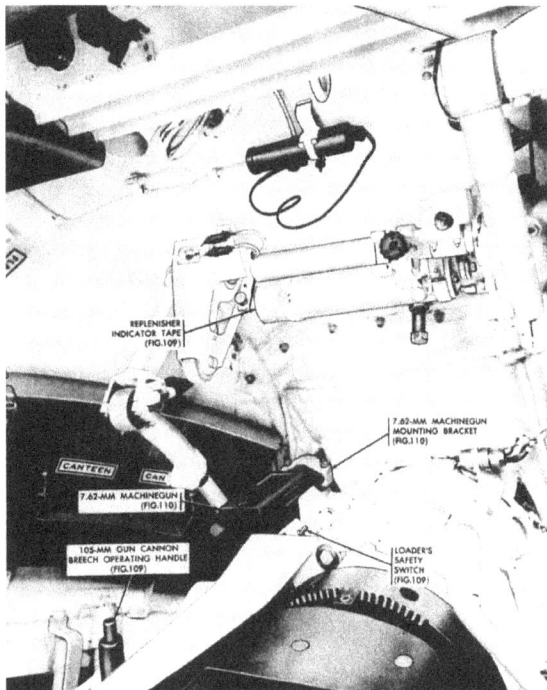

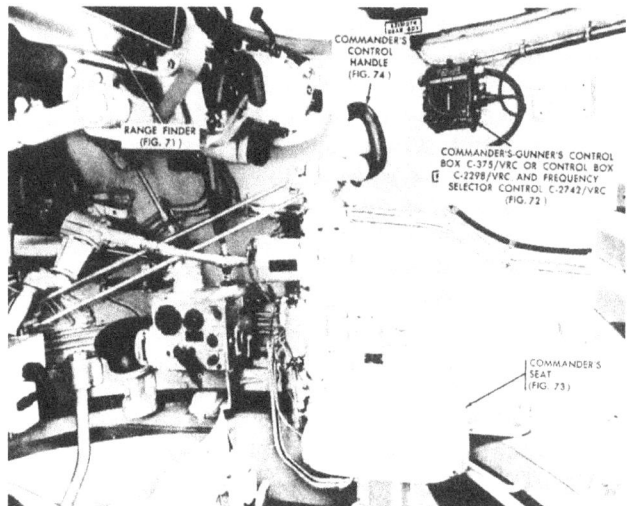

Labels: REPLENISHER INDICATOR TAPE (FIG.109) · 7.62-MM MACHINEGUN MOUNTING BRACKET (FIG.110) · 7.62-MM MACHINEGUN (FIG.110) · 105-MM GUN CANNON BREECH OPERATING HANDLE (FIG.109) · LOADER'S SAFETY SWITCH (FIG.109) · CANTEEN CAN

Labels: COMMANDER'S CONTROL HANDLE (FIG. 74) · RANGE FINDER (FIG. 71) · COMMANDER'S-GUNNER'S CONTROL BOX C-375/VRC OR CONTROL BOX C-2298/VRC AND FREQUENCY SELECTOR CONTROL C-2742/VRC (FIG. 72) · COMMANDER'S SEAT (FIG. 73)

The gunner's controls are shown at the right and below are two views of the 7.62mm M73 coaxial machine gun.

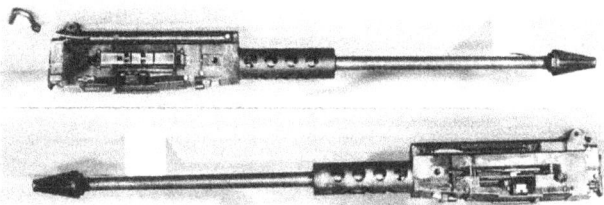

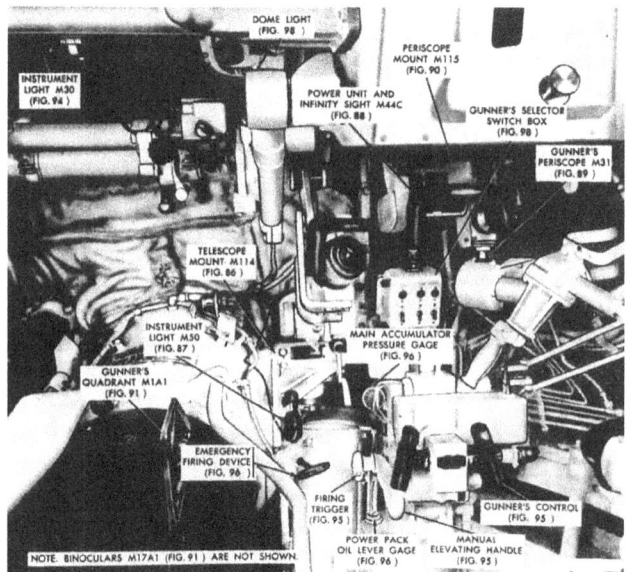

Labels: DOME LIGHT (FIG. 98) · PERISCOPE MOUNT M115 (FIG. 90) · INSTRUMENT LIGHT M30 (FIG. 94) · POWER UNIT AND INFINITY SIGHT M44C (FIG. 88) · GUNNER'S SELECTOR SWITCH BOX (FIG. 98) · GUNNER'S PERISCOPE M31 (FIG. 89) · TELESCOPE MOUNT M114 (FIG. 86) · INSTRUMENT LIGHT M50 (FIG. 87) · GUNNER'S QUADRANT M1A1 (FIG. 91) · MAIN ACCUMULATOR PRESSURE GAGE (FIG. 96) · EMERGENCY FIRING DEVICE (FIG. 96) · FIRING TRIGGER (FIG. 95) · POWER PACK OIL LEVER GAGE (FIG. 96) · MANUAL ELEVATING HANDLE (FIG. 95) · GUNNER'S CONTROL (FIG. 95) · NOTE: BINOCULARS M17A1 (FIG. 91) ARE NOT SHOWN.

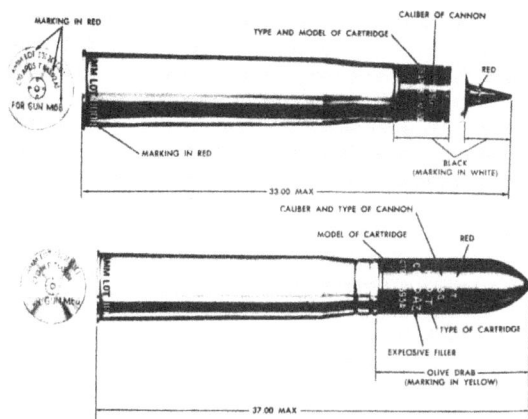

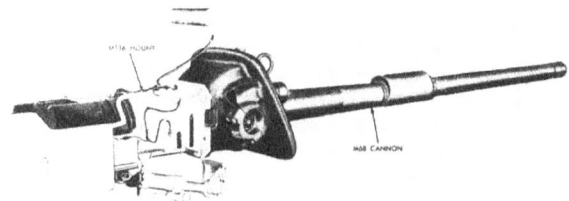

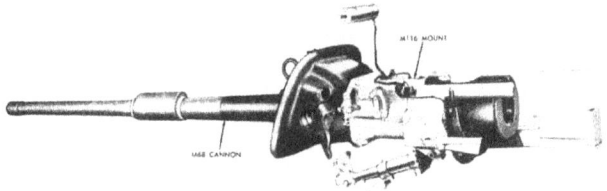

The standardized 105mm gun M68 (top right) retained the eccentric bore evacuator. It was installed in the mount M116 (right). Above are the APDS (upper) and HEP (lower) rounds of ammunition for the 105mm gun. The sketch below illustrates the ammunition stowage arrangement. In later M60s equipped with the smaller AN/VRC radio, eight rounds were carried in the turret bustle instead of six.

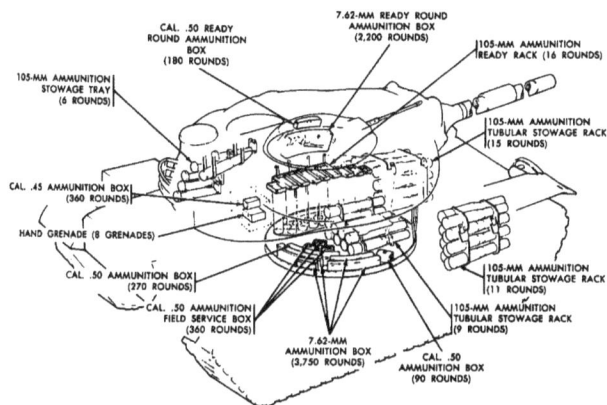

Below is the left rear of the turret showing the empty 105mm ammunition racks. At the right are views of the turret roof antennas (upper) and the AN/GRC radio (lower) in the turret bustle.

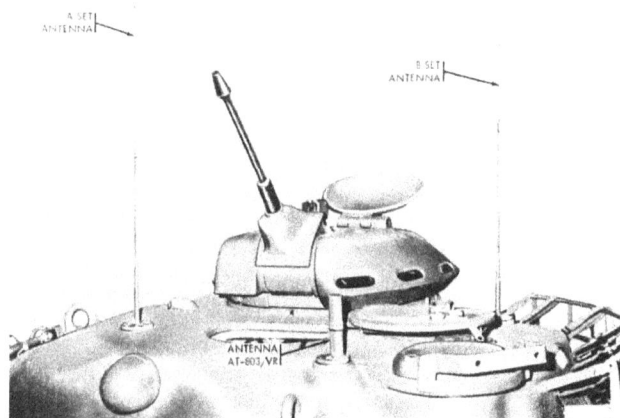

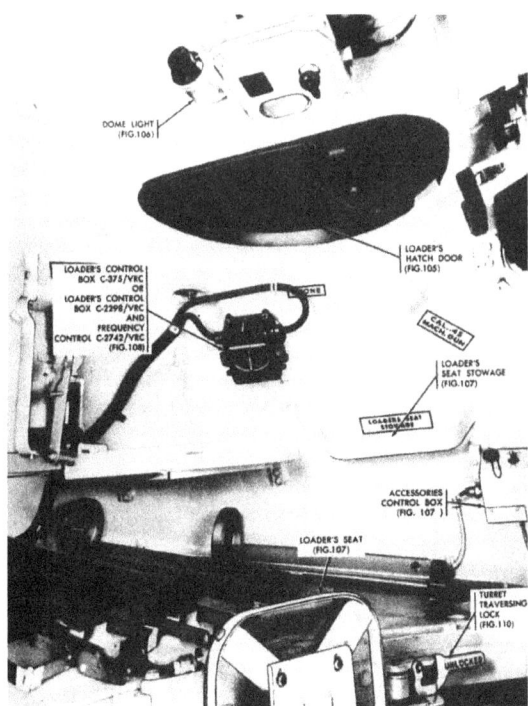

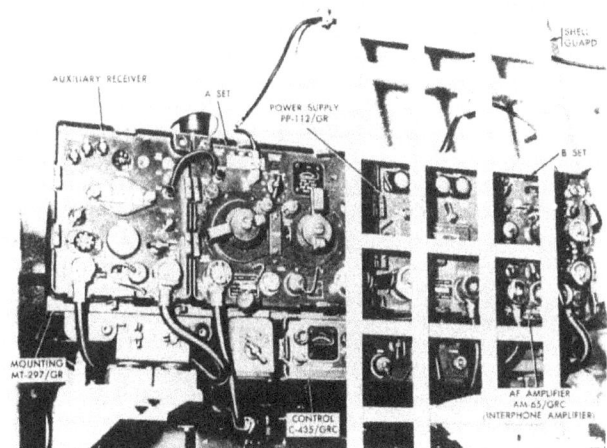

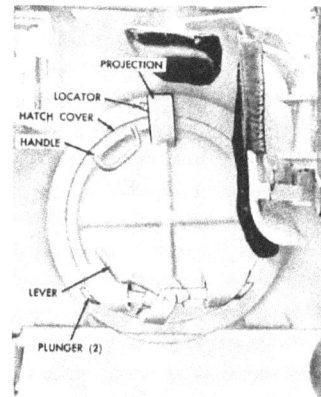

Above are the driver's controls. The periscopes are not mounted in this view. At the right, the driver's hatch (top) and the floor escape hatch (bottom) can be seen. The latter is open in the top view.

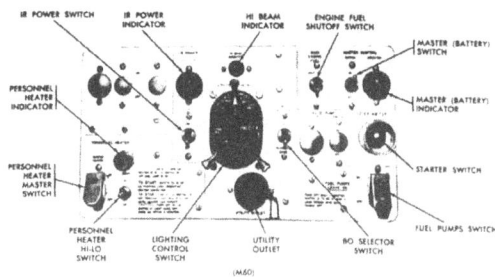

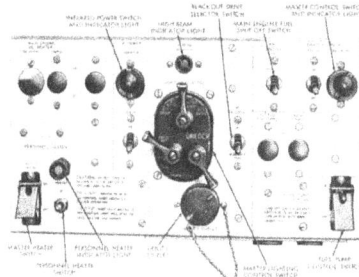

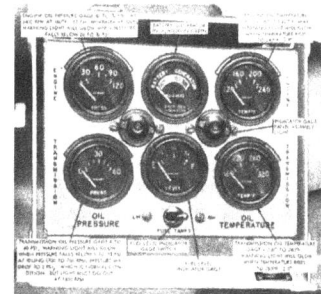

Above, the late M60 driver's master control panel (left) is compared with the early version (center). Note that a starter switch has been added to the later model. The driver's gage panel is at the right. Below are two views of the power pack consisting of the AVDS-1790-2 engine and the CD-850-6 transmission.

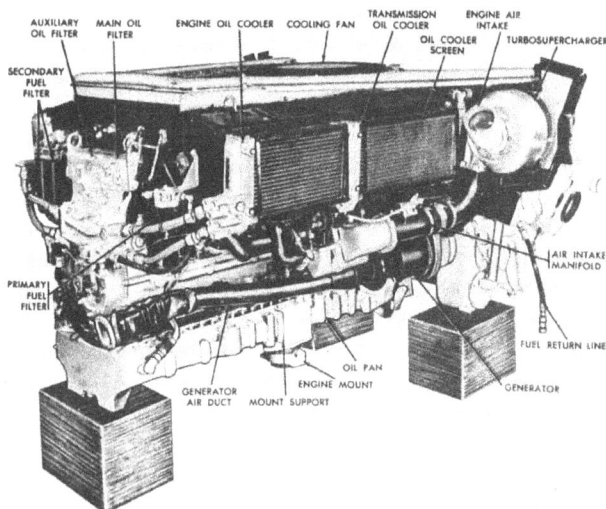

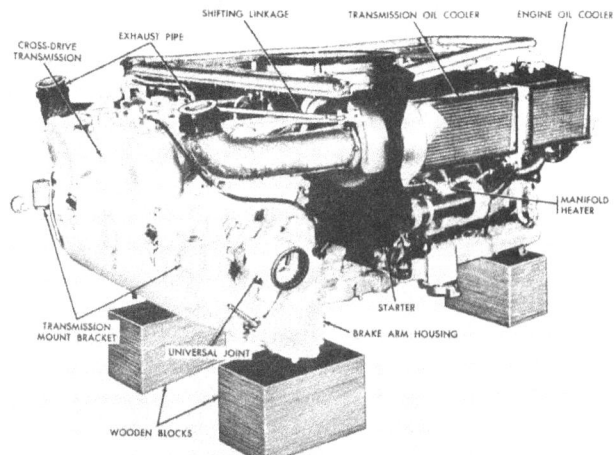

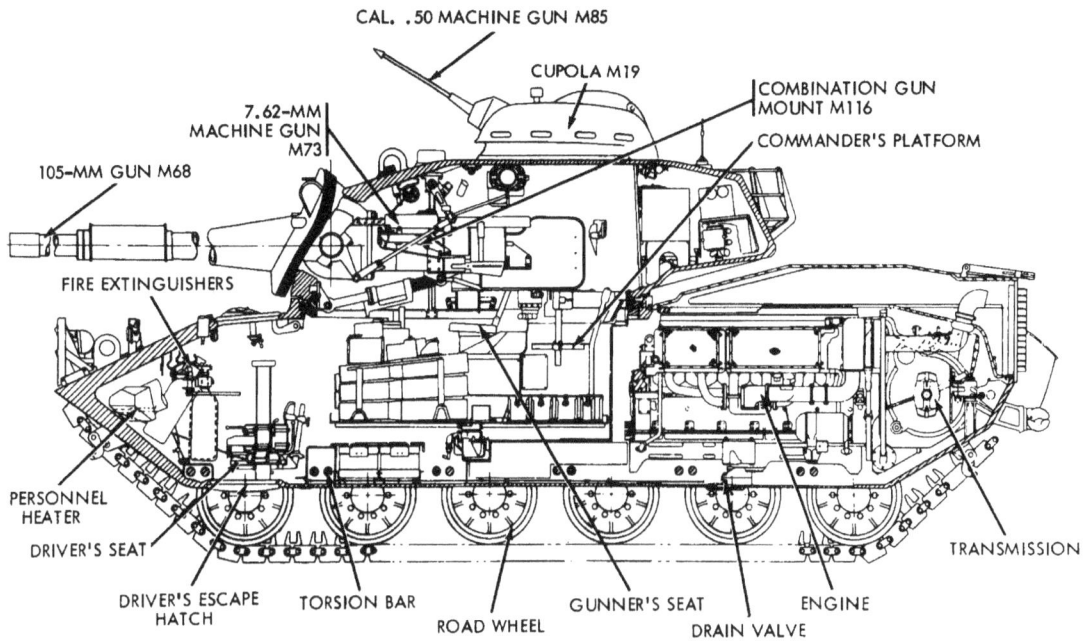

CAL. .50 MACHINE GUN M85

CUPOLA M19

COMBINATION GUN
MOUNT M116

COMMANDER'S PLATFORM

7.62-MM
MACHINE GUN
M73

105-MM GUN M68

FIRE EXTINGUISHERS

PERSONNEL
HEATER

DRIVER'S SEAT

DRIVER'S ESCAPE
HATCH

TORSION BAR

ROAD WHEEL

GUNNER'S SEAT

DRAIN VALVE

ENGINE

TRANSMISSION

The internal arrangement of the M60 is illustrated in the sectional drawing above. Note that the exhaust pipes extend back from the turbochargers without any mufflers. Compare with the M48A2C on page 120.

Tests at Aberdeen and Fort Knox showed that the free suspension was unsatisfactory both in firing stability and riding comfort. It was directed that shock absorbers be installed on all production tanks starting early in 1962. The tanks already built would have shock absorbers installed on the front and rear road wheel arms during overhaul. After the changes required by the test program, the M60 was considered satisfactory for troop use. The two features which proved to be the most popular with the user during the evaluation were the increased cruising range and the great improvement in firepower with the 105mm gun.

It was noted that in order to minimize the weight, the hull armor in some areas on the four pilot M60s and 15 of the early production tanks was too thin to provide adequate ballistic protection. This was corrected on subsequent vehicles and the light hull tanks were sent to the Armor School for use only in training.

The M60 on the railway car below is fitted with a protective cover for shipment. This photograph was taken at the Yuma Test Station on 20 May 1960. This tank has the original free suspension without snubbers or shock absorbers.

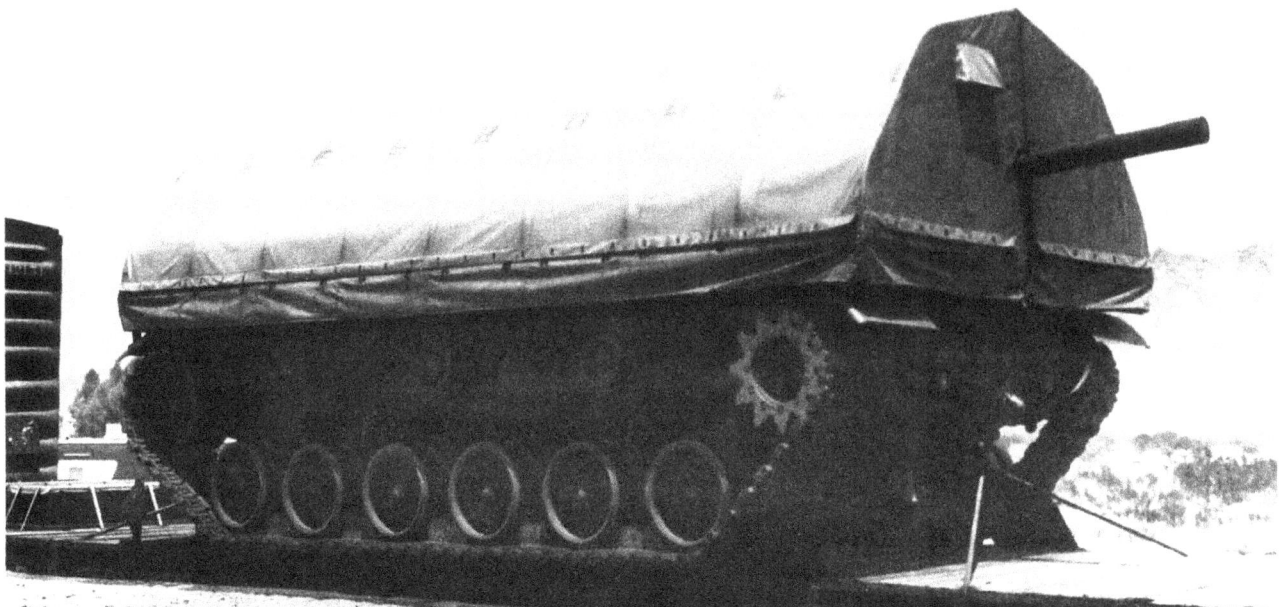

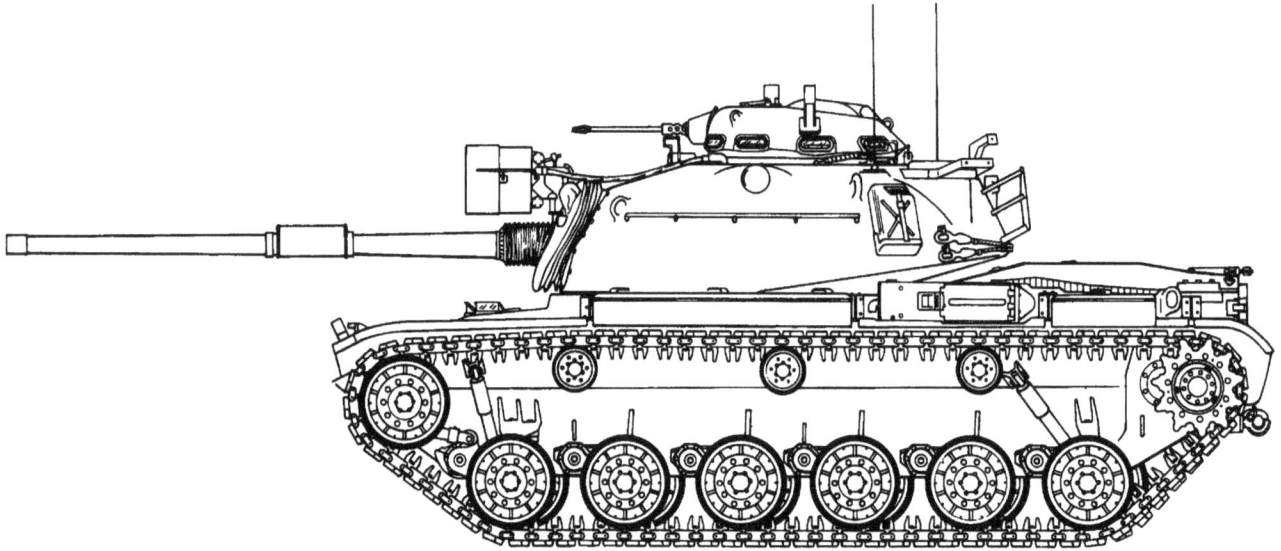

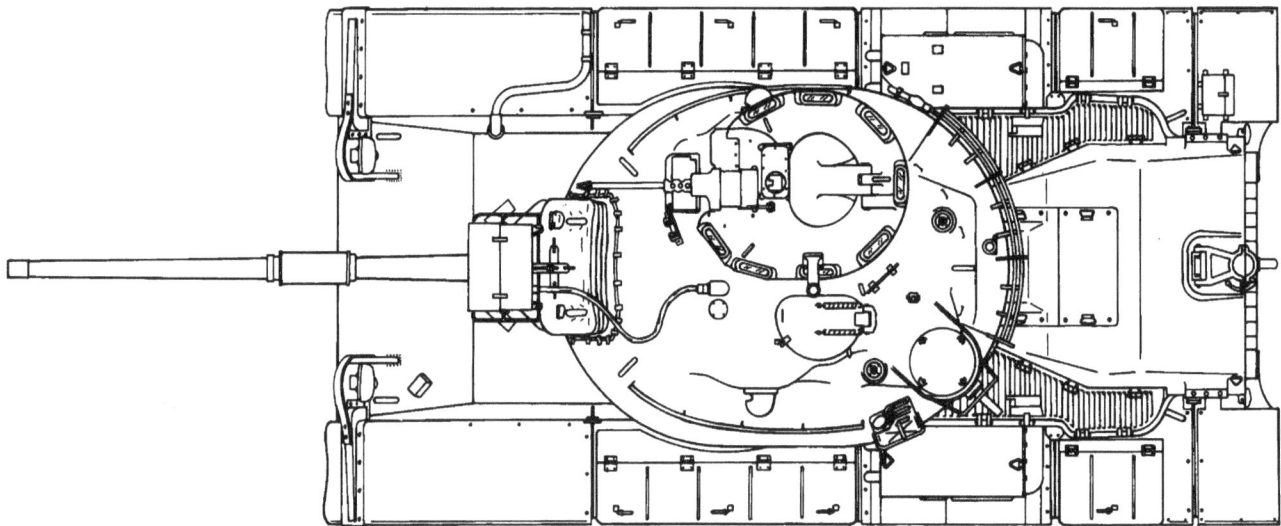

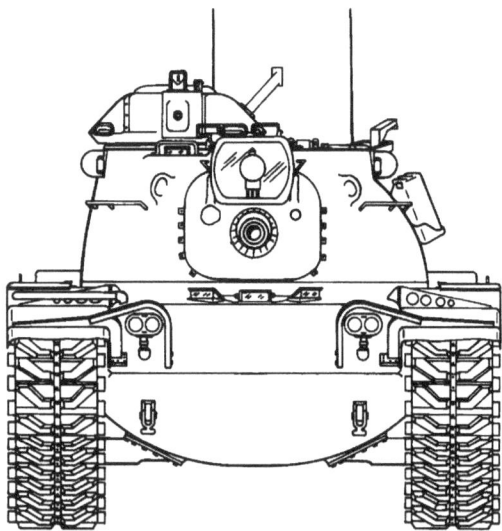

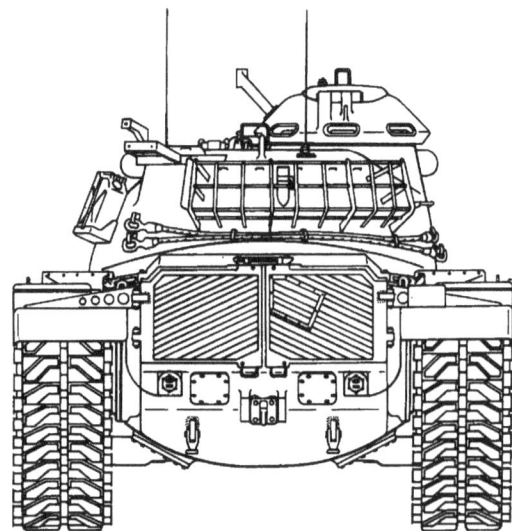

Scale 1:48

120mm Gun Tank M60

Late M60 production by Chrysler at the Detroit Tank Arsenal can be seen above during 1962. Turrets are being fitted out at the left and at the right, the hulls are moving along the assembly line.

Chrysler built the first 360 M60 tanks at the Newark, Delaware plant after which production was switched to the Detroit Tank Arsenal. Production continued until it was replaced on the line by the M60A1 in October 1962 after a total run of 2205 tanks. The first M60s were issued to the U.S. Army in Europe in December 1960 despite severe limitations on the supply of spare parts and ammunition.

Below is an M60 during test operations at Aberdeen Proving Ground on 23 March 1961. The commander's cupola is now armed with the new M85 .50 caliber machine gun.

This M60 has been fitted with the xenon searchlight kit having both infrared and visible light capability. These photographs were taken during the evaluation of the equipment at Fort Knox. Note that this tank has snubbers installed on the front and rear road wheel arms.

Above is a close-up view of the xenon searchlight installed on an M60 at Aberdeen Proving Ground on 25 March 1963. Below is a front view of the M60 test tank with the xenon searchlight kit at Fort Knox.

M60E1, pilot number 1, is shown above and at the bottom left on 19 May 1961 during preliminary tests at Detroit. The machine guns are not installed in the top photograph.

Although the M60 was fitted with the M48A2 type turret, work continued to develop the elongated turret based on the T95E7. Even without siliceous cored armor, this design provided superior ballistic protection and with the cannon mounted five inches farther forward, that much additional room was available for the turret crew. OTCM 37786, dated 29 June 1961, superseded earlier OTCMs and formally established the development project approving the technical characteristics for the 105mm gun tank M60E1. However, the project was well underway long before that date. The initial program had been approved on 21 March 1960 and Chrysler was directed to fabricate three pilot tanks. They consisted essentially of new turrets installed on reworked M60 chassis. The first pilot M60E1, registration number 9B3487, was completed on 6 May 1961 and after preliminary tests was shipped to Aberdeen Proving Ground. Pilot number 2, registration number 9B3488, was completed on 26 May and later shipped to the Maintenance Evaluation Group at Detroit Arsenal. The third M60E1, registration

Another view of M60E1 number 1 below at Aberdeen Proving Ground shows the snubbers on the front and rear road wheel arms.

168

M60E1, pilot number 1, appears above on 17 May 1961. The shape of the new turret with the elongated nose is obvious in this photograph.

number 9B3486, was finished on 30 June. After some tests, it was transferred to Fort Knox where it was received on 20 July.

In addition to the new elongated turret, the M60E1 incorporated many features shown to be necessary or desirable by the M60 test program. For example, friction snubbers were installed on the first and last road wheel arms. During the test program, the snubbers were replaced by hydraulic shock absorbers for comparison purposes. The driver's brake and accelerator pedals were rearranged for more comfortable and efficient operation. This was achieved by the substitution of an hydraulic actuating mechanism for the original mechanical linkage.

The steering wheel was replaced by a T bar steering control. The wire mesh seats in the M60 were considered uncomfortable and they were replaced by padded seats. The driver had a contoured bucket seat with an adjustable backrest. Two seats were provided in the new turret for the tank commander. An upper swivel type seat was mounted to a spindle attached to the turret roof at the right rear of the cupola. Rotated outward, it was used when riding with cupola open. The lower seat was adjustable in height along a post attached between the turret ring at the top and the basket at the bottom. The backrest of this seat could be folded forward to provide a platform for the tank commander when standing in the open cupola. The entire seat could be folded upward for stowage. The gunner also had a padded seat with an adjustable backrest attached to the turret basket to the right of the cannon. The loader's seat was removable and could be attached to the turret ring at the left of the gun or at the top of the turret lock to the left of the first position.

The rear of M60E1 number 1 during tests at Aberdeen is shown at the left. The external stowage rack can be seen on the rear of the turret.

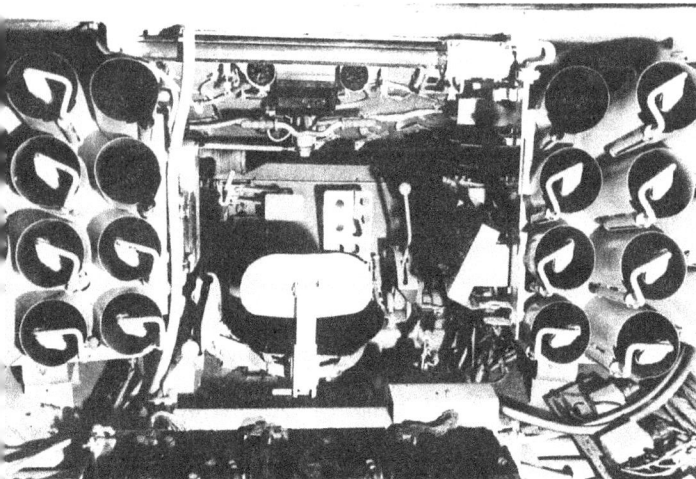

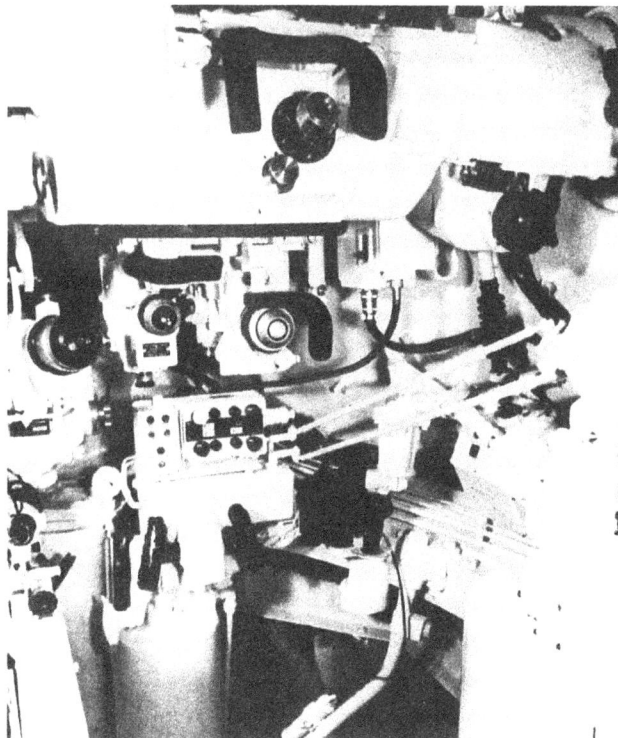

Above is the driver's compartment in the M60E1. Note the new T-bar control replacing the steering wheel of the M60. The 105mm ammunition racks can be seen on each side of the driver's seat. At the right is a view of the fire control equipment in the M60E1. The single eyepiece of the commander's coincidence type range finder appears at the top of the photograph above the gunner's periscopic and telescopic sights.

As originally built, the M60E1s were fitted with the XM16C electric computer and new vision devices for the gunner and tank commander. The gunner's primary sight was an XM35 periscope head containing a daylight periscope on the left and an infrared periscope on the right. This sight was coupled directly to the ballistic drive. An M105C articulated telescope was fitted on the right side of the cannon as a backup sight. The M17C coincidence type range finder provided the tank commander with ten power vision for both sighting and ranging a target. In the M19 cupola, the tank commander had either the XM34 daylight periscope or the XM36 combination day and night periscope. In addition to observation, both of these instruments were used to aim the M85 machine gun in the cupola. The XM34 contained two periscope heads for binocular vision. The XM36 also contained two heads, but each was for monocular vision only. The left head was for daylight use and the one on the right was an infrared unit for night vision. The periscope mount was fitted with a shield which could be controlled from within the cupola.

Below are the tank commander's XM34 daylight periscope (left) and XM36 day and night periscope (right).

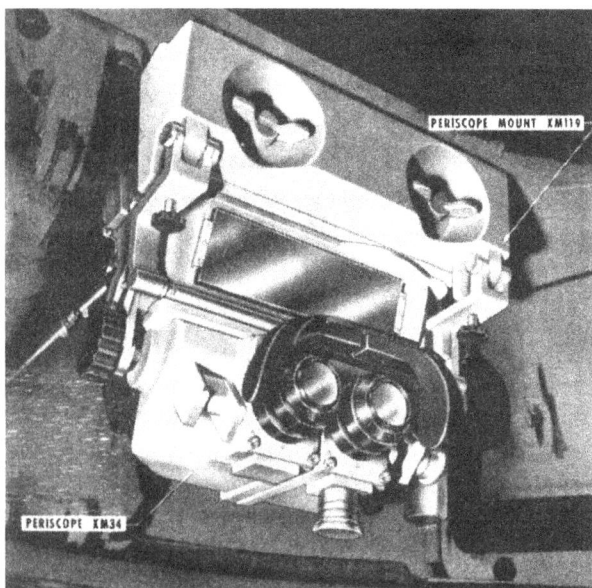

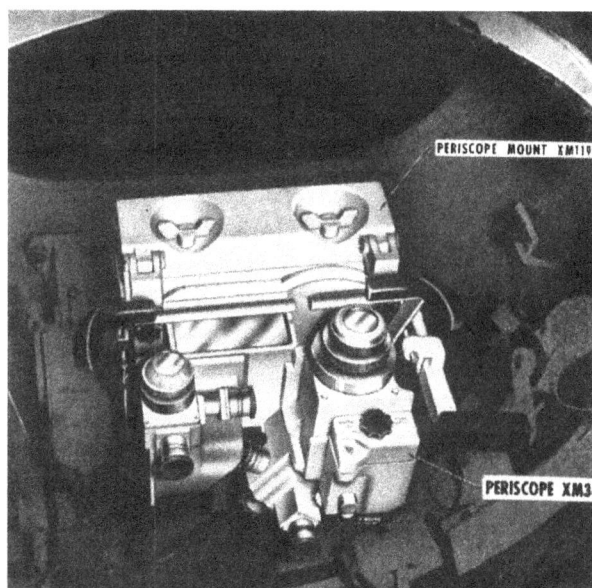

Both photographs on this page show a new M60A1 tank during tests at Fort Knox. The friction snubbers installed on the first two and last road wheel arms are readily visible.

The 105mm gun full tracked combat tank M60A1 was classified as Standard A by OTCM 37933 on 22 October 1961. The same action reclassified the M60 as Standard B effective upon the first delivery of production M60A1s. For the early production tanks, the M60 fire control equipment was specified. Provision was made for later substitution of the electric computer and the new periscopes when they became available. The M17A1 range finder replaced the M17C. In the M17A1, the boresight qualification position was 1200 meters instead of the 1500 meters in the earlier unit.

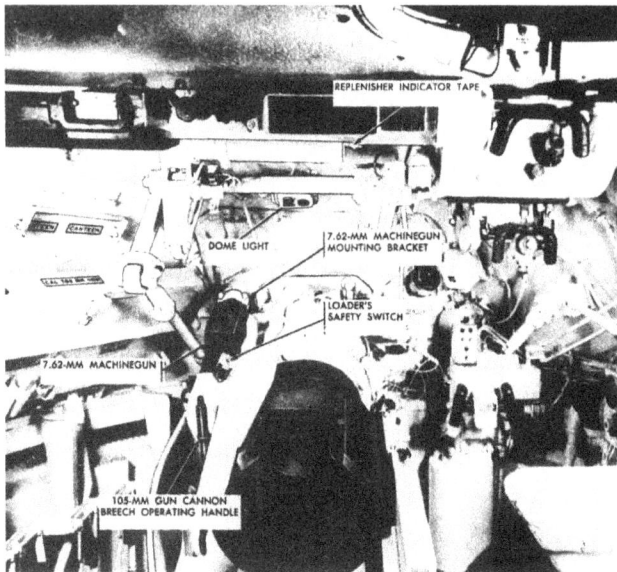

REPLENISHER INDICATOR TAPE

DOME LIGHT

7.62-MM MACHINEGUN MOUNTING BRACKET

LOADER'S SAFETY SWITCH

7.62-MM MACHINEGUN

105-MM GUN CANNON BREECH OPERATING HANDLE

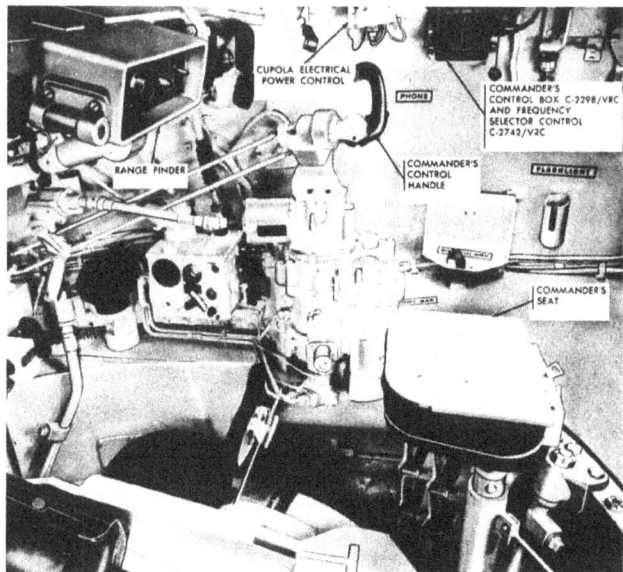

CUPOLA ELECTRICAL POWER CONTROL

PHONE

COMMANDER'S CONTROL BOX C-2298/VRC AND FREQUENCY SELECTOR CONTROL C-2742/V2C

RANGE FINDER

COMMANDER'S CONTROL HANDLE

COMMANDER'S SEAT

FIRE CONTROL SYSTEM
M60A1 TANK

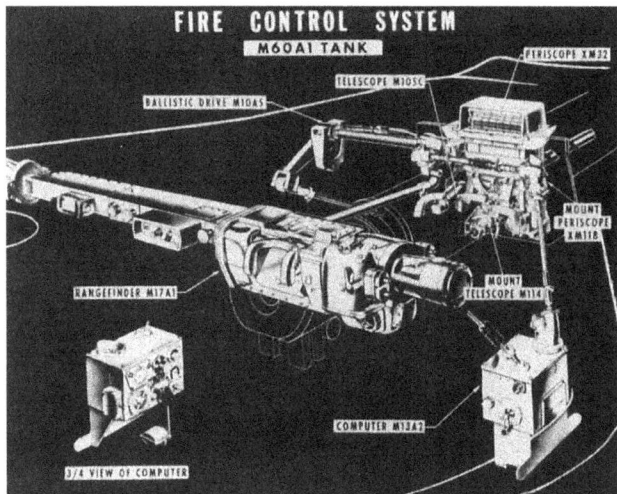

PERISCOPE XM32

TELESCOPE M105C

BALLISTIC DRIVE M10A5

MOUNT PERISCOPE XM118

RANGEFINDER M17A1

MOUNT TELESCOPE M114

COMPUTER M13A2

3/4 VIEW OF COMPUTER

The front interior of the M60A1 turret appears in the view at the top left. The range finder is at the top with the gunner's instruments and controls at the right. The coaxial machine gun is at the left of the cannon. Directly above is a view of the commander's controls on the right side of the turret.

The fire control system for the M60A1 can be seen at the left.

The bottom photographs show the turret bustle (left) without the radio and the AN/VRC-12,46, or 47 radio installation (right). The 105mm ammunition rack can be seen on the loader's side of the turret bustle.

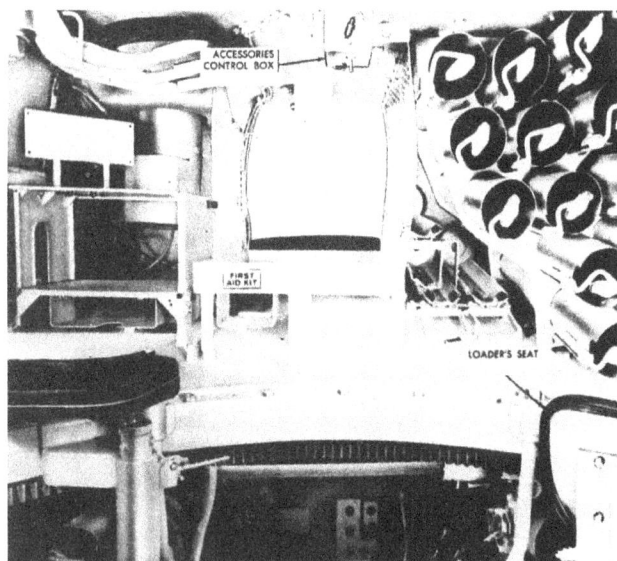

ACCESSORIES CONTROL BOX

FIRST AID KIT

LOADER'S SEAT

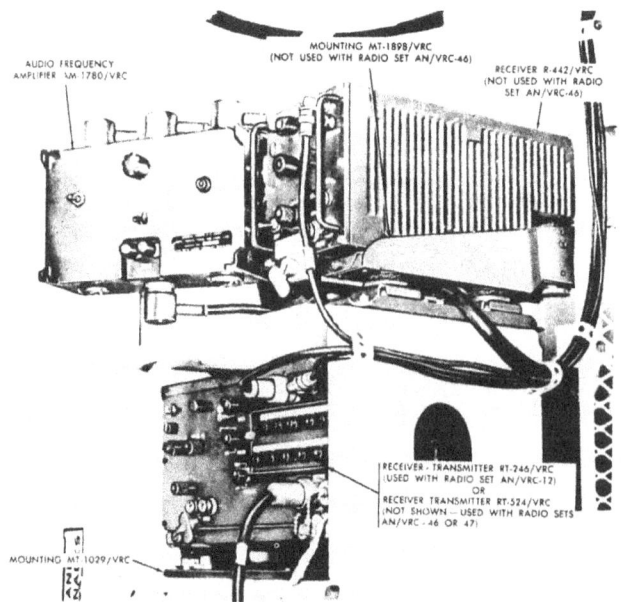

AUDIO FREQUENCY AMPLIFIER AM-1780/VRC

MOUNTING MT-189B/VRC (NOT USED WITH RADIO SET AN/VRC-46)

RECEIVER R-442/VRC (NOT USED WITH RADIO SET AN/VRC-46)

RECEIVER-TRANSMITTER RT-246/VRC (USED WITH RADIO SET AN/VRC-12) OR RECEIVER TRANSMITTER RT-524/VRC (NOT SHOWN — USED WITH RADIO SETS AN/VRC-46 OR 47)

MOUNTING MT-1029/VRC

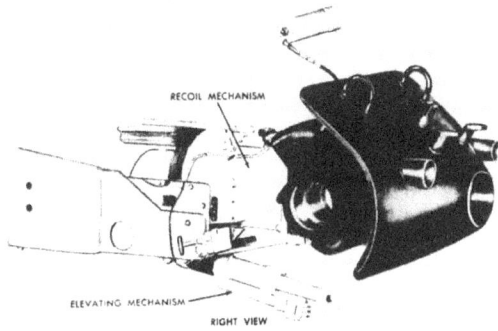

RECOIL MECHANISM

ELEVATING MECHANISM

RIGHT VIEW

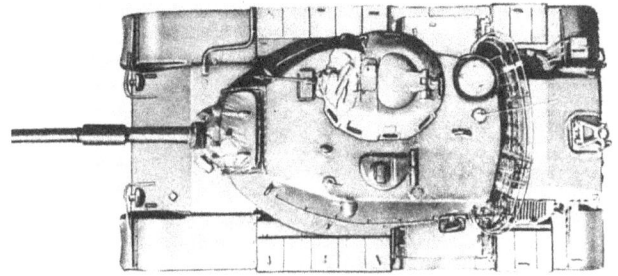

The M140 combination gun mount designed to fit the new long nosed turret is shown at the left (above and below). At the top right, the shape of the new turret can be seen. The ammunition stowage in the early M60A1s is illustrated in the sketch below at the right.

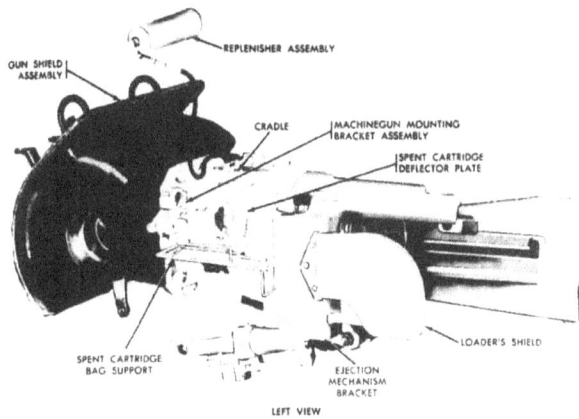

REPLENISHER ASSEMBLY

GUN SHIELD ASSEMBLY

CRADLE

MACHINEGUN MOUNTING BRACKET ASSEMBLY

SPENT CARTRIDGE DEFLECTOR PLATE

SPENT CARTRIDGE BAG SUPPORT

EJECTION MECHANISM BRACKET

LOADER'S SHIELD

LEFT VIEW

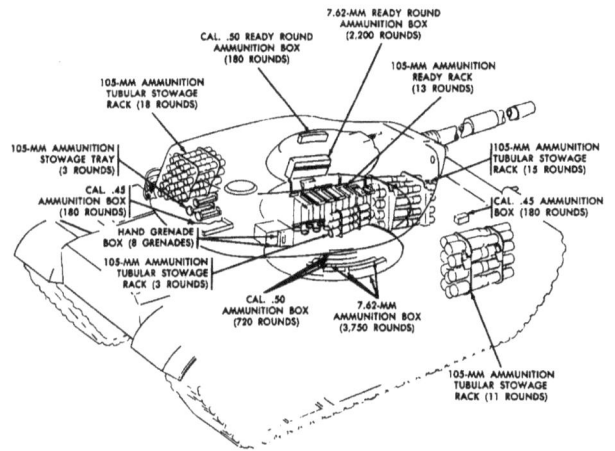

7.62-MM READY ROUND AMMUNITION BOX (2,200 ROUNDS)

CAL. .50 READY ROUND AMMUNITION BOX (180 ROUNDS)

105-MM AMMUNITION READY RACK (13 ROUNDS)

105-MM AMMUNITION TUBULAR STOWAGE RACK (18 ROUNDS)

105-MM AMMUNITION STOWAGE TRAY (3 ROUNDS)

105-MM AMMUNITION TUBULAR STOWAGE RACK (15 ROUNDS)

CAL. .45 AMMUNITION BOX (180 ROUNDS)

HAND GRENADE BOX (8 GRENADES)

105-MM AMMUNITION TUBULAR STOWAGE RACK (3 ROUNDS)

CAL. .45 AMMUNITION BOX (180 ROUNDS)

CAL. .50 AMMUNITION BOX (720 ROUNDS)

7.62-MM AMMUNITION BOX (3,750 ROUNDS)

105-MM AMMUNITION TUBULAR STOWAGE RACK (11 ROUNDS)

Components of the xenon searchlight kit appear at the left below. This kit with both infrared and visible light capability is installed on the M60A1 at the right below. This is the same kit fitted to the M60 on pages 166 and 167.

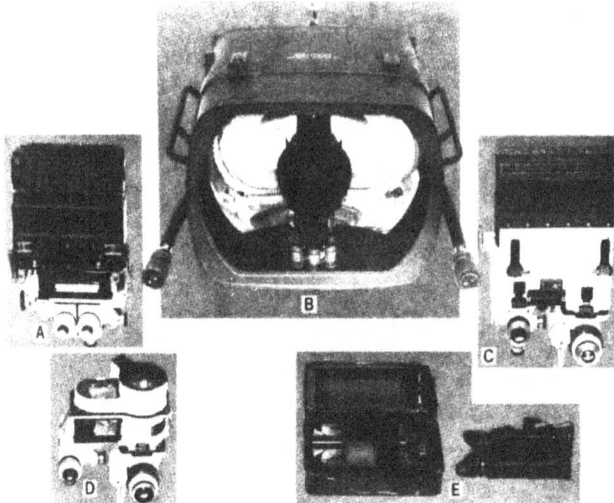

173

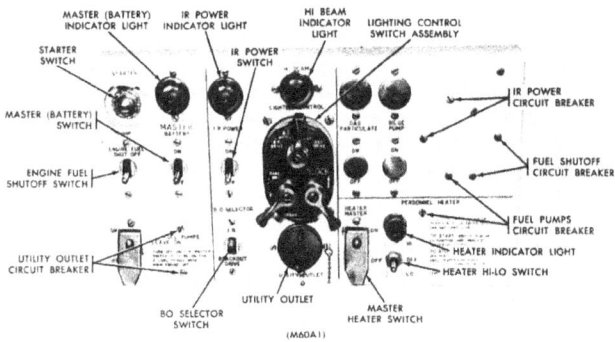

The driver's controls are at the right with a view of his master control panel above. The sketch below illustrates the steering of the M60A1 using the T-bar control.

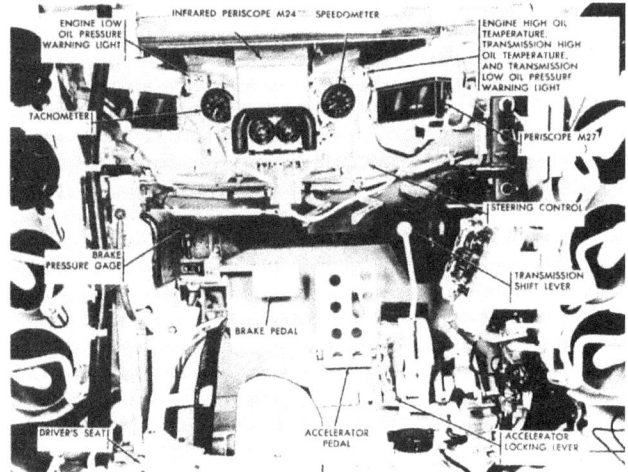

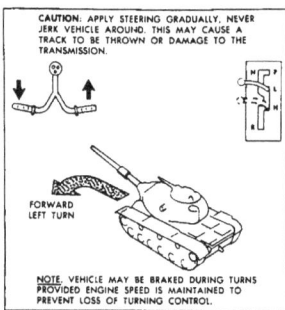

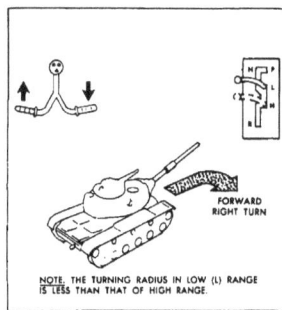

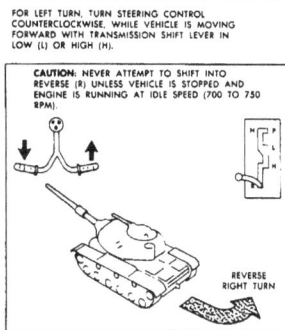

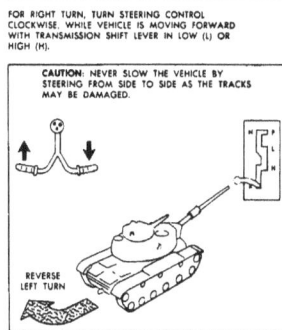

CAUTION: APPLY STEERING GRADUALLY, NEVER JERK VEHICLE AROUND. THIS MAY CAUSE A TRACK TO BE THROWN OR DAMAGE TO THE TRANSMISSION.

FORWARD LEFT TURN

NOTE. VEHICLE MAY BE BRAKED DURING TURNS PROVIDED ENGINE SPEED IS MAINTAINED TO PREVENT LOSS OF TURNING CONTROL.

FOR LEFT TURN, TURN STEERING CONTROL COUNTERCLOCKWISE, WHILE VEHICLE IS MOVING FORWARD WITH TRANSMISSION SHIFT LEVER IN LOW (L) OR HIGH (H).

FORWARD RIGHT TURN

NOTE. THE TURNING RADIUS IN LOW (L) RANGE IS LESS THAN THAT OF HIGH RANGE.

FOR RIGHT TURN, TURN STEERING CONTROL CLOCKWISE, WHILE VEHICLE IS MOVING FORWARD WITH TRANSMISSION SHIFT LEVER IN LOW (L) OR HIGH (H).

CAUTION: NEVER ATTEMPT TO SHIFT INTO REVERSE (R) UNLESS VEHICLE IS STOPPED AND ENGINE IS RUNNING AT IDLE SPEED (700 TO 750 RPM).

REVERSE RIGHT TURN

FOR RIGHT TURN WHILE VEHICLE IS MOVING REARWARD, TURN STEERING CONTROL COUNTER-CLOCKWISE WITH TRANSMISSION LEVER IN REVERSE (R).

CAUTION: NEVER SLOW THE VEHICLE BY STEERING FROM SIDE TO SIDE AS THE TRACKS MAY BE DAMAGED.

REVERSE LEFT TURN

FOR LEFT TURN WHILE VEHICLE IS MOVING REARWARD, TURN STEERING CONTROL CLOCKWISE WITH TRANSMISSION SHIFT LEVER IN REVERSE (R).

Although the M60E1 had the same hull armor thickness as the M60, this was increased on the production M60A1. The upper front plate went from 3.67 inches to 4.29 inches, both at 65 degrees from the vertical. Also, the sides of the crew compartment were 2.9 inches at 0 degrees at the apex compared to 1.9 inches at 0 degrees on the earlier M60. In August 1962, the Continental AVDS-1790-2A engine started production and was installed in the M60A1. It offered a significant reduction of exhaust smoke and an even lower fuel consumption. The M73 coaxial machine gun had never been fully satisfactory and it was superseded later by the 7.62mm M219 which, unfortunately, was not much better. The suspension of the production M60A1 differed from the M60E1 in that shock absorbers were installed on the first two and last road wheel arms on each side of the tank.

The introduction of the M60A1 began a production run that was to last with many modifications for over 20 years. At that time, it was considered to be an interim vehicle until the ideal tank could be designed. However, unforeseen difficulties with the various development programs extended the service of the M60A1 and resulted in its continued improvement. Fortunately, the basic design was sound and readily accepted the many modifications required.

Details of the headlight groups can be seen below (left) along with the M60A1 suspension (right). Note the three friction snubbers installed on each side compared to two on the M60.

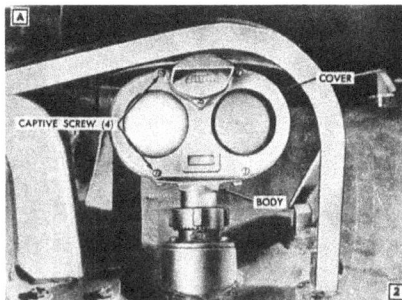

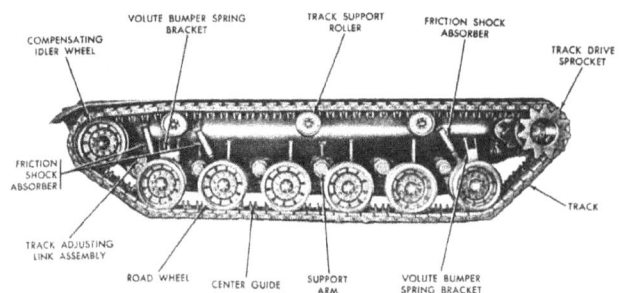

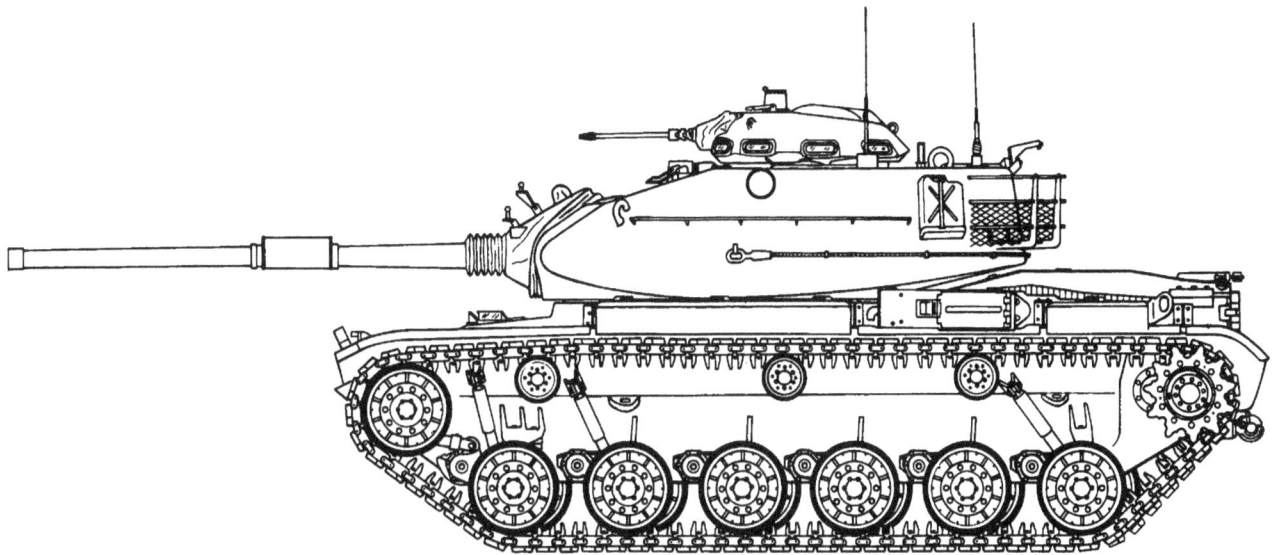

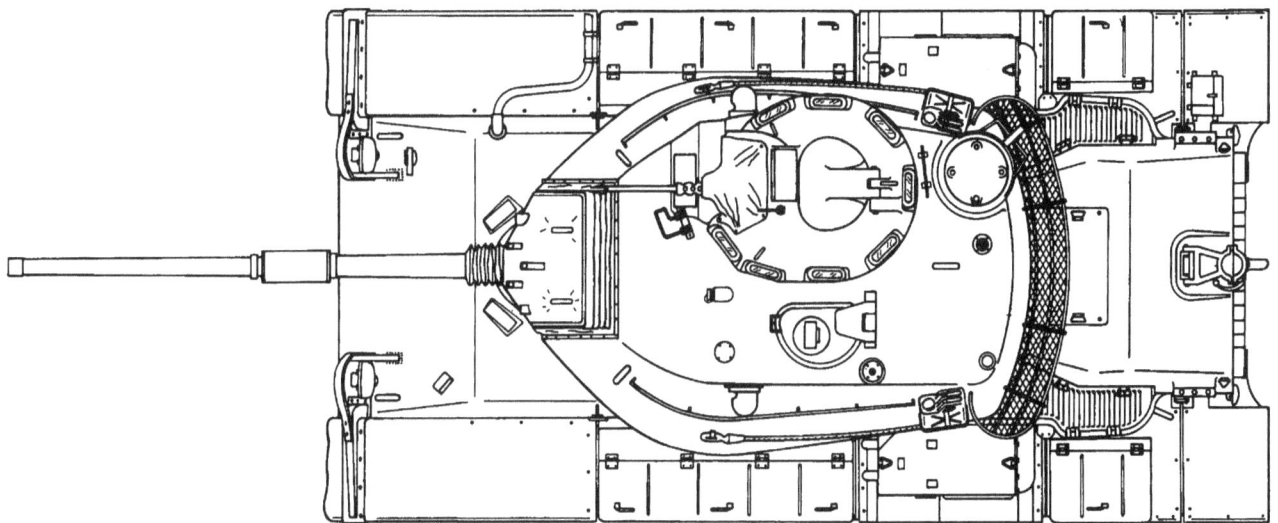

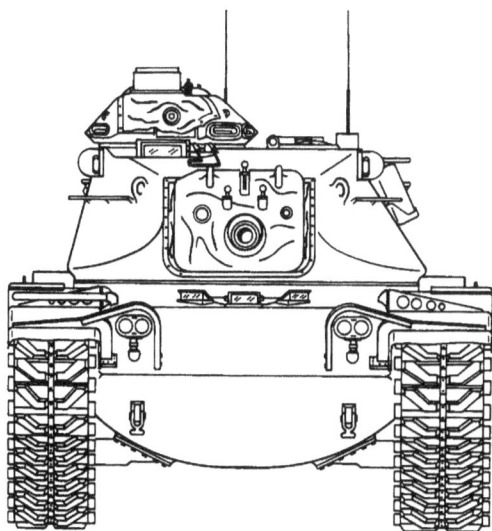

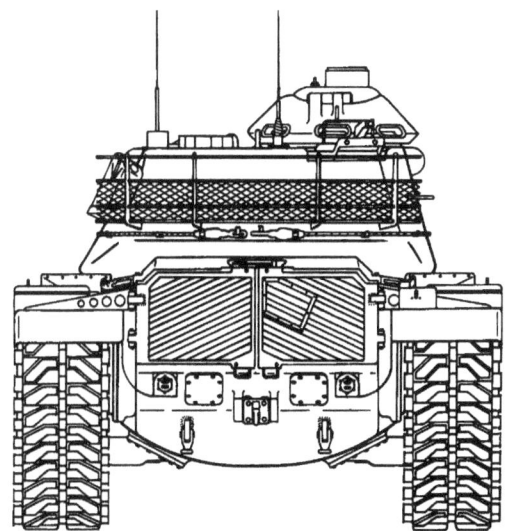

Scale 1:48

105mm Gun Tank M60A1

Fabrication of the turret (above) and hull (below) for the M60A1 at Detroit Tank Arsenal is illustrated in these photographs. Note the depression in the bottom of the turret bustle to provide extra clearance for the driver when the gun is locked in the travel position.

Above, the turret is mated to the hull during the assembly of an M60A1 at Detroit. Below, an M60A1 equipped with the xenon searchlight is shown at Fort Riley, Kansas in November 1971.

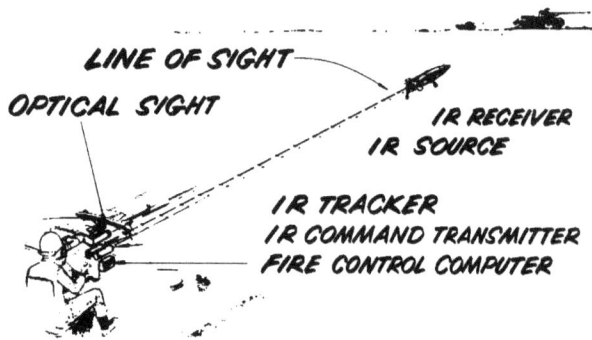

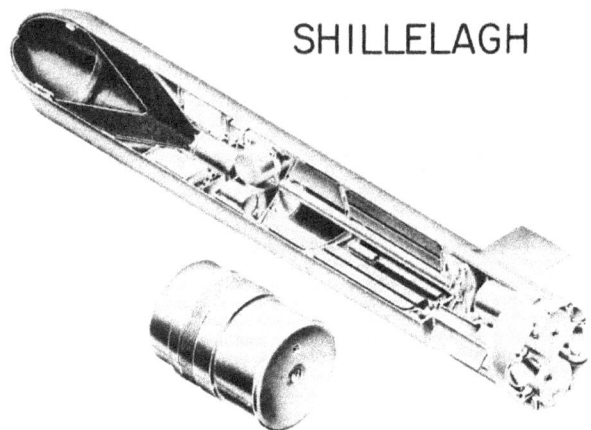

SHILLELAGH

The sketch at the left above depicts the operation of the Shillelagh missile using line of sight command guidance. At the right is the early XM13 missile.

The ARCOVE report published in January 1958 recommended that a major effort should be directed toward the development by 1965 of a guided missile weapon system for tanks. Such a system would have line of sight command guidance and would provide the main armament for future main battle tanks. To meet these requirements, a program was launched with the responsibility for the missile development being contracted on 11 June 1959 to the Aeroneutronics Division of Ford Motor Company. Named the Combat Vehicle Weapon System (CVWS) Shillelagh, it consisted of the 152mm XM81 gun-launcher which could fire a conventional projectile or launch the XM13 missile. The XM81 was approximately half the weight and length of the 105mm gun M68 making it very attractive for combat vehicle installation. Fitted with a separable chamber breech mechanism, it was designed to use conventional ammunition with a completely combustible case and primer. Such a system eliminated the problem of removing spent cartridge cases from the tank as well as the necessity of salvaging them for future use. Three types of conventional 152mm ammunition were under development for the XM81. These were the XM409 HEAT-MP, the XM410 white phosphorus, and the XM411 training round. As its designation indicated, the XM409 was a multipurpose projectile combining the armor defeating characteristics of the shaped charge with the blast effect of a standard high explosive round. Such a combination reduced the types of ammunition required. All of these rounds were spin stabilized by the rifling in the XM81. The XM409 had a muzzle velocity of 2260 feet/second and was estimated to penetrate 7 inches of rolled homogeneous armor at 60 degrees obliquity.

The XM13 Shillelagh missile was ejected from the gun-launcher at approximately 260 feet/second. The solid rocket motor then boosted its speed to about 1060 feet/second. The missile flew a line of sight trajectory to the target under the control of an infrared(IR) tracking and command system. As the gunner held the sight on the target, the infrared tracker measured the displacement of the IR source

The two views below show the XM81 gun-launcher originally designed for the test installation in the T95 turret.

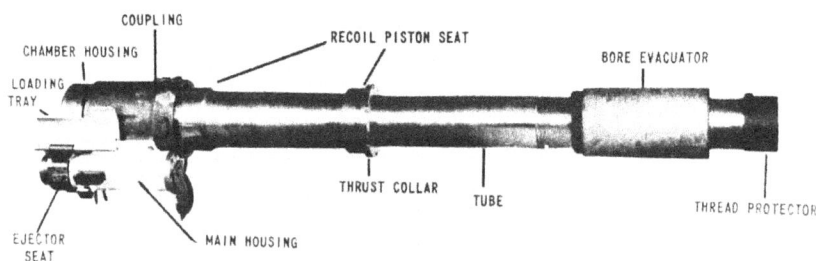

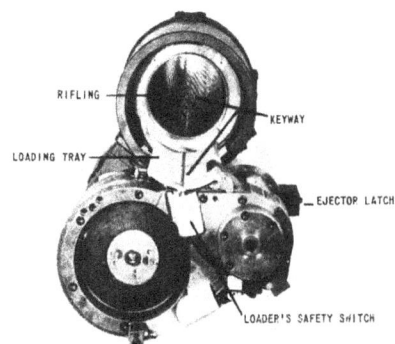

in the missile tail from the line of sight. This error signal was processed by the electronic computer and the correction command signal was transmitted to the missile by a xenon arc lamp. In the gun-launcher, the missile rode on top of the rifling lands and engaged a keyway in the bottom of the tube to prevent rotation. The missile itself was 43.25 inches long and its six-inch diameter shaped charge war head was capable of destroying any known armored vehicle at that time.

The Shillelagh was the preferred weapon system for both of the new tank type vehicles planned for the future army. As mentioned earlier, these were the airborne reconnaissance/airborne assault vehicle(AR/AAV), which eventually appeared as the Sheridan, and the new main battle tank (MBT). The latter was the subject of numerous design studies in which it was designated as the MBT(MR) referring to the medium range time period. Such a vehicle was expected to be ready for production in about five years.

Since the guided missile was envisiged as the future MBT armament, studies began on the upgrading of the existing tank fleet by the application of the new weapon system. In August 1961, tentative technical characteristics were prepared for an M60 with this installation. It was then planned to retrofit the Shillelagh system into the M60E1(M60A1) tanks at some future time. A mock-up had been constructed at the Detroit Arsenal model shop in the Spring of 1961 to study this arrangement. The only changes required were the installation of the gun-launcher, a new gun shield, a tracker telescope, and the missile guidance components. The stowage had to be rearranged to handle the new ammunition and missiles. At the time, the converted vehicle was referred to as the M60E2. Also, the M60E3 designation was suggested for future new production M60 series tanks incorporating the Shillelagh system. The M60E3 also was expected to carry a spotting rifle for use with the conventional ammunition. It was requested that the M60E1 turret casting be modified for production to permit the installation of the spotting rifle during a future retrofit program.

As part of the development project, three T95 turrets and two M60E1 turrets were requested for modification to carry the Shillelagh. The T95 turrets were to be installed on M48 chassis for test purposes. The M60E1 turrets were to be modified to the M60E2 configuration and installed on M60 chassis.

The designation 152mm gun-launcher XM81E5 was assigned to the weapon for the M60E2 while the XM81E6 referred to the gun-launcher in the proposed MBT(MR). Further studies noted that a new turret based on one of the MBT(MR) designs could be provided for the M60 Shillelagh program as quickly as the modified M60E1 turrets. This new design was ordered in June 1961. It was a three man version of the four man MBT(MR) turret. The latter included the driver who remained in the hull in the M60 version.

This same period saw the appearance of still another turret arrangement. Designed by Clifford Bradley and strongly supported by Joseph Williams, it also was proposed for the MBT(MR) program. The new turret was an extremely compact design with a greatly reduced frontal area. It was an outgrowth of earlier studies which featured a remote control cannon mounted above an armored hull crew compartment. Such an arrangement provided a low silhouette and excellent crew protection, but it introduced severe complications in providing adequate vision and in the remote control required for servicing the cannon. With the state of the art at that time, such a complex system would have created more problems than it would have solved. The new turret achieved many of the advantages of the remote control gun with a relatively simple design using proven gun and turret controls. The upper part of the compact turret was long and narrow with its width hardly exceeding that required for the gun mount. The tank commander was located directly behind the main weapon at a height giving excellent all round vision. The gunner and loader were placed low in the turret basket on the right and left sides of the gun respectively. The hatches in the turret roof over their positions were only slightly above the turret ring itself. The small frontal area of this design greatly reduced the probability of a hit and the weight saving from the decreased volume enclosed by the armor permitted a considerable improvement in the ballistic protection. With this turret, the driver remained in his usual location in the center front of the hull.

The model at the right illustrates the concept of the MBT (MR) with the compact turret. This particular version was armed with the 105mm gun.

The experimental test vehicle for the XM66 concept Type A appears above at the left. The model at the right illustrates Type B. Below are drawings of the Type B turret which was adopted for further development.

On 10 January 1964, the Department of the Army reviewed four MBT turret concepts which were armed with the Shillelagh weapon system. Two of these, types A and B, were based on the compact turret design and each was fitted with a remote control 20mm cannon as secondary armament. Type C was a conventional turret with a cupola mounted 20mm gun and Type D was basically the M60A1 turret with a shortened bustle. Types A and D had been tested experimentally and mock-ups were evaluated for the remaining two. Turrets A, B, and D were considered potentially suitable for adapting the Shillelagh system to the M60 series tanks and further evaluation selected the compact Type B for continued development. At that time, all four concepts were referred to as the XM66 main battle tank. However, Army Materiel Command Technical Committee(AMCTC) item 3100 assigned the nomenclature 152mm gun full tracked combat tank M60A1E1 on 1 February 1965. With the reorganization of the Army, the old OTCMs were now replaced by AMCTC items.

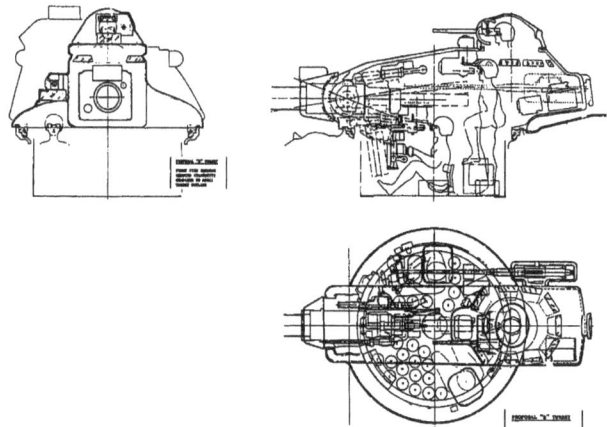

The M60A1E1 project was originally envisaged as a fairly low risk program to upgrade the standard production tanks. Unfortunately, these hopes were not realized. In AMCTC 3100, the 152mm gun-launcher XM81E10 was specified with a quick change tube mounted in a concentric recoil mechanism. Further development replaced this with the

The sketches below show the XM66 concepts Type C (left) and Type D (right). The latter was basically the version considered as the M60E2 or M60E3.

One of the M60A1E1 pilot tanks appears above and below after completion at Detroit. At the bottom right, one of the early pilots is at Aberdeen Proving Ground.

XM81E13 which was then redesignated as the 152mm gun-launcher XM162. Perhaps the E13 was considered to be unlucky. If so, it was an omen for the future.

M60A1E1 research and development pilots 1 and 2 were delivered for engineering and service tests during November 1965 and February 1966 respectively. After a limited program which included 12 missile firings, the vehicles were reworked and testing resumed in June. The original prototype was reassembled with final design components and joined the program in August as a third test tank. Severe problems with the combustible case ammunition now became obvious. These included misfires, flare back, and premature detonation. Smoldering residue from the incomplete combustion of a cartridge case would ignite a subsequent round, sometimes before the breech could be closed. This could set off other rounds of the combustible case ammunition in the turret. Further evaluation of the combustible case rounds showed them to be easily damaged by high humidity conditions and too fragile for rough handling. Corrective modifications solved most of these problems, but they resulted in delays in the test program and a loss of confidence by the troops in the new ammunition.

Another M60A1E1 pilot is shown here during evaluation at Fort Knox. Note that this tank utilized an M60 chassis with friction snubbers on just the front and rear road wheel arms.

Here are additional views of the M60A1E1 pilot at Fort Knox. The top view clearly shows the details of the compact turret. All of these photographs were dated 27 July 1966.

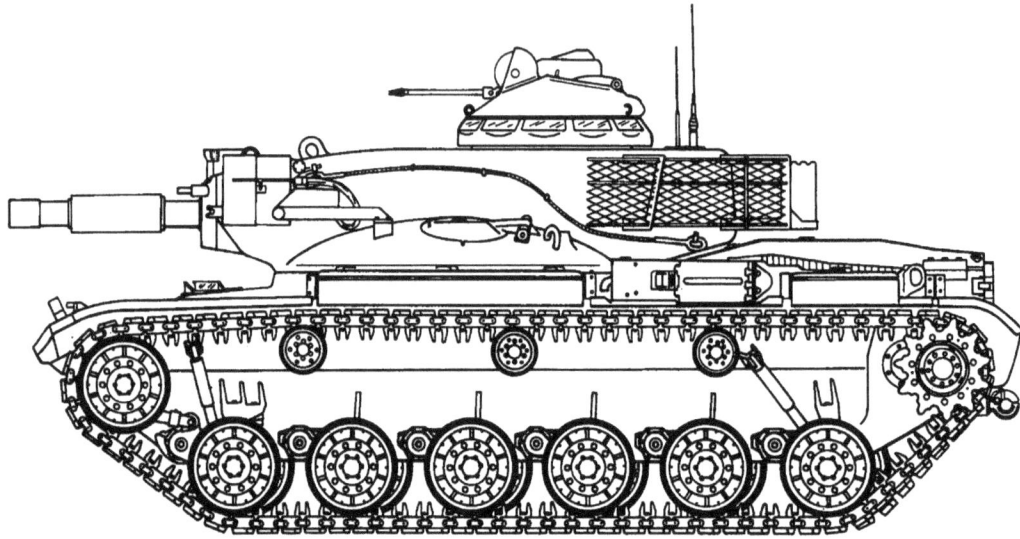

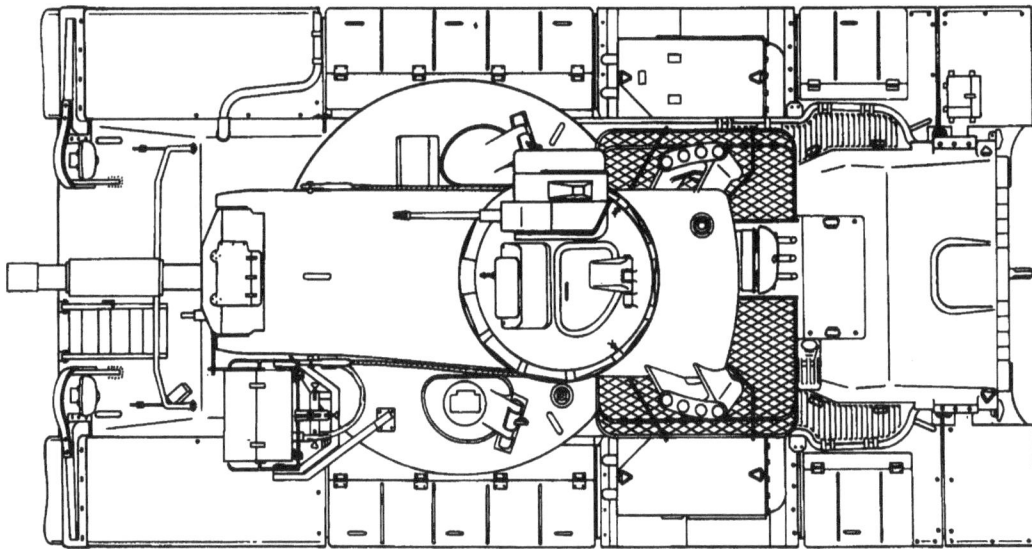

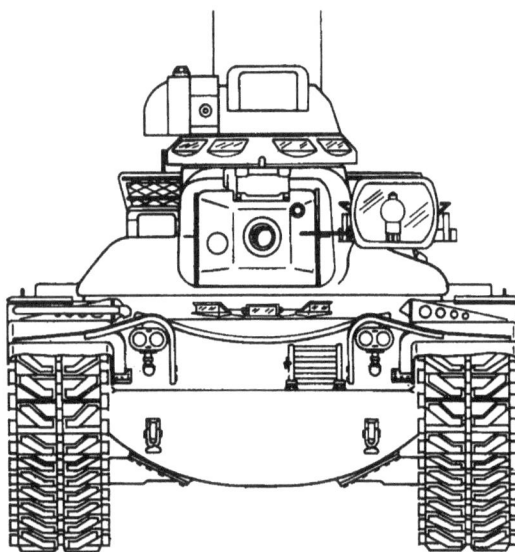

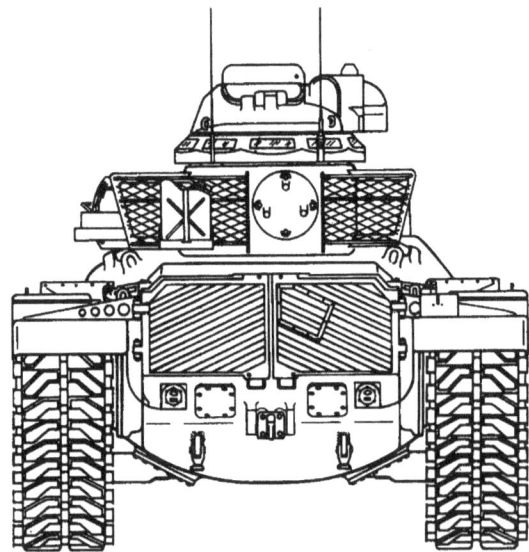

152mm Gun Tank M60A1E1

M60A1E1, Advance Production Engineering (APE) pilot number 1, appears here at Fort Knox on 4 April 1967. The xenon searchlight has not yet been fitted on this tank.

M60A1E1, APE pilot number 1 is shown here at Fort Knox on 30 October 1967. The xenon searchlight has now been installed. Also, note the changes in the infrared transmitter housing above the main gun mount.

The interior view above shows the turret front of M60A1E1 APE pilot number 1. The separable chamber breech of the 152mm gun-launcher can be seen along with the gunner's controls.

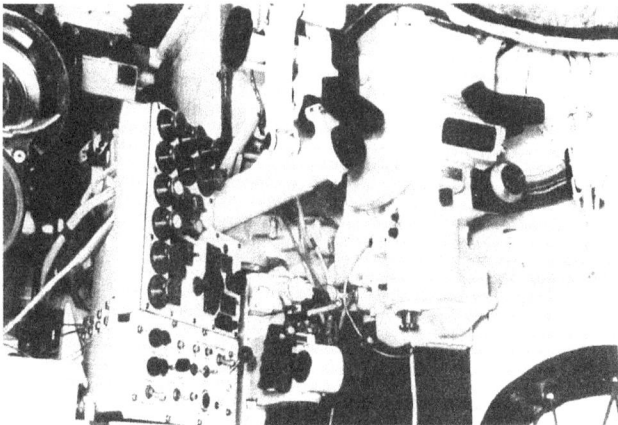

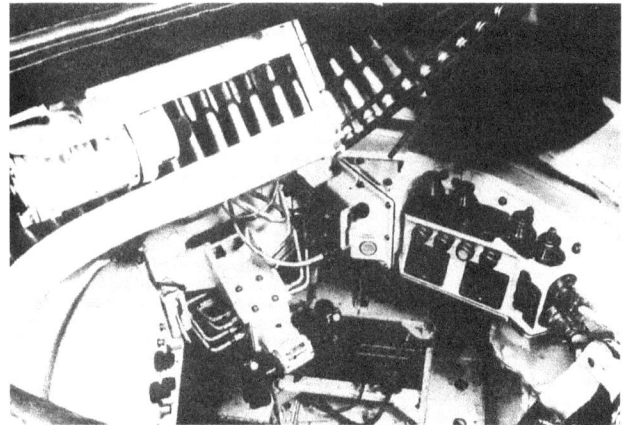

Above are the controls for the gunner (left) and the tank commander (right) in the M60A1E1 APE number 1 turret. At the right is a view of the radio installation in the turret bustle.

An M60A1E2 is shown here at Fort Knox during the 1968 test operations. This tank is not equipped with the closed breech scavenging system as can be seen by the normal lower rear hull in the view below.

On 3 July 1965, AMCTC 6325 clarified the type classification by designating the 152mm gun full tracked combat tank M60A1E2 as the vehicle based on the M60A1 chassis. The M60A1E1 nomenclature was retained for the tank based on the M60. AMCTC 4841 approved the procurement of 243 M60A1E1 turrets with fiscal year 1966 funds and 300 additional tanks with fiscal year 1967 funds. This was later amended to specify limited production M60A1E2 tanks for the latter 300 vehicles. These were later standardized as the 152mm gun full tracked combat tank M60A2. Chrysler data show that 526 vehicles were retrofitted with the new turret at Detroit during 1973 to 1975. However, U.S. Army figures indicate a total production of 540 M60A2s.

The additional photograph of the M60A1E2 at the left reveals its basic M60A1 chassis with three friction snubbers per side on the suspension.

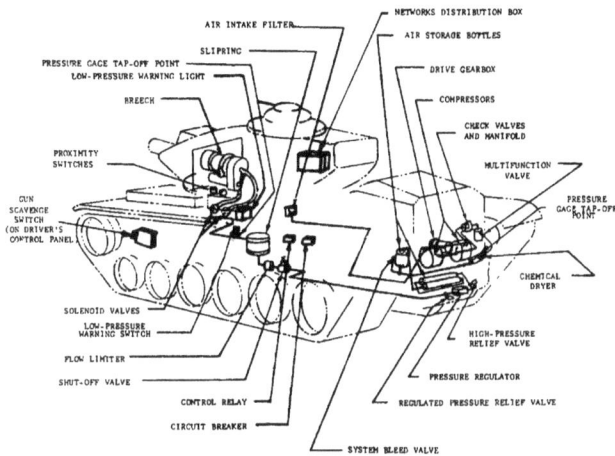

The sketch at the left above depicts the components of the closed breech scavenging system (CBSS). An M60A1E2 equipped with this system appears at the top right.

To insure the removal of any smoldering residue from the gun prior to the opening of the breech, a new closed breech scavenging system (CBSS) was developed. Including two gear driven compressors and two 3100 psi storage bottles installed in a bulge just below the louvered doors at the rear of the tank, this equipment provided three blasts of 1000 psi compressed air to clear any residue out of the gun bore. The first and third air blasts came from two nozzles located in the rear of the breech and the second was directed from two nozzles on the sides. Gun recoil automatically initiated the scavenging sequence prior to opening the breech. With the introduction of the CBSS, the bore evacuator was superfluous and it was eliminated from later production gun-launcher tubes.

The rear of the M60A1E2 below reveals the bulge in the lower hull below the exhaust grill doors required to house the CBSS. All of these photographs were taken during the 1971 test operations at Fort Knox.

The M60A2 above is armed with the late version of the 152mm gun-launcher without the bore evacuator. Another late M60A2 appears at the bottom of the page.

The primary fire control problems on the new tank were the limited night vision and the lack of a range finder. The narrow compact turret could not accommodate the standard range finder and some experiments were conducted using short base range finders mounted on the commander's cupola. However, three contracts were soon awarded for the development of a laser range finder which was to provide the ultimate solution to the problem. Among other new fire control components evaluated during the test program were an electronic azimuth indicator and two types of wind sensors.

The limited range of the night vision equipment also was an extremely serious problem as it restricted missile firing to daylight hours or to situations where the target could be illuminated by the white light xenon searchlight. The internal optics and image intensifier tubes in the periscopes at that time provided a passive recognition range of only 600 meters in ¼ moonlight. By using a "pink" filter on the xenon searchlight, the range could be extended to at least 1000 meters under all conditions of ambient illumination. This, of course, was inadequate for night missile firing, but it did permit the use of conventional ammunition in the dark at the shorter ranges. The use of these rounds was required in any case at close range, since the missile system was ineffective below approximately 1000 meters.

Consideration also was given to adapting the compact turret for the future installation of a long barreled gun-launcher capable of firing a kinetic energy armor piercing round in addition to its other ammunition. This weapon was the 152mm gun-launcher XM150 being developed for the future MBT.

The tank shown here represents the final version of the M60A2 fitted with the late model gun-launcher and the CBSS. Details of the turret and cupola are clearly visible in these photographs.

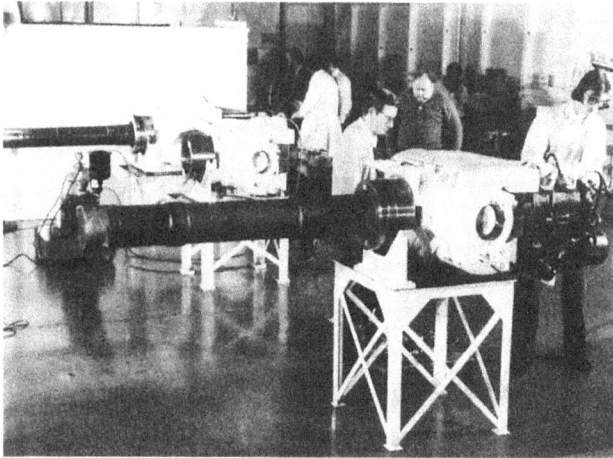

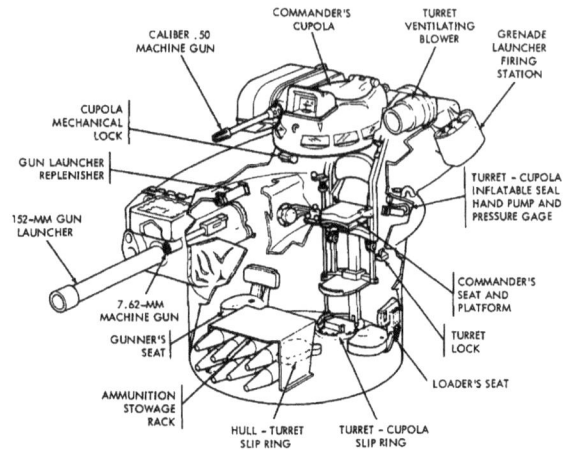

XM162 gun-launchers are being installed in the combination mount above at the left. The arrangement of the armament and other components in the M60A2 turret is illustrated in the sketch at the right above.

As standardized, the M60A2 was armed with the 152mm gun-launcher XM162E1. This weapon had a shallow keyway in the bore to improve tube life. It launched the latest version of the Shillelagh missile now designated as the MGM 51C. The earlier missiles with the large keys could not be fired from this launcher. Fifteen conventional rounds and seven missiles were stowed in the turret. An additional 18 conventional rounds and six missiles in the hull brought the total ammunition stowage to 33 conventional rounds and 13 missiles. All 13 missiles could be replaced by the smaller conventional rounds.

In addition to the gun-launcher, a coaxial 7.62mm machine gun was fitted on the left side of the combination mount. The M85 .50 caliber machine gun in the commander's cupola provided additional firepower for use against both ground and air targets. An hydraulic control system traversed the turret and elevated the main weapon as well as traversing the commander's cupola and elevating the .50 caliber machine gun. Both the main weapon and commander's cupola mount were stabilized in azimuth and elevation. However, they could be power operated without the stabilizer or manually operated.

The drawings below show the ammunition stowage (left) and the fire control instruments in the M60A2 turret (right).

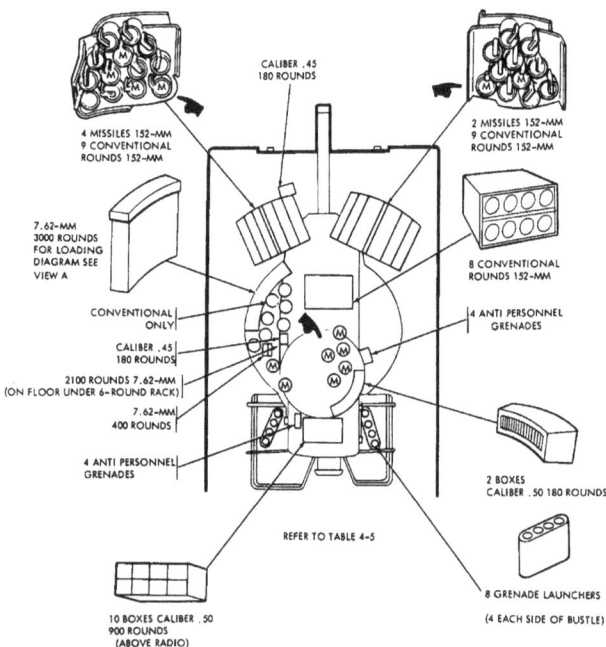

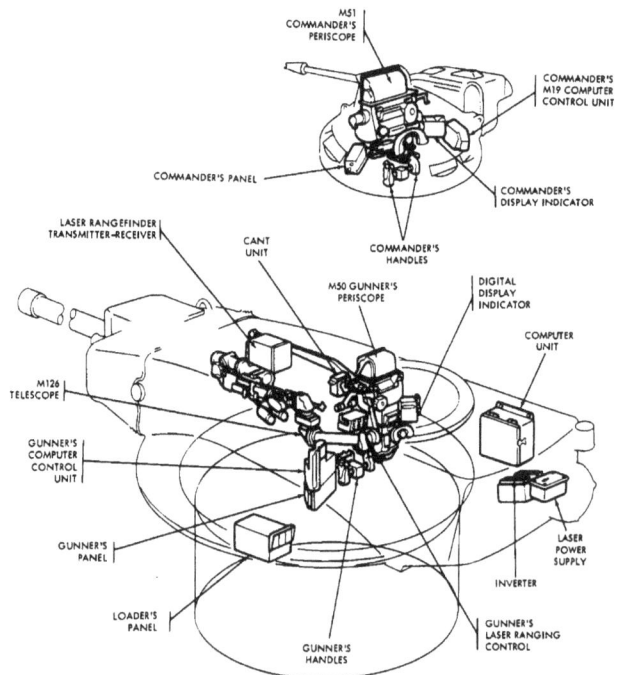

192

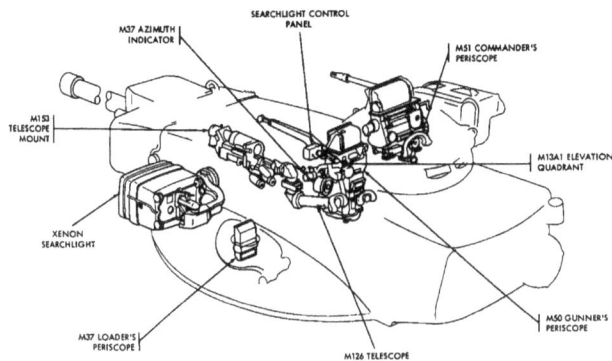

Another view of the fire control instruments in the M60A2 turret is sketched above. The radio installation appears at the right.

The commander's position in the compact turret provided excellent observation above the gun mount. Eleven vision blocks in the base of the cupola allowed an overlapping view in all directions. The XM51 periscope in the cupola top provided 10x magnification during daylight or 8x magnification under low light conditions. This periscope could be used by the tank commander to take over control from the gunner and to aim the main gun when firing conventional rounds. The gunner's XM50 periscopic sight was essentially the same as the commander's XM51. The gunner also was provided with an XM126 articulated telescope on the right side of the mount. Its primary function was to track the missile to the target. It also could be used as a backup sight for conventional ammunition. An M37 periscope pivot mounted in the loader's hatch provided observation on the left side of the tank.

The transmitter-receiver unit for the laser range finder was installed on the right side of the gun mount. When firing conventional ammunition, range data were transmitted to the XM19 ballistic computer which determined the necessary sight corrections to compensate for ammunition type, drift, gun jump, parallax, vehicle cant, and cross wind velocity. If required, it also could compute the proper lead for moving targets. These corrections were then transmitted to the XM50 or XM51 sights.

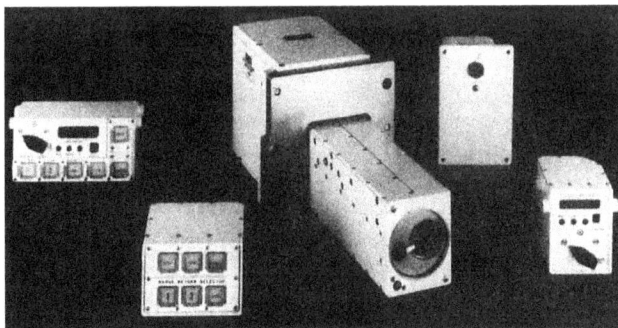

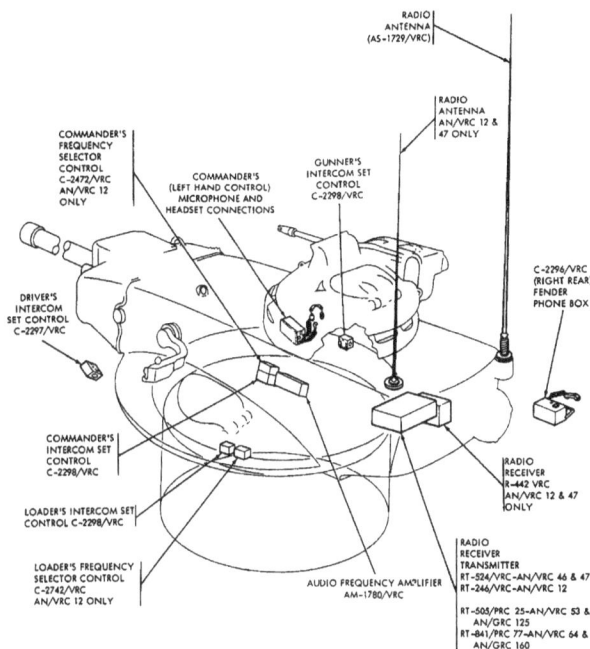

When firing the missile, the gunner established the line of sight by keeping the Shillelagh reticle in the XM126 telescope on the target. The infrared tracker, located in the right side of the gun mount, measured the deviation of the missile from the line of sight as it traveled toward the target. Correction commands were then sent to the missile by the infrared transmitter on the top center of the gun mount keeping the missile on the line of sight until it hit the target.

For indirect fire, the M60A2 was equipped with a mechanical XM37 azimuth indicator geared to the right side of the turret ring. Antipersonnel ammunition was developed for close range use with the gun-launcher. These were the XM617 and XM625 canister or flechette rounds. Also, eight XM176E1 grenade launchers were installed with four on each side of the turret. Each launcher carried two smoke grenades which were fired together.

Although eventually issued to the troops in 1974 and deployed to Europe the following year, the M60A2 was never a popular tank. Its complexity and the need for skilled maintenance and operation earned it the nickname "Starship". With the shift of interest away from missile armament to high performance kinetic energy weapons, the M60A2 was retired from service.

At the left is the AN/VVS-1 laser range finder which solved the range determination problem in the compact turret of the M60A2.

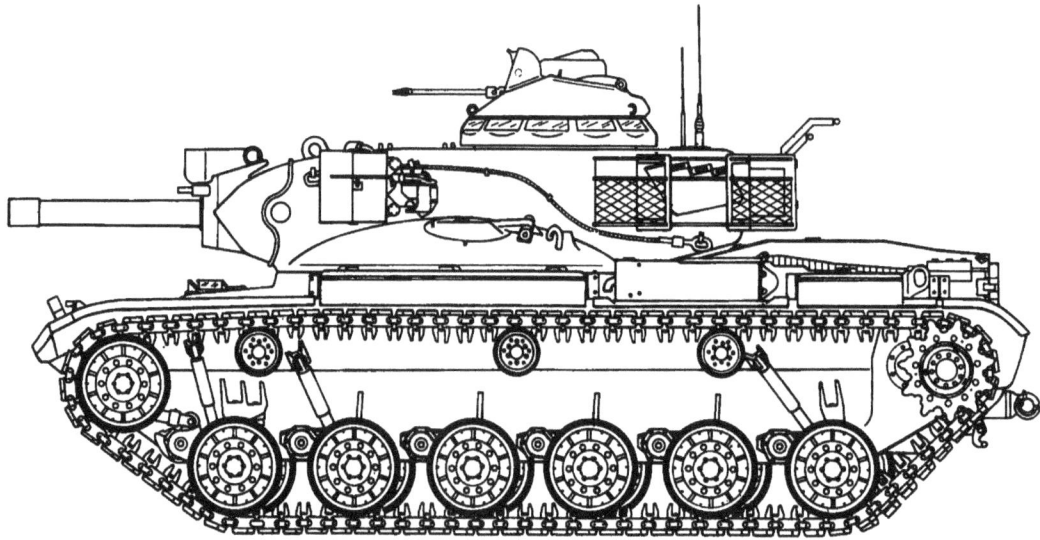

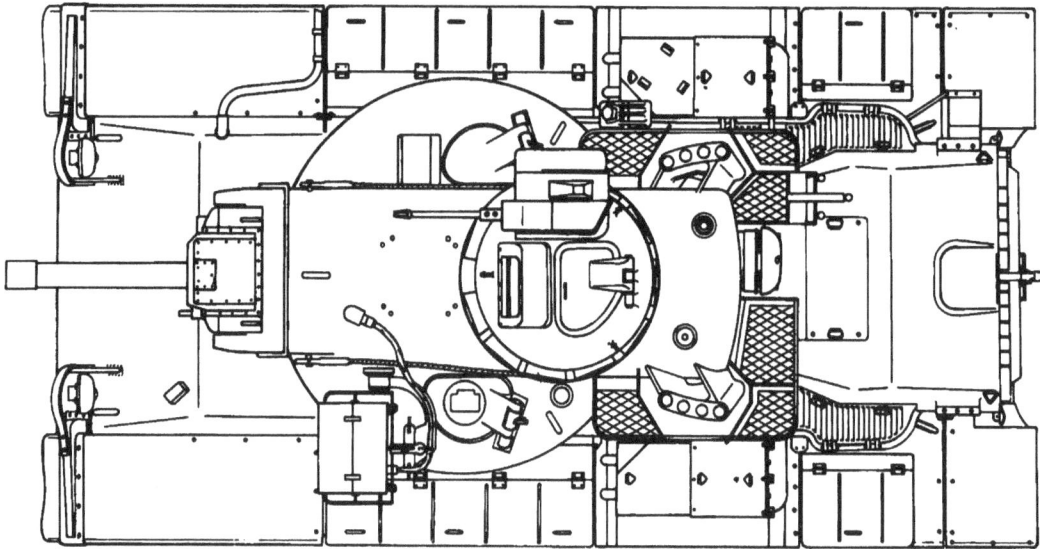

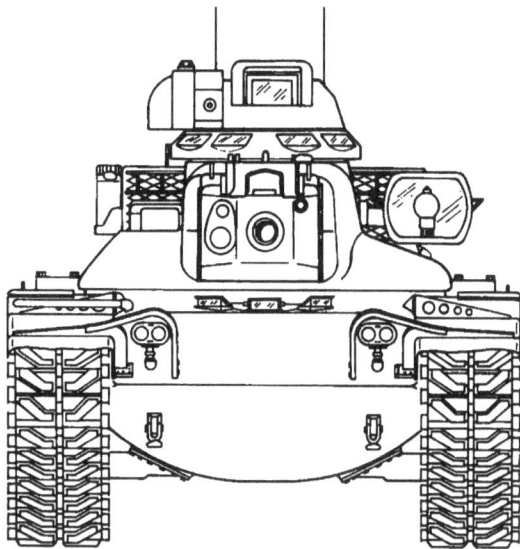

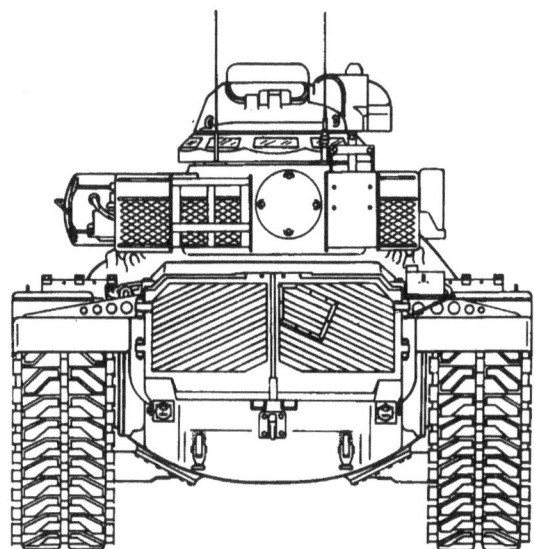

Scale 1:48

152mm Gun Tank M60A2

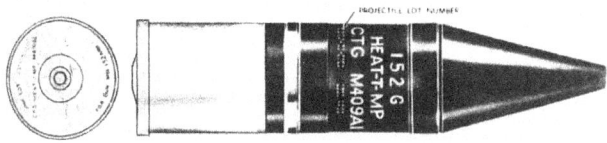

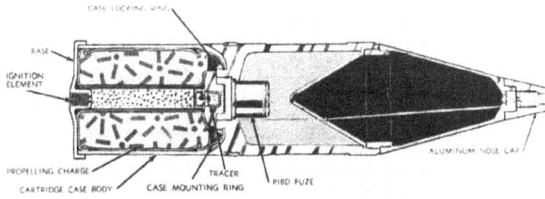

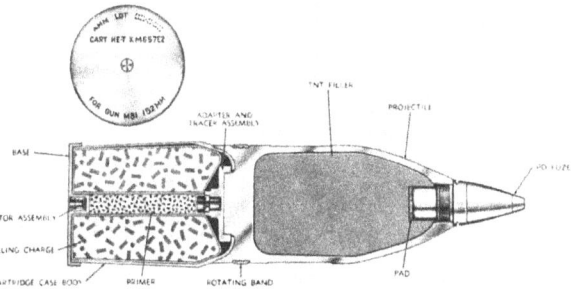

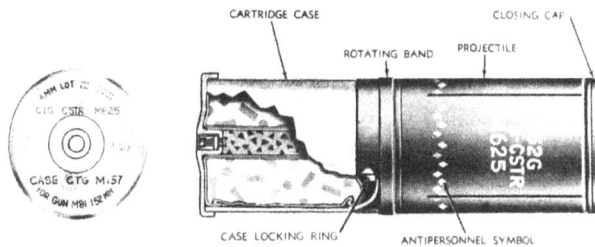

The combustible case conventional ammunition for the 152mm gun-launcher is shown here. At the top left is the M409A1 HEAT-T-MP round and the XM657E2 HE-T can be seen above. The M625 canister or flechette round is at the left. Compare the XM157 cartridge case on the XM657E2 with the later M157 case on the M409A1 and the M625.

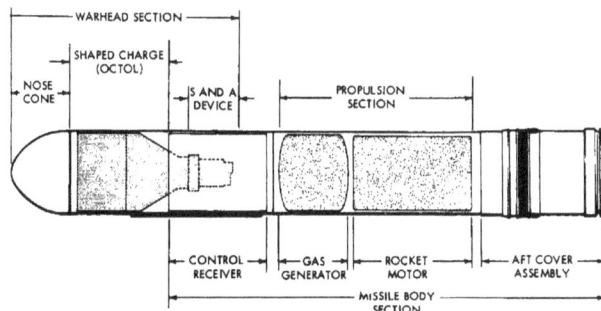

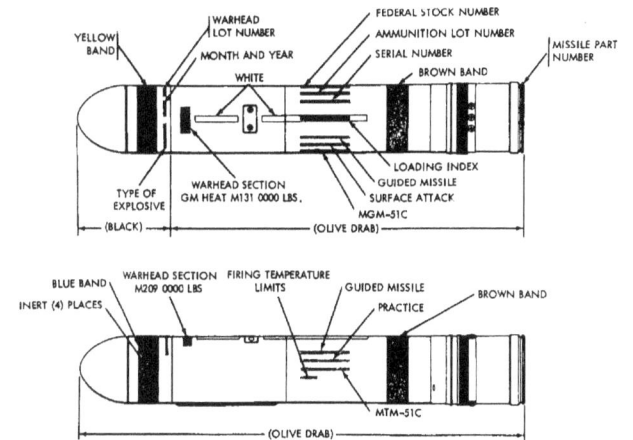

The layout of the various components in the MGM 51C Shillelagh missile appears in the drawing at the left above. Markings of the various missile types are shown at the right above. Below, a crew loads a Shillelagh into one of the M60A1E1 pilots (left) and the missile is launched at the White Sands Missile Range (right).

Above is Chrysler's proposed K tank with a wooden mock-up of the new turret installed on an M60A1 chassis. This turret also can be seen below at the left. In the same photograph is the mock-up of the semiautomatic loader. The operation of this loader is illustrated at the bottom right.

In late 1968, the development of the MBT70 under the joint project with the Federal Republic of Germany was in full swing and the M60A1E2 was still being tested and modified. On 29 November, Chrysler presented a third alternative as an unsolicited proposal to the Army for a product improvement of the M60 series. It involved the installation of a new turret and weapon system. To support this

proposal, a mock-up was constructed and fitted to an M60A1 chassis. Designated by Chrysler as the K tank, the new concept was armed with a modified version of the MBT70's 152mm gun-launcher XM150. As mentioned earlier, the XM150 was a long barreled gun-launcher with the capability of firing a kinetic energy armor piercing round (APDS) in addition to the projectiles already available. Fitting the standard 85 inch diameter ring, the new turret was designed to be fabricated from rolled armor plate

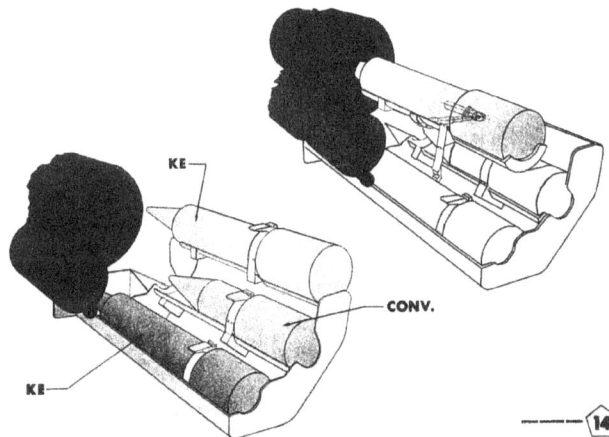

The views of the proposed K tank above clearly show the highly sloped sides of the new turret. Note the remote control .50 caliber machine gun on the commander's cupola.

with a highly sloped front, sides, and rear. The XM150 gun-launcher was mounted with a 7.62mm coaxial machine gun in the turret front and utilized the same fire control equipment as on the M60A1E2. The gunner and the loader were seated on the right and left sides of the cannon respectively. A three round semiautomatic loader was attached to the breech of the gun for use with conventional ammunition. Its location did not interfere with the loading of the missile. The semiautomatic loader would move any one of the three rounds it carried to the ram position. The human loader would then unclip and manually ram the round. He also kept the three round magazine loaded. This system was expected to increase the rate of fire while on the move. The

tank commander was located behind the cannon under a cupola fitted with a remote control .50 caliber machine gun. This machine gun was fed through the hollow pedestal mount and could be reloaded, aimed, and fired without opening the cupola. Although the main gun mount was fully stabilized, the commander's cupola was not, in order to simplify the system.

The proposed K turret offered improved protection for about an 1800 pound weight increase over the M60A1E2. The cupola, however, weighed less than that on the M60A1E2. Main armament ammunition stowage was increased from 46 to 57 rounds. This could include 22 missiles, 14 HEAT-MP, and 21 APDS.

The drawings at the left below show the crew location and the ammunition stowage in the K tank. The side views of the tank show the normal M60A1 suspension on the mock-up.

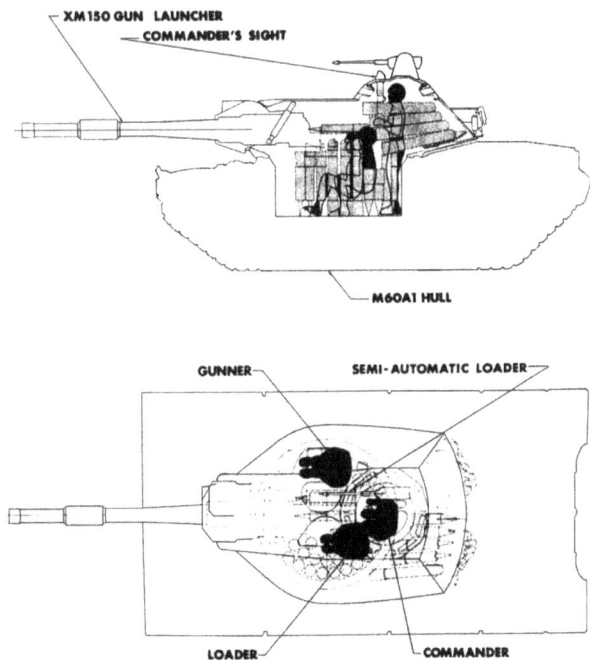

The K tank armed with the 120mm Delta gun is illustrated by the mock-up above. Note the length of this cannon compared to the 152mm gun-launcher. Below is a sketch of the tube-over-bar suspension proposed for the K tank. However, it was not installed on the mock-up.

A mock-up also was prepared replacing the gun-launcher with the 120mm Delta gun. This was a hypervelocity weapon developed from the earlier smooth bore 90mm T208 and 105mm T210 guns and it also fired a fin stabilized projectile. The 120mm gun installation added about 1000 pounds to the weight of the tank.

It also was proposed to improve the mobility of the vehicle by replacing the torsion bar suspension with the new tube-over-bar(TOB) design then being considered for the product improved version of the M60A1. With this arrangement, the road wheel arm was attached to a solid torsion bar that was enclosed in a tube to which it was attached at the opposite end. The tube was then anchored to the tank on the road wheel side. Thus, the effective length of the torsion spring was doubled, permitting about a 45 per cent increase in wheel travel and a lower spring rate. Rotary shock absorbers also were proposed for installation next to the hull. Such an arrangement allowed the heat generated in the shock absorbers to be dissipated in the steel hull which was much more

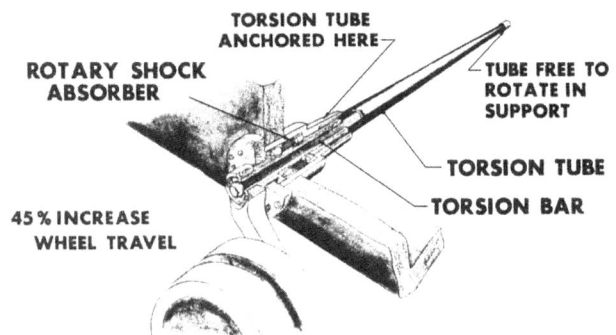

efficient than dissipation in air as with the standard shock absorbers.

Although the proposed K tank had considerable merit, all available resources at that time were being expended upon other projects and to finance the war in Vietnam. Under these conditions, it did not receive any support and further studies on this concept were dropped.

The 120mm Delta gun with combustible case ammunition is sketched below. The fin stabilized armor piercing round for this weapon is illustrated at the left.

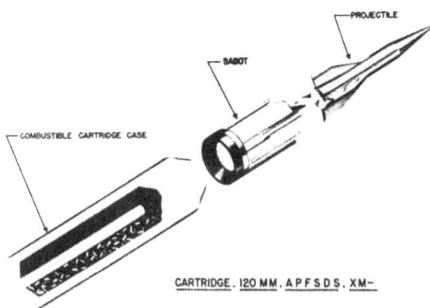

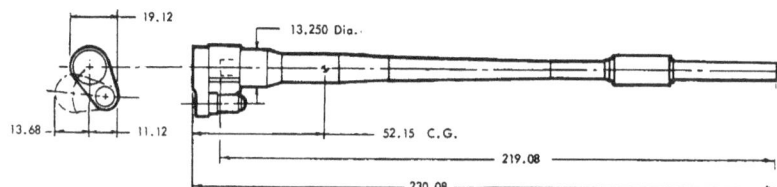

The new hydraulically powered commander's cupola is shown above before being shipped to Fort Knox. With the .50 caliber machine gun offset to one side, the weapon was easier to service, but it did exert a torque on the cupola when fired. It was not adopted for service use.

As development problems and rapidly escalating costs continued to push the date for fielding a new main battle tank further into the future, it became obvious that the M60A1 would have to serve far longer than originally intended. Like the Sherman of World War II, it became the subject of numerous product improvement programs that would greatly increase its effectiveness. The introduction of the tank into widespread use had revealed the usual need for numerous modifications to improve its performance and reliability.

One problem area was the commander's cupola. Although the M19 cupola was considered generally satisfactory, studies continued to develop a better commander's station. One such project at Chrysler produced a new cupola incorporating a ring of 11 vision blocks, each of which was larger than those in the M19. The M85 .50 caliber machine gun was offset to the right side of the cupola with easier access for operation and maintenance. The ammunition capacity was increased from 180 to 270 rounds. The new design had improved ballistic protection and hydraulic power was provided to traverse the cupola and to elevate the .50 caliber gun. Including

the machine gun, the cupola weighed about 4130 pounds. Installed on an M60A1 tank, the prototype cupola was shipped to Fort Knox for test on 27 March 1967.

The report of the Senior Officers Materiel Review Board(SOMRB) dated 22 December 1969 recommended an extensive product improvement program for the M60A1. Intended to upgrade the tank and keep it current with the state of the art, this program was approved by the Chief of Staff in January 1970. It was expected that many new components could be retrofitted to both the M60 and M60A1, but the major interest was in the latter and the upgraded vehicle was initially designated as the M60A1(PI) indicating product improvement.

The first of the new components appeared in 1971. This was the top loading air cleaner(TLAC) which reduced dirt and dust ingestion increasing the engine life. The top loading arrangement also made it easier to service. The TLAC was followed by add-on stabilization(AOS) which was released for production tanks in late 1972. This work had begun with several company sponsored programs which developed add-on stabilizers for the M48 series

The new top loading air cleaner is open at the left below with the filter element exposed. The various components of the add-on stabilization system are sketched below at the right as they were installed in the tank.

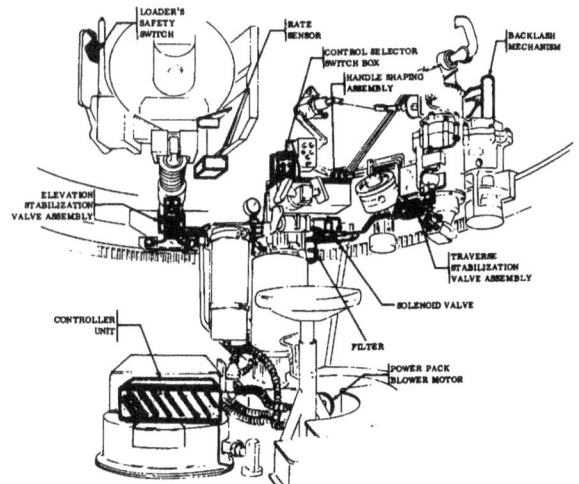

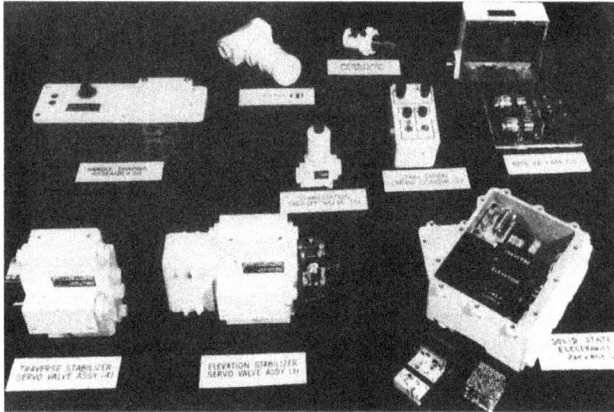

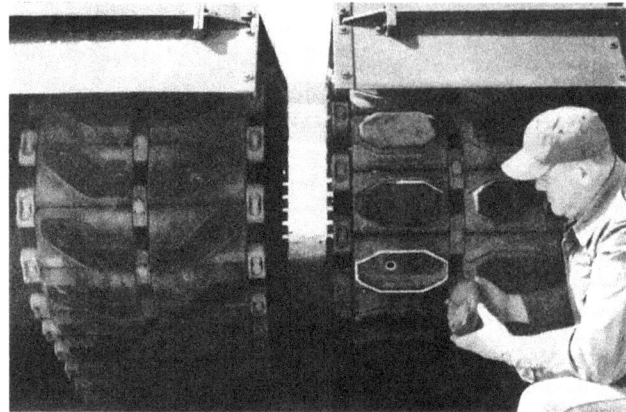

The component parts of the add-on stabilization system can be seen at the left above. The T142 track with the replaceable pads is compared at the top right with the earlier T97 track. The various parts of the new T142 track shoe are shown below.

tanks during 1962–1964. In 1964–1965, two of these companies modified their stabilizers for use on the German Leopard tank and later they were adapted for the M60A1. After competitive tests, one was selected for installation on the production tanks. The add-on stabilizer kit was designed to fit into the existing tank hydraulic gun control system with a minimum of modification. It consisted of the rate sensor package, a control selector box, the electronics package, a shut-off valve, the traverse servo-valve assembly, the elevation servo-valve assembly, the handle shaping assembly, an hydraulic filter, and an antibacklash cylinder. When installed in the tank, the equipment could be operated fully stabilized, non-stabilized with power, or in the manual mode. Test results from Aberdeen showed that hit probabilities from a moving M60A1 were better than 50 per cent at short to medium ranges. Without the stabilizer, the hit probability on the move was essentially zero.

Another feature applied to the product improved M60A1 was the new T142 steel track. This 28 inch wide double pin track had a greatly improved life and the rubber track pads were replaceable. It could be substituted directly for the T97 track. When equipped with the top loading air cleaner, the add-on stabilizer, and the T142 track, the tank was now designated as the M60A1(AOS). This essentially completed phase I of the program to upgrade the M60A1 tank.

The second phase of the improvement program covered longer range development items. These included a more reliable engine and electrical system, a new solid state computer, a laser range finder, and an improved suspension system. A program to increase the life of engine components showing the highest failure rate had resulted in an engine with

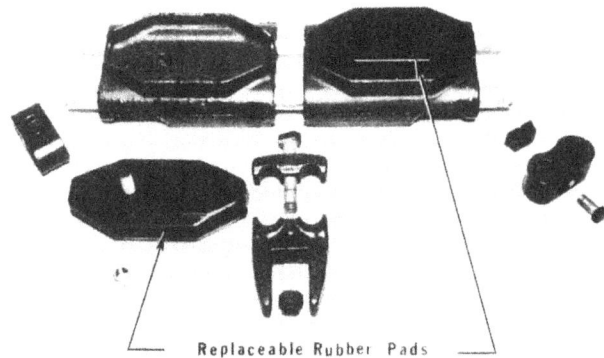

Replaceable Rubber Pads

greater reliability. Referred to as the reliability-improved-selected-equipment (RISE) engine, it averaged over 5000 miles in test operation before requiring replacement. The efficiency of the new top loading air cleaner as well as improved starters, fuel injector lines and nozzles, stronger cylinders, and better turbosuperchargers contributed to the increased service life of the new engine. The RISE engine with the 300 ampere generator and the original electrical system was designated as the AVDS-1790-2D. A new electrical system incorporated a 650 ampere oil-cooled alternator, a solid state regulator, and a new wiring harness providing more accessible quick disconnects. When this electrical system was

At the right are the 650 ampere oil-cooled alternator and the solid state regulator used on the AVDS-1790-2C engine.

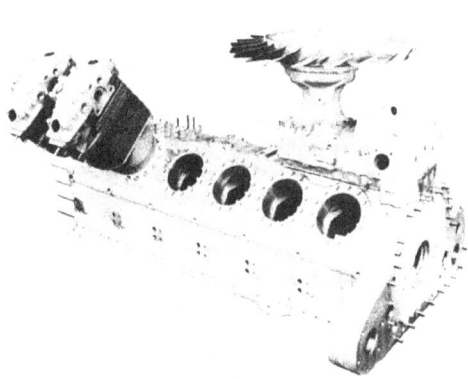

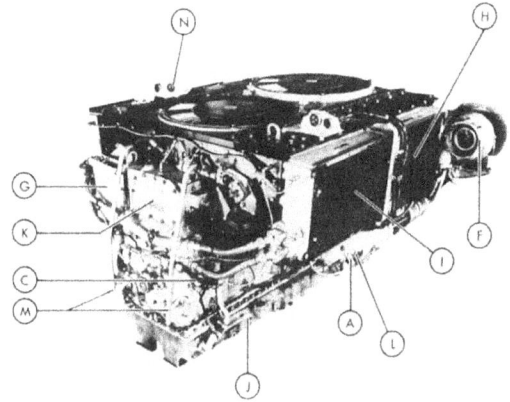

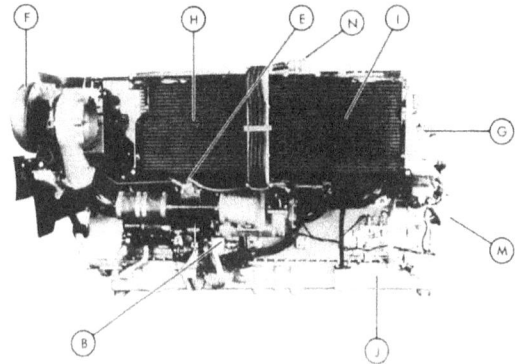

The basic components of the AVDS engine are shown at the left above during assembly. The attachment of the separate air-cooled cylinders to the aluminum block and the cooling fan arrangement can be clearly seen in these views. Details of the AVDS-1790-2C appear at the right above. The various components are as follows: A. Generator assembly and drive, B. Starter low voltage protection, C. Oil pressure switch and temperature transmitters, D. Exhaust manifold, E. Manifold flame heater system, F. Turbosupercharger, G. Fuel filter system and automatic water drain, H. Transmission oil cooler, I. Engine oil cooler, J. Oil pan and oil pump, K. Damper housing and oil filter system, L. Oil-cooled alternator, M. Front protective guard, N. Top-side electrical quick disconnects.

applied to the RISE engine, it was designated as the AVDS-1790-2C. With the installation of the new engine in the M60A1(AOS), it became the M60A1 (RISE) which was introduced in 1975.

AMCTC 9197, dated May 1972, outlined many of the proposed improvements for the M60A1. It also designated prototypes which included all of the features of the M60A1(RISE) plus the laser range finder, the solid state computer, and a tube-over-bar (TOB) suspension as the 105mm gun full tracked combat tank M60A1E3. The TOB, described earlier, was installed with its rotary shock absorbers only on the first two and last road wheel stations (numbers 1, 2, and 6) on each side. By March 1974, 12

The two views below show the M60A1E3 after assembly at Detroit. Note the thermal shroud on the main gun to minimize barrel distortion due temperature gradients.

These additional views of the M60A1E3 show the tube-over-bar suspension with its rotary shock absorbers. Note the lack of the usual tubular shock absorbers or snubbers. The wind sensor can be seen on the turret roof.

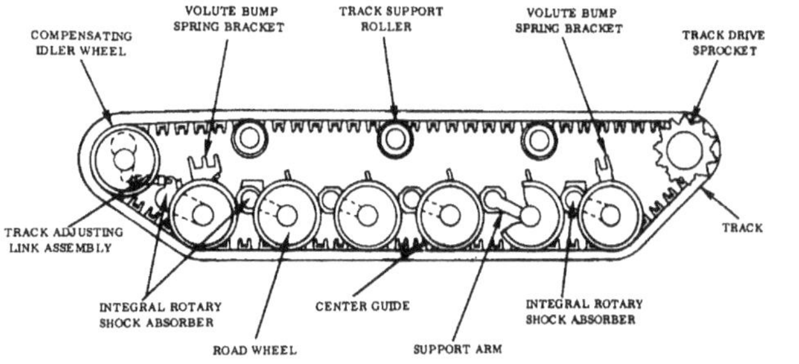

The arrangement of the M60A1E3 suspension is sketched above showing the rotary shock absorbers at road wheel stations 1, 2, and 6. The drawing below depicts the M60A1E3 turret control system incorporating stabilization. Compare with the turret control system for the M48A2 on page 114.

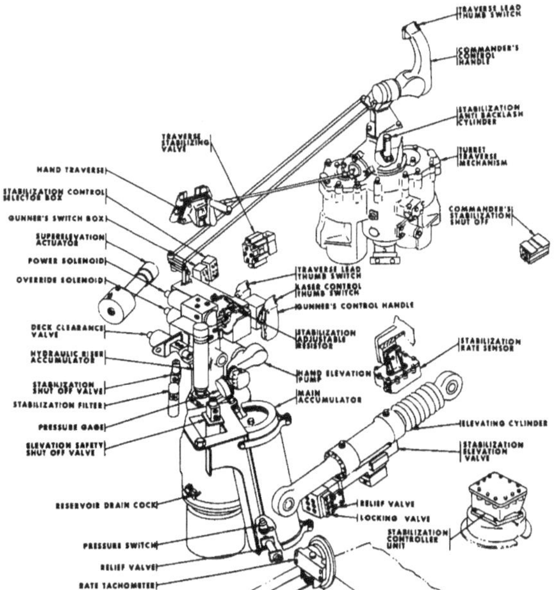

M60A1E3s had been assembled and the development tests already had begun in January with the contractor's tests being completed in April. Another item included on the test tanks was the new AN/VSS-3A searchlight. Smaller in size than earlier models, it produced both white and infrared light. The infrared was the so-called "pink" light very near the visible spectrum. New passive night sights were installed for both the gunner and the tank commander. The barrel of the 105mm gun M68 was enclosed in a thermal shield to minimize distortion resulting from uneven heating or cooling. Development tests of the M60A1E3 were carried out at the Yuma and Aberdeen Proving Grounds as well as at Fort Knox. As the tests proved the relative value of the new components, many were applied to the production tanks. In 1977, these included the improved passive periscopic sights for the tank commander and gunner, a new night vision device for the driver, and a new deep water fording kit. When these components were added to the M60A1(RISE), it became the M60A1(RISE)(PASSIVE).

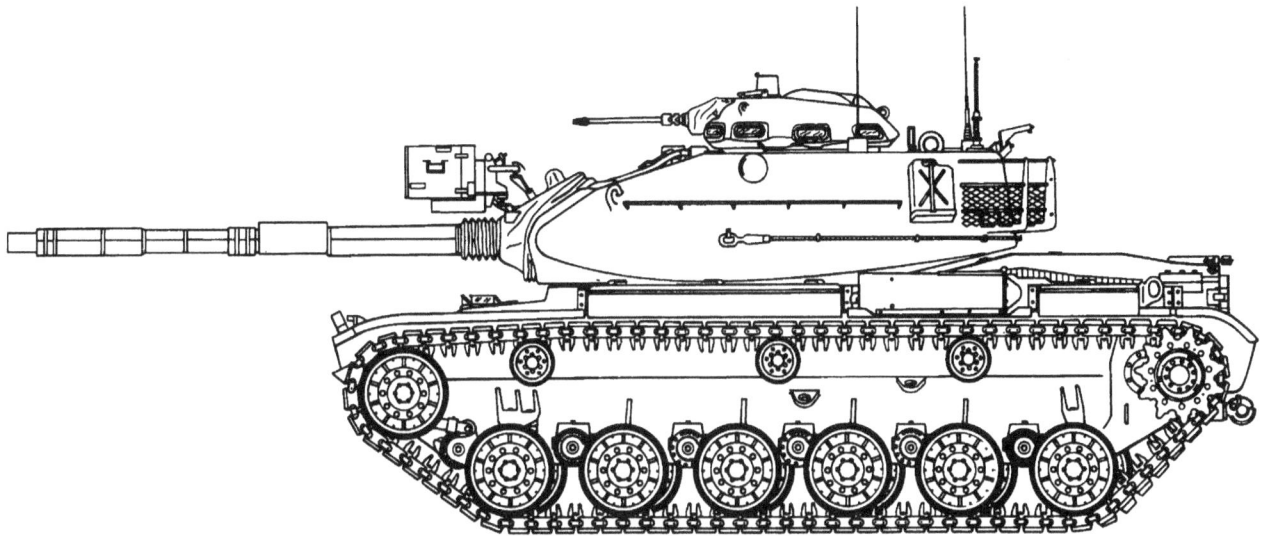

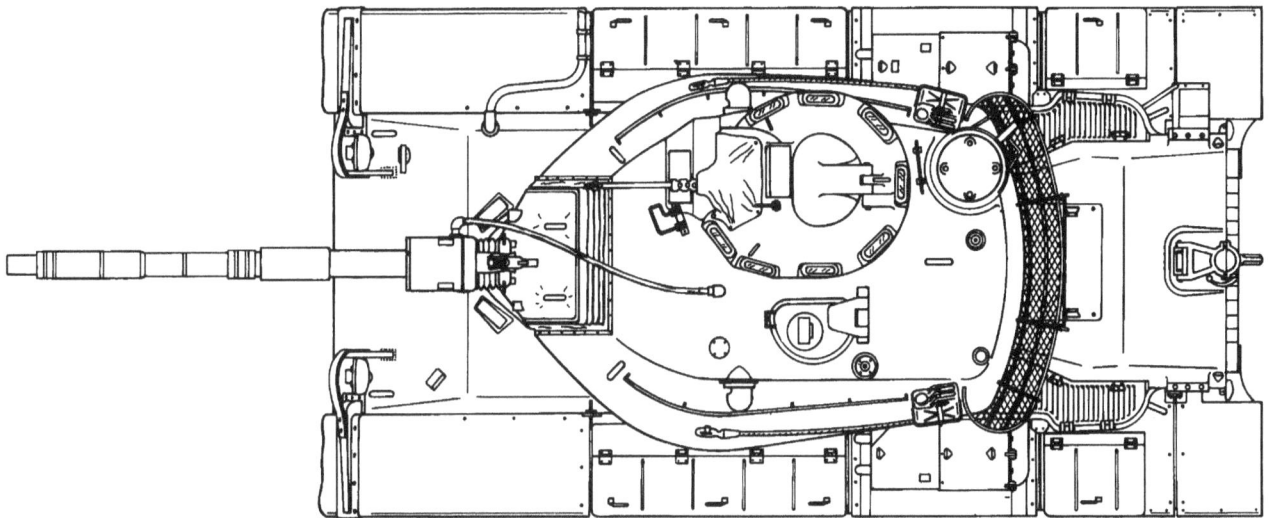

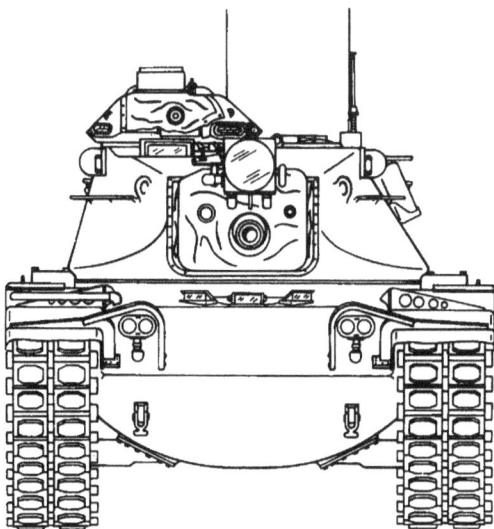

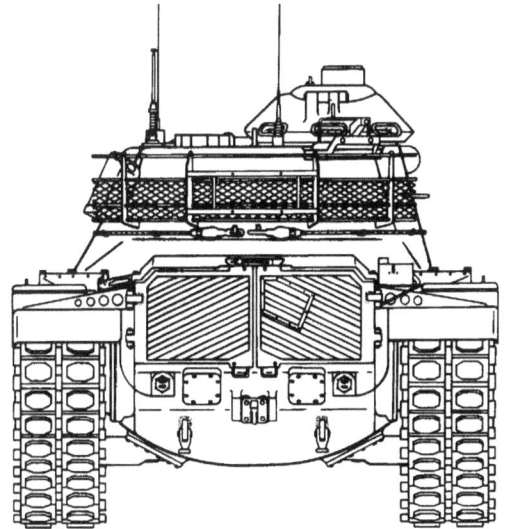

Scale 1:48

105mm Gun Tank M60A1E3

Labels (left photo): COMMANDER'S LASER CONTROL PANEL, GUNNER'S AMMUNITION SELECTOR BOX, GUN REPLENISHER, BALLISTICS DRIVE M10A3, DOME LIGHT, TELESCOPE M105D, GUNNER'S LASER CONTROL BOX, GUNNER'S CONTROL, PERISCOPE M35, GUNNER'S GUN CONTROL BOX, GUNNER'S STABILIZATION CONTROL BOX

Labels (right photo): LASER RANGEFINDER, CUPOLA POWER CONTROL, COMMANDER'S COMMUNICATION CONTROL BOXES, SEARCHLIGHT CONTROL BOX, COMMANDER'S CONTROL HANDLE, COMMANDER'S AMMO SELECT, COMPUTER, TURRET TRAVERSE GEAR BOX ASSEMBLY, GUNNER'S CONTROL UNIT, COMMANDER'S SEAT

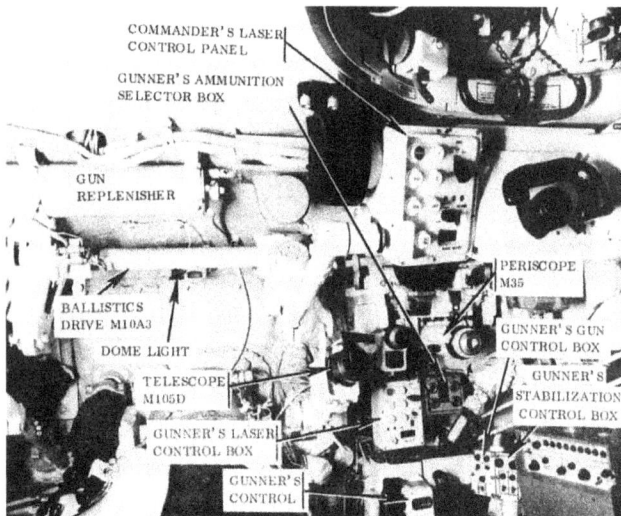

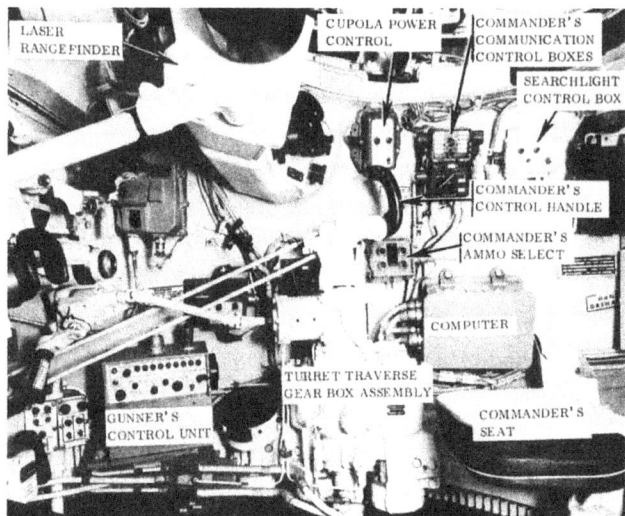

Above are the controls in the front (left) and right side (right) of the M60A1E3 turret. Note that the optical range finder has been removed and replaced on the right side by the AN/VVG-2 laser system. The arrangement of the M60A1E3 sighting and fire control equipment is sketched at the right below.

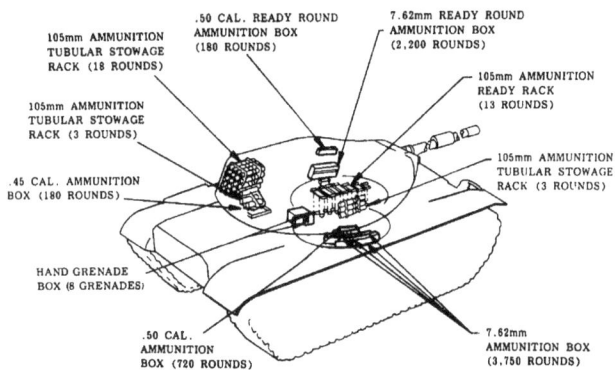

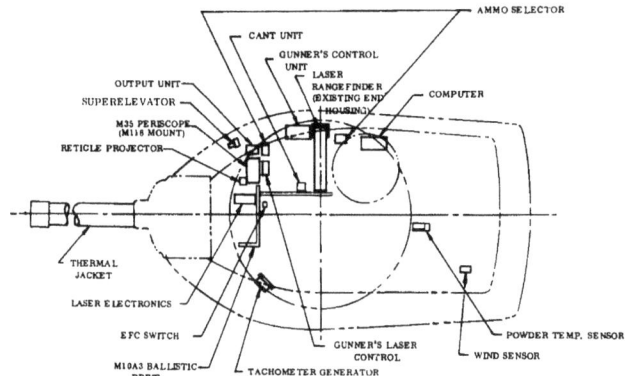

Labels (left diagram): 105mm AMMUNITION TUBULAR STOWAGE RACK (18 ROUNDS), .50 CAL. READY ROUND AMMUNITION BOX (180 ROUNDS), 7.62mm READY ROUND AMMUNITION BOX (2,200 ROUNDS), 105mm AMMUNITION READY RACK (13 ROUNDS), 105mm AMMUNITION TUBULAR STOWAGE RACK (3 ROUNDS), 105mm AMMUNITION TUBULAR STOWAGE RACK (3 ROUNDS), .45 CAL. AMMUNITION BOX (180 ROUNDS), HAND GRENADE BOX (8 GRENADES), .50 CAL. AMMUNITION BOX (720 ROUNDS), 7.62mm AMMUNITION BOX (3,750 ROUNDS)

Labels (right diagram): AMMO SELECTOR, CANT UNIT, GUNNER'S CONTROL UNIT, LASER RANGEFINDER (EXISTING END HOUSING), COMPUTER, OUTPUT UNIT, SUPERELEVATOR, M35 PERISCOPE (M118 MOUNT), RETICLE PROJECTOR, THERMAL JACKET, LASER ELECTRONICS, EFC SWITCH, M10A3 BALLISTIC DRIVE, TACHOMETER GENERATOR, GUNNER'S LASER CONTROL, POWDER TEMP. SENSOR, WIND SENSOR

The 105mm ammunition stowage in the M60A1E3 was the same as the late M60A1 tanks. The turret stowage is sketched above at the left. It differs from the early M60A1 arrangement by replacing the three tray holders in the turret bustle with a three round tubular rack. Compare with the drawing on page 173. Below, M60A1E3s are firing on the Cedar Creek tank range at Fort Knox during service tests on 30 December 1974.

An M60A1 fitted with the tube-over-bar suspension is shown in these photographs during the evaluation at Fort Knox. The absence of the tubular shock absorbers or snubbers is obvious and the rotary shock absorbers can be seen on the first two and last road wheels.

Tests on the M60A1E3 confirmed that the TOB suspension greatly improved the cross-country mobility of the tank. The doubled length of the spring element permitted a large increase in the wheel travel. From the static position to the bump stop, the travel distance was approximately 11 inches compared to 6.5 inches for the original torsion bar suspension. The greater road wheel travel permitted faster cross-country speeds with reduced shock loading of the hull. This in turn, improved the gun platform stability and reduced crew fatigue.

Although the TOB suspension provided a significant improvement in performance, it was not released for production. By this time, two new suspensions had appeared on the scene. One of these was the hydropneumatic suspension system (HSS),

Here, the hydropneumatic suspension system is installed on this M60A1 at Fort Knox. Once again, the absence of the tubular shock absorbers is readily apparent and the individual HSS units can be seen in the side view.

other versions of which had been evaluated on the T95 tank and the MBT70. Consisting of independent suspension units installed on the outside of the hull at each road wheel station, the HSS did not occupy any interior hull space. It provided a road wheel travel of about 13.5 inches from the static position to the bump stop. The HSS units also combined the spring and shock absorber functions in a single device.

The second new suspension system tested along with the HSS was the advanced torsion bar (ATB). This consisted of new single torsion bars fabricated from high strength H-11 electroslag refined steel (ESR). With a spring rate roughly halfway between standard torsion bars and the TOB system, the ATB permitted a wheel travel of slightly over 12 inches from the static position to the bump stop. The splines in the new torsion bars were compatible with the original design to permit the retrofit of existing M60 series tanks with only minor modification during overhaul. The new torsion bars also were fitted with aluminum protective covers

The advanced torsion bar suspension is fitted to the M60A1 above at Fort Knox. Rotary shock absorbers are installed for road wheels 1, 2, and 6. Note that this tank is fitted with the low silhouette commander's cupola intended for the new M1 tank.

and rubber seals to exclude water or other corrosive materials. The ATB system utilized rotary shock absorbers at the first, second, and sixth road wheel stations.

Starting in October 1976, tests at Aberdeen and Fort Knox compared the performance of the hydropneumatic and advanced torsion bar systems with that on the standard M60A1. Little difference was noted between the two new suspensions in stabilized gun performance while firing on the move and both were much superior to the original system. However, the non-linear load-deflection characteristics of the HSS did provide a somewhat smoother

ride permitting a slightly higher cross-country speed than with the ATB.

Later a hybrid vehicle was tested by replacing the HSS units at the third, fourth, and fifth road wheel stations with advanced torsion bars. Retaining the HSS units at stations 1, 2, and 6 eliminated the requirement for rotary shock absorbers which were the most expensive components of the full ATB system. Testing of the hybrid suspension was initiated at Fort Knox in February 1977 in competition with the full hydropneumatic and advanced torsion bar systems. However, none of these were applied to production tanks.

At the right is a close-up view of the advanced torsion bar suspension on the M60A1. The rotary shock absorber housing for road wheel number 2 is visible here. The configuration of the ATB suspension is sketched below.

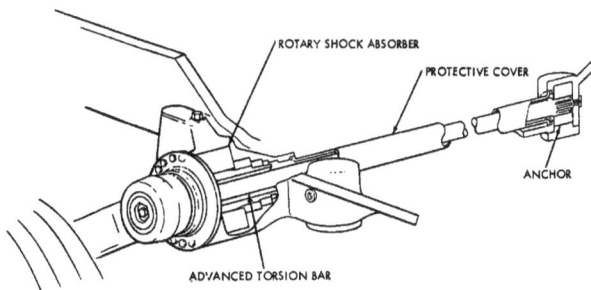

POWER SUPPLY SYNCHRONIZER

RECEIVER/ TRANSMITTER

COMPUTER UNIT

GUNNER'S CONTROL UNIT

AMMUNITION SELECTORS

CROSSWIND SENSOR

RATE TACHOMETER

CANT UNIT

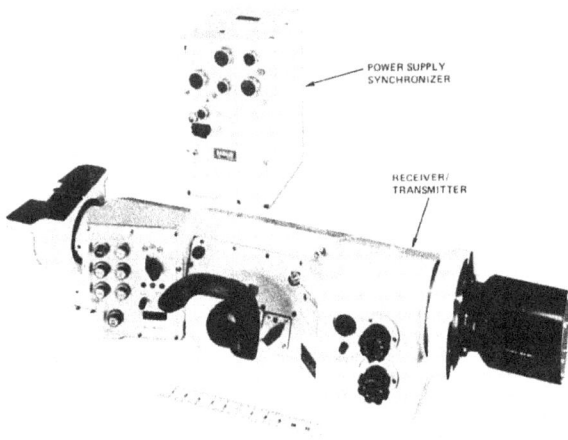

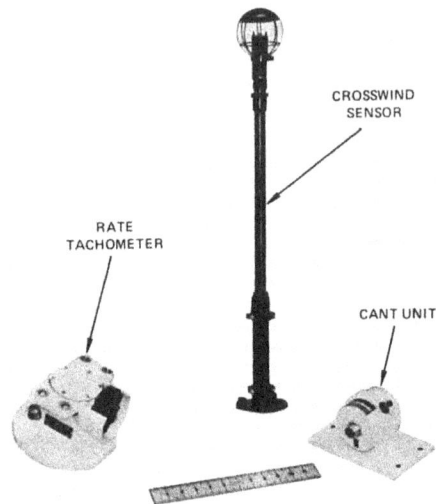

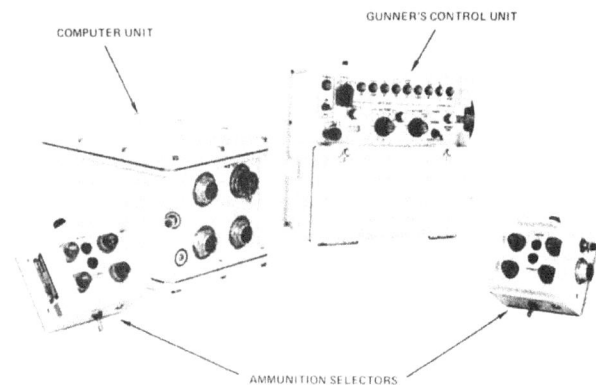

Some of the new fire control components in the M60A1E3 are illustrated here. At the top left is the AN/VVG-2 laser range finder and its power supply. Above are the sensors which supplied data to the XM21 solid state ballistic computer seen at the left.

The next major items from the M60A1E3 program to be released for production were the ruby laser range finder and the solid state ballistic computer. The AN/VVG-2 range finder was installed in the upper right side of the turret using the right blister previously occupied by the optical range finder. With the new XM21 computer, it was a major element of the primary fire control system. The dual power (6x–12x) telescope in the range finder served as the tank commander's primary sight for directing the fire of the main gun. Operating from 200 to 5000 meters, the range finder had a specified accuracy of plus or minus 10 meters and the capability of resolving two targets as close together as 20 meters. Range data were fed into the ballistic computer which provided aiming information compensating for drift, cross wind, horizontal target motion, altitude, gun tube wear, cant, sight parallax, and gun jump. Wind conditions were detected by a sensor mounted on a flexible support on the left rear of the turret roof.

The various new features standardized in the M60A3 are shown on the sketch at the right indicating the areas of performance to which they applied.

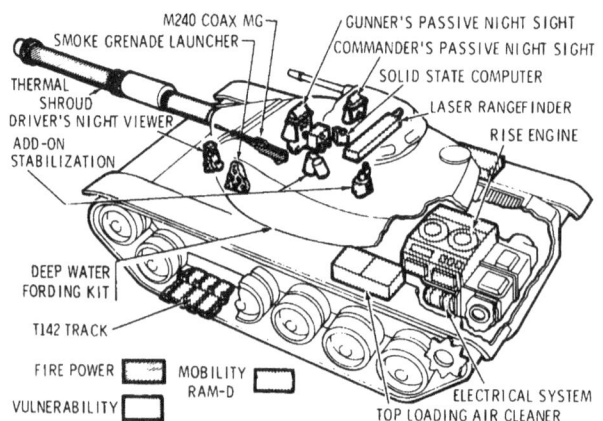

The new fire control system was released for production tanks in 1978 along with the British designed M239 smoke grenade launcher. A fully satisfactory coaxial machine gun finally appeared when the new 7.62mm M240 replaced the M219. The M240, which was the U.S. designation for the Belgian MAG-58, was considered to be at least five times as reliable as the earlier machine gun. The significant improvements resulting from all of these modifications required a new designation for the vehicle and on 10 May 1979, it was standardized as the 105mm gun full tracked combat tank M60A3. At the same time, it was approved for full production. This, of course, was somewhat after the fact and on 26 May, the 1st Battalion, 32nd Armor, 3rd Armored Division in Europe received 54 of the new tanks.

M240 COAX MG
SMOKE GRENADE LAUNCHER
THERMAL SHROUD
DRIVER'S NIGHT VIEWER
ADD-ON STABILIZATION
GUNNER'S PASSIVE NIGHT SIGHT
COMMANDER'S PASSIVE NIGHT SIGHT
SOLID STATE COMPUTER
LASER RANGEFINDER
RISE ENGINE
DEEP WATER FORDING KIT
T142 TRACK
FIRE POWER
VULNERABILITY
MOBILITY RAM-D
ELECTRICAL SYSTEM
TOP LOADING AIR CLEANER

208

Diagram labels:
TRIGGER SAFETY
FIRE POSITION
COVER ASSEMBLY
CHARGER MOUNTING STUD
CHARGER HANDLE
CHARGER ASSEMBLY
FEED TRAY GROUP
JACKET ASSEMBLY WITH BEARING GROUP
CAUTION: MAKE SURE PARTS ARE ASSEMBLED FOR LEFT HAND FEED AND THE HAND CHARGER ASSEMBLY IS INSTALLED ON THE LEFT SIDE.

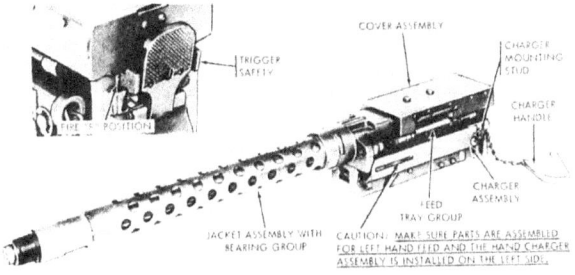

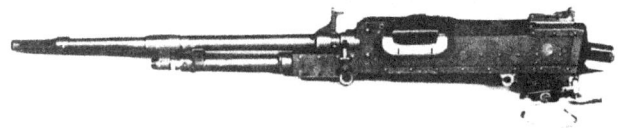

The ill-fated 7.62mm M219 machine gun at the left is compared with its replacement, the M240 above. The M219 was a modified version of the earlier M73.

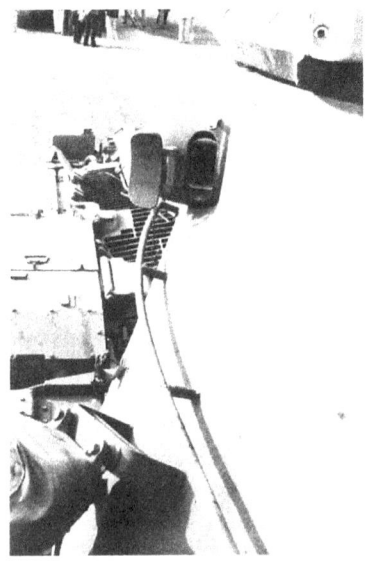

At the left and below are views of the production M60A3. The tubular shock absorbers of the standard M60A1 suspension can be seen in these photographs as well as the new M239 smoke grenade launchers on each side of the turret. Above, is a close-up showing the small door covering the port for the laser range finder.

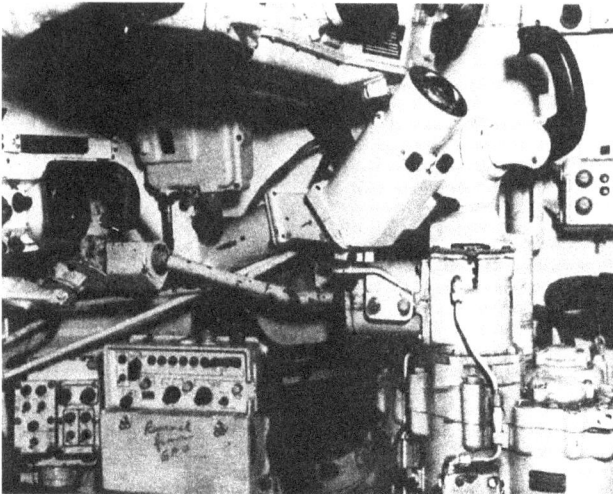

The experimental version of the Texas Instruments tank thermal sight is installed at the left above in an M60A1 during tests at Aberdeen. Designated as the AN/VSG-2(XE-2), it provided the basis for the production sight. At the right above is a view of another tank as seen through the thermal sight.

A new sighting device for the gunner was standardized in June 1978. Designated as the tank thermal sight(TTS), it replaced the gunner's M35E1 passive image intensification sight. The new device was independent of ambient light and utilized heat emitted by the target to form an image on a screen. More than twice as effective as the passive sight it replaced, the TTS could also sense targets through smoke, fog, or rain. The thermal sight consisted of a head assembly, the gunner's viewer, and a light pipe which enabled the tank commander to view the image on the gunner's screen. The new sight was installed in selected tanks starting in August 1979. These vehicles were then designated as the M60A3(TTS).

Another new feature applied to the M60A3 was the vehicle engine exhaust smoke system(VEESS).

Developed by Teledyne-Continental Motors, it consisted of an arrangement to inject diesel fuel into the exhaust system. Vaporized and expelled, the fuel mixed with the outside air and condensed to form a thick white cloud. During tests, the smoke cloud effectively screened the tank from visual observation and detection by infrared devices.

In addition to the new components from the development programs, numerous other improvements were applied to the production tanks. In January 1977, an armored top loading air cleaner(ATLAC) was released. This modification reduced the vulnerability of the air cleaners by replacing the aluminum alloy housing with one of steel. It also eliminated problems with broken cover hinges on the aluminum model. To reduce the hinge breakage on the tanks in the field, a kit was prepared which

Below, the M239 smoke grenade launcher can be seen at the left installed on the turret. The grenade storage box is on the upper outside of the turret wall behind the range finder blister. At the right below is a schematic drawing of the vehicle engine exhaust smoke system.

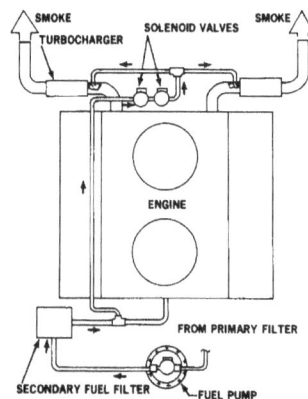

EARLY M60A1 MODELS   LATER M60A1 MODELS

provided a guard to protect the cover hinges. Other changes included the replacement of the aluminum track return rollers with ones of steel. This resulted in a weight increase of about 40 pounds per tank, but reduced procurement and maintenance costs. Beginning in May 1980, steel road wheels replaced the aluminum models on the new production tanks. Unlike their aluminum alloy counterparts, the steel road wheels and track support rollers did not require wear plates and they were less expensive. Steel wheels also were purchased as spares for future use on tanks in the field.

An improved driver's escape hatch was installed on the later production tanks with a better seal and locking device. Also, to provide sufficient power for the stabilized turret, the 5 horsepower turret motor was replaced by a 10 horsepower model on all new vehicles.

Since the start of production in 1962, a total of 7948 M60A1 tanks of various models were produced. These included 578 for the U.S. Marine Corps and 874 for foreign military sales. The last M60A1 was completed in 1980 and it was replaced on the line by the M60A3 which had started production during the 1978 fiscal year. In addition, a program was initiated to upgrade large numbers of M60A1s to the M60A3 standard.

The plan for product improvement of the M60A1 as outlined in AMCTC 9197 in May 1972 proposed replacing the tank's engine and transmission. Later studies at Detroit Arsenal considered the application of several new power packs to the vehicle. The engines included higher powered versions of the AVDS-1790 such as the AVDS-1790-5A and the AVDS-1790-7A. At 2400 rpm, the gross output of these engines was 908 horsepower and 950 horsepower respectively. Also considered was the new

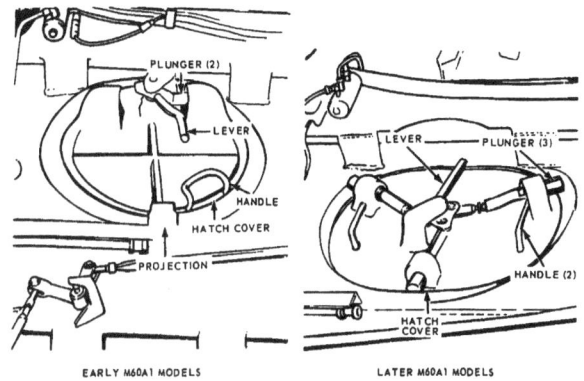

AVCR-1790-1 which developed a maximum of 1200 gross horsepower at 2400 rpm. This engine retained the same displacement as the original AVDS-1790, but it was fitted with variable compression ratio pistons, hence the new designation. These pistons had an outer shell which was extended by hydraulic pressure to increase the compression ratio. The high compression necessary for good starting decreased as the firing pressure increased under load. This maintained a controlled firing pressure within the structural design limits over a wide load range. On shutdown, the piston shell automatically returned to the high compression ratio position necessary for the next start. The VCR design allowed a very high horsepower output per cubic inch of displacement. It did, of course, increase the fuel consumption. At 2400 rpm, the AVCR-1790-1 consumed 504 pounds of fuel per hour compared to 296 pounds per hour for the AVDS-1790-2 under the same conditions.

The higher powered engines also required a new transmission, since the CD-850 was close to the top of its performance range with the 750 horsepower AVDS-1790. Other transmissions considered included the X-700, the X-1100, and the Renk. The first two were built by Allison and the latter was a German design under license to Teledyne-Continental Motors.

Below, the AVCR-1790-1 diesel engine appears at the left with the Renk transmission. At the right is a sectional drawing illustrating the double shell construction of the variable compression ratio piston.

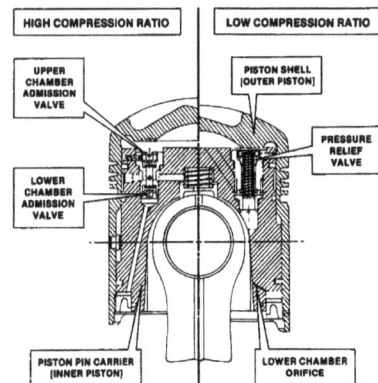

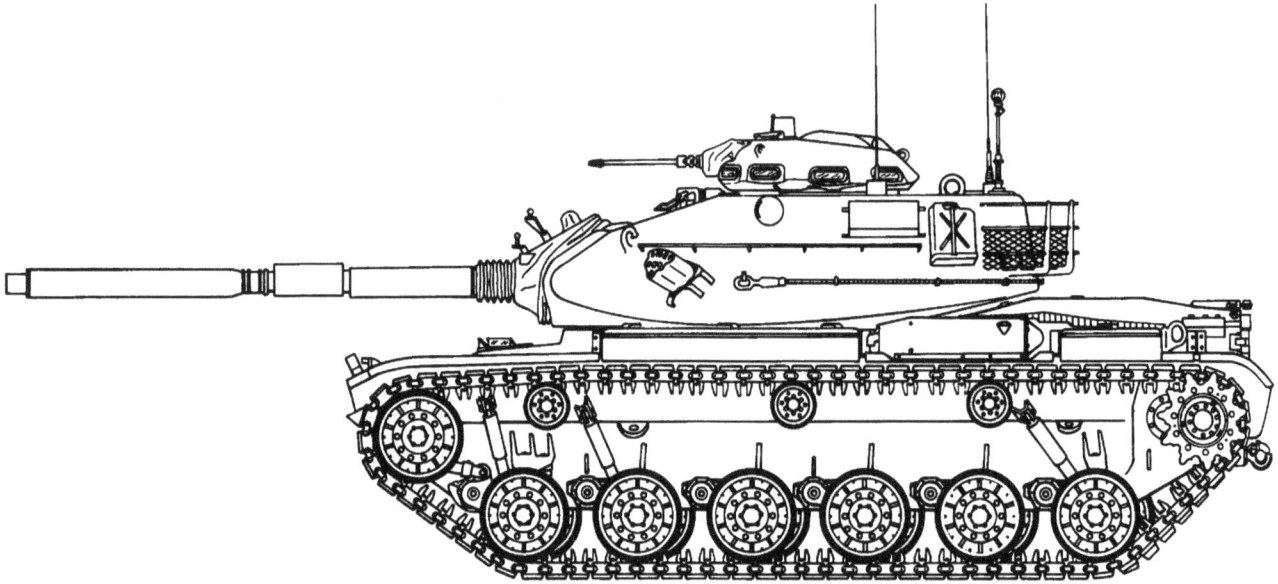

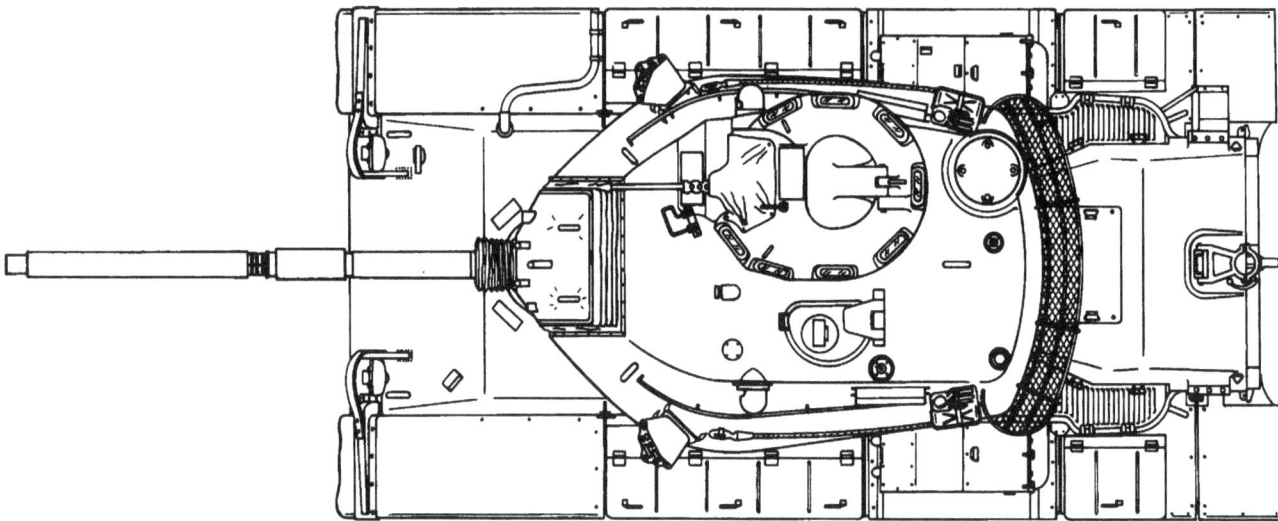

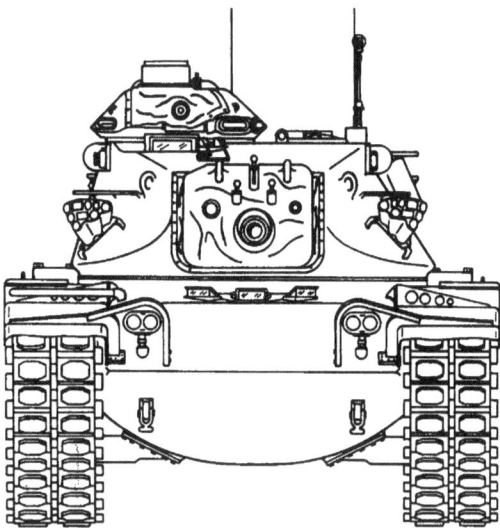

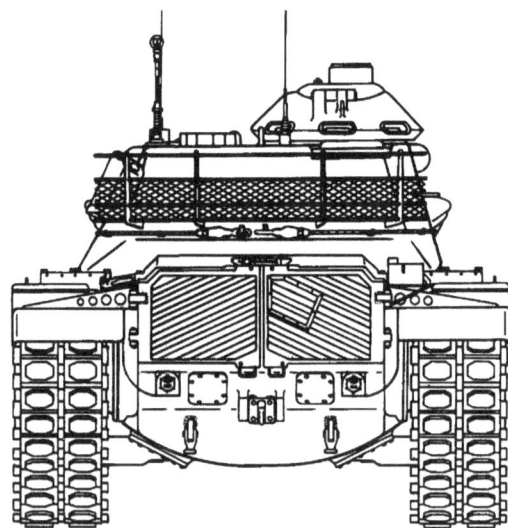

Scale 1:48

105mm Gun Tank M60A3TTS

These views of an M60A3 from the 194th Armored Brigade at Fort Knox show the details of the thermal shroud on the 105mm gun. The .50 caliber machine gun is not mounted in the cupola and the protective shields are closed over the periscopes for the gunner and the tank commander.

On the M60A3 above, the protective shields are open exposing the periscopes for the gunner and tank commander. Neither the .50 caliber machine gun nor the xenon searchlight is installed. The deflector to protect the latter from the machine gun can be seen folded forward alongside the gunner's periscope. Below, the M60A1 type suspension retained on the M60A3 is clearly visible with the three shock absorbers on each side. These photographs were taken by Major Fred Crismon in February 1984.

M60 AN TURRET
LK 10602

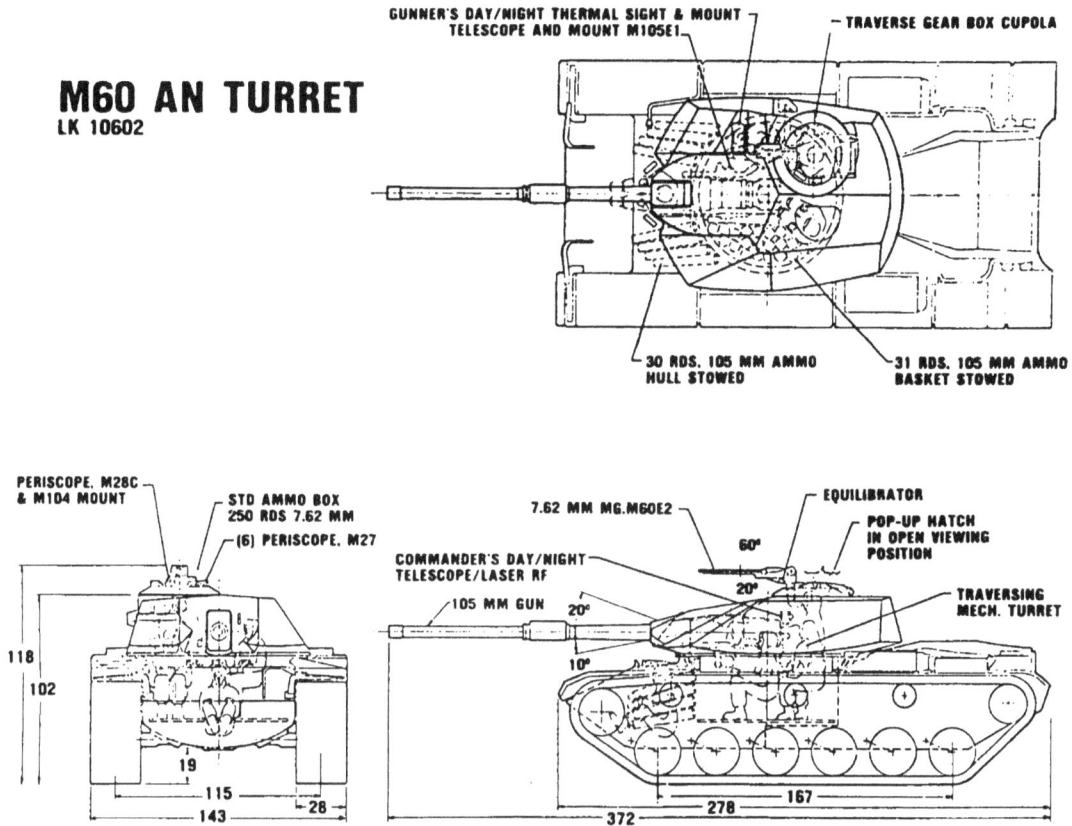

GUNNER'S DAY/NIGHT THERMAL SIGHT & MOUNT
TELESCOPE AND MOUNT M105E1

— TRAVERSE GEAR BOX CUPOLA

30 RDS. 105 MM AMMO HULL STOWED

31 RDS. 105 MM AMMO BASKET STOWED

PERISCOPE. M28C & M104 MOUNT

STD AMMO BOX 250 RDS 7.62 MM

(6) PERISCOPE. M27

7.62 MM MG.M60E2

EQUILIBRATOR

POP-UP HATCH IN OPEN VIEWING POSITION

COMMANDER'S DAY/NIGHT TELESCOPE/LASER RF

TRAVERSING MECH. TURRET

105 MM GUN

118
102
19
115
143
28
60°
20°
20°
10°
372
278
167

The Tank Automotive Command design study above included an improved turret with special armor for the M60 series tanks. It also incorporated such features as a hydropneumatic suspension and a more powerful engine. The muzzle reference system proposed for the M60A3 is sketched at the right below.

The studies at Detroit concluded that several types of engines and transmissions would be suitable for the M60A1, but preference was given to the combination specified for the new M1 tank. This consisted of the Avco-Lycoming AGT-1500 gas turbine and the Allison X-1100 transmission. Although the fuel consumption was higher, the development potential was considered to be greater and such a selection would simplify the logistics by providing a common power train for the M60 series and the M1.

Although the proposed modifications greatly increased the effectiveness of the M60A1-M60A3, the new M1 tank now entered the picture. It was obvious that an extensive product improvement program for the earlier tanks would be in direct competition for funds with the M1 itself. In April 1977, the Army Vice Chief of Staff directed the formation of a task force to develop a program for the best available tank fleet balanced between the product improved M60 series and the new M1. The Army Staff, the Army Materiel Development and Readiness Command, the Training and Doctrine Command, as well as other major commands all contributed to this task force.

Considering the expected availability of funds, the task force concluded that the most effective combination for use into the 1990s would be a force consisting of at least 7000 M1s and 3600 M60A3s. Based on the combat effectiveness of such a fleet, the only additional improvements recommended for the M60A3 were the tank thermal sight, a muzzle reference system, a Halon fire extinguisher system, and hardware to permit the installation of devices such as a chemical alarm. The recommendations of the task force were reviewed by the Army System Acquisition Review Committee during the late Summer of 1977 and subsequently approved by the Chief of Staff.

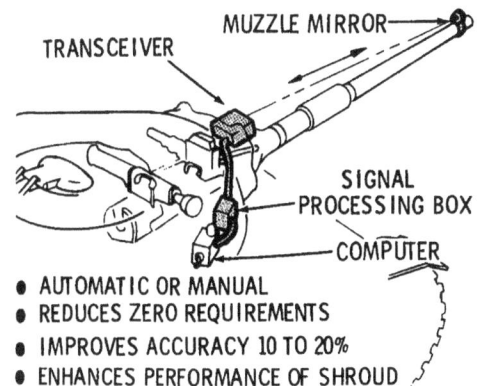

MUZZLE MIRROR

TRANSCEIVER

SIGNAL PROCESSING BOX

COMPUTER

● AUTOMATIC OR MANUAL
● REDUCES ZERO REQUIREMENTS
● IMPROVES ACCURACY 10 TO 20%
● ENHANCES PERFORMANCE OF SHROUD

The Teledyne-Continental Motors M60AX appears in the photographs above and below showing the special armor added to the hull and turret.

Although the Army did not pursue the installation of a new power train and suspension system in the M60 series, a private venture by Teledyne-Continental Motors featured the AVCR-1790-1B engine with the Renk RK-304 transmission. Unofficially dubbed the M60AX, it also was equipped with the National Waterlift hydropneumatic suspension system (HSS). After initial tests, additional modifications were applied which included extra armor particularly intended to increase the protection against shaped charge projectiles. The M19 cupola was replaced by a low silhouette type with a pop-up hatch for the tank commander. Although the modifications increased the weight of the tank by about 9500 pounds, the new power train and suspension system provided greatly increased performance. The top speed on level ground increased from 30 to 45 miles/hour. The HSS permitted cross-country speeds of 24 miles/hour compared to 9 miles/hour for the M60A1 under similar conditions. The cruising range was approximately 280 miles.

The M60AX prototype illustrated the potential for upgrading the M60 tank fleet and even the M48 series as well. Such a modernization program would follow the precedent already set with the earlier tanks.

SPACED APPLIQUE ARMOR
EQUALS 16 INCHES (40.64 CM) RHA

SPACED APPLIQUE
ARMOR EQUALS
16-INCHES RHA

SPALL LINER

SKIRTING PLATE

CAST ARMOR

30°

ATTACK HORIZONTAL

COMPOSITE MODULE

30° ATTACK

FRONT HULL SECTION APPLIQUE ARMOR
EQUALS 14.5 IN (36.83 CM) RHA

COMPOSITE MODULE

CAST HULL ARMOR

HIGH HARD SKIRTING PLATE

SPALL LINER

73°

SHOT LINE

0.88 IN (2.23 cm)

45°

10.5 IN (26.67 cm) CAST ARMOR

2.36 IN (5.99 cm) X 1.4 IN 3.55 cm)

1.0 IN (2.54 cm)

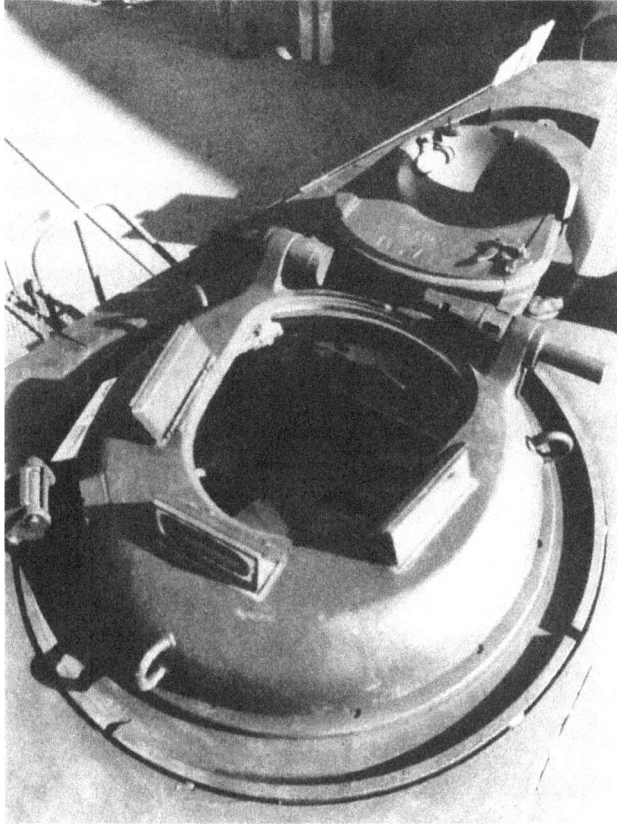

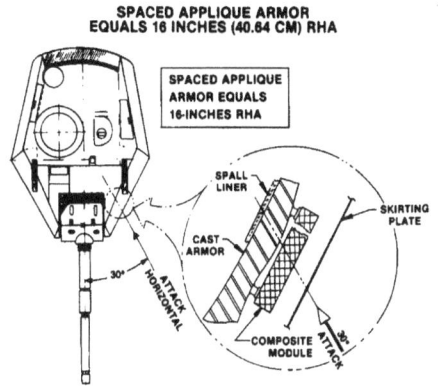

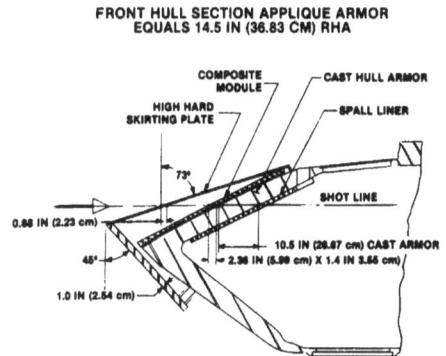

The low silhouette commander's cupola and hatch on the M60AX can be seen in the photograph by Steven Zaloga above at the left. Details of the armor added to the turret and hull appear at the right. The sectional view below shows the various modifications applied to the M60AX by Teledyne-Continental Motors.

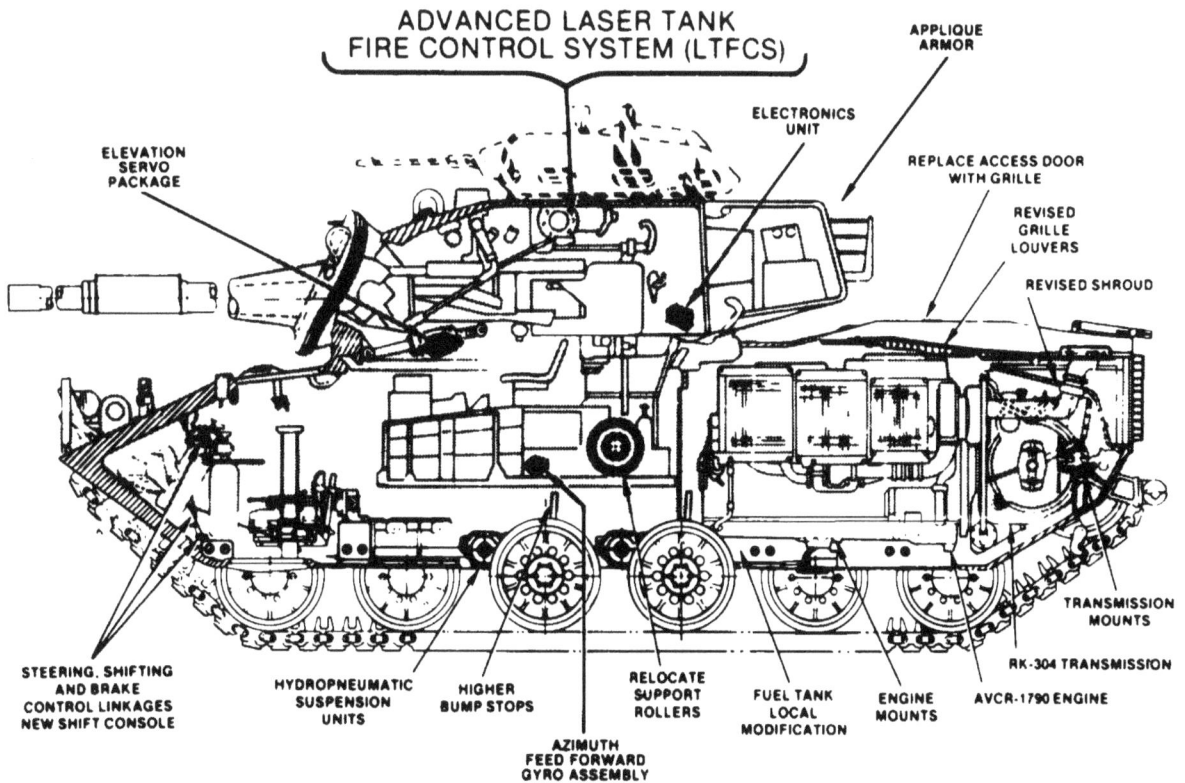

ADVANCED LASER TANK
FIRE CONTROL SYSTEM (LTFCS)

APPLIQUE ARMOR

ELECTRONICS UNIT

REPLACE ACCESS DOOR WITH GRILLE

REVISED GRILLE LOUVERS

REVISED SHROUD

ELEVATION SERVO PACKAGE

TRANSMISSION MOUNTS

RK-304 TRANSMISSION

AVCR-1790 ENGINE

ENGINE MOUNTS

FUEL TANK LOCAL MODIFICATION

RELOCATE SUPPORT ROLLERS

AZIMUTH FEED FORWARD GYRO ASSEMBLY

HIGHER BUMP STOPS

HYDROPNEUMATIC SUSPENSION UNITS

STEERING, SHIFTING AND BRAKE CONTROL LINKAGES NEW SHIFT CONSOLE

217

Interior details of the M60AX turret front can be seen in the wide angle photographs above. Note that the coaxial machine has not been installed at the left of the cannon. The sketch at the right shows the fire control equipment provided in the M60AX.

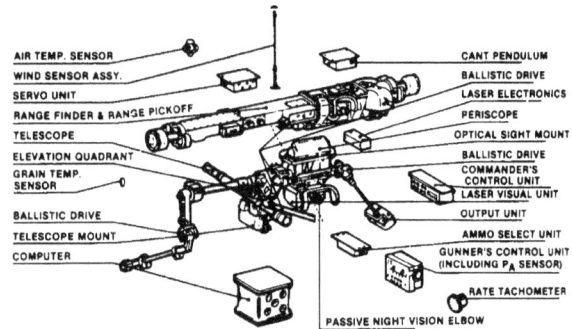

AIR TEMP. SENSOR
WIND SENSOR ASSY.
SERVO UNIT
RANGE FINDER & RANGE PICKOFF
TELESCOPE
ELEVATION QUADRANT
GRAIN TEMP. SENSOR
BALLISTIC DRIVE
TELESCOPE MOUNT
COMPUTER

CANT PENDULUM
BALLISTIC DRIVE
LASER ELECTRONICS
PERISCOPE
OPTICAL SIGHT MOUNT
BALLISTIC DRIVE
COMMANDER'S CONTROL UNIT
LASER VISUAL UNIT
OUTPUT UNIT
AMMO SELECT UNIT
GUNNER'S CONTROL UNIT (INCLUDING P_A SENSOR)
RATE TACHOMETER
PASSIVE NIGHT VISION ELBOW

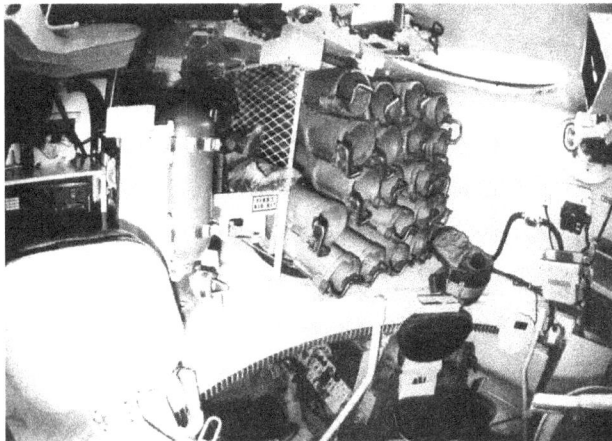

The rear of the M60AX turret interior is shown above. Obviously, these photographs were taken with the gun traversed to the rear as the driver's compartment can be seen below the turret ring. At the left is a wide angle view of the driver's compartment in the M60AX. All of the photographs on this page were taken by Steven Zaloga.

The pilot M48A1E1 is shown above and below at Fort Knox. Like on the other diesel powered tanks, the personnel heater exhaust extends to the right side of the hull and the fender mounted dry type air cleaners have been installed.

UPGRADING THE TANK FLEET

The development of the M60 tank with the 105mm gun and the diesel engine provided a combat vehicle with greatly increased firepower and cruising range. Application of these two features to the M48A1s already in the inventory was expected to give them comparable performance. In December 1958, the Chief of Staff requested information on the feasibility of such a conversion. It then was concluded that a retrofit of the M48 series tanks was technically feasible if sufficient funds were available after meeting the production requirement for 60 new M60 tanks per month. On 10 June 1959, a memorandum from the Chief of Ordnance to OTAC referred to the M48A1 retrofitted with M60 components as the M48A1E1. This designation was officially assigned as the 105mm gun full tracked combat tank M48A1E1 in OTCM 37387 on 21 April 1960. Six pilot M48A1E1s were authorized by this action. The first vehicle was delivered to Aberdeen Proving Ground in March 1960 and the second arrived at Fort Knox on 6 April. Pilot number 3 went to OTAC for field

service maintenance evaluation and number 4 was used for studies at Detroit Arsenal. The fifth M48A1E1 was delivered to the Arctic Test Board for evaluation during the 1960–61 season and the sixth was used for climatic tests at Yuma, Aberdeen, and the arctic.

Further details of the M48A1E1 at Fort Knox can be seen in these views. Above, the hull surfaces where the second and fourth track return rollers were removed are obvious. Note in the front view that a shield has been installed around the personnel heater exhaust pipe.

Like the M60, the M48A1E1 was powered by the Continental AVDS-1790-2 engine using the Allison CD-850-6 transmission. It also was fitted with the M60 aluminum fuel tanks and the same type of top rear deck and grills as well as the dry type air cleaners. The suspension retained most of the M48A1 components, but it was upgraded to include the new compensating idler assembly, track adjusting link, double bump springs, and heavy front road wheel arms. The track tension idler was removed and the number of track return rollers was reduced from five to three. Friction snubbers replaced the hydraulic shock absorbers at road wheel stations 1, 2, and 6. The M60 fire control system was installed. This included the coincidence type range finder, a modified ballistic computer, and an articulated telescope. The 90mm gun was replaced by the 105mm gun M68 and the stowage was rearranged to carry 57 rounds of 105mm ammunition. The modified M87 combination gun mount was fitted with the 7.62mm M73 machine gun to the left of the cannon. The original M1 commander's cupola was retained along with its .50 caliber M2 machine gun. The combat loaded weight of the M48A1E1 was about 52.6 tons.

The test program report concluded that the M48A1E1 was comparable in practically all respects to the M60 and was definitely superior to all other tanks of the M48 series because of the great increase in cruising range and firepower. It was noted that the M48A1E1 did not have the growth potential of the M60 in either daylight or infrared vision capability for the tank commander. This was due to the limited space in the M1 cupola which would not permit the installation of the new vision devices planned for the M60.

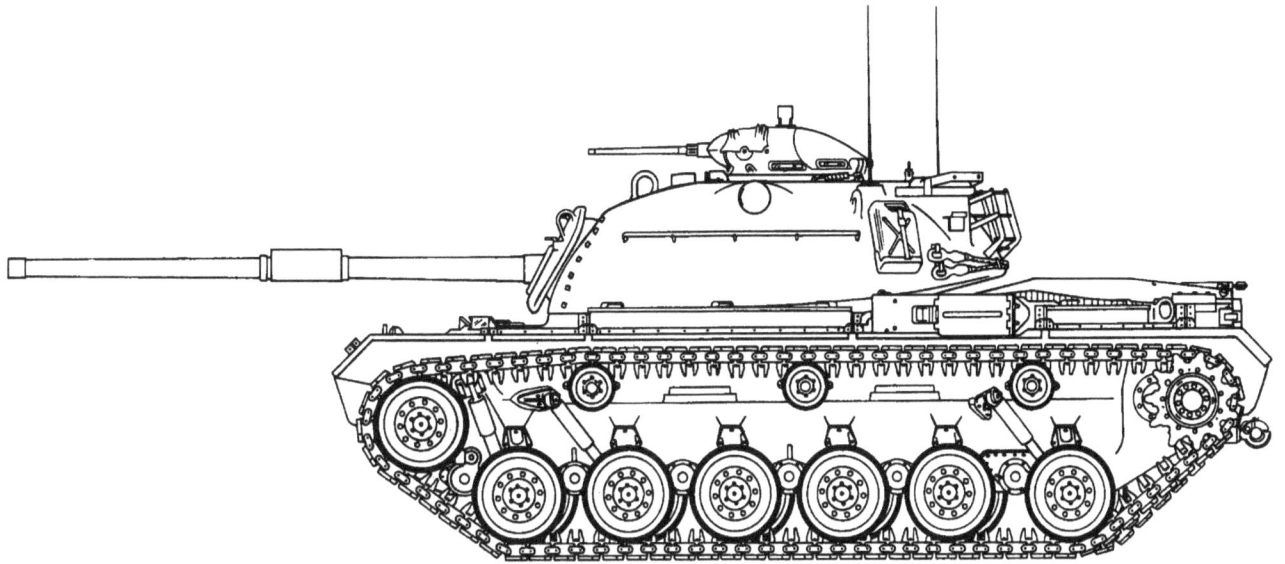

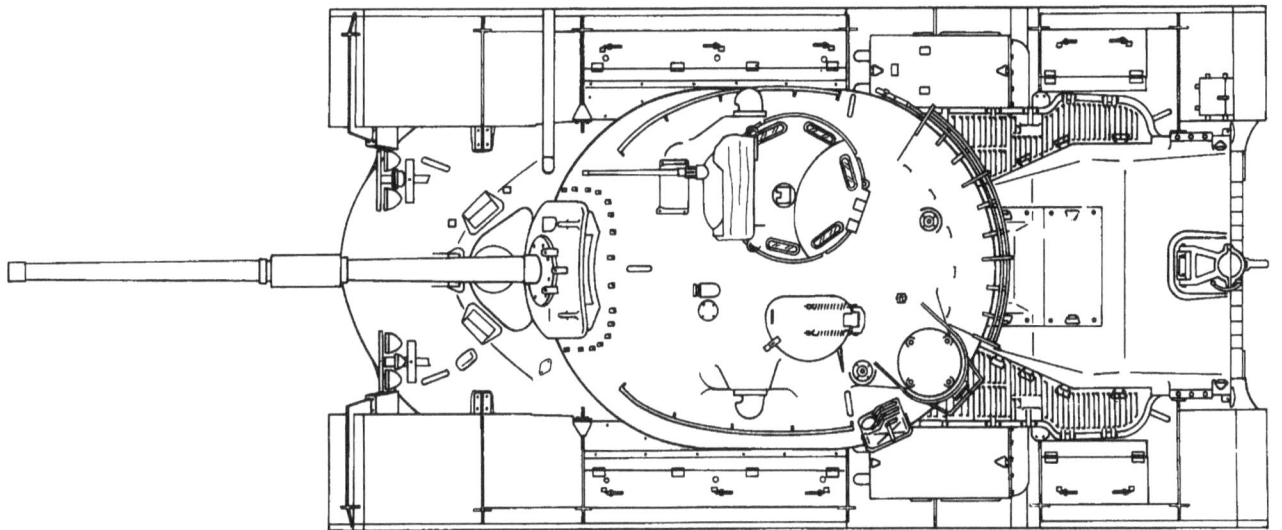

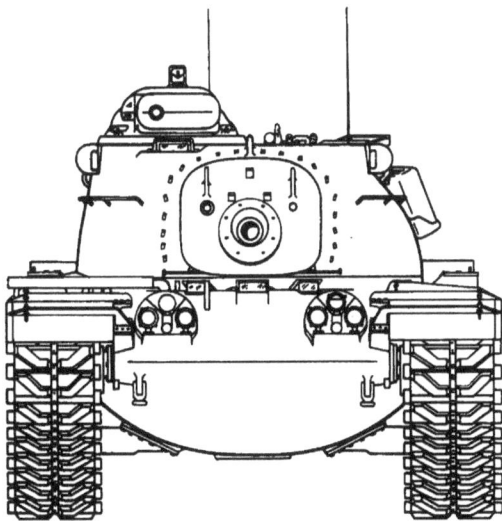

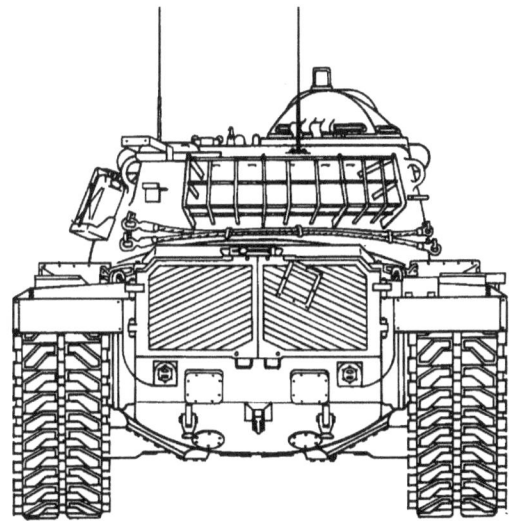

105mm Gun Tank M48A1E1

M48A1E2, registration number 9A 9551, appears in the two views above. Below, the same tank is shown after its transfer to the Marine Corps in June 1964. Note that the USA has been converted to USMC, but the same number is retained. The three track return rollers are obvious in all three photographs.

In November 1960, a study by OTAC concluded that the M48A1 modernization program could be accomplished more economically at Ordnance Depot facilities than by using a commercial contractor. A subsequent recommendation by the Chief of Ordnance that 600 M48A1s be converted at Ordnance Depots was approved in February 1961. However, the 105mm gun was deleted from the program and the modernized tanks were to retain the 90mm gun. This was mainly due to the large stocks of 90mm ammunition on hand and budgetary limitations in supplying sufficient 105mm ammunition for the converted tanks. Two of the M48A1E1 pilots were rearmed with the 90mm gun and they were then redesignated as the 90mm gun full tracked combat tank M48A1E2. These vehicles were used in production engineering activities for the final version of the modernized tank which was standardized as the 90mm gun full tracked combat tank M48A3. The two M48A1E2s, modified from M48A1E1s,

could be readily identified by the three track return rollers on each side. The M48A3s converted from M48A1s on a production basis retained all five of the track return rollers.

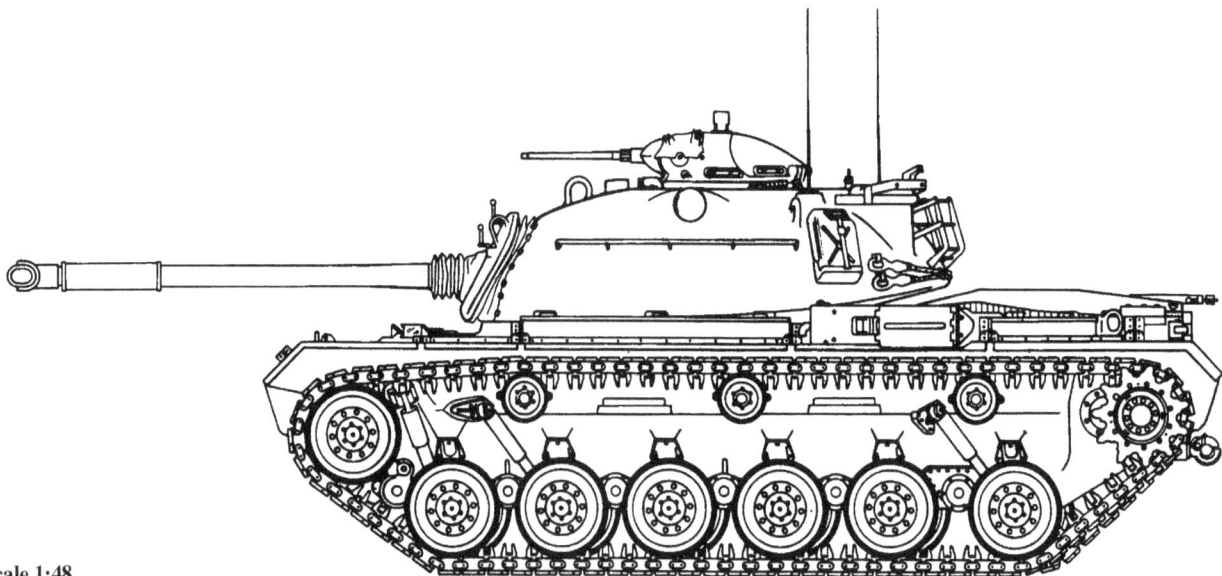

Scale 1:48

90mm Gun Tank M48A1E2

222

A pilot M48A3 at Chrysler is shown above and at the left below. This tank, registration number 9A 6270, retains all five track return rollers and a xenon searchlight has been installed. At the right below, a load of M48A1s is enroute to the Red River Army Depot for conversion to M48A3s. The blast deflectors and bore evacuators have been removed and the gun tubes are sealed for shipment.

The conversion work began at the Anniston and Red River Army Depots with the first tank being accepted by the Army in February 1963. This program provided 600 M48A3 tanks for the Army and 419 for the U.S. Marine Corps by late 1964. These tanks were deployed later to Vietnam where they were used in battle by both United States and South Vietnamese forces.

The M48A1 was converted to the M48A3 configuration by replacing the original power train with the AVDS-1790-2A engine and the CD-850-6A transmission. This required major rework of the hull to provide the M60 type top deck and rear grill doors. The modernized tanks incorporated improved driver's controls and a new fuel and electrical system. Since it was no longer required with the fuel efficient diesel engine, the auxiliary generator and engine were removed. The interior oil bath air cleaners were replaced by the M60 type fender mounted dry units. The M48A1s manufactured by Chrysler, Ford, and General Motors differed in minor details of fenders, grills, and stowage box arrangements requiring different conversion kits for these vehicles. Other differences in stowage and equipment requirements between the Army and the Marine Corps required additional variations in the conversion process. On the suspension, the track tension idler was removed and, as mentioned earlier on the M48A1E1, the suspension was upgraded, but all five track return rollers were retained. The M60 type personnel heater was installed with its exhaust pipe extending out to the right side of the hull. The two shot carbon dioxide fire extinguisher system replaced the earlier model and the M2A2 CBR (chemical, biological, radiological) equipment was provided.

Above is another view of the M48A3 pilot at Chrysler. Below are the controls for the gunner (left) and tank commander (right). At the bottom, the ammunition stowage in the M48A3 is sketched at the left with an interior view of the tank commander's cupola at the right.

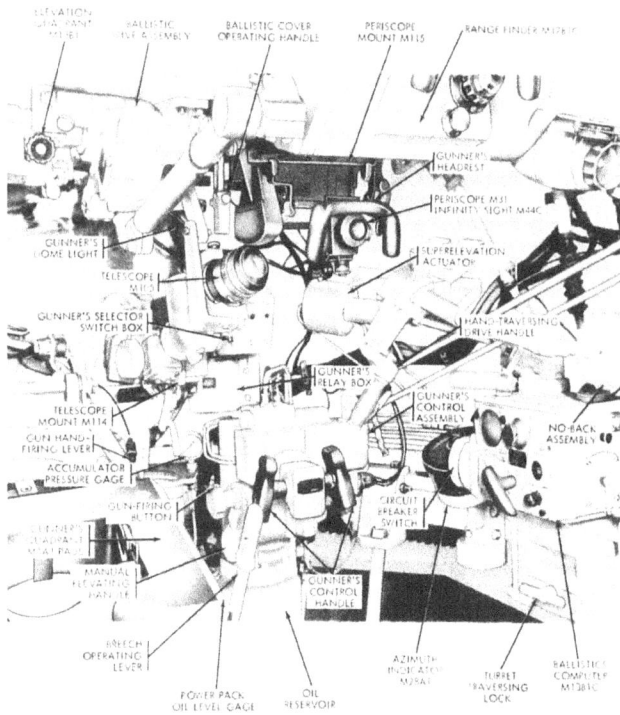

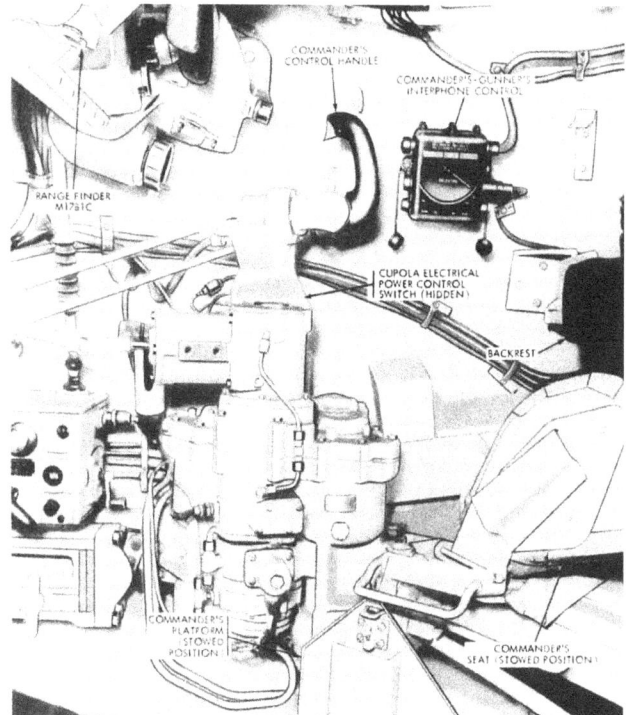

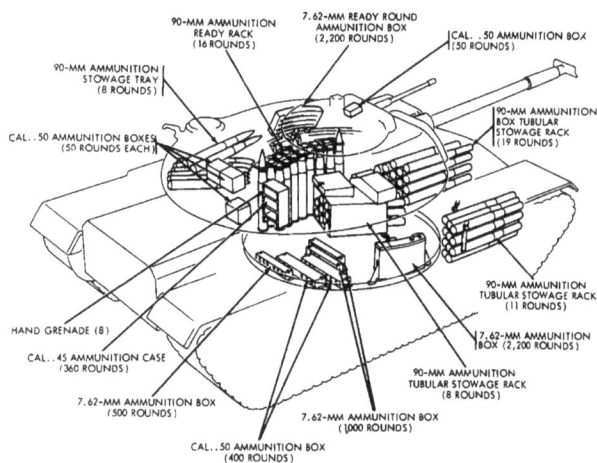

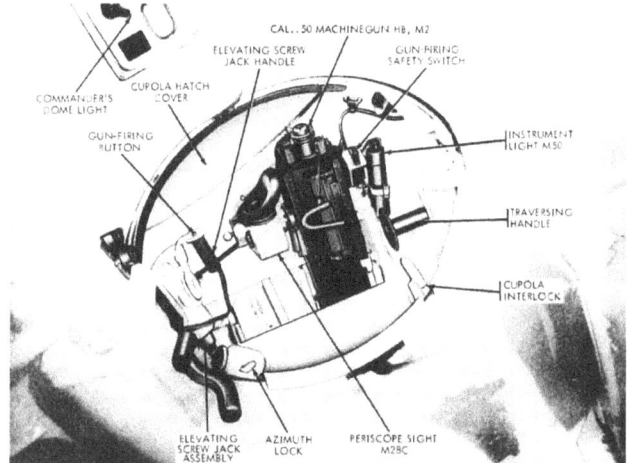

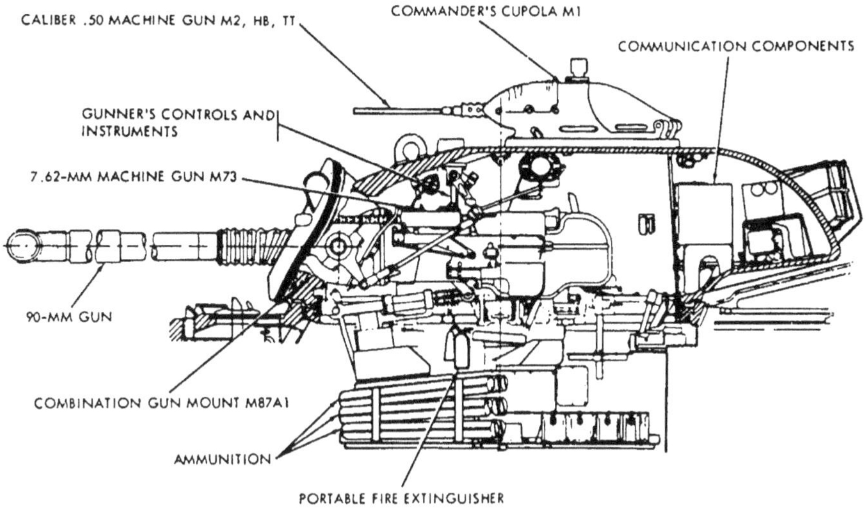

CALIBER .50 MACHINE GUN M2, HB, TT

COMMANDER'S CUPOLA M1

COMMUNICATION COMPONENTS

GUNNER'S CONTROLS AND INSTRUMENTS

7.62-MM MACHINE GUN M73

90-MM GUN

COMBINATION GUN MOUNT M87A1

AMMUNITION

PORTABLE FIRE EXTINGUISHER

The interior arrangement of the M48A3 turret is shown in the sectional drawing above. Details of the gun shield cover and the studs for mounting the xenon searchlight can be seen at the right.

Although the 90mm gun M41 was retained as main armament, the ammunition stowage was revised to include two additional rounds bringing the total carried to 62. The turret and turret platform stowage was rearranged and the early Oilgear turret control was replaced by the Cadillac Gage system. Mounts were added for the 2.2 kilowatt xenon searchlight. A new waterproofed gun shield cover and a nylon inner ballistic shield were provided. An inflatable seal was installed between the turret and hull and the entire vehicle was modified for deep water fording. Improved fire control equipment included the M17B1C coincidence range finder, the M10B1 ballistic drive, the M13B1 ballistic computer, the M105 telescope and mount, and the M31 gunner's periscope. The M1 commander's cupola was retained. By this time, the limited space inside the cupola had resulted in a further reduction in the size of the .50 caliber ammunition box. It now held only 50 rounds and the space available was still inadequate for operation to be fully satisfactory. This would become clearly apparent when the M48A3 was committed to battle in Vietnam.

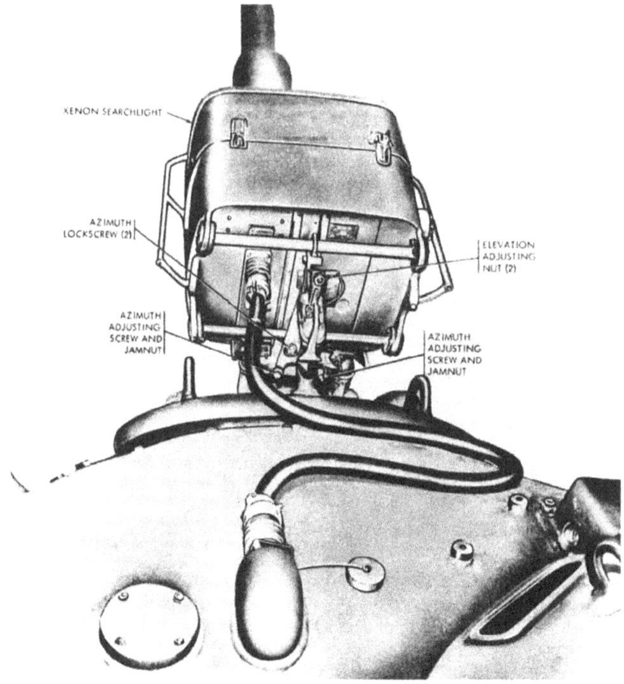

The 2.2 kilowatt xenon searchlight is installed in its operational position on the gun mount at the right.

225

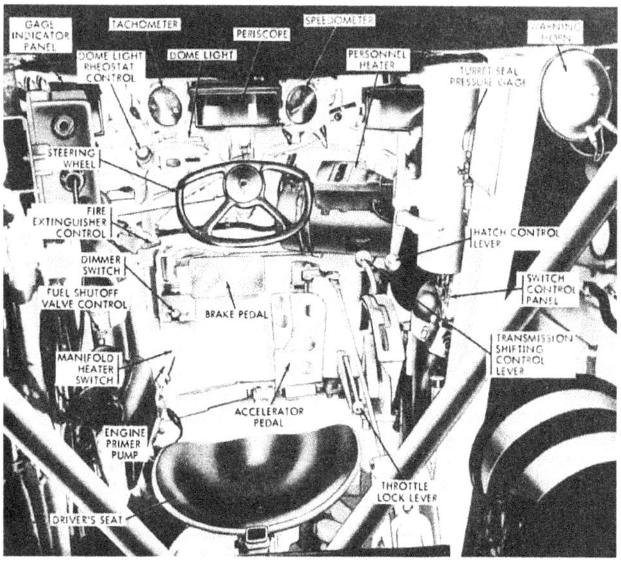

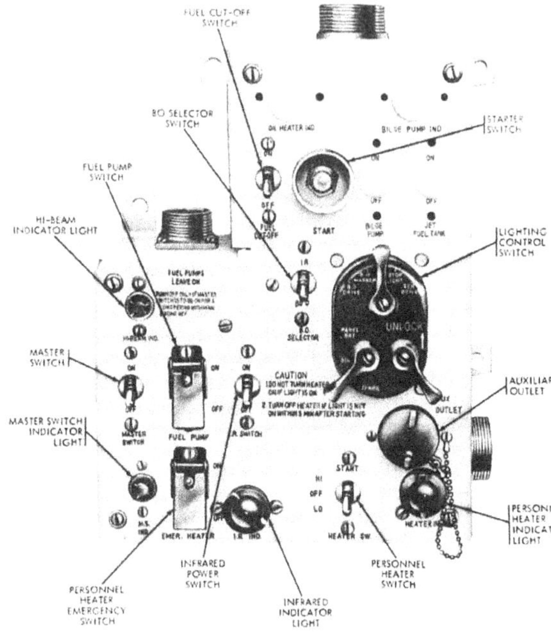

Above, the driver's controls appear at the left with a view of his switch control panel at the right. The latter was located at the right of the driver's seat.

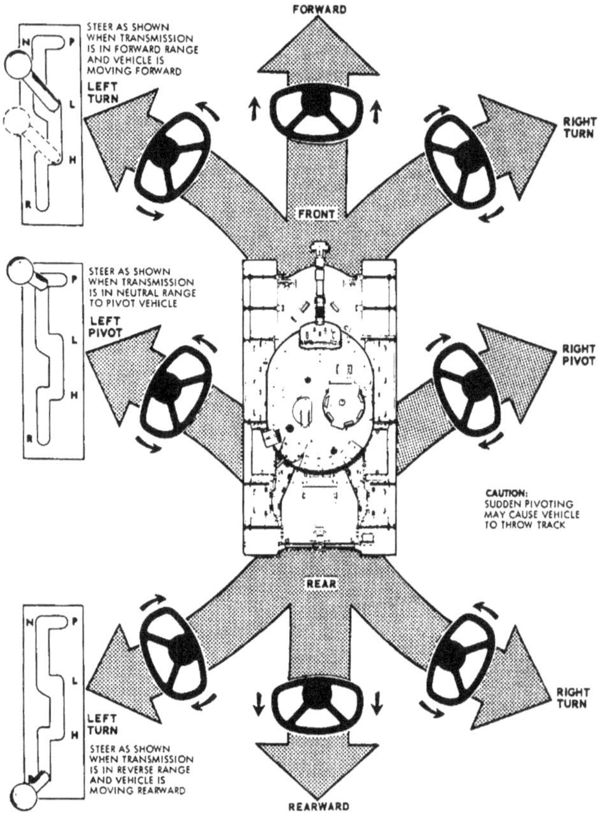

Steering the M48A3 was the same as for the earlier M48 series tanks as illustrated by the steering chart above.

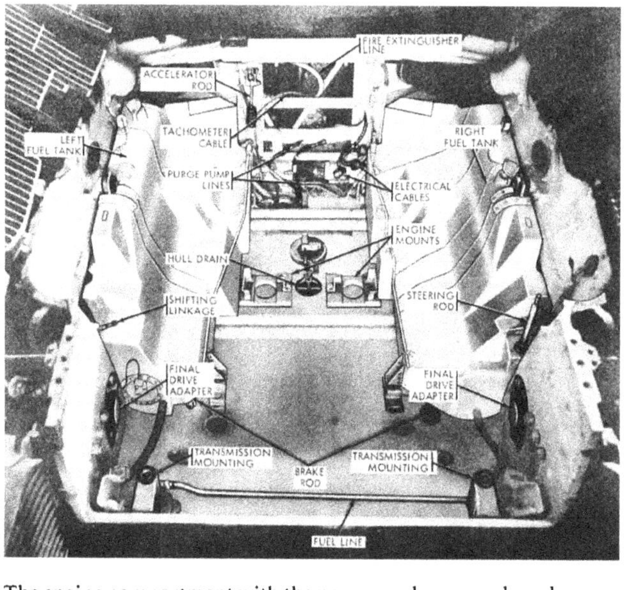

The engine compartment with the power pack removed can be seen above. Note how the fuel tanks are fitted into the space available to obtain the maximum volume. Below is a sketch of the intake and exhaust system for the AVDS-1790 diesel engine.

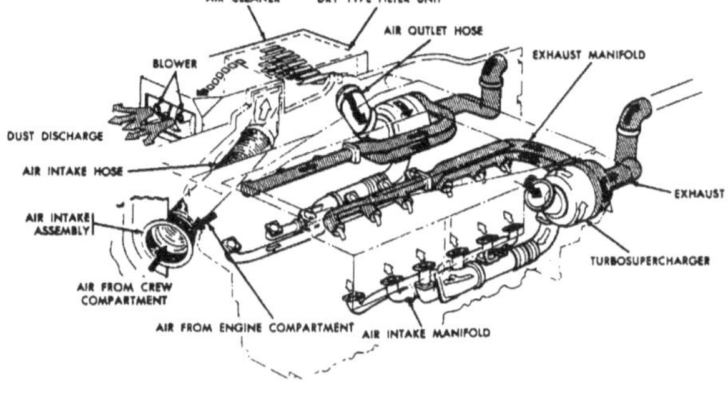

The M48A3 (Mod B) above was converted from an M48A1 at Bowen-Mclaughlin-York Company. Note the late design headlight arrangement. Although neither the .50 caliber machine gun nor the searchlight is installed, the gun deflector is fitted on the turret roof.

On 7 December 1965, a new program was approved to rearm the M48A3 fleet by installing the 105mm gun M68. However, before the program could be implemented, all funds available were required for higher priority projects, primarily the war in Southeast Asia. Also, by this time, many of the M48A3s intended for rearming were already earmarked for shipment to Vietnam. In fact, as the war increased in intensity and the use of armor became more widespread, it was obvious that additional tanks would be required. Because of the work load at the Army Depots, this project was to be accomplished at a commercial facility. On 14 April 1967, a contract was awarded to the Bowen-McLaughlin-York Company of York, Pennsylvania to convert additional M48A1 tanks to the latest M48A3 standard.

The tanks modernized under the fiscal year 1967 contract incorporated numerous product improvements not available under the fiscal year 1962 program. To differentiate between the two, the later vehicles were referred to as the M48A3 (Mod B).

These tanks included many of the latest M60A1 components such as the driver's controls and instruments. The fuel lines were relocated from the hull floor to the apex of the side armor. Two new suspension system kits provided a torsion bar knock-out hole in the suspension housing and revised the track support roller mud shield. The rear grill doors and tail lights were modified to make them more rugged and resistant to damage during jungle operations. New detachable composite headlights were introduced and a raised fender telephone box was installed.

Infrared fire control equipment was provided for the gunner in the M48A3 (Mod B), but the most obvious change was in the commander's cupola. A spacer ring with nine large vision blocks was installed between the cupola and the turret. This greatly improved the all round vision as well as the head room inside the cupola. The latter was further increased by a new bulged hatch cover. Unlike the original hatch, the new cover did not incorporate any vision blocks.

At the top left, the M19 cupola is mounted on the M48A3 turret using the adapter ring. At the right above and below, the M1 cupola is fitted with the nine block vision ring and the bulged hatch cover.

The necessity for the cupola modifications had been emphasized by combat reports from Vietnam. Here the troops had removed the .50 caliber machine gun and mounted it on the cupola roof forward of the tank commander's hatch. The commander then fired the weapon standing in the open hatch. The problem was clearly described in a report from the 3rd Tank Battalion, 3rd Marine Division which concluded that "extensive use of this cupola in Vietnam discloses poor visibility, general unreliability, and inherent mechanical failures that are unacceptable in combat." The report indicated that in addition to poor visibility, the major deficiencies were insufficient headroom, inadequate ammunition space (50 rounds), link chute jamming, and extreme difficulty in reloading. The latter was significant, since the 50 round ammunition box had to be frequently reloaded. All of these problems resulted in the gun being remounted outside and fired by the tank commander from the open hatch position.

Two solutions to the cupola problem were studied at Detroit. The first approach was to install the M19 cupola from the M60 tank on the M48A3. However, the M19 had a 34 inch diameter mounting ring compared to the 29.75 inch ring on the M1 cupola, thus an adapter ring was required. Such an adapter ring was fabricated and the M19 cupola was installed with a height increase of about five inches. However, it was noted that the rear overhang of the M19 interfered with the loader's hatch when the cupola was rotated 90 degrees to the right. Also, the larger M19 blocked the use of the turret lift eye and had to be removed before the turret could be lifted. These problems combined with a shortage of M19 cupolas ruled out its use on the M48A3.

An interior view of the M1 cupola fitted with the vision block ring is at the right.

The second solution which was applied to the problem has already been mentioned. It utilized the vision ring and the bulged hatch cover to increase the visability and headroom with the M1 cupola. Unfortunately, it did not solve the link jamming problem, the limited ammunition supply, or the reloading difficulties. As a result, many tank commanders continued to remount the .50 caliber machine gun on the cupola roof and use it in the open.

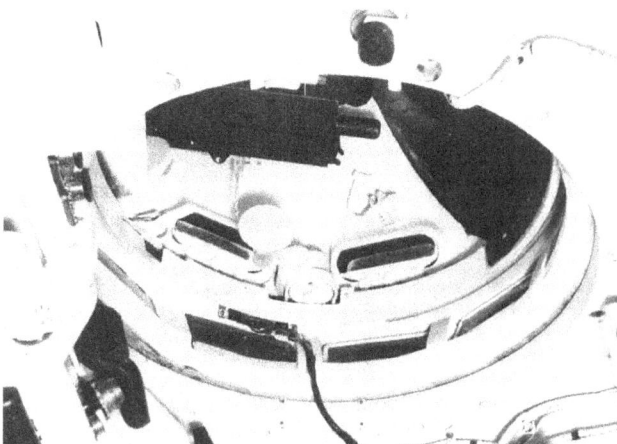

This M48A3 (Mod B) was used during test operations at Fort Knox. The headlight arrangement can be seen in the upper photograph. Note the protection added for the taillights and the grill doors in the rear view below. A platform for special equipment is attached to the gun mount and the .50 caliber machine gun deflector has been enlarged.

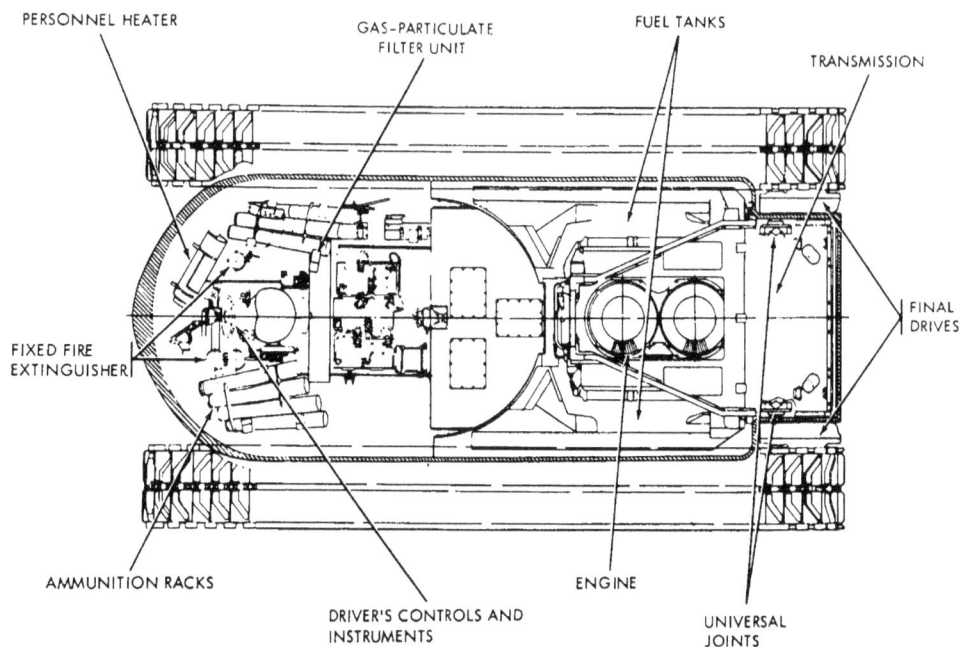

PERSONNEL HEATER
GAS-PARTICULATE FILTER UNIT
FUEL TANKS
TRANSMISSION
FIXED FIRE EXTINGUISHER
FINAL DRIVES
AMMUNITION RACKS
DRIVER'S CONTROLS AND INSTRUMENTS
ENGINE
UNIVERSAL JOINTS

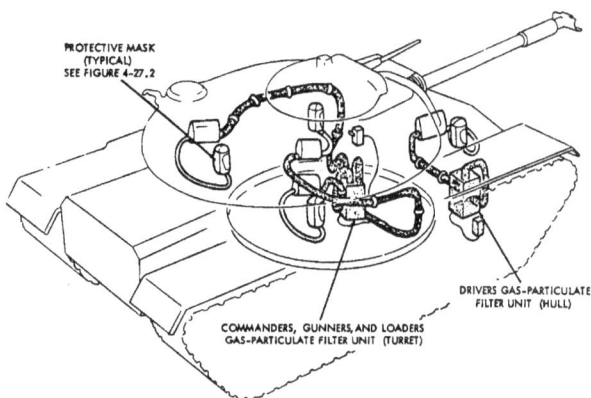

PROTECTIVE MASK (TYPICAL) SEE FIGURE 4-27.2

DRIVERS GAS-PARTICULATE FILTER UNIT (HULL)

COMMANDERS, GUNNERS, AND LOADERS GAS-PARTICULATE FILTER UNIT (TURRET)

An horizontal section drawing of the M48A3 hull appears above. The six batteries connected in series-parallel can be seen just behind the driver's seat. The sketch at the left shows the location of the gas-particulate filter units and the protective masks.

Bowen-McLaughlin-York converted a total of 578 tanks to the M48A3(Mod B) configuration under their contract. The vision block ring and other late modifications were eventually retrofitted to the earlier M48A3s. After that program was complete, the Mod B designation was dropped.

Below is a newly converted M48A3 (Mod B). With no gun shield cover fitted, the details of the searchlight mount are visible.

The M48A3 in these photographs has been prepared for shipment. The blast deflector has been removed and the muzzle of the 90mm gun sealed. Although it is equipped with the vision block ring and the bulged cupola hatch cover, this vehicle retains the early headlight arrangement and lacks the protective shields around the taillights and on the grill doors.

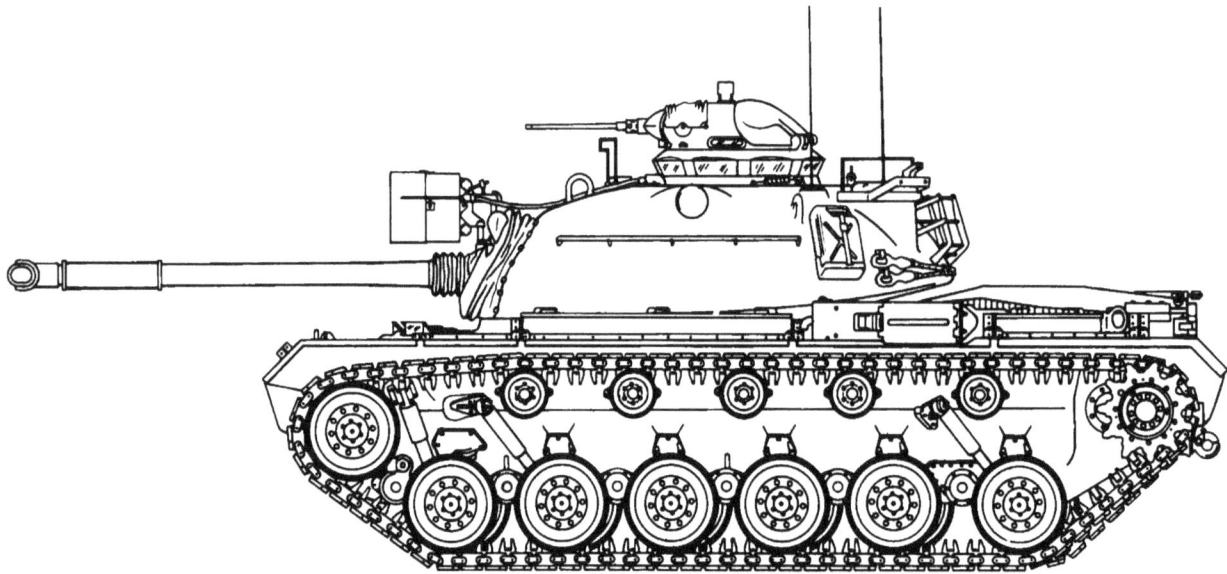

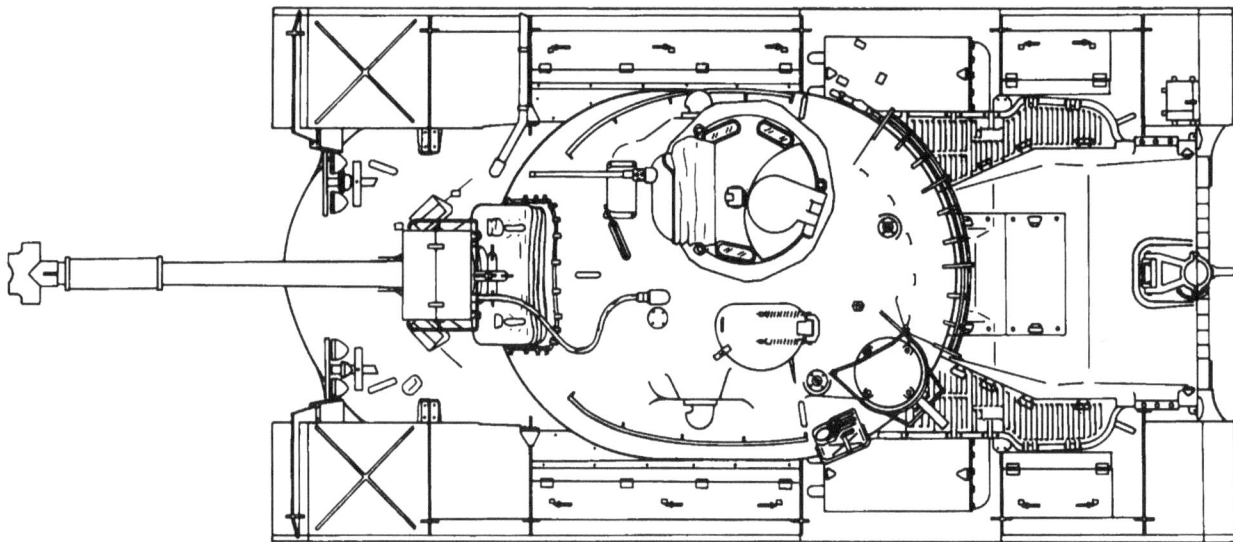

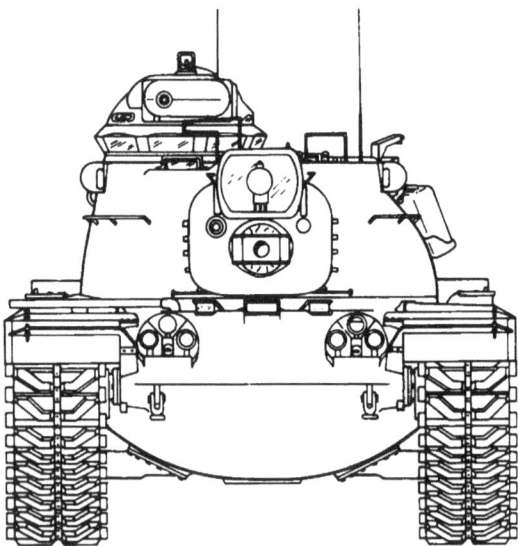

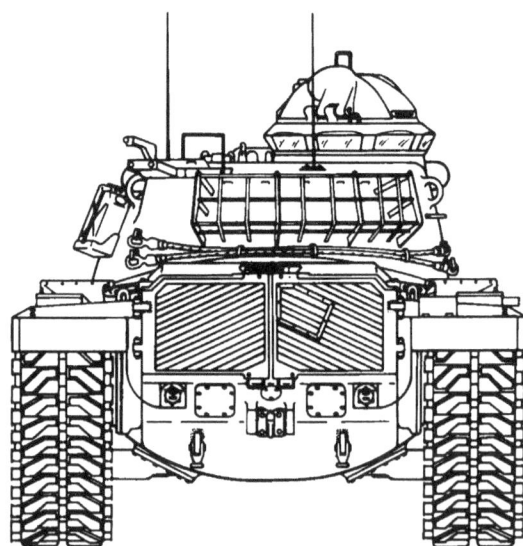

Scale 1:48

90mm Gun Tank M48A3

The Chrysler pilot M48A3 here has been fitted with the vision block ring and bulged cupola hatch cover. It also has a xenon searchlight, but appears to lack a deflector for the .50 caliber machine gun.

Above is the M48A1E3 after assembly at Chrysler. Although the chassis utilized late M48A3 components, it retained the early headlights and lacked the ruggedized grill doors and taillights. Below, during the assembly of the M48A1E3, details of the hull can be seen with the turret removed.

the XM735 tank, in December 1965 using many of the M48A3 conversion kits. M60 turrets were installed on two tanks which subsequently were designated as the 105mm gun full tracked combat tank M48A1E3. One of these, registration number 9A9849, was shipped to Fort Knox during the Spring of 1967. It consisted essentially of an M48A3(Mod B) chassis with the stowage rearranged to accommodate the 105mm ammunition. Inner and outer spacer rings raised the turret two inches to provide clearance for the M60 turret basket. The ballistic outer ring was welded to the hull around the turret opening and the inner ring was machined to fit over the M48 bolt circle. M60A1 ammunition racks were installed with new hardware to fit the elliptical front hull.

A program was approved to retrofit 243 M48A1 tanks to the new configuration with the M60 turret. They were to be classified as Standard B and designated as the 105mm gun full tracked combat tank M48A4. However, with the decision to limit the numbers of M60s converted to the Shillelagh, the program was cancelled and no additional tanks were modified.

When it was planned to retrofit the M60 tanks with the Shillelagh turret under the M60A1E1 program, it appeared that large numbers of 105mm gun turrets would become available. To utilize these assets, a project was initiated to install the turrets on M48A1 chassis modernized to the M48A3 standard. Chrysler began the project, originally referred to as

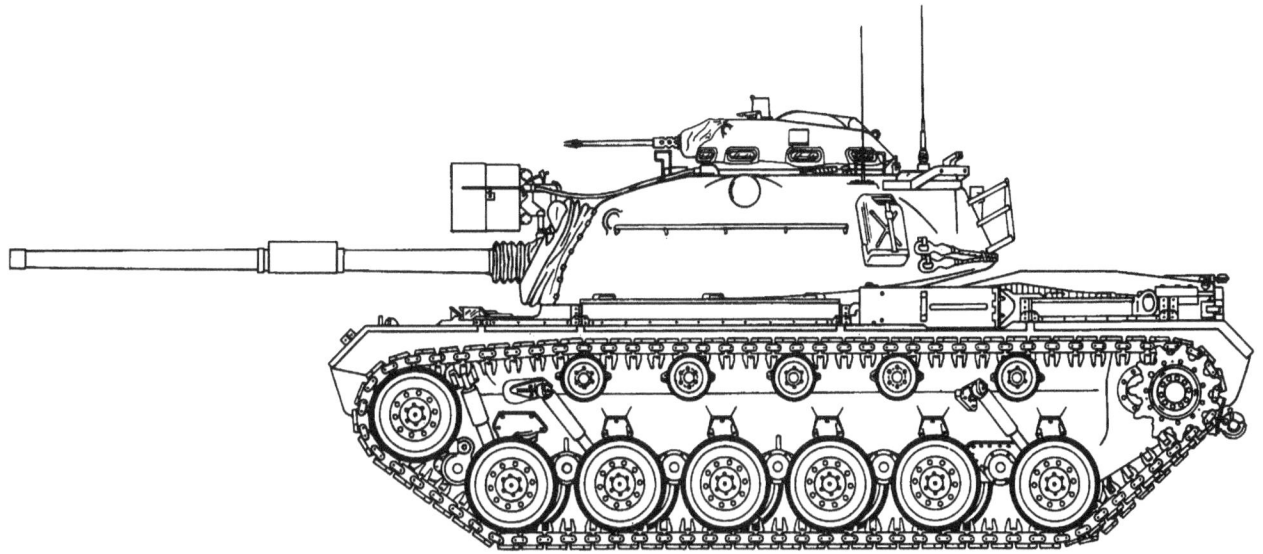

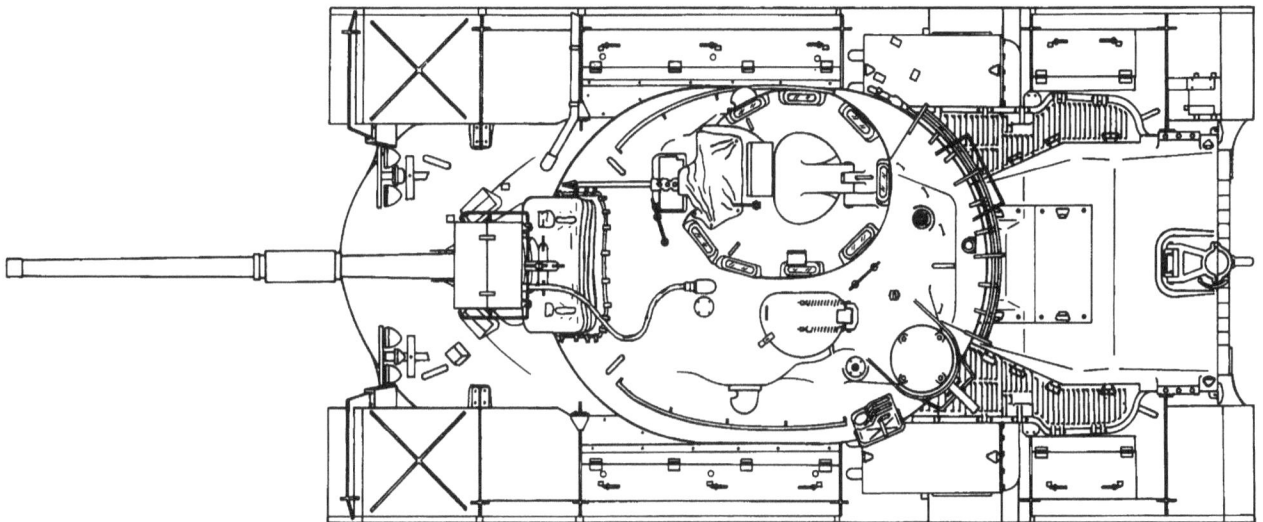

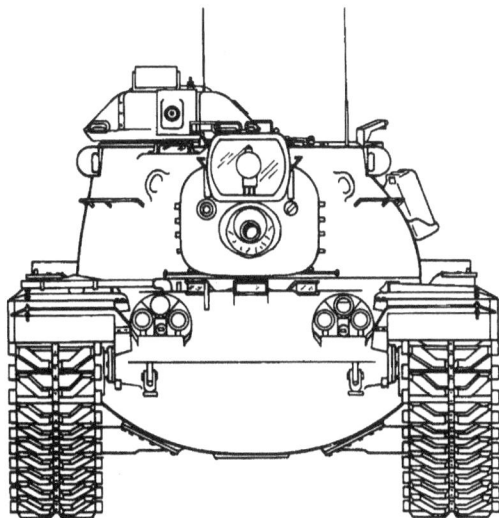

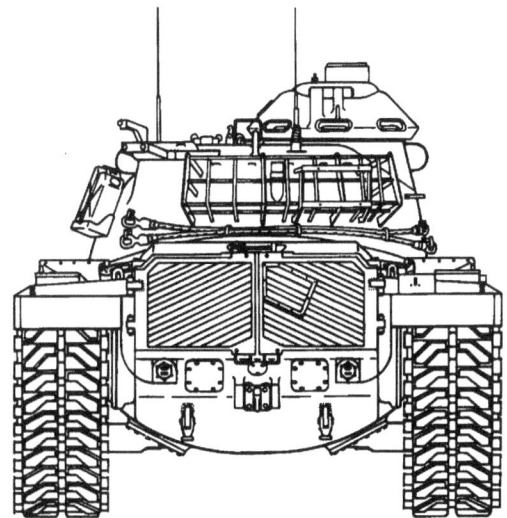

Scale 1:48

105mm Gun Tank M48A1E3(M48A4)

This is the M48A1E3 as it appeared during tests at Fort Knox on 4 May 1967. The new T142 tracks have been installed with the replaceable track pads.

These are additional views of the M48A1E3 at Fort Knox. All of the optical equipment is in place. The telescopic sight and the range finder lenses can be seen in the front view as well as the gunner's and driver's periscopes. The stowage bracket for the xenon searchlight is visible on the left rear of the turret.

The tanks on this page are typical of the early M48A5 with the modified M1 cupola and vision block ring. The xenon searchlight is installed with a deflector for the .50 caliber machine gun. Note the late M48A3 type headlights and the protected taillights and grill doors.

For operations in Vietnam, the 90mm gun in the M48A3 was more than adequate. However, compared to the main armament of the new MBTs, it was clearly obsolete and the project to replace it with the 105mm gun was revived once again. Originally, the subject of this latest program was designated as the XM736 tank, but this was changed later to the M48A3E1. After standardization in May 1975, the official nomenclature became the 105mm gun full tracked combat tank M48A5.

The initial version of the M48A5 was essentially an M48A3(Mod B) rearmed with the 105mm gun. It utilized as many M60A1 components as possible and retained the late version of the M1 commander's cupola with the vision block ring and the bulged hatch cover. Five test tanks were delivered, two in June and three in July 1975. A contract was let to the Annistion Army Depot to convert 501 (including the five test tanks) M48A3s to the M48A5 configuration. The first of the production conversions was delivered in October 1975 and the last of the batch of 501 was completed in December 1976.

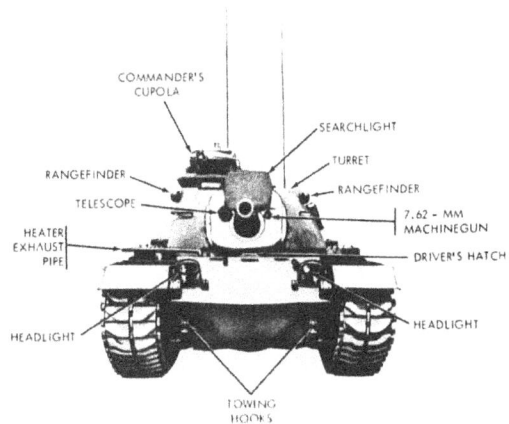

The elevated external interphone box can be seen below on the right rear fender. At the left, the whip type radio antennas are tied down.

Above, the low profile cupola is shown with the hatch fully open (left) and in the pop-up position (right). The action of the hatch and the scissors type machine gun mount is illustrated in the sketch below.

In preparation for the project in October 1974, the Department of the Army directed that maximum use be made of the Israeli experience in upgrading their M48 series tanks. In line with these instructions, a number of product improvements were developed and introduced into the production line as soon as possible. The most important of these modifications were an increase in 105mm ammunition stowage from 43 to 54 rounds, a 7.62mm M60D machine gun on the turret roof for the loader, and a low profile Israel Defence Forces (IDF) type cupola for the tank commander. This cupola mounted a second 7.62mm M60D machine gun for the tank commander's use when the hatch was open.

The low profile cupola had a pop-up hatch which allowed a 360 degree view for the tank commander with little exposure. Three periscopes were installed, one in front and one at each side, for vision when the hatch was closed. The cupola was fitted with a two position scissors type mount for the commander's machine gun. The M60D weapon was originally developed as a helicopter door gun. Two fixed sockets were located on the turret roof for the loader's gun, one in front and one at the left of the loader's hatch. The weapon could easily be switched between the two firing positions. These new features were incorporated in the conversion program during August 1976 and for a time, the designation M48A5PI was applied to the product improved tanks. After the earlier vehicles were retrofitted to the new standard, the PI was dropped.

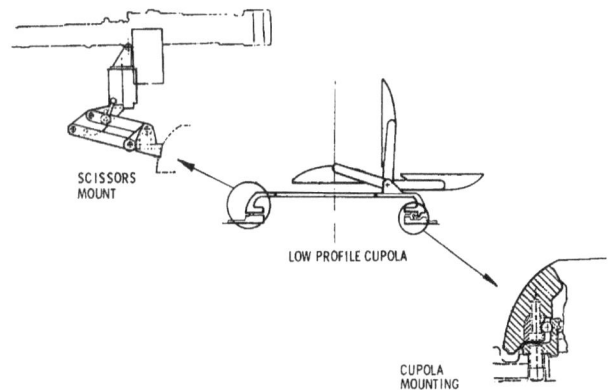

Below is the ammunition stowage on the product improved M48A5 providing space for 54 rounds of 105mm ammunition.

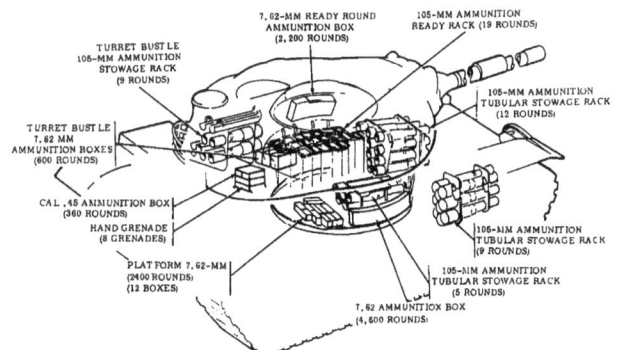

Details of the product improved M48A5 are illustrated in these photographs. Note that there are no deflectors to protect the xenon searchlight from the two M60D machine guns on the turret roof. A little discretion is required from the machine gunners.

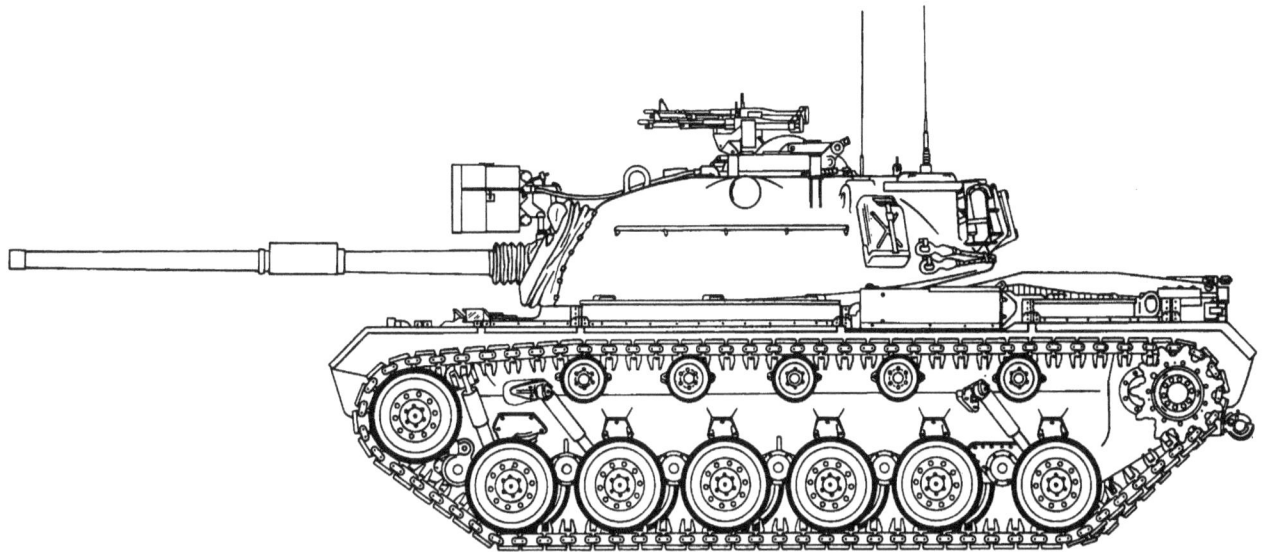

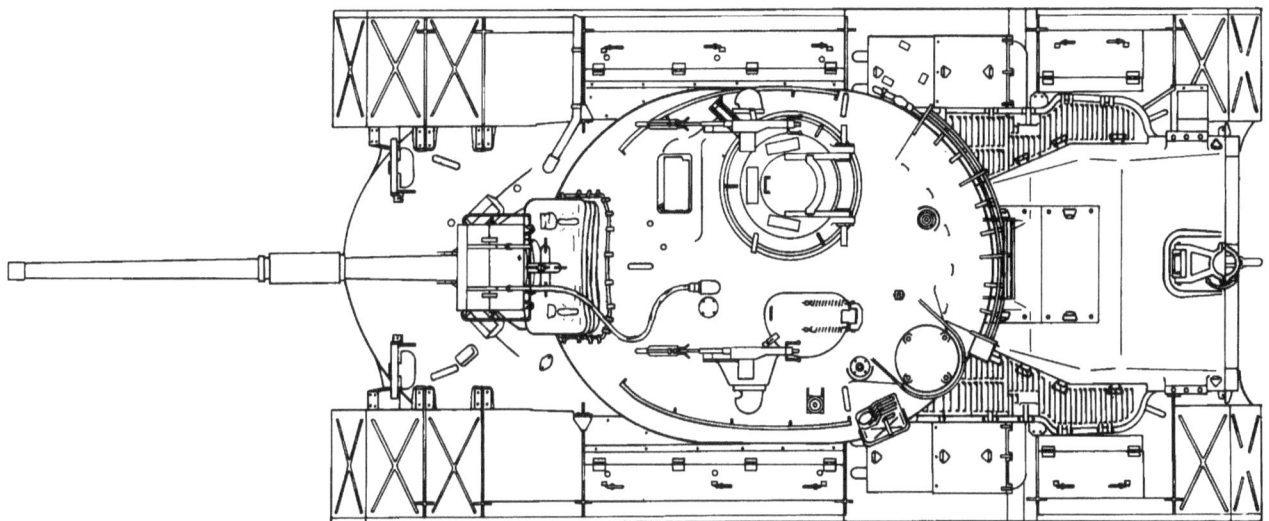

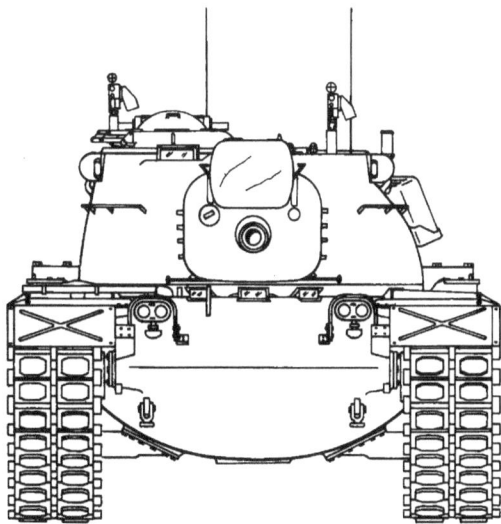

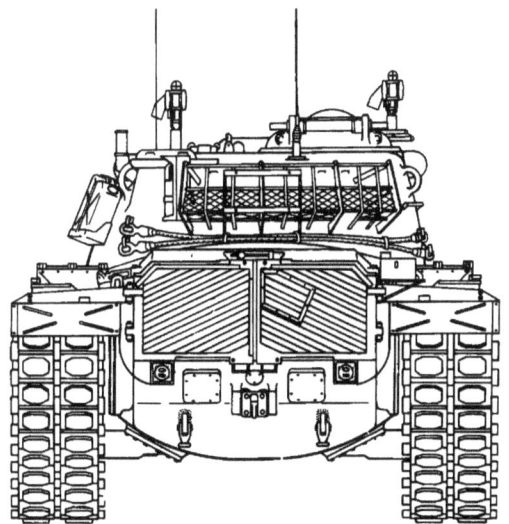

Scale 1:48

105mm Gun Tank M48A5

241

The M48A5PI can be compared with the earlier M48A5 in the view at the left. The elevated interphone box on the right rear fender and other details can be seen in the rear view above.

Shortly after work began on the M48A3 to M48A5 conversions, a far more extensive project was started to rework the M48A1 to the new configuraton. While only 11 conversion kits had to be applied to the M48A3, 67 such kits were required to bring the M48A1 up to the M48A5 standard. Two pilot M48A1s were converted at Anniston and delivered in August 1976 followed by full production in October. This work continued until March 1978 and completed a batch of 708 M48A5s. Further work under this conversion program at Anniston continued until December 1979 ending with a grand total of 2069 M48A5s.

The majority of the M48A5 tanks were destined for U.S. Army Reserve and National Guard units. However, 140 M48A5s were deployed to Korea to replace the M60A1 tanks of the U.S. 2nd Infantry Division. The first two of these tanks arrived at Pusan, Korea on 8 June 1978 and were issued to the units eight days later. The final tanks were handed over on 21 July. The M60A1s were then returned to the United States for a badly needed overhaul. The M48A5s deployed to Korea had the tank commander's 7.62mm M60D machine gun replaced by a .50 caliber M2 weapon with appropriate changes in ammunition stowage.

The modifications required to bring the M48A1 or M48A3 up to the M48A5PI standard are listed in the sketch below.

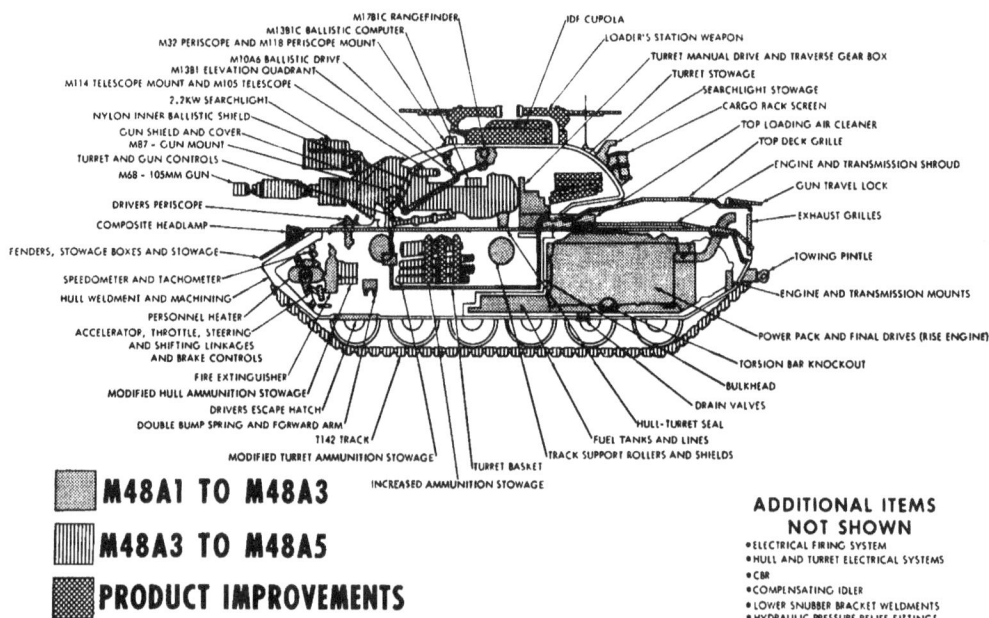

M17B1C RANGEFINDER
M13B1C BALLISTIC COMPUTER
M32 PERISCOPE AND M118 PERISCOPE MOUNT
M10A6 BALLISTIC DRIVE
M13B1 ELEVATION QUADRANT
M114 TELESCOPE MOUNT AND M105 TELESCOPE
2.2KW SEARCHLIGHT
NYLON INNER BALLISTIC SHIELD
GUN SHIELD AND COVER
M87 - GUN MOUNT
TURRET AND GUN CONTROLS
M68 - 105MM GUN
DRIVERS PERISCOPE
COMPOSITE HEADLAMP
FENDERS, STOWAGE BOXES AND STOWAGE
SPEEDOMETER AND TACHOMETER
HULL WELDMENT AND MACHINING
PERSONNEL HEATER
ACCELERATOR, THROTTLE, STEERING AND SHIFTING LINKAGES AND BRAKE CONTROLS
FIRE EXTINGUISHER
MODIFIED HULL AMMUNITION STOWAGE
DRIVERS ESCAPE HATCH
DOUBLE BUMP SPRING AND FORWARD ARM
T142 TRACK
MODIFIED TURRET AMMUNITION STOWAGE
INCREASED AMMUNITION STOWAGE
TURRET BASKET

IDF CUPOLA
LOADER'S STATION WEAPON
TURRET MANUAL DRIVE AND TRAVERSE GEAR BOX
TURRET STOWAGE
SEARCHLIGHT STOWAGE
CARGO RACK SCREEN
TOP LOADING AIR CLEANER
TOP DECK GRILLE
ENGINE AND TRANSMISSION SHROUD
GUN TRAVEL LOCK
EXHAUST GRILLES
TOWING PINTLE
ENGINE AND TRANSMISSION MOUNTS
POWER PACK AND FINAL DRIVES (RISE ENGINE)
TORSION BAR KNOCKOUT
BULKHEAD
DRAIN VALVES
HULL-TURRET SEAL
FUEL TANKS AND LINES
TRACK SUPPORT ROLLERS AND SHIELDS

M48A1 TO M48A3
M48A3 TO M48A5
PRODUCT IMPROVEMENTS

ADDITIONAL ITEMS NOT SHOWN
• ELECTRICAL FIRING SYSTEM
• HULL AND TURRET ELECTRICAL SYSTEMS
• CBR
• COMPLENSATING IDLER
• LOWER SNUBBER BRACKET WELDMENTS
• HYDRAULIC PRESSURE RELIEF FITTINGS

PART IV

SPECIALIZED ARMOR BASED ON THE PATTON

Above, a flame thrower is installed on the medium tank M26 chassis. The tank turret was removed and replaced by the low silhouette welded structure.

FLAME THROWER TANKS

During World War II, the use of tanks armed with flame throwers proved to be extremely effective, particularly in the Pacific theater of operations. Unfortunately, the value of such weapons was not recognized in the United States until late in the war and no regular production flame thrower tanks were available in time for use in combat. The urgent requirement for these weapons was seen much more clearly out in the Pacific and a project in Hawaii converted a number of Sherman tanks by replacing the cannon with a main armament flame thrower. A few were successfully used by the U.S. Marine Corps on Iwo Jima, but the first full battalion of these tanks was employed on Okinawa. Here, the successful operations of the Army's 713th Tank Battalion finally convinced the War Department of the value of main armament flame thrower tanks and resulted in a high priority program for their production. Designated as the flame thrower tanks M42B1 and M42B3, they were based on the medium tanks M4A1 and M4A3. All of these flame throwers were designed to use thickened gasoline fuel usually referred to as Napalm because of the napthenic and palmitic acid thickening agents. The surrender of

Japan during the Summer of 1945 brought production to a halt and limited further work to a few development projects also based primarily on the Sherman tank.

The Army's latest production tank at the end of the war was the M26 General Pershing and on 11 October 1945, OCM 29326 approved the military characteristics for the flame thrower tank T35 based on this vehicle. However, the lack of urgency and reduced development funds during the postwar period slowed the rate of progress on the project.

A variety of concepts were studied to meet the flame thrower tank requirement and some did not follow the characteristics outlined for the T35. One of these eliminated the turret with its cannon and replaced it with a flame gun in a small turret on a large fixed superstructure. Such a design was expected to increase the capacity for the flame thrower fuel to about 900 gallons. Another study installed a flame gun in a low silhouette mount on the M26 chassis for test purposes. However, the major part of the investigation was directed toward the addition of the flame gun to the M26 while retaining its main armament.

Below are the concept drawings for the proposed high capacity flame thrower tank based on the medium tank M26 chassis.

The view above shows the flame gun mock-ups installed at the five locations proposed for the T35 flame thrower tank. At the right, plans 4 and 5 illustrate the two most desirable arrangements. Although poor in quality, these photographs are included as they appear to be the only surviving views of this installation.

Studies at Edgewood Arsenal indicated that the original concept of the T35 retaining its full five man crew was impractical and a revised list of characteristics was prepared. The new concept reduced the crew to four and replaced the assistant driver with a large tank of flame thrower fuel. A remote controlled .30 caliber machine gun was specified to replace the bow machine gun. The 90mm gun remained, although some consideration was given to replacing it with a 76mm weapon to obtain extra space in the turret. A major effort was devoted to finding the proper location for the flame gun. Five different arrangements were investigated. Plan number 1 mounted the flame gun on the front hull, either on the ventilator blister or on the assistant driver's left periscope mount. He would then operate the weapon, but the traverse would be limited to about 150 degrees. This approach was dropped when it became necessary to eliminate the assistant driver to obtain adequate fuel stowage. The second design mounted the flame gun coaxially with the cannon in the combination mount. Unfortunately, this arrangement limited the range of elevation to that of the cannon unless use was made of a complicated and cumbersome externally mounted device. Further study showed that this approach was impractical. The third concept replaced the loader's periscope in the turret top with the flame gun. An advantage of this location was that the cannon and the flame gun could be used independently. Unfortunately, in this position, the flame gun fuel tended to drip all over the vehicle and thus it was considered less desirable than the two following arrangements.

Plan number 4 moved the flame gun to the left top front of the turret. Here, the desired elevation range of +45 to −15 degrees could be obtained by using a simple linkage to the cannon mount. Also, the use of quickly interchangeable 1:1 and 1:2 linkages between the cannon mount and the gunner's periscope would permit its use to sight either weapon. The chief disadvantage of this location was the long nozzle extension required to prevent fuel from dripping on the vehicle. Although this design was satisfactory, number 5 was considered to be superior. It located the flame gun just to the right of the gun shield. In this position, the hazard from the dripping fuel was minimized and the flame gun was easily connected by 2:1 linkage to the gun shield, thus obtaining the +45 to −15 degree elevation range. As with the fourth design, the gunner's periscope was used to sight the flame thrower by using the appropriate linkage. Mock-ups were constructed to study each of these locations on the M26. The report from Edgewood Arsenal, dated 25 April 1947, concluded that the development of the T35 carrying 300 gallons of flame thrower fuel was technically feasible with the four man crew. However, by this time the M26 was becoming obsolete and the whole concept of main armament flame thrower tanks was being reconsidered.

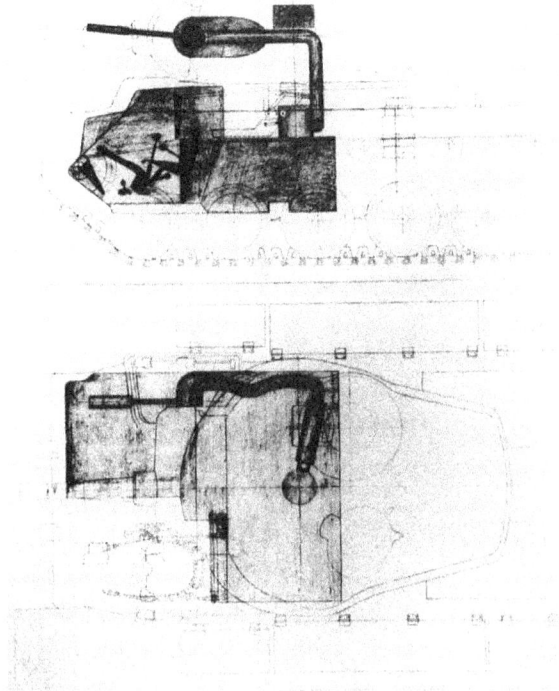

On 7 July 1948, a conference on mechanized flame throwers was convened at Fort Monroe, Virginia to review the tactical concept of flame thrower tanks and to recommend a course for future development. This conference concluded that a specialized main armament flame thrower tank was uneconomical from a tactical, logistical, training and procurement standpoint. It recommended that the T35 project be cancelled and that future efforts be directed toward the development of large capacity auxiliary flame throwers which could be attached to standard combat vehicles. This concept envisaged a trailer installation to obtain the required fuel capacity. Following this recommendation, one study applied the design of the British Crocodile to the M26. The flame gun fuel and its pumping equipment were carried in a trailer towed behind the tank. The fuel was pumped through a protected line to the flame gun mounted near the assistant driver's position and he fired the weapon. Ordnance Committee action on 6 October 1948 followed the recommendation of the mechanized flame thrower conference and cancelled the T35 project. The trailer concept was recommended in its place.

The plan 5 arrangement for the flame thrower equipment in the T35 can be seen in the drawing above. Note how the flame gun fuel tank extended into the space previously occupied by the bow gunner. Below is a drawing of the British Crocodile design with the flame gun fuel trailer applied to the M26 tank.

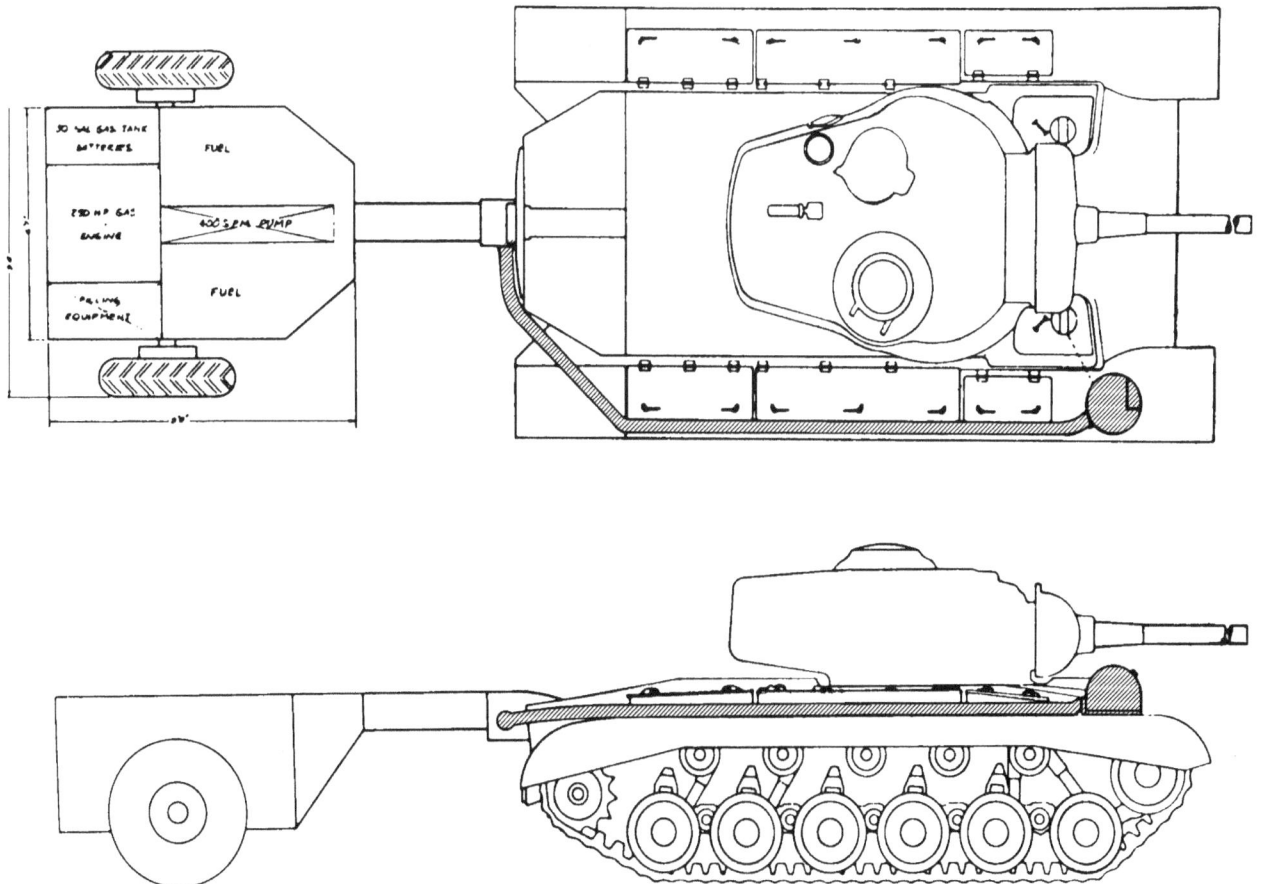

Above is the flame thrower tank T66 based on the M47 tank. The pistol port on the early T42 turret used for this pilot has been welded up.

The decision to drop the main armament flame thrower tank reflected the view that it was inefficient to have a specialized tank that was limited to the close infantry support role. However, the U.S. Marine Corps viewed the situation from a different standpoint. The primary use of Marine Corps tanks was close infantry support and their combat experience showed the great value of the main armament flame thrower tank in this application. With the outbreak of the Korean War, the situation for the Corps became urgent. The only flame thrower tanks available were the M42B1 and M42B3 built at the end of World War II as well as the Shermans converted in Hawaii for the invasion of Japan. With nothing better at hand, the Marines made use of these vehicles in the Korean fighting, but they continued to request the development of main armament flame throwers based on late model tanks.

To meet the Marine Corps requirement, the Chemical Corps initiated a project to install a main

armament flame thrower in the T42 tank. An early T42 turret was modified with a flame gun replacing the 90mm cannon. By the time this was ready, the T42 was no longer being considered for production and the modified turret was installed on the chassis of the M47 tank. This combination was designated as the flame thrower tank T66. However, only one pilot was completed since the M47 itself was now being replaced by the newer M48 and work continued to adapt the flame equipment to the later vehicle.

On 13 October 1954, the Chemical Corps Technical Committee classified the M7-6 main armament mechanized flame thrower as a standard type and a component of the flame thrower tank T67 based on the M48. As its Chemical Corps designation indicated, the M7-6 consisted of the fuel and pressure unit M7 and the flame gun M6. Prior to standardization, this was the E28-30R1 referring to the experimental E28 fuel and pressure system and the E30R1

The pilot flame thrower tank T67 below is at Aberdeen Proving Ground on 5 November 1953. This pilot was based on the M48 tank with the low silhouette Chrysler commander's cupola.

The view above of the pilot T67 at Aberdeen shows that it was based on one of the very early M48s. Note the thin (½ inch) transmission inspection cover plates on the rear hull. These were thickened on the later vehicles to equal the protection of the remainder of the rear hull plate.

flame gun. The complete assembly, including the turret itself, was designated in the Ordnance records as the flame thrower turret T7. The pilot tank was assembled using an M48 type turret with the low silhouette Chrysler cupola and the external .50 caliber machine gun mount. Since the flame thrower tank had no cannon, the loader was eliminated reducing the crew to three. The tank commander and the gunner remained in their usual positions with the same seats as in the M48. The M6 flame gun was installed in a shroud designed to resemble the 90mm gun, although it was shorter and slightly larger in diameter. This dummy gun tube also had holes in the side to permit the entrance of air for combustion and holes and drip shields in the bottom

for drainage. A removable cover on the top of the tube allowed access to the ignition components. The complete assembly was attached to the gun shield and pivoted on trunnions in the front of the turret opening. A .30 caliber coaxial machine gun was mounted to the right of the flame gun. The muzzle heavy gun assembly was balanced by an hydraulic equilibrator throughout its elevation range of +45 to −12 degrees. Many standard M48 components were retained such as the hydraulic power traverse unit, the radio, and the ventilating equipment. The controls for the commander and the gunner were similar in appearance to those in the M48, but they were modified for operation of the flame gun. The gunner aimed both the flame gun and the .30 caliber coaxial

The small driver's hatch of the M48 is apparent in the front view of the T67 at the right. Note also the flattened tops of the headlight brush guards to permit clearance of the flame gun at maximum depression. Below, the T67 is firing during a field test on 27 November 1953.

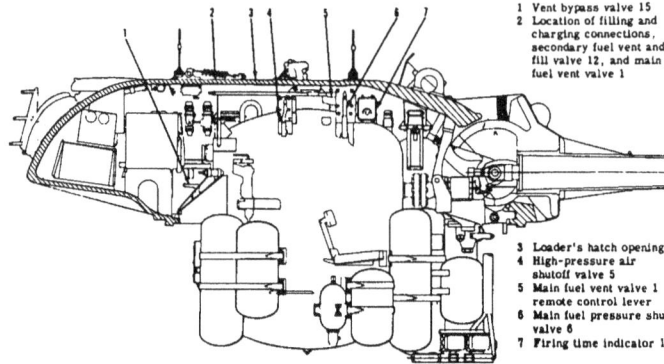

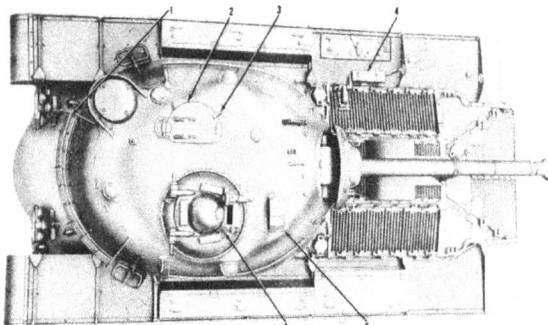

1 Stowage rack   3 Padlock   5 Gunner's periscope
2 Loader's hatch   4 Auxiliary engine exhaust muffler   6 Padlock

A top view of the T67 flame thrower tank is above at the left. As can be seen here, the muffler for the auxiliary generator engine has been relocated to the right fender. Above at the right is a sectional view showing the internal arrangement of the left side of the T67 turret.

machine gun using the T39 periscopic sight. This was a unity power instrument with a wide angle field of view designed as the primary sighting device for the flame thrower tank. It was later standardized as the periscopic sight M21.

The only topside entrance to the turret was through the commander's cupola as the loader's hatch was used to provide access to the flame thrower fueling and charging controls. The thickened fuel for the flame gun was stored in the main fuel container located in the left center of the turret. A 10.2 gallon secondary fuel container supplied unthickened gasoline to the atomizer for starting, and to the flame gun where it was used to coat the main fuel stream for ignition under cold weather conditions. Both the main and secondary fuel systems

were pressurized to 325 psi. The main fuel container had a total capacity of 398 gallons, but the necessity of leaving room for expansion and other losses reduced the usable capacity to 365 gallons. This permitted total average firing times of 55 and 61 seconds with the ⅞ inch and ¾ inch nozzles respectively. The fuel ignition system consisted of two high voltage (approximately 24,000 volts) spark plug igniters in front of the nozzle inside the dummy gun tube. A carbon dioxide snuffer system also was included to extinguish any residual fuel burning inside the shroud after the gun itself was shut off.

Below is the T67 gunner's station with a close-up of the gunner's controls at the right. The .30 caliber coaxial machine gun is not mounted in this photograph.

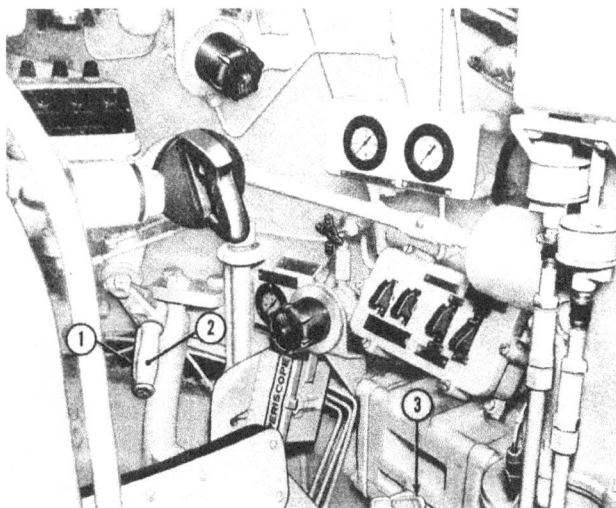

1 Release lever   2 Manual traversing control handle
                       3 Turret traverse lock

1 Handgrip   3 Supercharge accumulator
2 Periscope linkage   4 Valve block

1 Gunner's control handle
   assembly switches
2 Indicator lights
3 Fuel air pressure gage 4
4 Pilot fuel pressure regulator 3
5 Gunner's control handle
6 Fuel pressure setting gage 7
7 High-pressure air gage 12
8 Commander's control handle
9 Test panel
10 Secondary fuel pressure
   shutoff valve 9

11 Secondary fuel pressure
   regulator 8
12 Secondary fuel pressure
   gage 10
13 Manual elevating control
14 Manual traversing control
   handle
15 Equilibrator pressure gage
16 Trigger of CO_2 fire extinguisher 11
17 Hand supercharging pump
18 Foot firing switch
19 Turret traverse lock

1 Mount attaching nut   3 Elevation boresight knob
2 Azimuth boresight   4 Boresight locking lever
  knob   5 Head catch

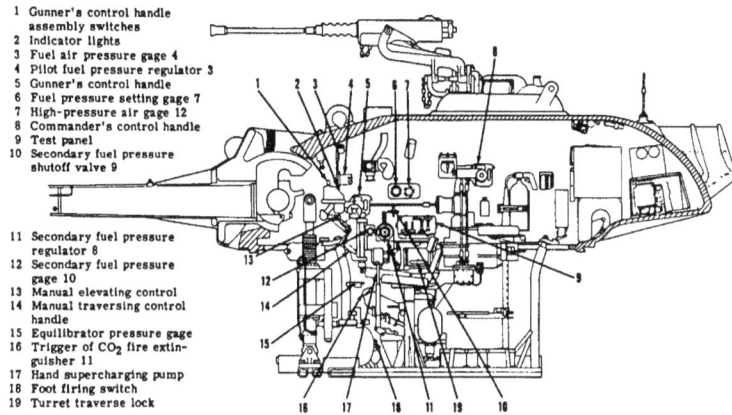

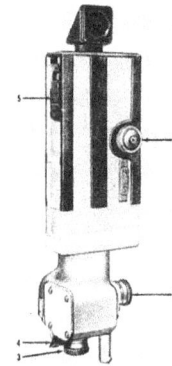

The right side of the T67 turret interior can be seen above at the left. The T39 periscopic sight designed for use with the flame gun appears above at the right. The special travel lock required for the flame gun is shown in the view of the rear deck below.

1 Gun tube   2 Gun traveling lock

Certain modifications to the M48 hull were required when the flame thrower turret was installed. Since the maximum depression for the flame gun was greater than that for the cannon, the tops of the headlight brush guards were flattened to provide adequate clearance. On the M48, the auxiliary engine muffler was mounted on the rear deck. On the T67, it was relocated to the right rear fender to prevent interference with the flame thrower fuel tank vent which protruded through the left bottom of the turret bustle. The 90mm ammunition racks were removed from each side of the driver's position and replaced by stowage racks for flame thrower spare parts, a tool box, and machine gun ammunition. A new electrical slip ring was required to handle the circuits for the flame gun.

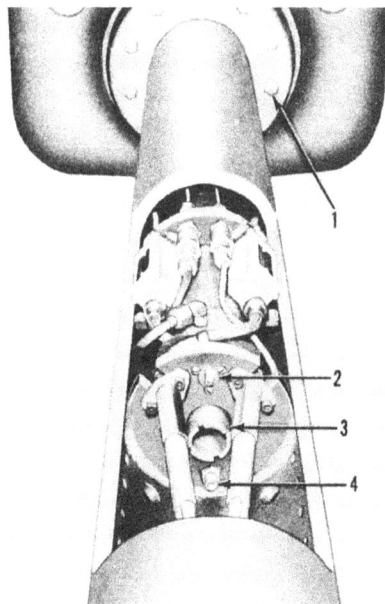

At the left is the flame gun with the access cover removed from the dummy gun tube. Below, the driver's seat in the T67 can be seen between the stowage racks which replaced the 90mm ammunition racks.

1 Gun shield capscrews   3 Main fuel nozzle extension
2 Atomizer nozzle   4 CO_2 snuffer outlet

Above is the standardized M67 flame thrower tank (left) and the pilot M67A1 flame thrower tank at Chrysler (right). Note that the M67 is based on the M48A1 with the large driver's hatch and M1 cupola for the tank commander. The M67A1 uses the chassis of the M48A2. Below is an additional view of the M67A1 at Chrysler.

After tests by the Marine Corps, a total of 56 complete T67 tanks were procured in addition to 17 of the T7 flame thrower turrets. All of these were based on the M48A1 tank with the M1 commander's cupola. The 17 turrets were then installed on modified M48A1 hulls and the pilot T67 was reworked to the M48A1 standard bringing the total inventory to 74 tanks. On 1 June 1955, OTCM 35901 standardized the T67 as the flame thrower tank M67. The same action designated the T7 turret as the flame thrower tank turret M1.

After the appearance of the M48A2 tank, the M1 flame thrower turret was fitted to the new chassis. The M7 fuel and pressure system was modified to U.S. Army requirements and redesignated as the M7A1. Thus under Chemical Corps terminology, the complete flame thrower now became the M7A1-6. On 8 January 1959, OTCM 36947 classified the M48A2 chassis with the new turret mounted flame thrower as Standard A and designated it as the flame thrower tank M67A1. The new vehicle incorporated features of the later model tanks such as the Cadillac Gage constant pressure hydraulic turret control system. It also was fitted with the new XM30 periscopic sight for the gunner. This sight had a magnification of 1.5x with a 48 degree field of view. Thirty-five of the M67A1s were built by Chrysler at the Delaware plant in 1955–56. This was the only version of the flame thrower tank to see U.S. Army service and that was only for a limited period.

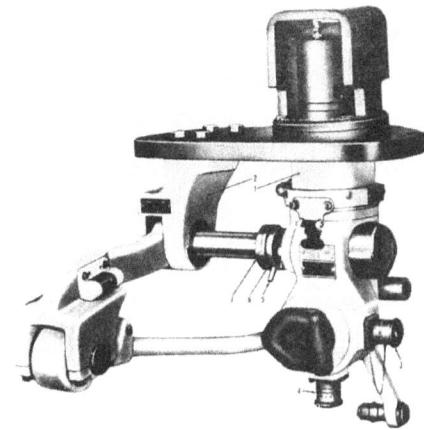

1 XM30 periscope	5 Coupling lever
2 XM113 periscope mount	6 Input coupling
3 Azimuth (deflection) boresight knob	7 Output coupling
4 Elevation boresight knob	

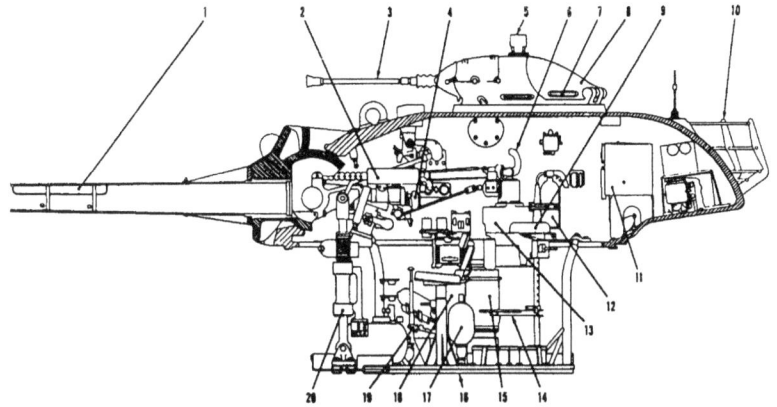

1 Barrel cover	8 Hatch door	15 Power pack assembly
2 Cal. .30 machinegun	9 Commander's seat	16 Turret platform
3 Cal. .50 HB Browning M2 machinegun	10 Stowage rack	17 Equilibrator accumulator
4 Riser and gunner's control	11 Radio sets	18 Main accumulator
5 M28C periscope	12 Portable fire extinguisher	19 Hand pump
6 Commander's control handle	13 Turret traversing mechanism	20 Elevating mechanism
7 Direct view vision blocks	14 Commander's platform	

Sectional views of the M67A1 turret are shown above and below at the right. At the top left is the XM30 periscopic sight. Below at the left is the tank assembly for the flame thrower fuel.

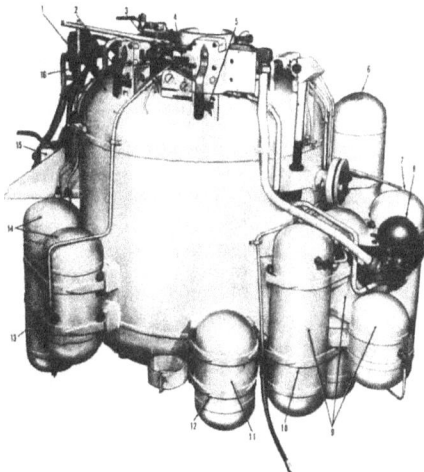

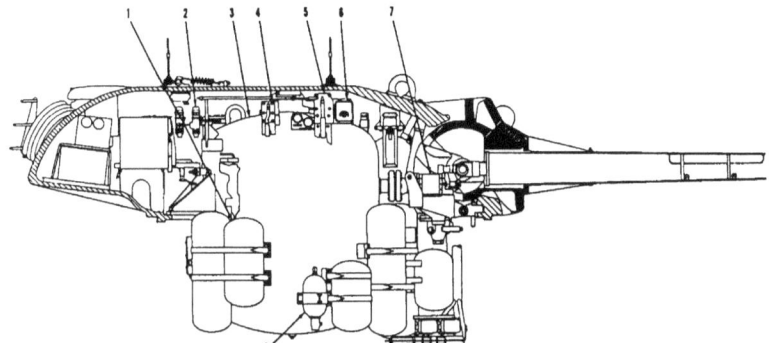

1 Air pressure container	4 High-pressure air shutoff valve 5	7 Equilibrator
2 Relief valves	5 Main fuel pressure shutoff valve 6	8 Equilibrator accumulator
3 Main fuel container	6 Firing time indicator 14	

1 Main fuel container	9 Pressure container
2 Name plate	10 Pin and strap
3 Air charging header assembly	11 Vertical pressure container
4 Nameplate	12 Pin and strap
5 Nameplate	13 Pin and strap
6 Pressure container	14 Pressure container
7 Pressure container	15 Bypass vent valve 15
8 Large volume regulator	16 Relief valves

At the right are the gunner's controls in the M67A1. Note the use of the Cadillac Gage turret control system. Below is a top view showing the general arrangement and stowage of the M67A1.

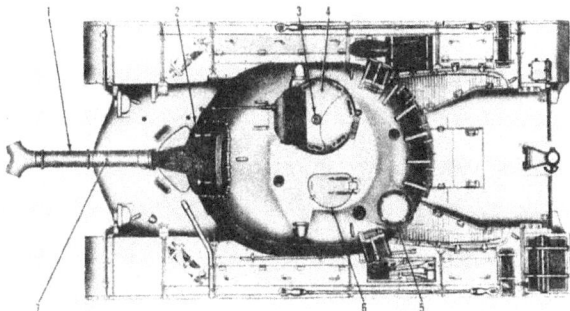

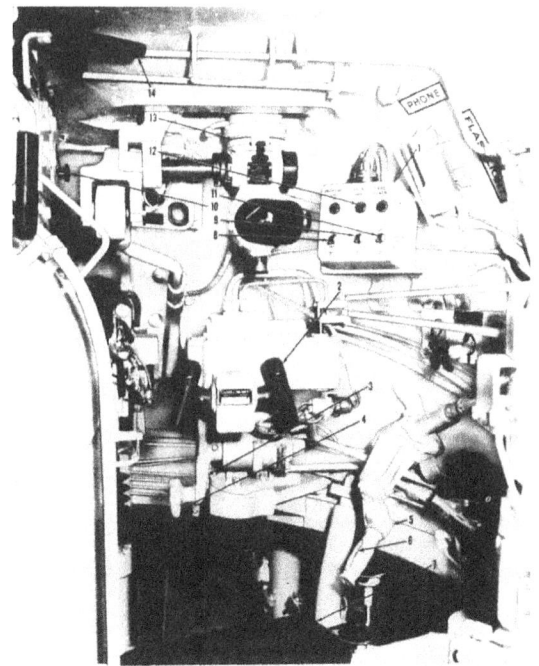

1 Flame thrower gun	4 Cal. .50 machinegun mount	6 Loader's hatch
2 Combination gun mount	5 Ventilating blower	7 Barrel cover
3 M28C periscope		

1 Gunner's control switch box	6 Release lever	11 Turret power switch
2 Gunner's control handle	7 Gunner's seat	12 Indicator lights
3 Hand elevation pump handle	8 Machinegun firing switch	13 XM30 periscope
4 Cal. .50 machinegun firing switch	9 Main gun switch	14 Main fuel valve remote control handle
5 Hand traversing control handle	10 Firing time indicator	

The flame thrower tank M67A2 above is firing during field trials. Based on the M48A3, this tank can be identified by the five track return rollers, the personnel heater exhaust extending to the right side of the tank, and the fender mounted dry type air cleaners.

After the introduction of the diesel engine and other new features in the M48A3, the Marine Corps requested that their M67 flame thrower tanks be modernized to the same standard. Funds were provided in late 1961 to convert 35 of the M67s to the M48A3 hull configuration. The modernized vehicle was designated as the full tracked combat tank, flame thrower, M67E1 by OTCM 37996 on 1 February 1962. A pilot tank, which was part of the first 35 to be converted, was completed at Detroit Arsenal to check all of the modifications before they were released to production. The converted hull was essentially the same as the M48A3 with the ammunition racks replaced by the appropriate stowage. The turrets were brought up to the later standard of the M67A1 including the new fire control equipment, a nylon ballistic shield, a new gun shield cover, and installation of the 7.62mm M73 coaxial machine gun. On 25 June 1962, the M67E1 was standardized as the full tracked combat tank, flame thrower, M67A2 by AMCTC 128. Seventy-three of the M67A2s were converted in parallel with the 1963–64 M48A3 program at the Anniston and Red River Army Depots. This was the version of the flame thrower tank later employed by the Marine Corps in Vietnam.

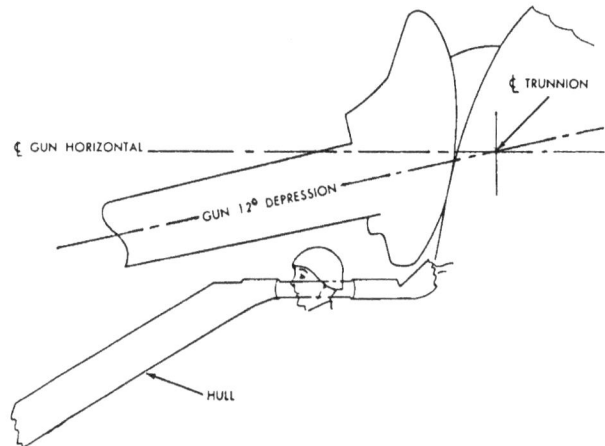

The sketch at the right above illustrates the limited head clearance for the driver when the flame gun was fully depressed. The two views below show an M67A2 converted by Chrysler from an M67A1 in January 1968. Note the three track return rollers on each side.

The pilot 155mm self-propelled gun T97 is shown in the two photographs on this page. The early 155mm gun T80 is fitted with a muzzle brake.

SELF-PROPELLED ARTILLERY

Both the original Equipment Review Board in mid-1945 and the later Stilwell Board recommended the further development of self-propelled artillery. During World War II wide use was made of various tank chassis to provide the basis for artillery motor carriages. Although many of these were highly successful, it was noted that the use of a tank chassis frequently resulted in greater weight than necessary for self-propelled guns and howitzers. An objective of the postwar development program was to minimize this weight penalty, but still retain as many common components as possible between the various vehicles to simplify the logistics problem. The light tank chassis provided a suitable basis for weapons such as the 155mm howitzer or smaller, but it was inadequate for heavier pieces such as the 155mm gun.

A meeting between the Ordnance Research and Development Division and the Army Ground Forces at Detroit Arsenal in July 1946 produced tentative military characteristics for a new 155mm gun motor carriage. This design was intended to use as many power train and running gear components from the medium tank as possible. Studies at Detroit indicated that it was feasible to mount either the 155mm gun or the 8 inch howitzer on the new chassis using the same mount. This, of course, had already been done on the 155mm gun motor carriage M40 and the 8 inch howitzer motor carriage M43.

Reflecting the limited funds available during the postwar period, the development work was contracted out a piece at a time. Pacific Car & Foundry Company of Renton, Washington received a fixed price contract on 9 April 1948 to complete the design and build a full scale wooden mock-up of the proposed vehicle. This was followed on 13 April 1950 by another contract to construct a prototype armed with the 155mm gun. A supplemental agreement on 11 April 1951 covered the manufacture of an additional prototype armed with the 8 inch howitzer. These vehicles were originally designated as the 155mm gun motor carriage T97 and the 8 inch howitzer motor carriage T108. The lightweight

255

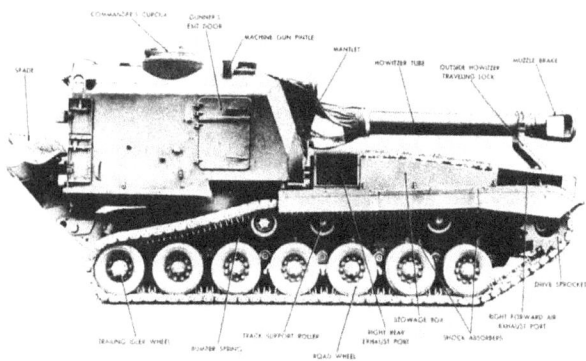

The pilot 8 inch self-propelled howitzer T108 appears above. Like the 155mm gun T80, the early 8 inch howitzer T89 is equipped with a muzzle brake on the pilot vehicle.

weapons developed for the vehicles were the 155mm gun T80 and the 8 inch howitzer T89. They were interchangeable in the mount T58. In this vehicle, the engine and transmission were in the front of the hull driving the sprockets at the front of each track. This arrangement left the rear of the chassis free for mounting the turret carrying the cannon. The turret was operated by hydraulic power and was limited to a total traverse of 60 degrees, 30 degrees to the right or left of center. Elevation ranged from +65 to −5 degrees with both weapons. A crew of six manned the vehicle with the driver in the left front of the turret.

The first pilot T97 was delivered in April 1952 followed by the T108 in July. Both weapons were fitted with muzzle brakes, but this feature was dropped for the production models. A so-called "ultimate" fire control system was installed in the first pilot. However, further study selected a simplified system for the second pilot and for the production vehicles. The initial production contract called for 30 T97s and 70 T108s and the first delivery was made in August 1952. However, production orders for both vehicles were increased and they continued to come off the line until April 1955. Like the tank

program, the rush to produce these vehicles resulted in a long list of required modifications after they were tested by the user. To make these changes, a modification program was set up at Pacific Car & Foundry on 1 July 1955 and the work was completed on 30 November 1956.

The two vehicles were standardized as the 155mm self-propelled gun M53 and the 8 inch self-propelled howitzer M55. The lightweight cannon were standardized as the 155mm gun M46 and the 8 inch howitzer M47, both in the mount M86. The stowage racks for the two sizes of ammunition were removable and interchangeable. When fitted for the 155mm gun, 20 complete rounds were carried. This was reduced to 10 complete rounds with the 8 inch howitzer. In early 1956, the Army started a program to convert all of its M53s to the M55 configuration dropping the 155mm gun. The Marine Corps, however, retained the M53 in service.

The development and production of the M53 and M55 paralleled rapid changes in the medium tank program. Since maximum interchangeability of parts was desirable, numerous changes were required during this period. However, the running gear based on the road wheels and the 23 inch wide

The production model 155mm self-propelled gun M53 is shown below. The transmission access port covers on the front armor plate differ from those on the early vehicles. Compare with the M55 on the opposite page.

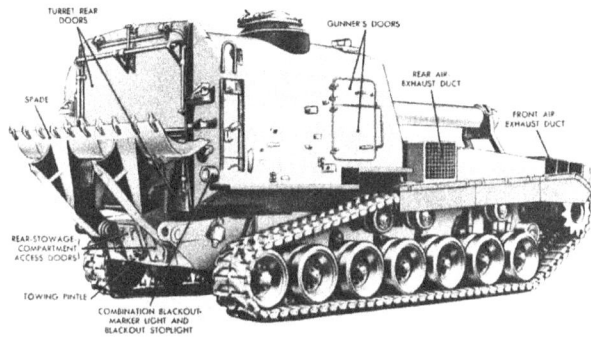

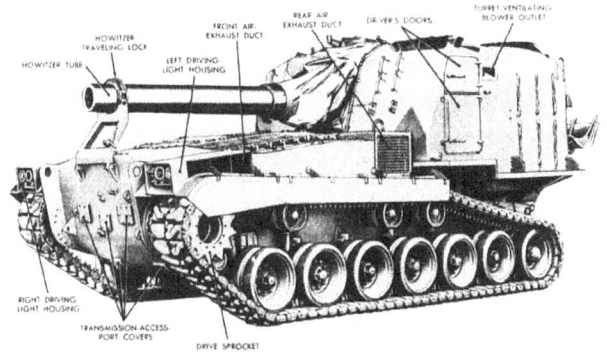

The muzzle brake has been eliminated on the 8 inch self-propelled howitzer M55 (T108) above, but this early vehicle retains the transmission access port configuration of the pilot. Later production vehicles had the same arrangement as the M53 on the previous page.

tracks of the M46 and M47 tanks was retained throughout the production run. The early vehicles were powered by the AV-1790-5B engine with the CD-850-4 transmission. Later in production, these were replaced by the AV-1790-7B and the CD-850-4B. Likewise, the M47 wobble stick steering was superseded in later vehicles by the steering wheel of the M48. Two vehicles were fitted with the unit cooled fuel injection engine and designated as the 8 inch self-propelled howitzer M55E1. Other experimental installations continued throughout the service life of these vehicles.

A top view of the early 8 inch self-propelled howitzer M55 (T108) is at the right. Except for the cannon and the traveling lock, the M53 (T97) was identical. A sectional drawing of the 8 inch self-propelled howitzer M55 (T108) appears below.

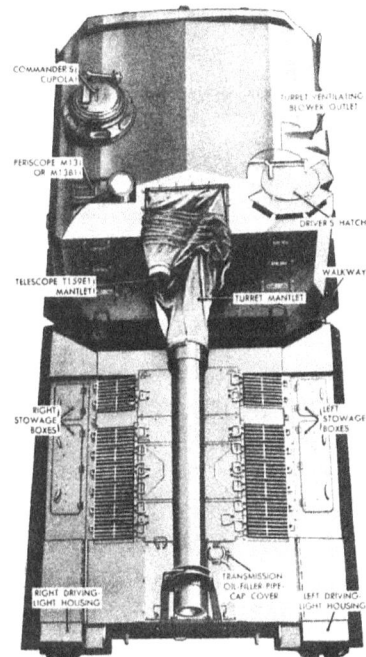

A—Engine
B—Muffler
C—Personnel heater
D—Equilibrator vertical-adjusting cylinder
E—Main instrument panel
F—Driver's seat
G—Recoil cylinders
H—Vertical equilibrator cylinder
J—Commander's seat
K—Projectile stowage rack
L—Projectile rammer and spade hoist
M—Turret bearing
N—Horizontal equilibrator cylinders
P—Equilibrator gas bottles
Q—Equilibrator accumulator bottle
R—Bilge pump
S—Auxiliary engine muffler
T—Carburetor
U—Oil cooler
V—Transmission

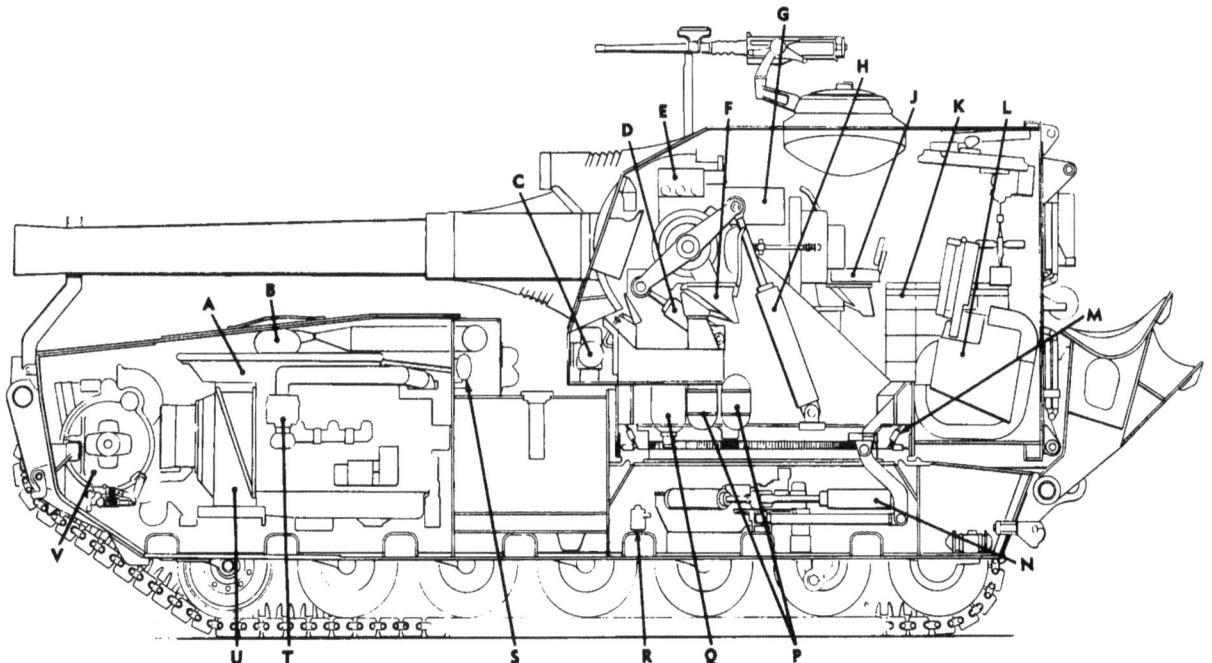

257

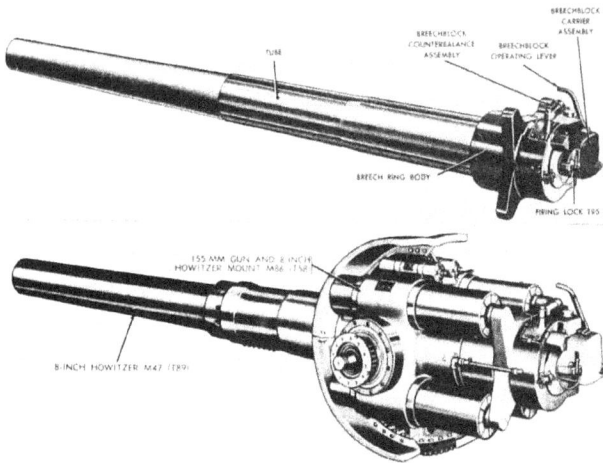

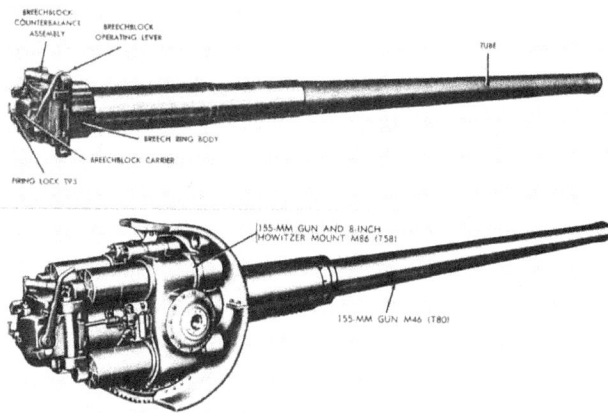

Above, the 8 inch howitzer M47 (left) and the 155mm gun M46 (right) are shown separately and installed in the mount M86. Below, the details of the shield and turret front can be seen at the left and the sighting and fire control instruments are at the right.

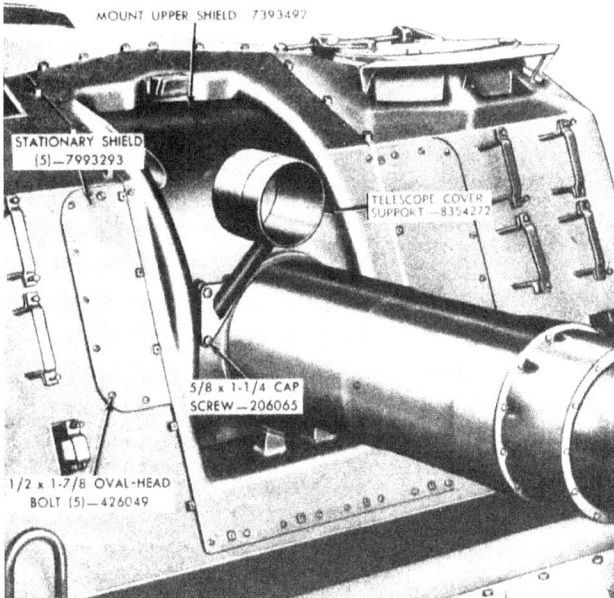

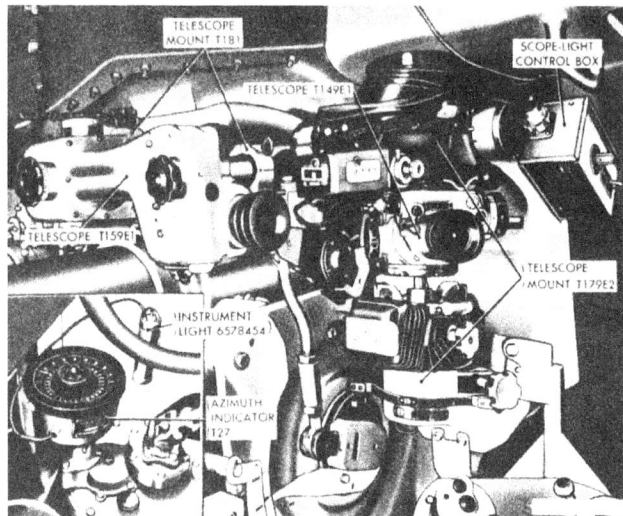

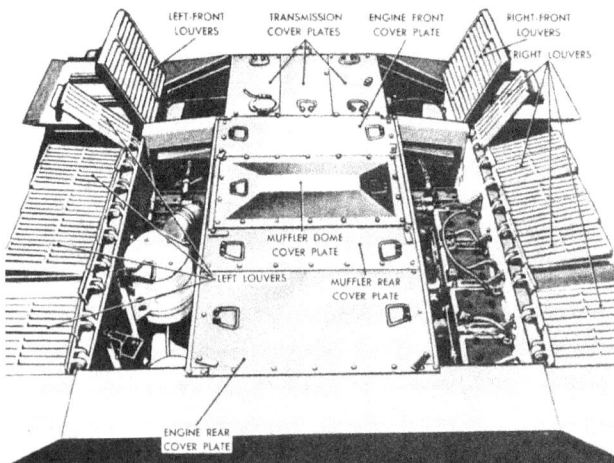

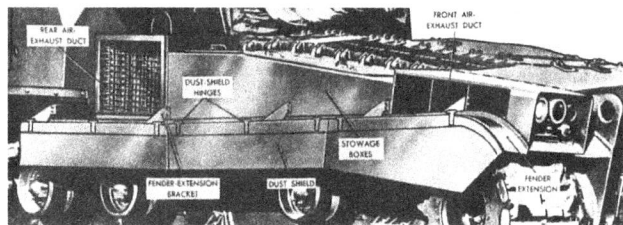

Details of the right front fender and the suspension system are shown above and below respectively. At the left is the engine deck with all of the louvered grills open.

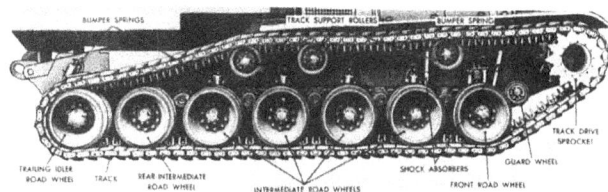

Above, a 155mm self-propelled gun M53 is operating with the 2nd Marine Division near Vieques, Puerto Rico during 1953. Below, a Marine Corps 8 inch self-propelled howitzer M55 is photographed at Camp Lejeune, North Carolina in 1961. Note that both of these vehicles have the late configuration for the transmission access port covers.

These views show the 8 inch self-propelled howitzer M55E1. Only two pilots were built by the Pacific Car and Foundry Company. With a combat weight of 98,000 pounds, this vehicle was powered by the AVI-1790-8 engine using the CD-850-5 transmission.

Above is an early artist's concept of a new self-propelled 175mm gun. Even at this early stage, there is a considerable resemblance to the T97 and T108. Note that the gun is equipped with a bore evacuator and a muzzle brake.

Although the Army converted its M53s to the howitzer armed M55, the need was recognized for a high powered long range gun. The Army Equipment Development Guide of 29 December 1950 recommended that a gun approximately 170mm in caliber be developed to replace the older 155mm weapon.

To meet this requirement, the 175mm gun T181 was developed and Pacific Car & Foundry received a contract to adapt this weapon to a self-propelled mount. Designated as the 175mm self-propelled gun T162, it was based on the T97-T108 design then in production. Although similar in appearance to the earlier vehicles, the T162 differed in several important respects. The most obvious difference was the

long barreled 175mm gun. The size and weight of this weapon required changes in the self-propelled mount. Installed on a wider hull, the suspension was strengthened and fitted with the 28 inch wide tracks of the M48. The T162 was powered by the supercharged Continental AVSI-1790-6 gasoline engine which developed 1000 gross horsepower at 2800 rpm. This was a unit cooled engine with fuel injection and its high output required the use of the Allison XT-1400-3 transmission. The driver's wobble stick or wheel control on the M53 or M55 was replaced by a steering bar in the T162. Like the earlier vehicles, it also was fitted with a compressed air scavenging system to remove powder fumes from

Below is the pilot 175mm self-propelled gun T162 as completed by the Pacific Car and Foundry Company. Unlike the T97 and T108, this vehicle used the 28 inch wide tracks of the M48 tank.

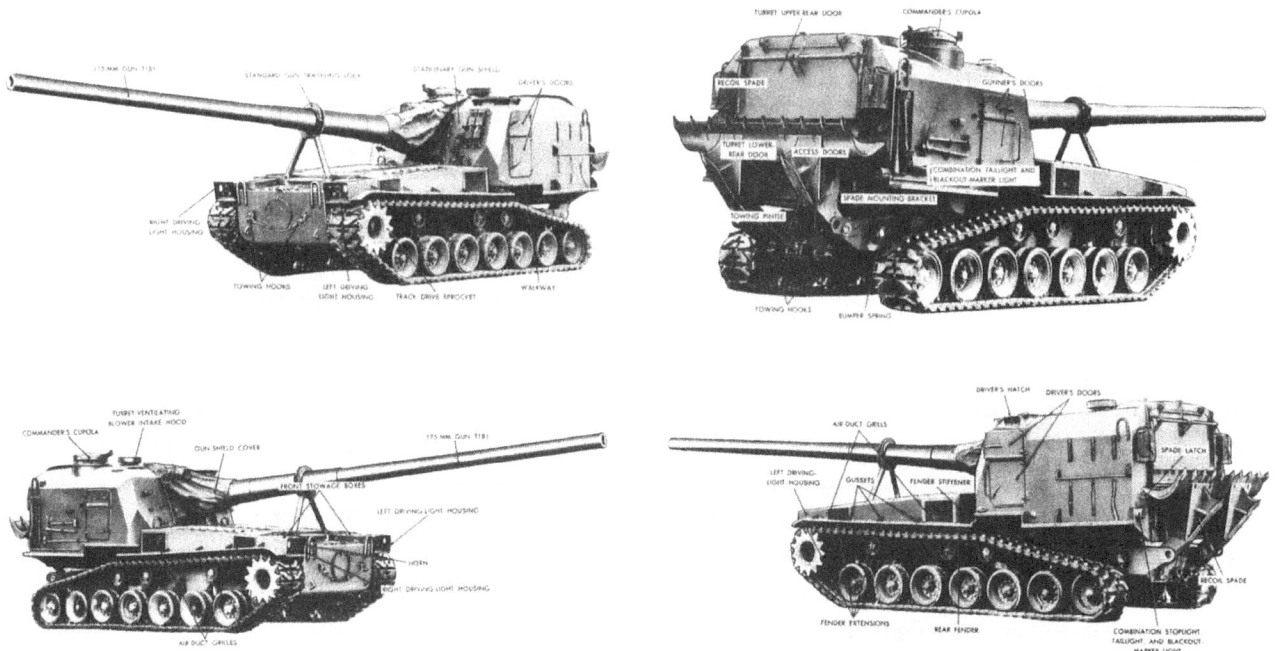

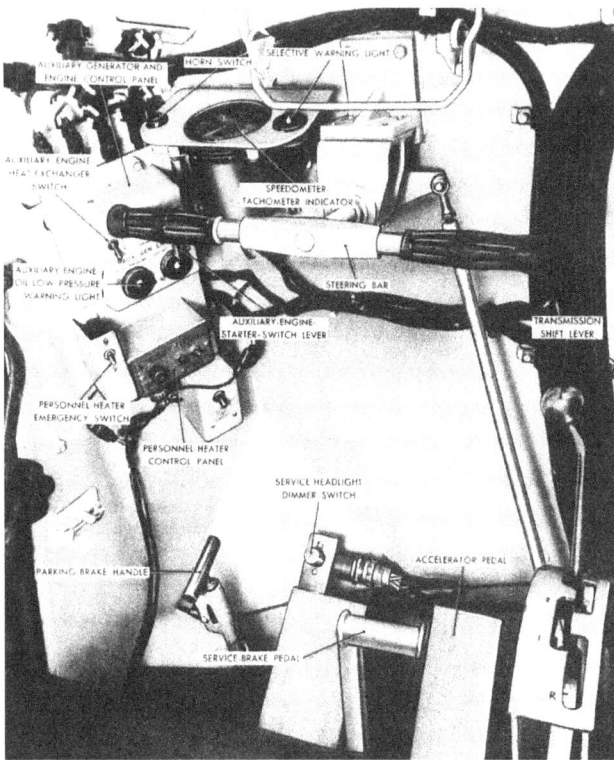

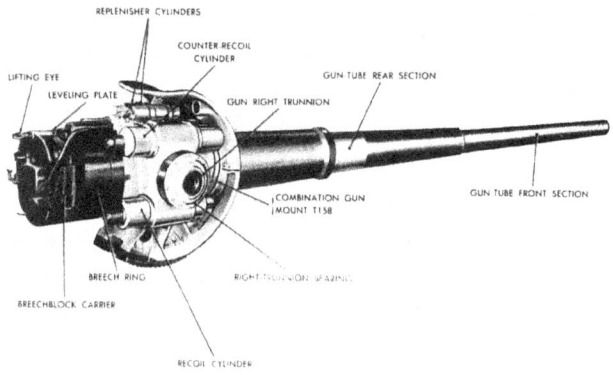

The driver's controls in the T162 appear at the left. Note that T-bar steering has replaced the earlier wobble stick or steering wheel. Above, the 175mm gun T181 is shown installed in the mount T158.

the turret. The hull frontal armor was reduced from 1 inch on the M53 and M55 to ¾ inches on the T162, but ½ inch thick plate was retained for the hull sides and the turret.

Three pilot T162s were built by Pacific Car & Foundry Company. Pilots 1 and 2 used hydraulic power to move the turret in traverse and the gun in elevation. On number 3, this was replaced by an electrical system. During the test program, the T162 had the usual number of problems expected from a new piece of equipment. However, the greatest obstacle to its adoption was its size and weight. By the middle of the 1950s, the Army policy was to make as

much of its equipment as possible transportable by air. This was not practical for medium tanks at that time, but self-propelled artillery was another matter and several projects were launched to develop such lightweight weapons. With this shift in interest, the T162 program was cancelled.

If the T162 was considered too big and heavy, the situation could have been much worse if some earlier ideas had ever seen the light of day. In the late 1940s concepts were considered for even heavier self-propelled artillery. The designations T146 and T147 were assigned for self-propelled versions of a 240mm howitzer and an 8 inch gun respectively. Both were dropped early in the planning stage when it became obvious that the maximum weight could not be limited to the specified 60 tons.

The engine deck on the T162 and the traveling lock for the 175mm gun can be seen at the right. Below is the suspension using many components from the M48 tank. However, note the very large drive sprocket with 13 teeth.

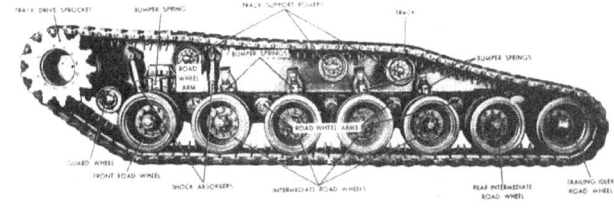

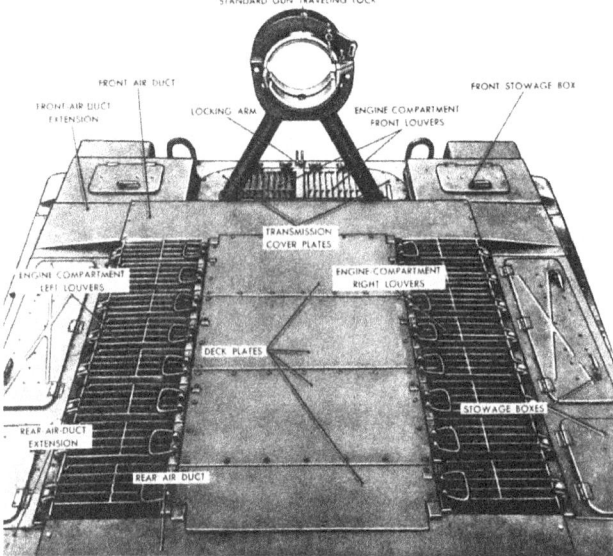

Above is the concept drawing of an antiaircraft tank as proposed by Colonel John Berres early in 1975. Note the similarity of this vehicle to the later development of the DIVAD gun.

ANTIAIRCRAFT TANKS

During the early 1970s, there was concern in many quarters of the U.S. military regarding the inadequate state of short range air defense in the Army. In December 1972, an all-weather short range air defense weapon was specified as a requirement by the Field Army Air Defense Study. Subsequent studies at Stanford Research Institute indicated that a 35mm gun would be two to three times more cost effective than the 20mm Vulcan cannon then in use. In any case, the Vulcan was limited to ranges below 2000 meters, half of that desirable for an effective weapon. The research also indicated that a gun and missile combination would provide the most effective defense. These conclusions were supported by the Short Range Air Defense (SHORAD) study of August 1973 and the Division Air Defense study in April 1974. Both of these recommended the development of an air defense gun to complement the antiaircraft missile.

In May 1974, the U.S. Army Air Defense School proposed a requirement for an air defense gun. This was approved by the Vice Chief of Staff and in November, the U.S. Army Materiel Command and the Training and Doctrine Command agreed that the weapon should be 30mm to 40mm in caliber. Previous weapons of this type, such as the 40mm twin self-propelled gun M42 had mounted their guns in open turrets on lightly armored chassis. Although they lacked the modern fire control equipment necessary to be effective in the antiaircraft role, these self-propelled guns proved to be extremely useful against ground targets in Korea and Vietnam. However, such use frequently exposed

them to ground fire revealing the vulnerability of the lightly armored vehicles.

The value of self-propelled guns against both air and ground targets was clearly recognized by Colonel John Berres, president of the U.S. Army Armor and Engineer Board. In an article in Armor magazine early in 1975, Colonel Berres proposed the development of an upgraded version of the M42 with radar controlled guns utilizing a fully armored turret on the chassis of a dieselized M48 or M60 tank. This concept clearly indicated the trend of future events.

After presentations to the Army Vice Chief of Staff and the Secretary of Defense, the Defense Department approved the operational requirement for the new self-propelled weapon. Originally referred to as the Advanced Radar-directed Gun Air Defense System (ARGADS), it later became the Division Air Defense (DIVAD) gun system. At this time, the caliber had not been selected and the request for proposal only specified 30mm to 40mm leaving the choice of the gun to the industry. Studies at the Ballistic Research Laboratory, the Air Defense School, as well as other agencies had listed five weapons as candidates to arm the new system. These were the 30mm Mauser F gun then under development, the General Electric 30mm GAU-8 multi-barrel cannon used in the Air Force A10 attack plane, the Oerlikon 35mm twin guns in production for the German Gepard antiaircraft tank, a 35mm version of the Sperry 37mm T250 Vigilante Gatling type gun, and an adaptation of the Bofors 40mm L/70 twin guns.

The request for proposal was issued to industry on 26 April 1977 and by July responses had been received from the Aeroneutronics Division of Ford Aerospace & Communications Corporation, the Pomona Division of General Dynamics, the General Electric Company, the Raytheon Company, and the Sperry Division of Sperry Rand Corporation. Two contractors were to be selected to develop prototypes during a 29 month accelerated program. It was intended that each contractor would utilize as many items of standard equipment as possible to minimize cost and improve reliability. In both cases, the gun system was to be installed on government-furnished modified M48A5 tank chassis.

The Raytheon proposal adapted the turret from the Dutch version of the German flakpanzer Gepard to the M48A5 chassis. This turret, used on the Dutch CA-1 Cheetah antiaircraft tank, was armed with twin Oerlikon KDA 35mm cannon. Fire control equipment utilized the Hollandse Signaalapparaten radar and a Contraves computer. The Raytheon study showed that the turret and fire control systems were compatible with the M48A5 chassis and that it had 94 per cent maintenance line replaceable unit commonality with its European counterpart in NATO. Raytheon and their subcontractors were licensed to manufacture the system in the United States.

Not surprisingly, the General Electric Company proposal adapted their successful 30mm GAU-8 seven barrel cannon to an antiaircraft turret. In this mount, the externally powered GAU-8A was manned by a crew of two. Both the commander and the gunner had firing controls and either could operate the entire system. Nicknamed the Avenger, the proposed weapon had an effective kill range of 4000 meters as a design goal. Originally, the radar was an improved version of the one developed for the earlier Vulcan air defense system with which the fire control equipment had some items in common. Later, a more advanced radar system was proposed.

Below is a model of the DIVAD system proposed by the General Electric Company armed with the 30mm GAU-8A cannon. Above, an artist depicts the vehicle in action against enemy aircraft.

The Sperry short range air defense gun system is illustrated above.

The Sperry entry in the DIVAD competition made use of their previous experience in developing the Vigilante antiaircraft weapon system. It utilized basically the same Gatling type gun modified from its original 37mm caliber to fire the 35mm NATO round. A two man turret crew, seated side by side, manned the weapon system which, like the other entries, was fitted to the M48A5 tank chassis. The 35mm gun mounted at the left side of the aluminum armor turret had two rates of fire. These were 3000 rounds/minute for engaging aircraft and 180 rounds/minute for use against ground targets. The weapon was fed from a magazine containing 1464 rounds of ammunition. In addition to the Sperry integrated radar/IFF (Identification, Friend or Foe) system, other sensors provided independent 360 degree direct or television viewing for the crew.

General Dynamics armed their vehicle with twin Oerlikon KDA 35mm cannon. These weapons could be fired in either the automatic or semiautomatic mode and their combined rate of fire was 1100 rounds/minute. The magazines held 600 ready rounds and were accessible for rapid reloading. The twin gun mount was installed in the front center of the welded aluminum armor turret. A high performance electrohydraulic turret drive and stabilization system provided a shoot-on-the-move capability. The radar fire control was derived from the Phalanx system already in production. Independently stabilized optical sights, including a laser range finder, provided a separate means of target engagement for the two man turret crew seated directly behind the gun mount.

The arrangement of the early General Dynamics DIVAD gun proposal is illustrated by the drawing below. At the left is an artist's concept of the vehicle in action.

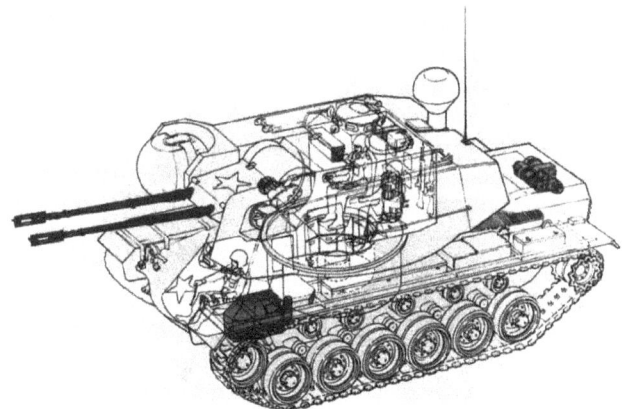

Above, the General Dynamics DIVAD gun moves at high speed over dusty terrain during test operations. Further details of this vehicle are illustrated in the views below.

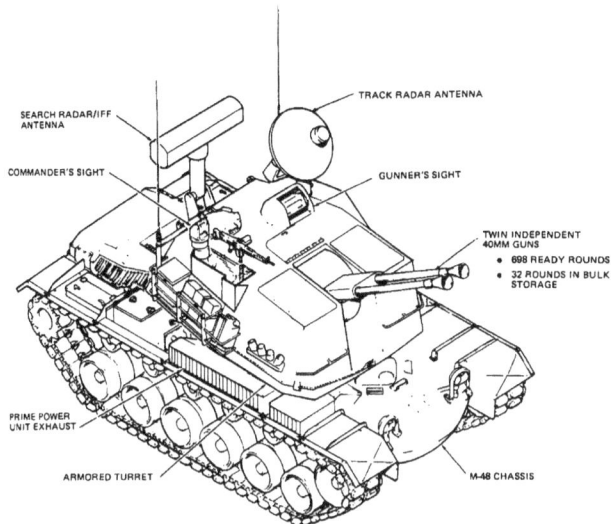

The Ford Aerospace DIVAD gun is shown in the concept stage above. Note the early version of the smoke grenade launchers on the turret.

The Ford Aerospace proposal, nicknamed Gunfighter, was armed with two 40mm Bofors L/70 guns, the largest caliber permitted under the specification. Although the combined firing rate for the twin guns was only 600 rounds/minute, the lethality of the larger rounds was increased by the availability of a prefragmented projectile fitted with a proximity fuze. The radar fire control was a modified version of the Westinghouse equipment developed for the F16 fighter. This system included the IFF apparatus as well as the search and track radar. The two man turret crew was seated side by side at a control console just behind the gun mount. The turret was welded from homogeneous steel armor and, like all of the other designs, it was installed on the M48A5 chassis with the driver retained in his usual hull position.

Following review by the Source Selection Evaluation Board, General Dynamics and Ford Aerospace were chosen to develop competitive prototypes. On 13 January 1978, contracts were signed with each company to build two vehicles. The General Dynamics and Ford Aerospace designs were designated as the 35mm air defense artillery self-propelled gun XM246 and the 40mm air defense artillery self-propelled gun XM247 respectively. These prototypes were delivered to the Army in June 1980 for the development and operational test (DT/OT) program. Operated by army personnel on over 200 missions, the two weapon systems engaged targets ranging from helicopters to jet aircraft flying at high subsonic speeds. These targets were taken under fire both from stationary positions and while the DIVAD was on the move. The evaluation

The Ford Aerospace DIVAD system as submitted for test can be seen below. M239 smoke grenade launchers have now been installed on the turret.

267

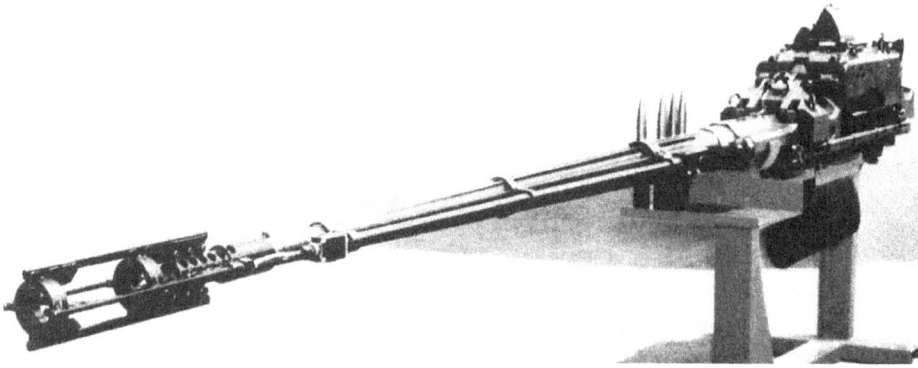

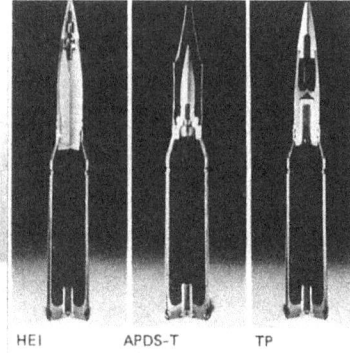

The 35mm Oerlikon gun used in the General Dynamics DIVAD system is shown above along with sectional views of its ammunition.

concluded in mid-November with war games involving three separate 72 hour periods of continuous operation to evaluate the reliability, maintainability, and general effectiveness of the two systems.

The competition was essentially between a weapon with a smaller projectile having a higher rate of fire versus one with a larger more lethal projectile, but with a lower rate of fire. The 35mm round carried 112 grams of explosive compared to 120 grams in the 40mm projectile. With respective firing rates for the twin guns of 1100 and 600 rounds/minute, a two second burst threw about 4.1 kilograms of explosive in 35mm rounds compared to approximately 2.4 kilograms with the 40mm ammunition. Although this appeared to give the advantage to the smaller caliber, the 35mm projectiles were detonated by contact while the 40mm rounds were prefragmented and fitted with a proximity fuze. Thus the latter did not have to actually hit the target

to destroy it. At a late date, a proximity fuze was developed for the 35mm round, however, the smaller size of the projectile did not permit the effective prefragmentation obtained with the 40mm caliber.

Although both weapon systems performed well in the test program, the Army decided in favor of the 40mm gun and on 7 May 1981, Ford Aerospace was selected to produce the new DIVAD gun system. The initial program called for 618 fire units mounted on the M48A5 chassis and 13 million rounds of ammunition. Deployment of the new weapon to Europe was scheduled for late 1984. In mid-1982, the XM247 was named the Sergeant York in honor of Sergeant Alvin C. York, the Medal of Honor winner in World War I.

Below is a drawing of the Bofors 40mm gun modified for use in the Ford Aerospace DIVAD system. The ammunition illustrated from the left is the prefragmented round with the proximity fuze, the high explosive point detonating shell, and the target practice round which simulated the high explosive. At the far right below is a close-up of the prefragmented projectile with the proximity fuze. The damage to an aircraft from this round is shown at the right.

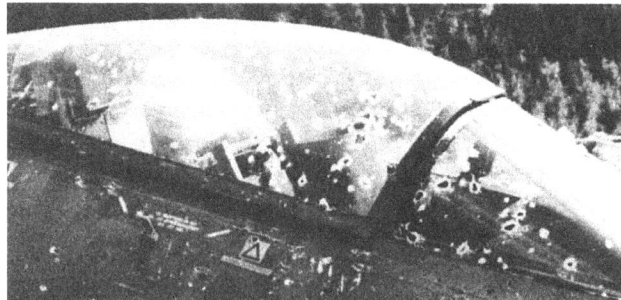

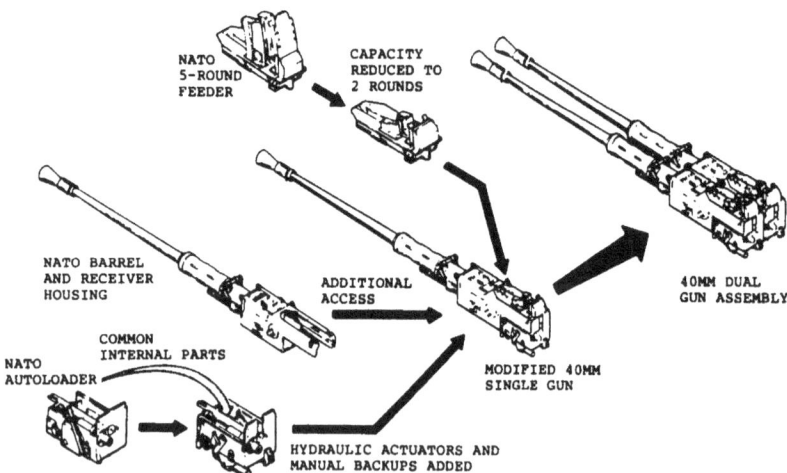

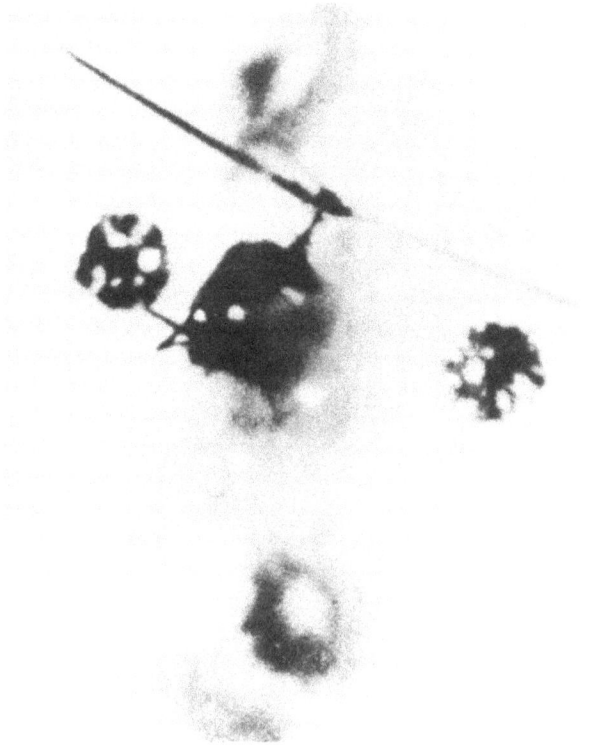

At the left, a helicopter is under fire from the proximity fuzed 40mm rounds showing the lethal effect of this ammunition. The crew stations in the Ford Aerospace DIVAD system are visible in the drawing above and the general arrangement of the vehicle can be seen in the cutaway view below.

☐ ARMAMENT

▨ GUNSIGHT

■ FIRE CONTROL AND DISPLAY

▦ SEARCH AND TRACK RADAR

▦ COMMUNICATIONS AND POWER

▤ SQUAD LEADER'S PERISCOPE

The Ford DIVAD gun during test operations is shown in these two photographs. Note the empty cartridge cases being ejected while firing below.

Details of the Ford Aerospace DIVAD system can be seen in these views of the vehicle during the test program.

Concept No. 1

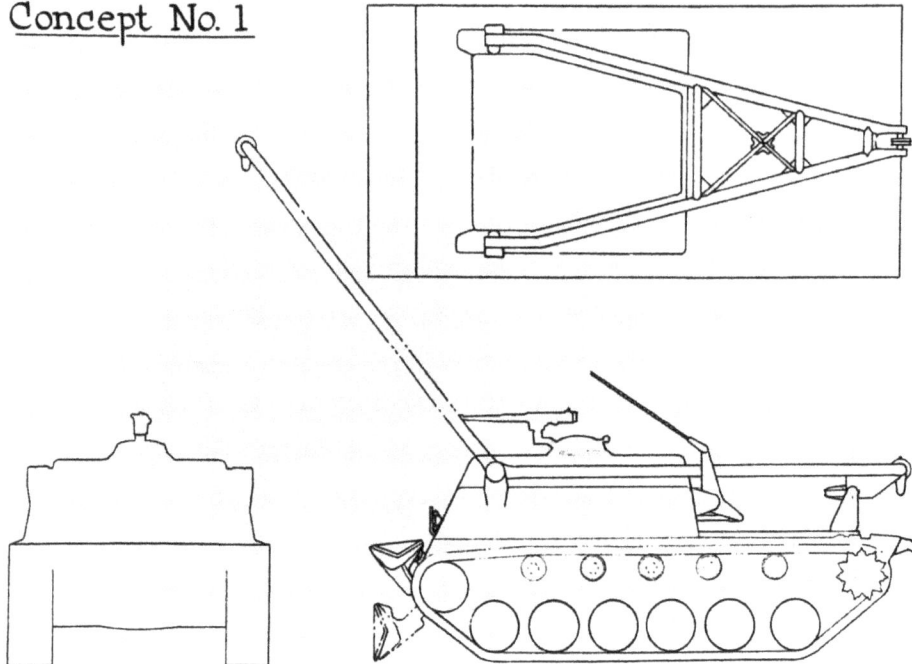

Above is the concept proposed by Bowen-McLaughlin, Inc. for the conversion of the M48 tank to a medium recovery vehicle.

TANK RECOVERY VEHICLES

The requirement for an armored vehicle capable of recovering disabled tanks from the battlefield dated from World War II. Experience in that conflict revealed that the ability to recover and repair damaged combat vehicles was frequently the deciding factor between victory or defeat. In late 1942, the T2 tank recovery vehicle was converted from the M3 medium tank. It was later standardized as the M31 series and some of these served until the end of the war. The M31 was soon superseded by an improved recovery vehicle based on the Sherman. This was the T5 which was equipped with an "A" frame type boom. Standardized as the M32 series, these vehicles served through both World War II and the Korean War. However, urgent requirements of the later conflict resulted in the conversion of additional late model M4A3 tanks with the horizontal volute spring

suspension. Standardized as the tank recovery vehicle M74, they were greatly improved over the earlier M32 series. Although the M74 was successful, the M4A3 tank was rapidly becoming obsolete and an urgent requirement existed for a tank recovery vehicle based on the contemporary medium tank. The

The model at the right is the medium recovery vehicle proposed by Bowen-McLaughlin, Inc.

development of such a vehicle was initiated by OTCM 35540, dated 7 October 1954. The same action outlined the desired military characteristics, assigned the designation medium recovery vehicle T88, and authorized the manufacture of three pilots.

A conference was held at Detroit Arsenal on 16 December 1954 to review various design proposals for the new medium recovery vehicle T88. Three concepts were considered with the first presented by Bowen-McLaughlin, Inc. describing the conversion of the standard M48 tank to a medium recovery vehicle. Such a conversion required the removal of the turret and the forward elliptical section of the hull to provide adequate space for stowage and for mounting a fixed cab. Extension of the hull also was required to incorporate a new power plant necessary to meet the performance specification. It was proposed that the vehicle width be reduced to 138 inches by installing 23 inch wide tracks from the M47 tank. An "A" frame type boom of tubular steel construction was mounted on the forward portion of the fixed cab. The boom did not traverse, but it was elevated by hydraulic cylinders and had a lifting capacity of 25 tons at a clear reach of 84 inches and a hook height of 230½ inches. The winch system consisted of a 25 ton boom or hoisting winch and a 50 ton tow winch, both of which were powered and controlled hydraulically. A cable operated spade was mounted on the hull front and a pair of spade stabilizers were installed at the rear. The cab was constructed of one inch armor on the front, sides, and rear and had a ¾ inch thick roof. It provided space for four crew members, one of which operated both the driving and winching controls.

Although the Bowen-McLaughlin concept had many desirable features, the conversion of M48 tanks to the recovery vehicle configuration was not recommended. In the first place, the availability of tanks for conversion was not assured and it would be more economical to design and manufacture a completely new hull. However, it was recommended that many of the features proposed in this concept be adopted for the new vehicle.

The second and third concepts reviewed, utilized the same basic chassis with one inch armor and a new flat track suspension eliminating the track return rollers. Concept number 2 featured a low silhouette cab constructed of ½ inch armor plate with structural reinforcements to support the hydraulically operated "A" frame boom mounted at the top front. Concept number 3 was similar, but it featured a high fixed cab with ¾ inch thick sides to provide maximum personnel and stowage space. However, it did conform to the railway height and width limitations (Berne International). Both vehicles had seats for the four man crew plus two additional seats for extra personnel. Concepts 2 and 3 were equipped with two winches for towing and hoisting and number 3 had an additional auxiliary winch to operate the spade and elevate the boom. A variety of power train combinations were considered, but the AVI-1790-8 engine and the XT-1400 transmission appeared to be the most suitable. Provision was made on designs 2 and 3 for jettisonable

Below is the second concept considered for the medium recovery vehicle. Note the flat track suspension and the jettisonable fuel drums along the sides.

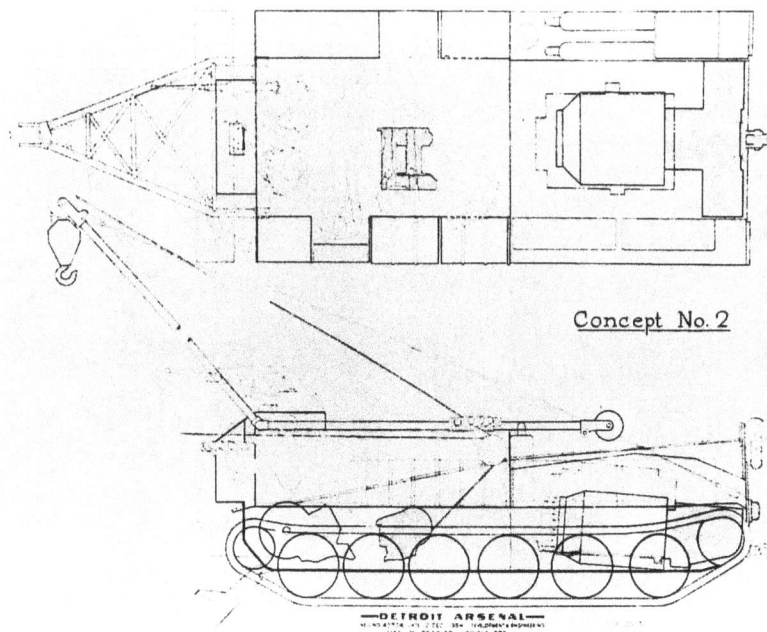

Concept No. 2

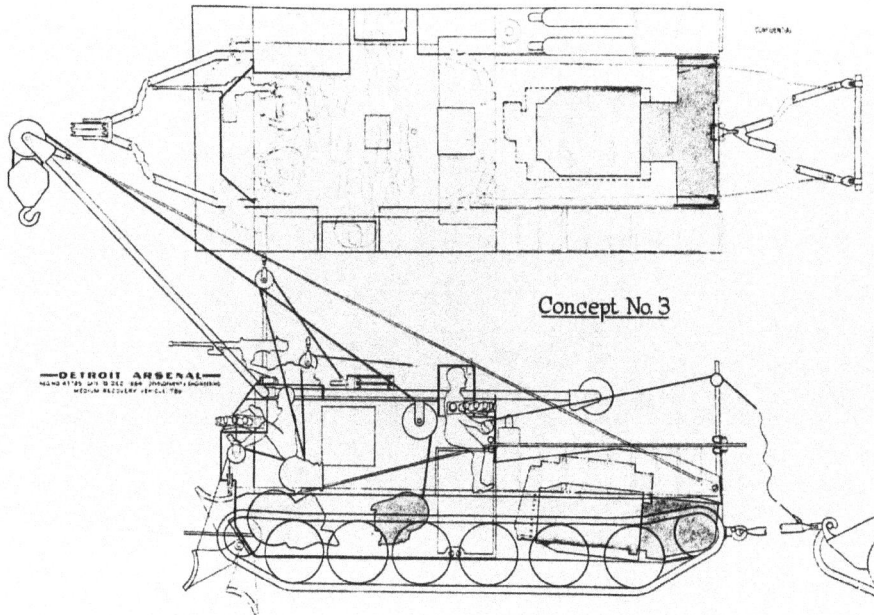

Concept No. 3

Concept number 3 (above) for the medium recovery vehicle is similar to number 2, but it has a much higher silhouette to provide the maximum personnel and stowage space within the limitation of the Berne International railroad clearance diagram.

fuel drums along the sides of the vehicle to increase the cruising range.

After considering the concepts presented, the conference recommended that the new recovery vehicle closely follow the future tank family to permit maximum interchangeability and standardization of components. However, the M48 series tanks were not to be converted to this application. Subsequently, two study contracts were let for the competitive evaluation of recovery vehicle designs. One of these went to Bowen-McLaughlin, Inc. and it resulted in the proposal that all winching, hoisting, and bulldozing operations be accomplished from the front end of the vehicle. The other study performed by Pacific Car & Foundry Company recommended that some of these operations be performed at the rear. Mock-ups were prepared for both concepts and Food Machinery and Chemical Corporation also submitted a mock-up of their own design which featured winching operations at the rear of the vehicle. After review of these mock-ups and a strong recommendation from the Continental Army Command (CONARC), it was decided that all winching operations would be accomplished from the front in the direction of the boom to facilitate operator visibility. With this decision, the equipment to winch and to provide spade emplacement from either end of the

vehicle was dropped. In October 1955, authorization was given to proceed with the manufacture of three pilots featuring all recovery operations from the front of the vehicle. This followed the Bowen-McLaughlin design and they received the contract to perform the work on 26 April 1956.

Considerable delay in the manufacture of the pilots resulted in reconsideration of the entire T88 program before any test results were available. A meeting at the Ordnance Tank Automotive Command on 29 January 1957 noted that the vehicle design was only 90 per cent complete, although orders had been placed for all major components. One problem had been to obtain a sub-contractor for the hull castings. Also, at that time, it was expected that the future medium tank would be one of the T95 series and it appeared that it would be entering

At the right is the Bowen-McLaughlin, Inc. mock-up of the T88 medium recovery vehicle. Note the changes in the cab and boom configuration compared to the earlier model.

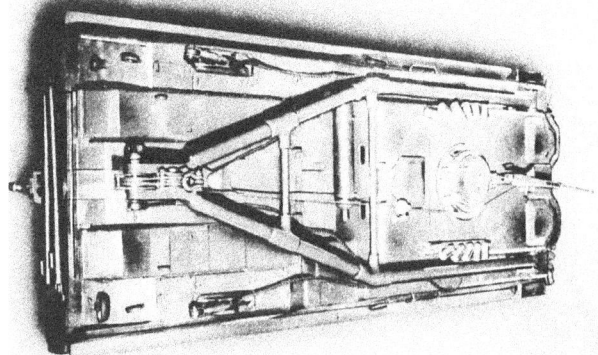

The options remained open for the suspension design on the T88 at this stage. All bases were covered on the mock-up with an M48 type suspension on the left side and a flat track design on the right.

production concurrently with the T88. In view of this, it was suggested that the T88 be cancelled and a new recovery vehicle be designed based on the lighter T95 components. The proposed designation for such a vehicle was the T88E1.

Since there were large numbers of M48 tanks in service, it was suggested that an interim recovery vehicle could be produced for use with them by applying a modification kit to the M48 itself. At this same time, it was proposed that the future recovery vehicle and the combat engineer vehicle be combined in one design. Since the T95 was expected to be the future production tank, the combined concept also was to be based on this chassis.

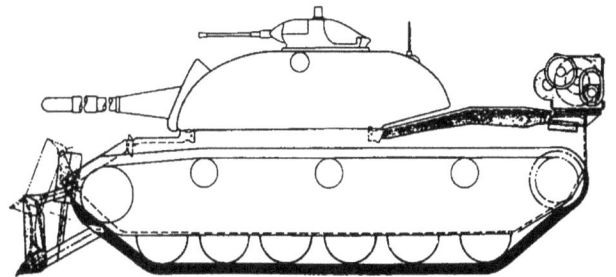

Above at the right is a drawing of the proposed interim recovery vehicle converted from the M48A2 tank. The dimensional data for the medium recovery vehicle T88 are presented below.

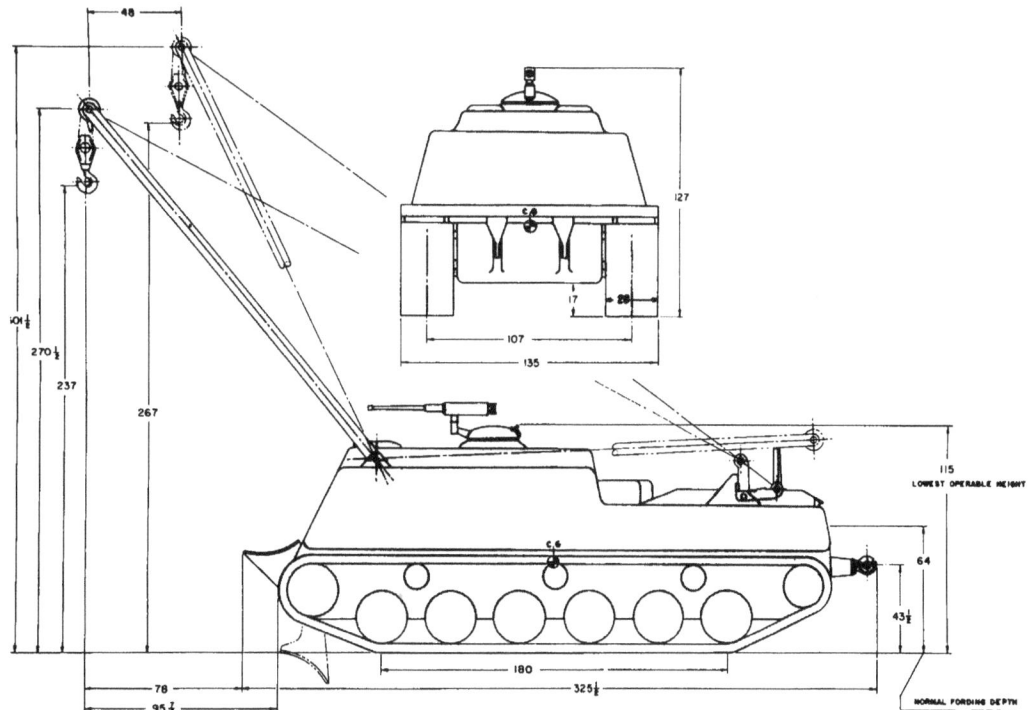

The silhouette of the medium recovery vehicle T88 is compared at the right with that of the earlier M74 converted from the Sherman tank.

The delivery of the pilot T88s and the first tests did not improve the situation. Initial studies had indicated that the AVI-1790-8 engine would provide sufficient power to obtain the maximum drawbar pull in a vehicle of the T88's weight. This selection also was desirable since it would mean using the same engine as in the M48A2. However, tests of the pilot recovery vehicles showed only marginally acceptable performance when towing the 52 ton M48. Since some weight increases were expected in the future, the T88 with AVI-1790-8 was rejected and it was recommended that the more powerful AVSI-1790-6 be used in future tests. Considerable experience with this engine and the XT-1400 transmission was already available from operations of the heavy recovery vehicle M51. This was the most powerful engine available to Ordnance at that time and a meeting on 20 October 1958 agreed that it would be installed in the engineering and user pilot T88s for retest. Because of the urgency of the requirement and the experience with the AVSI-1790-6 and the

XT-1400 in the M51, OTCM 36994, dated 19 February 1959, approved the manufacture of three production pilots and standardized the T88 with the new power train as the full tracked medium recovery vehicle M88.

By this time, the plans were for the XM60 to replace the T95 as the Army's future tank thus firmly establishing the requirement for a recovery vehicle in the M88's weight class. Bowen-McLaughlin, now renamed Bowen-McLaughlin-York Company (BMY), received a production contract for the M88 which ran from 1960 until 1964 with a total output during this period of 1075 vehicles.

Below, the medium recovery vehicle M88 is at Aberdeen Proving Ground on 3 April 1961. The greater road wheel spacing compared to the M48 series tank can be seen in this photograph.

A new medium recovery vehicle M88 is shown here at Bowen-McLaughlin-York Company. Details of the external stowage are visible in these views.

US ARMY
9B 3295

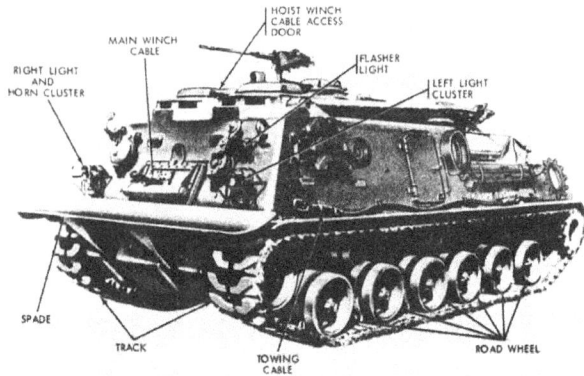

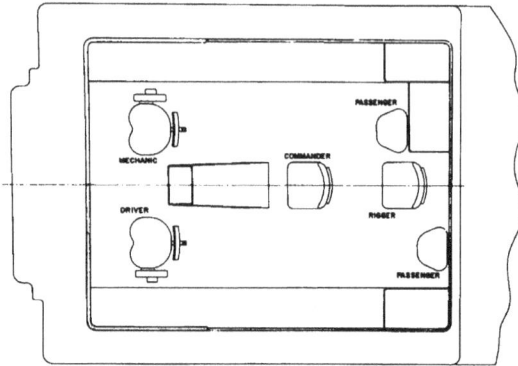

The left side external stowage can be seen above (left) and the diagram (right) indicates the crew stations inside the cab.

The production M88 was powered by the Continental AVSI-1790-6A engine with the Allison XT-1400-2 transmission. The engine was rated at 980 horsepower at 2800 rpm. It was estimated to deliver about 690 horsepower to the drive sprocket compared to 580 horsepower for the AVI-1790-8 with the same transmission. This provided adequate power for satisfactory performance. The four man crew consisted of the commander, driver, mechanic, and rigger. The driver and mechanic were located at the front of the cab with the driver on the left. The commander was in the center under the cupola just in front of the rigger. The vehicle was armed with a .50 caliber machine gun externally mounted on the commander's cupola. In addition to the cupola, a hatch was provided in the cab roof for each of the other three crew members as well as a door in each side of the vehicle. The recovery equipment consisted of an hydraulically powered and controlled spade, main winch, boom, and hoist winch. The spade, mounted at the front, was operated by the driver to anchor the vehicle or to support it when lifting a load with the boom. It also could be used like a dozer blade for light clearing or leveling. The main winch, mounted beneath the crew compartment, had a capacity of 90,000 pounds and was provided with 200 feet of 1¼ inch diameter cable. The exit for the cable was through a hole in the hull front. The "A" frame type boom was attached to the vehicle by a trunnion lever on each side of the cab. The levers extended downward and were connected to the hydraulic boom cylinders which were anchored to the hull in the crew compartment. The driver operated the boom from his normal position. The hoist winch, mounted below the cab, had a capacity of 50,000 pounds using a four-part ⅝ inch diameter line. This line passed through a cable chute between the floor of the crew compartment and the cab roof. A gasoline fired personnel heater was installed at the right rear of the crew compartment and the radio sets were mounted on a shelf along the right cab wall.

The details on each side of the M88 are shown below. Again, note the wide spacing between the road wheels resulting in the long ground contact.

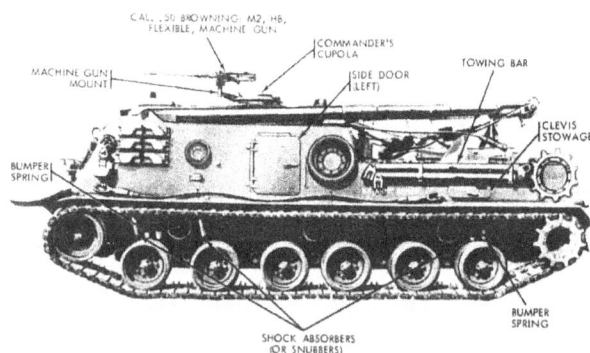

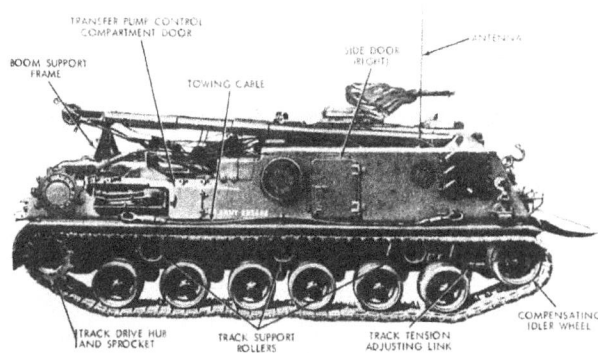

278

DRIVER'S HATCH
MECHANIC'S HATCH
RIGGER'S HATCH
SPOTLIGHT
BOOM
LIFTING EYE
LIFTING EYE
AUXILIARY POWER SLAVE RECEPTACLE
RIGGER'S LIGHT
RIGGER'S INFRARED LIGHT
EXHAUST DEFLECTOR
REAR TAILLIGHT
REAR TAILLIGHT
LIFTING CHAIN
TOWING PINTLE
TOWING BAR

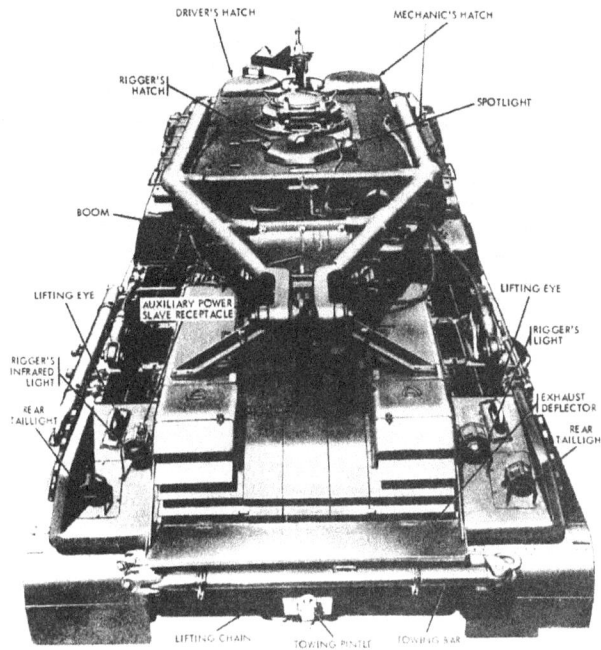

The front and rear interiors of the M88 cab are shown above and below respectively. The top rear of the vehicle and the power pack appear at the left. At the bottom is a sectional drawing of the M88.

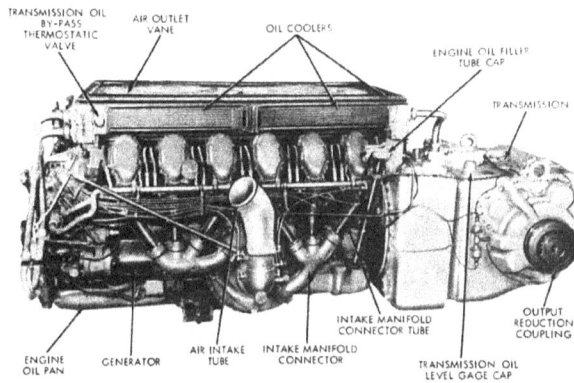

TRANSMISSION OIL BY-PASS THERMOSTATIC VALVE
AIR OUTLET VANE
OIL COOLERS
ENGINE OIL FILLER TUBE CAP
TRANSMISSION
ENGINE OIL PAN
GENERATOR
AIR INTAKE TUBE
INTAKE MANIFOLD CONNECTOR
INTAKE MANIFOLD CONNECTOR TUBE
TRANSMISSION OIL LEVEL GAGE CAP
OUTPUT REDUCTION COUPLING

STEERING AND SHIFTING CONTROLS
SEATS
BOOM SUPPORT AND RELATED PARTS
HOISTING BOOM
PERSONNEL HEATER
HYDRAULIC SUB-PLATE ASSEMBLY
FRONT
REAR
HOIST WINCH
DRAIN VALVES
ENGINE AND TRANSMISSION
MAIN WINCH AND SPADE
MECHANICAL TRANSMISSION AND HYDRAULIC PUMP
ENGINE MOUNT

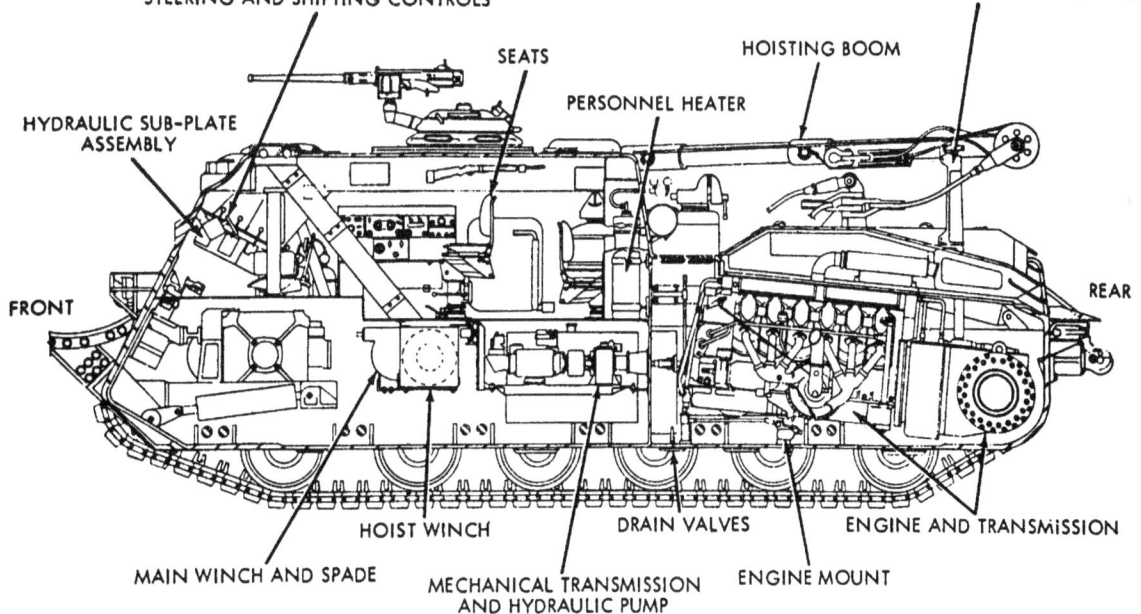

279

The medium recovery vehicle M88A1 is shown above and below. M239 smoke grenade launchers are installed in the top photograph and the mounts for these launchers can be seen in the lower view. In the latter photograph, the boom is in the operating position.

Although the AVSI-1790-6A provided sufficient power, the fuel consumption with the gasoline engine was high and by the time the M88 reached the troops, the M60 and other diesel powered vehicles were coming into service. An obvious answer to the problem was to adapt a diesel engine to the recovery vehicle. Eventually, this was done, but production of the diesel powered version, designated as the M88A1, did not begin at Bowen-McLaughlin-York until June 1975. Prior to that time, the M88 served with U.S. forces all over the world and was highly successful during operations in Vietnam.

The M88A1 was quite similar to the M88, except that the original power train was replaced by the Continental AVDS-1790-2DR diesel engine and the Allison XT-1410-4 transmission. An auxiliary generator driven by a 10.2 horsepower diesel engine replaced the gasoline powered unit in the earlier vehicle. Also, a new personnel heater operating on diesel fuel was installed in place of the gasoline model.

These two photographs clearly reveal the exterior details of the M88A1. Once again, the M239 smoke grenade launchers are installed on the front of the cab.

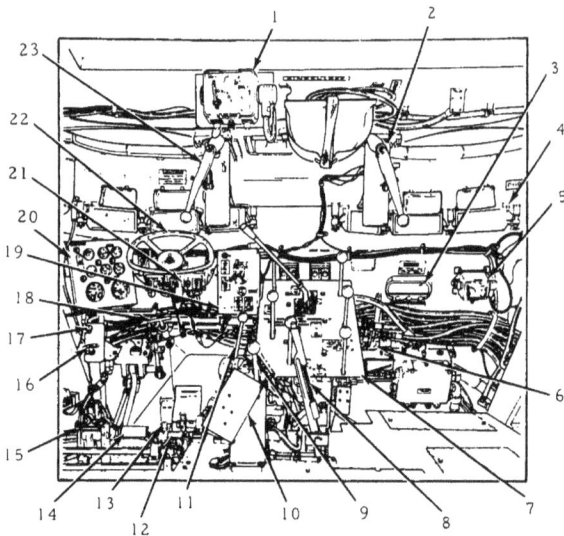

CAB INTERIOR, FRONT
1.Commander's interphone box, 2.Mechanic's hatch control, 3.Front vision blocks, 4.M17 periscopes(6), 5.Mechanic's interphone box, 6.APU emergency winch control valve, 7.Hydraulic controls panel, 8.Drain valve lever, 9. Main winch shift lever, 10.Auxiliary power unit control box, 11.Transmission shift lever, 12.Accelerator, 13.Dimmer switch, 14.Brake pedal, 15.Purge pump, 16.Manual fuel shutoff handle, 17.Engine hand throttle, 18.Spade locking release handle, 19.Switch panel, 20.Gage panel, 21.B.O. receiver light switch and indicator light, 22.Steering wheel, 23.Operator's hatch control

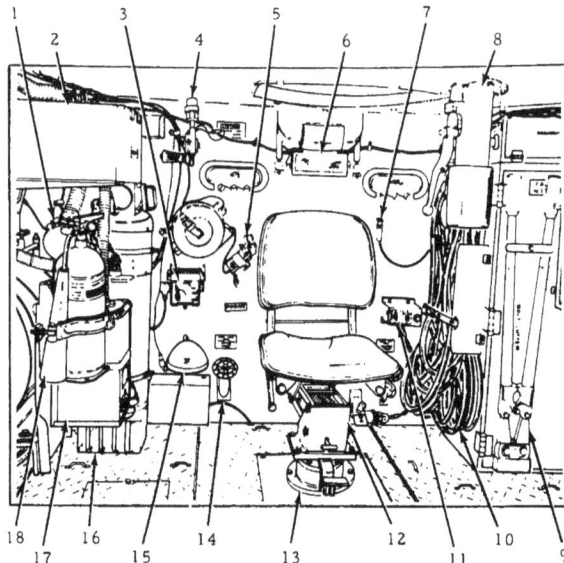

CAB INTERIOR, REAR
1.Auxiliary power unit(APU) air filter housing, 2.M24 periscope stowage box, 3.Rigger's interphone box, 4.Rigger's spotlight control, 5.Rigger's portable spotlight handle, 6.M17 periscope, 7.M24 periscope cable receptacle, 8.Rigger's hatch control, 9.Bolt cutter, 10.Oxygen and acetylene hoses, 11.Rigger's lights, 12.Generator blower air intake grill(hidden behind rigger's seat), 13.Rigger's seat, 14.Fuel control valve(4), 15.Rigger's portable spotlight, 16.Personnel heater, 17.Gas-particulate filter unit, 18.Portable fire extinguisher

The interior of the M88A1 cab is illustrated in these four sketches. Above are the front (left) and rear (right) and below the left and right sides are shown at the left and right respectively.

CAB INTERIOR, LEFT SIDE
1.LAW rocket(2), 2.Operator's air purifier control switch and dome light, 3.Caliber .45 machine gun, 4.Fixed fire extinguisher, 5.Water can(2), 6.Tool box(3), 7.Track link adjusting wrench, 8.Ammunition stowage rack

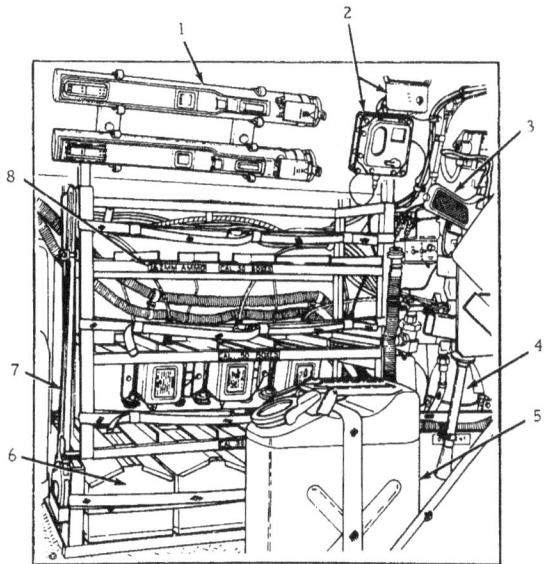

CAB INTERIOR, RIGHT SIDE
1.M16 rifle(M14 optional), 2.Mechanic's air purifier control switch and dome light, 3.Communication amplifier AM1780/VRC, 4.Radio equipment, 5.M16 rifle (M14 optional), 6.Spare barrel for .50 caliber MG, 7.Oil can(2), 8.LAW rocket(8), 9.Oddment compartment

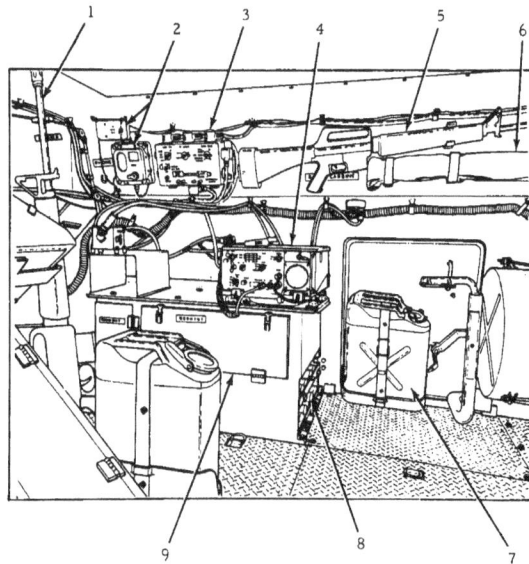

The top of the M88A1 is detailed in the drawing at the right with the numbered items listed below.

1.Auxiliary power slave receptacle, 2.Vise, 3.Oxygen cylinder, 4.Rigger's hatch, 5.Operator's hatch, 6.Hoist winch cable access door, 7.Mechanic's hatch, 8.Spotlight, 9.Tow cable(2), 10.Exhaust deflector

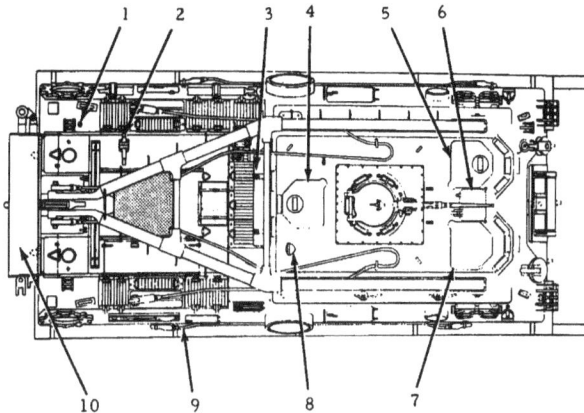

Above, the medium tank M46 (left) is equipped with the M3 tank mounting bulldozer. At the right is a similar view of the 90mm gun tank M47 fitted with the M6 bulldozer.

COMBAT ENGINEER VEHICLES

The value of a tank mounted bulldozer for clearing obstacles or debris under fire was clearly established by the end of World War II and the M1 and M2 series of bulldozers had been standardized for use on various models of the Sherman tank. These were employed by regular armored units as well as engineer troops. During the postwar period, this trend continued with the development of the M3 tank mounting bulldozer for use on the medium tank M46. The M3 was subsequently modified as the M3E1 for installation on the M47 tank. However, it

was considered to be unsatisfactory and it was redesigned to reduce installation time and to restore lost ammunition stowage space inside the tank. The redesigned version was standardized as the tank mounting bulldozer M6.

With acceptance of the M48 tank series, a project was initiated to adapt the T18 bulldozer, designed for the heavy tank T43, to the medium tank. The new version, designated as the T18E1, was originally intended to have maximum interchangeability of parts with the T18 and M3 bulldozers. However,

Below, the T18E1 bulldozer is installed on the 90mm gun tank M48A1 at Chrysler on 18 January 1955. The dozer blade is raised in the traveling position in this photograph.

The 90mm gun tank M48A2 is equipped with the T19 bulldozer kit in the photographs on this page. The blade is raised for traveling in the upper view and lowered for operation below.

further study modified the design to reduce field installation time and to eliminate the need to cut holes in the hull for the hydraulic lines. Subsequently, the T18 itself was redesigned to maintain the interchangeability of parts. On 10 January 1957, the modified T18E1 was standardized as the tank mounting bulldozer M8 for use on the M48 and M48A1 tanks.

Further changes were required with the introduction of the M48A2 tank. Modified for attachment to this vehicle, the bulldozer became the M8E1. This was later standardized as the M8A1. Continued evolution of the design to keep up with new tank models resulted in the introduction of the M8A3 bulldozer for use with the M48A3 tank in the late 1960s. The new M60 tank series also required changes in the bulldozer design and the tank mounting bulldozer M9 standardized for use with these vehicles was basically a modified version of the M8A1.

The installation of the M8A3 bulldozer on the 90mm gun tank M48A3 is illustrated by the drawing at the right. Note the protective covers for the hydraulic lines and control linkage.

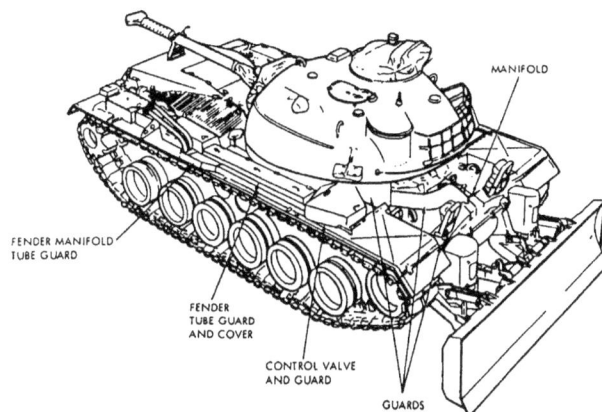

MANIFOLD

FENDER MANIFOLD
TUBE GUARD

FENDER
TUBE GUARD
AND COVER

CONTROL VALVE
AND GUARD

GUARDS

M60 SHOWN

M60 SHOWN

The M9 bulldozer is shown above fitted to the M60 tank. Below, the same equipment is installed on the M60A1 at Fort Knox. Note the elevated headlights required for operation with the dozer blade in the travel position.

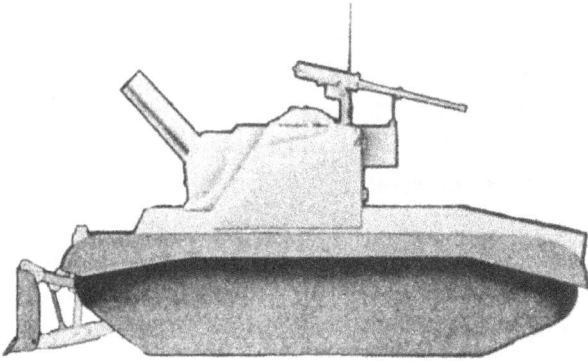

At the left is a proposal sketch for the pioneer tank T39.

The concept of an armored vehicle dedicated specifically for use by engineer troops doing demolition work under fire or other battlefield tasks originated in Great Britain during World War II. Based on the heavily armored Churchill tank and designated as the Armored Vehicle Royal Engineers (AVRE), they proved to be extremely effective during the latter part of the war. Efforts in the United States to produce a similar vehicle were limited to the conversion of some Sherman tanks for the combat engineer role, but the war ended before any of these saw active service. Early in 1945, a program was initiated for the development of an armored engineer vehicle designated as the demolition tank

T31. With the end of the war, this project was cancelled early in 1946, but it was recommended that work be continued to develop a new demolition tank. This was in line with the report of the Stilwell Board which included a requirement for "a special tank for engineer demolition work".

The Office of the Chief of Engineers conferred with Ordnance, the Army Ground Forces, the Army Service Forces, and the U.S. Marine Corps on 12 February 1946 to review proposed military characteristics of a new engineer armored vehicle. These characteristics specified that the vehicle be based on the chassis of the medium tank recommended by the Army Ground Forces Equipment Review Board. However, this tank would not be available until a much later date and on 7 August 1947, a meeting was held at Detroit Arsenal to select a chassis for initial studies. It was agreed that a medium tank M26, with the turret and stowage removed, would be immediately made available for the construction of a mock-up of the proposed vehicle. OTCM 31974 on 2 October 1947 outlined the characteristics, recommended the manufacture of two pilots, and assigned the designation pioneer tank T39.

Below is the mock-up of the pioneer tank T39 on the medium tank M26 chassis at Detroit Arsenal on 14 April 1948.

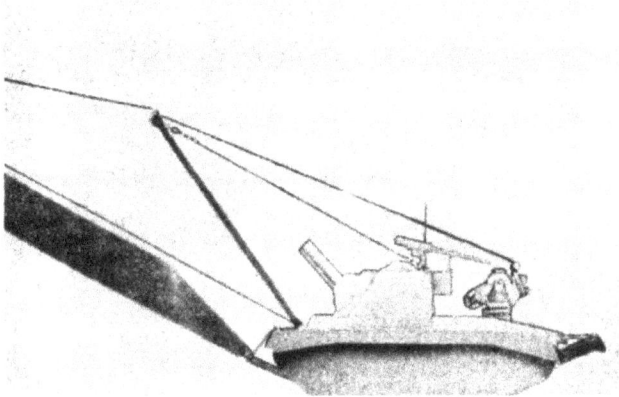

At the left is a sketch of the original concept of the T39 with an assault bridge.

The original specifications for the T39 called for a five man crew with space for two additional men inside the vehicle. Exit doors for the demolition personnel were required on each side and, if possible, in the rear of the superstructure. Based on the standard medium tank, the vehicle was to have comparable armor and the main armament was to be a launcher for demolition charges with an effective range of 500 yards. Initially, this weapon was envisaged as a breech loading rocket launcher 6 to 8 inches in caliber similar to the 7.2 inch T105 launcher installed experimentally in the Sherman tank. Thirty demolition charges, in addition to 2000 pounds of explosives, were to be stowed inside the vehicle. If possible, the tank's .30 caliber bow machine gun was to be retained in addition to a .50 caliber machine gun on top of the superstructure. Consideration was to be given to possible remote control of these machine guns.

The pioneer tank was to have attachments for mine detection and mine clearing equipment as well as a rack for placing 2000 to 6000 pounds of demolition charges against obstacles without exposing the crew. A dozer blade was required and was expected to be capable of operation simultaneously with the mine detection equipment. Attachment points at the front were to handle either an "A" frame or the launching device for an assault bridge at least 40 feet in length. This bridge was to be able to carry a 60 ton tank with a clear road width of at least 162

inches. The vehicle also was to have a 20 ton winch with a minimum of 200 feet of wire rope to handle various attachments. It was obvious that too many tasks were being assigned to a single type of vehicle. Despite this, a later requirement was added for the inclusion of a flail type mine exploder.

The pioneer tank, without the mine clearance or bridge launching attachments, was optimistically expected to have a maximum weight of about 45 tons and performance similar to that of the medium tank. Two pilots were authorized for construction based on the M26 tank chassis. The first vehicle was to use the standard M26 power plant, but the second was to be fitted with the new Continental AV-1790 engine, then under development. However, in August 1948, the medium tank M46 (T40) was standardized and the M26 was reclassified as Limited Standard. In view of this change, OTCM 32681, on 21 January 1949, recorded the decision to manufacture both pilot T39s based on the chassis of the new M46 (T40). This action also eliminated the requirement for the flail type mine exploder and recommended that it be replaced by fittings to permit the use of the demolition snake M3. It also recommended that one medium tank T40 and one medium tank M46 be diverted from production for conversion as pilots of the pioneer tank T39.

Another major modification was the replacement of the main armament rocket launcher by the British 6.5 inch demolition gun Mk 1. With minor breech ring modifications, it was designated as the 165mm gun T156 in U.S. service. The demolition gun had an effective range of about 1000 yards, twice that expected from the rocket launcher it replaced. It fired the 30 pound T237 high explosive plastic (HEP) projectile using a small propellant charge in a perforated case permanently attached to the base of the projectile. The more descriptive British designation for this type of round was high

Below are side views of the pilot tank now designated as the engineer armored vehicle T39. This is the pilot, registration number 30162865, converted from the tenth medium tank T40.

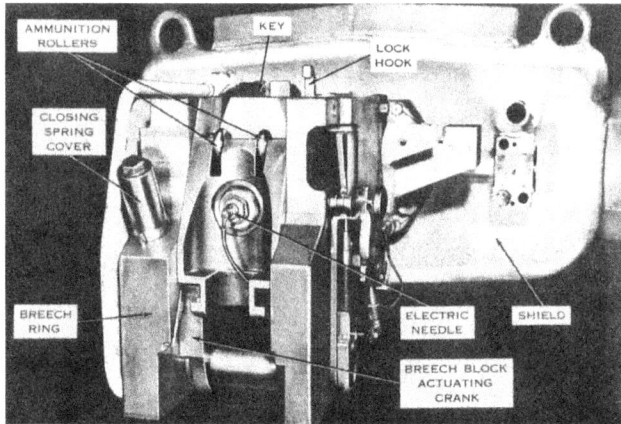

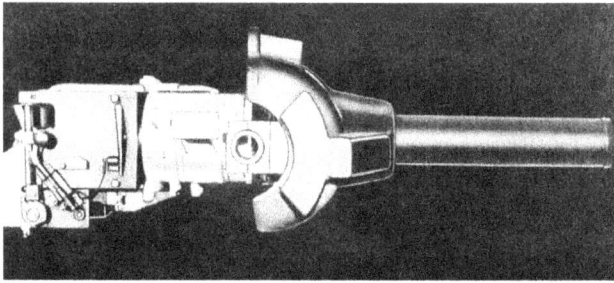

Details of the 165mm (6.5 inch) demolition gun and mount can be seen in these photographs. The mount is shown here before the installation of the shield over the telescope aperture.

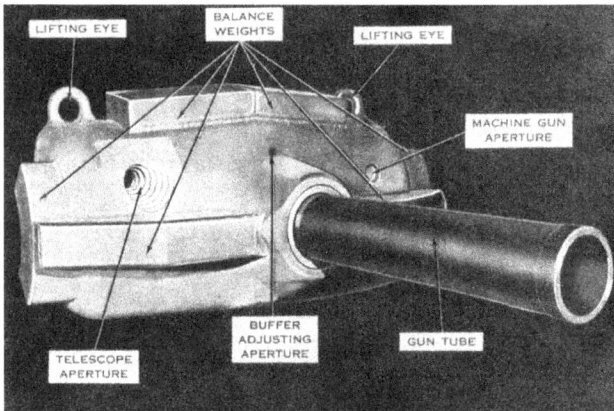

explosive squash head (HESH) referring to its behavior on striking the target. When fired, the powder gases escaped into the chamber through the holes in the case and both projectile and case went out of the barrel. The round was designed primarily to breach concrete obstacles. The T156 gun consisted of a rifled tube with external threads on the breech end to accept a chamber bushing. This was a collar also threaded externally to receive the breech ring which was screwed and shrunk onto it. The bushing and breech ring provided the entire

chamber volume as when the cartridge was seated in the rifling, the case extended out of the tube. The rear of the breech ring carried the obliquely sliding breechblock which was semiautomatic in operation. The breechblock was fitted with an obturator pad to prevent gas leakage and an electric firing needle to carry the current to the electric primer. A low chamber pressure of 3600 psi resulted in a muzzle velocity of about 850 feet/second.

Some realistic specification changes for the T39 eliminated the requirement for the assault bridge, reduced the crew to four men, but added a requirement for a mine laying device. On 11 May 1950, Ordnance Committee action dropped the pioneer tank designation and assigned the nomenclature, engineer armored vehicle T39. Both pilots were manufactured at Detroit Arsenal and number 1 was completed in June 1951 followed by number 2 in July. Both vehicles were shipped to Aberdeen Proving Ground for engineering tests.

As completed, the T39 differed considerably from the original concept. It replaced the fixed superstructure with the rotating M46 tank turret mounting the 165mm demolition gun in a concentric recoil mechanism coaxially with a .30 caliber machine gun. An M76E5 telescope was located on the right side of the gun and steel counterweights were

At the left is a sectional drawing of the breech mechanism and chamber area of the 165mm (6.5 inch) gun. Note the outline of the cartridge case extending out of the tube into the large diameter chamber. Below is a drawing of the HEP M123 round for use in the demolition gun. The removeable handle used to load the round can be seen on the base of the cartridge case.

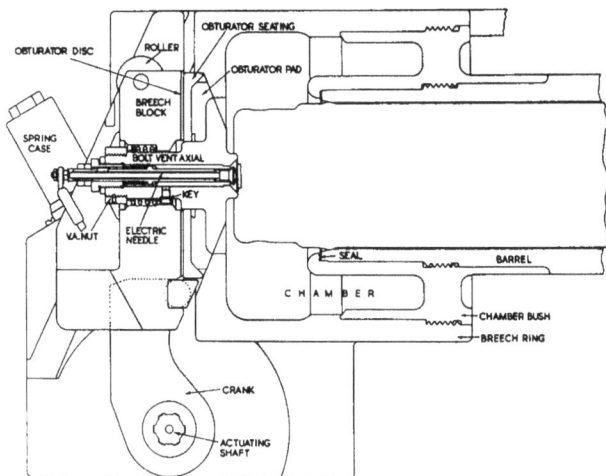

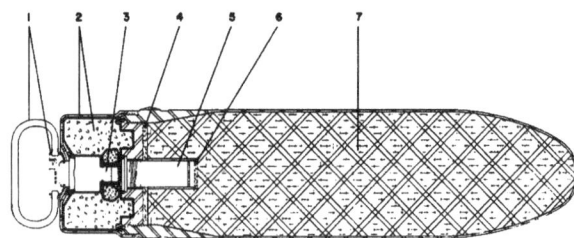

The first pilot engineer armored vehicle T39 is shown above during evaluation at Aberdeen Proving Ground. Note the shield installed around the large telescope aperture in the gun mount.

welded to the shield to balance the mount. Both manual and hydraulic power traverse systems were retained from the M46 and a manual-hydraulic elevating mechanism was provided. An access door in the left side of the turret permitted rapid exit and entry for the demolition crew. A .50 caliber machine gun was mounted on the turret roof near the commander's cupola. The bow machine gun was eliminated, the assistant driver's hatch was replaced by a

steel plate, and the space in the right front hull was used for stowage. Twenty-six rounds of 165mm ammunition were carried and space was allocated for 280 pounds of M5 demolition blocks inside the hull. The vehicle was equipped with the M3 tank mounting bulldozer, a 20 ton winch and "A" frame installed above the rear deck, and a front boom for obstacle removal and emergency assistance. Fully loaded, the T39 weighed about 52 tons.

The large winch housing on the rear deck of the T39 can be seen in the photograph below. The hold-out bars used for towing are stowed on the bracket over the winch power take-off on the hull rear.

Above at the left, the T39 has the forward boom erected in the operating position. The photograph at the right shows the first pilot T39 during tests with a new winch and "A" frame installation.

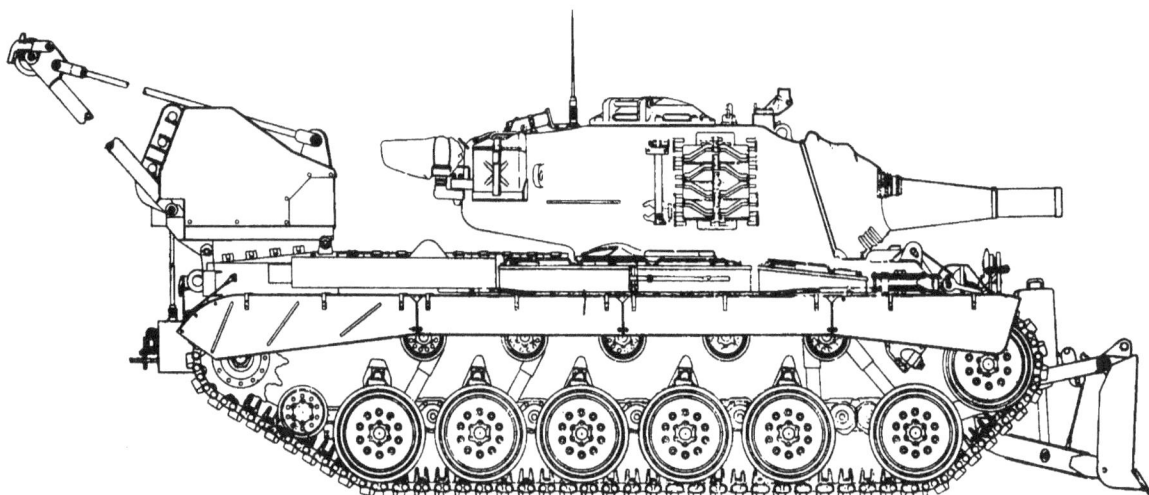

These drawings and those on the opposite page were reproduced from the design layouts of the engineer armored vehicle T39. They do not reflect changes introduced during the test program.

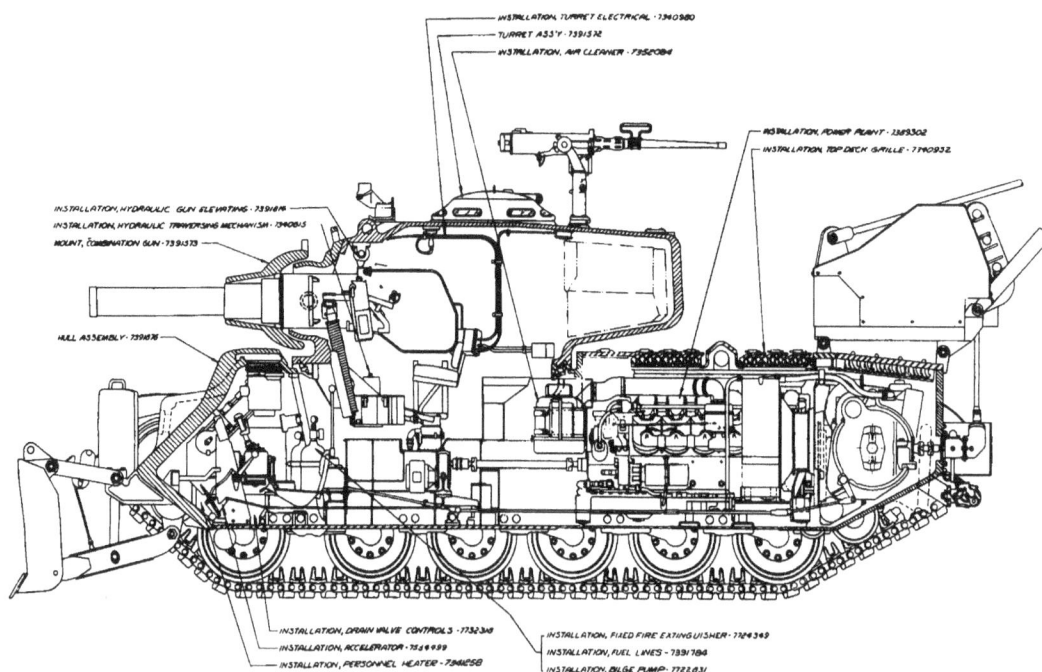

290

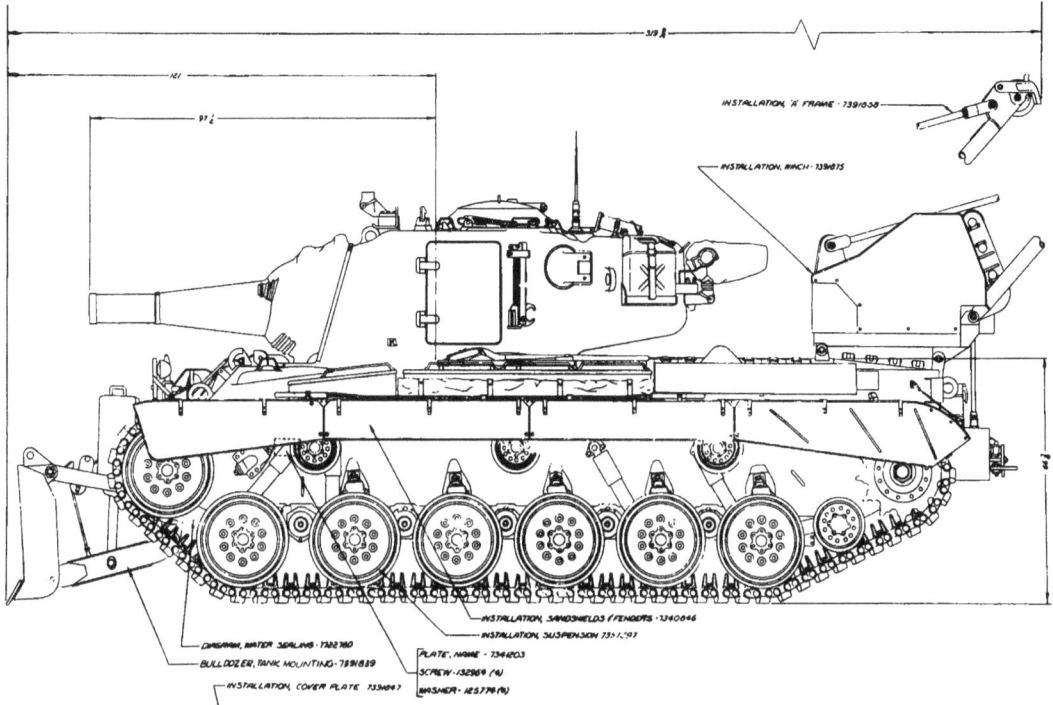

INSTALLATION, 'A' FRAME · 7391658

INSTALLATION, WINCH · 7394875

DIAGRAM, WATER SEALING · 7322760
BULLDOZER, TANK MOUNTING · 7891889
INSTALLATION, COVER PLATE · 7334047

INSTALLATION, SANDSHIELDS / FENDERS · 7340846
INSTALLATION, SUSPENSION · 7351797
PLATE, NAME · 7344203
SCREW · 132869 (4)
WASHER · 125778 (4)

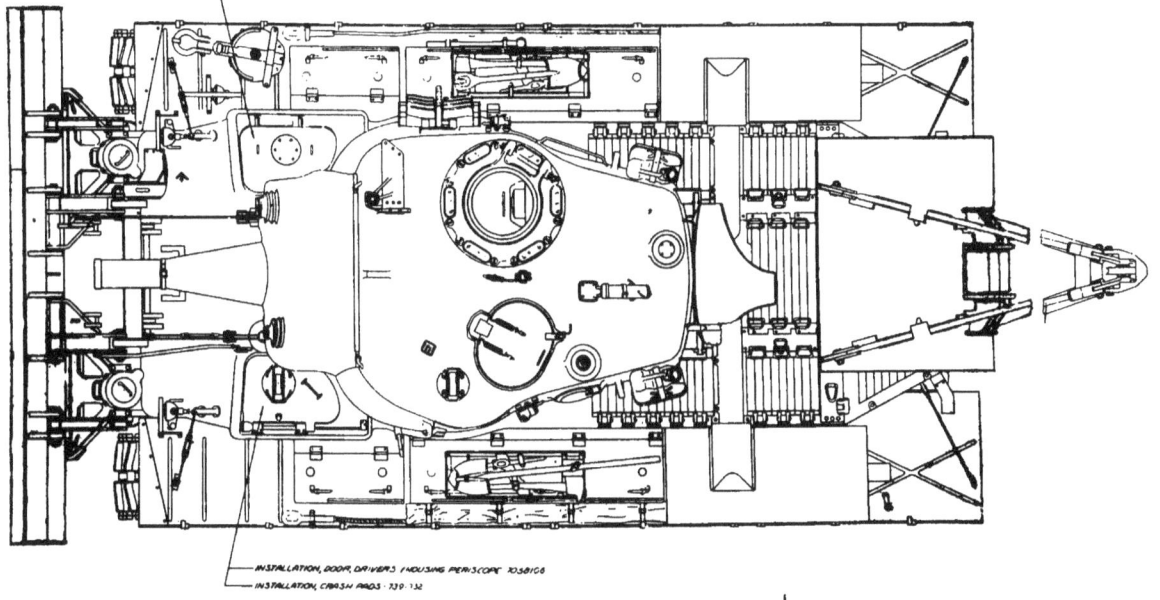

INSTALLATION, DOOR, DRIVERS / HOUSING PERISCOPE · 7058108
INSTALLATION, CRASH PADS · 739-132

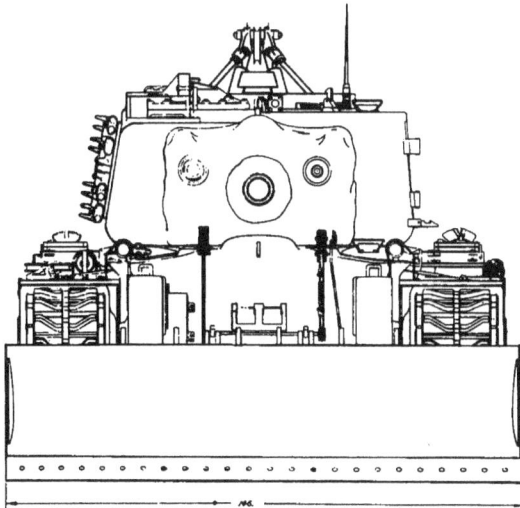

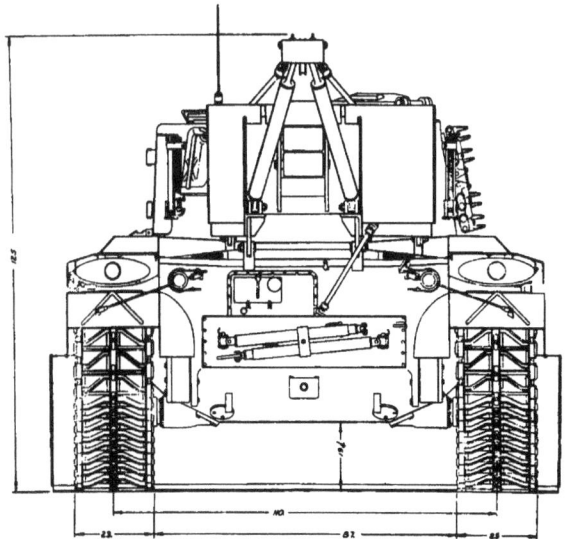

The sketch at the right depicts the proposed engineer armored vehicle T39E1. Although based on the M47 tank, it retained the same winch and boom arrangement as the T39.

Long before delivery of the two T39s, the decision was made to initiate the development of two additional pilots utilizing the chassis of the new M47 tank. Preliminary design began at Detroit Arsenal in July 1951 under the tentative designation engineer armored vehicle T39E1. However, manufacture was suspended in March 1952 pending further study to better define the requirements and to determine the most suitable tank chassis. Studies during this period evaluated the use of the basic chassis from the medium tanks M4A3, M46, M47, and M48. The outcome was the selection of the M47 as the basis for the interim engineer armored vehicle now designated as the T39E2.

Two T39E2s were built by Detroit Arsenal and the first pilot was shipped to Aberdeen Proving Ground for test arriving on 6 June 1955. Except for the fact that it was based on the M47 tank, the T39E2 was similar in many respects to the T39. The 165mm demolition gun in its concentric recoil mechanism was installed with a coaxial .30 caliber machine gun in a modified version of the tank's M78 combination mount. Twenty-four rounds of 165mm ammunition were carried in the hull in addition to six in the turret ready racks bringing the total to 30 compared to 26 in the T39. A .50 caliber machine gun was installed in the commander's cupola and four grenade launcher assemblies were attached to the outside of the turret. Each launcher assembly consisted of six tubes, except for that on the left front which had only five to allow clearance for the

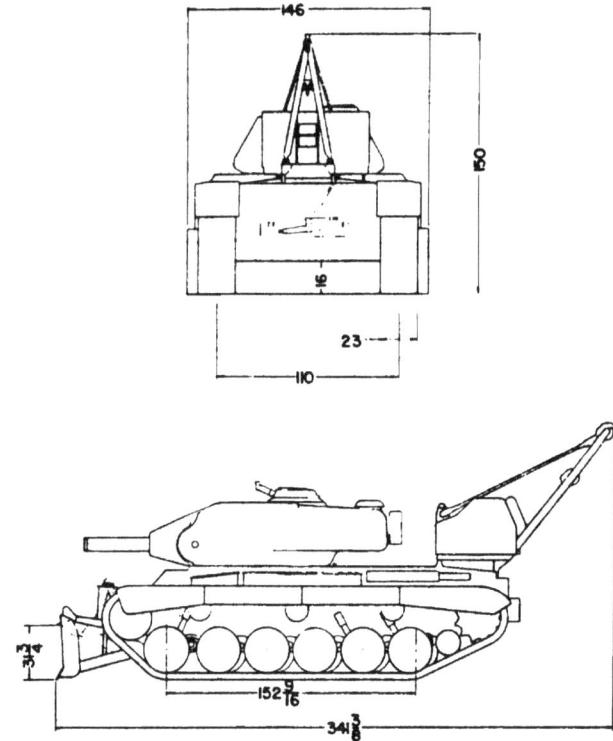

access door in that side of the turret. The new vehicle had the same four man crew as the T39 consisting of the commander, gunner, loader, and driver. The tank's M12 range finder was retained as the primary sighting device for the gunner. Provision was made on the front of the vehicle to mount a high "A" frame boom and a modified M6 hydraulically operated bulldozer. A 20 ton capacity winch and a low "A" frame boom were located at the rear of the chassis. The assistant driver's seat and the bow machine gun were eliminated to provide space

The engineer armored vehicle T39E2 based on the M47 tank is illustrated below with both the front and rear "A" frame booms erected. Note the smoke grenade launcher assemblies on each side of the turret.

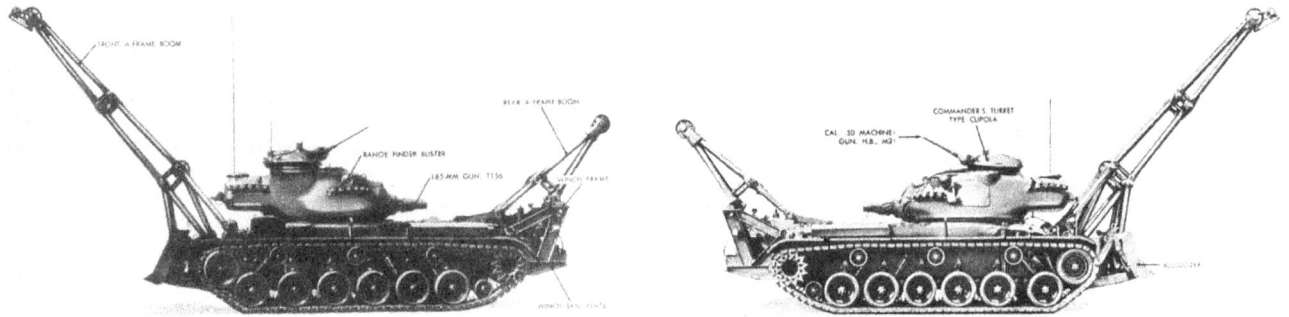

Further details of the T39E2 pilot can be seen in the side views above. Below, the same vehicle, registration number 30186149, is under test at Aberdeen Proving Ground on 25 August 1955.

Above is the T39E2 during the test program at Aberdeen. The .50 caliber machine gun has not been installed in the commander's Aircraft Armaments model 108 type cupola. Below, the "A" frame booms are stowed in this view of the T39E2.

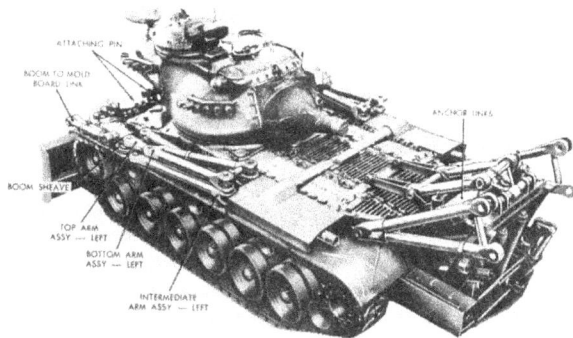

for the winch and bulldozer hydraulic system. The assistant driver's hatch allowed access for servicing the equipment. Attachments were provided for operating the demolition snake M3 and a mine clearing device. Internal hull stowage included 280 pounds of M5 demolition blocks. The vehicle was fitted with the M47 suspension and power train with some minor modifications. These included the power take-off shaft connected to the transmission and an increase in the final drive gear ratio from 4.47:1 to 5.47:1.

The engineering tests on the T39E2 showed generally satisfactory results. It was noted that supporting the front boom on the bulldozer resulted in early failure of the hydraulic system cylinder heads. Other suggested modifications included relocation of the rear winch to improve the access for brake adjustment and to increase the angle of departure. The T39E2 also suffered from the same front spindle arm and bearing failures as the M47 and the modifications recommended for the tank were even more important here because of the vehicles heavier weight. Combat loaded, the T39E2 weighed about 57 tons. As would be expected, the weight increase and higher final drive ratios with the same power plant reduced the performance compared to the M47.

After completion of the test program, standardization was recommended for the T39E2. This was accomplished by Ordnance Committee action on 1 March 1956, but once again the name was changed. It was first intended to use the term combat engineer tank, but the final official designation was the 165mm gun full tracked combat engineer vehicle M102. About this time, the T156 demolition gun and its T237E4 high explosive plastic round were standardized as the 165mm gun cannon M57 and the 165mm HEP M123 respectively. When standardized, the production of 122 of the new vehicles was planned for the 1957 fiscal year. However, time had now run out for the long development program. The M47 tank, upon which the M102 was based, was out of production and no longer considered as first line equipment. Thus the production plans were cancelled and later model tanks were than considered for adaptation to the combat engineer vehicle (CEV) role.

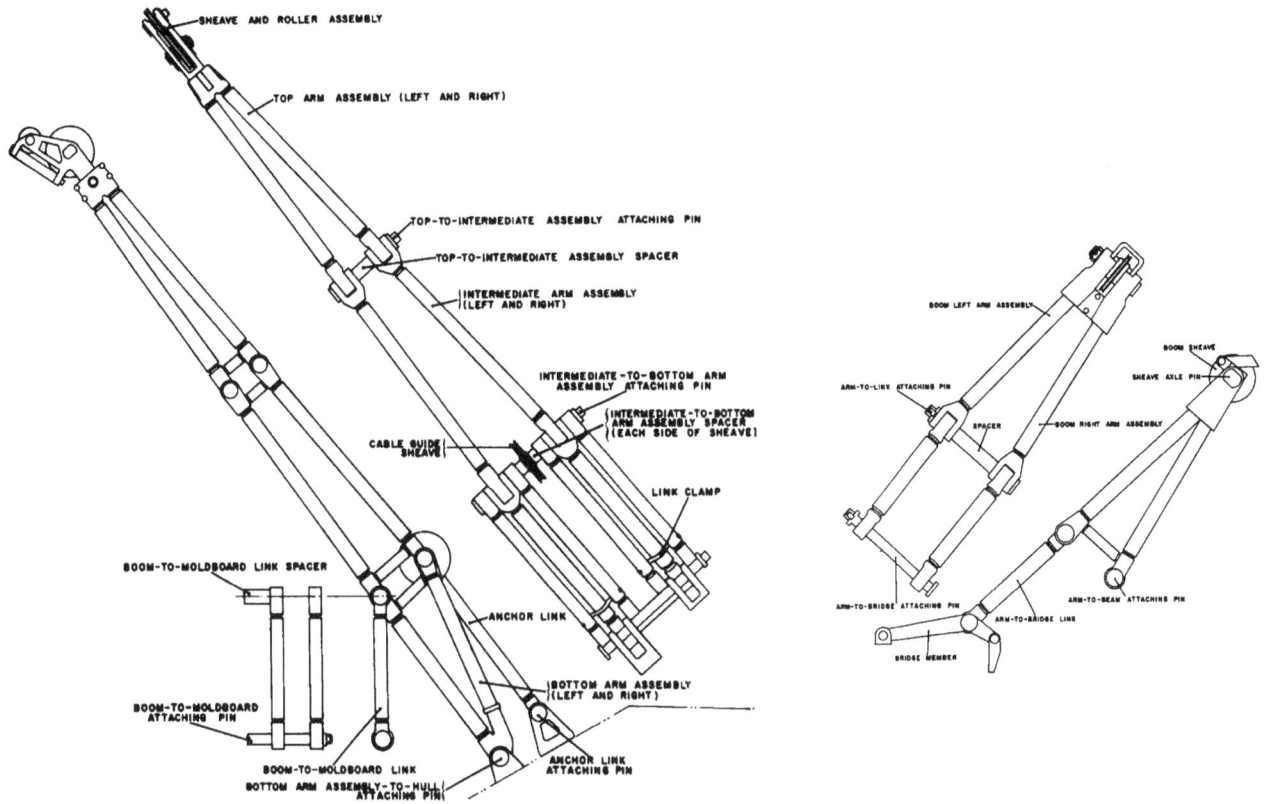

Details of the front (left) and rear (right) "A" frame booms on the T39E2 are sketched above. Below is a side view drawing showing the general arrangement of the vehicle.

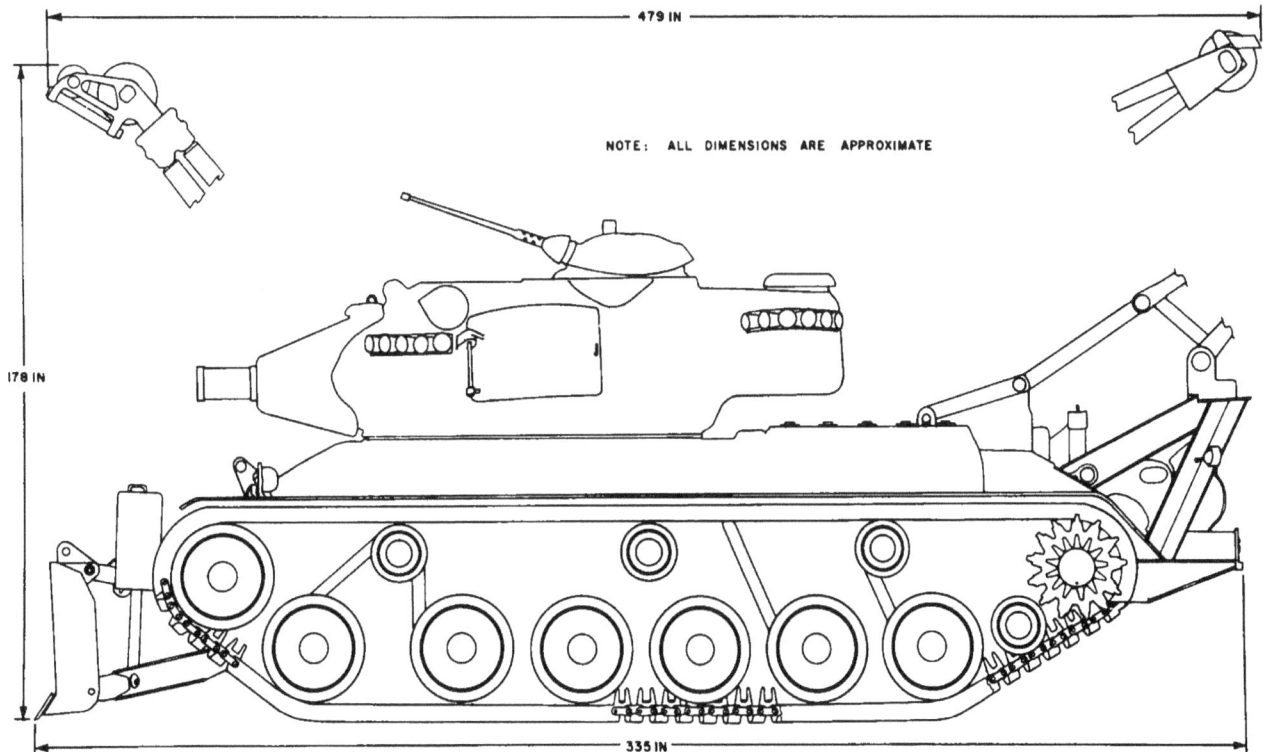

NOTE: ALL DIMENSIONS ARE APPROXIMATE

295

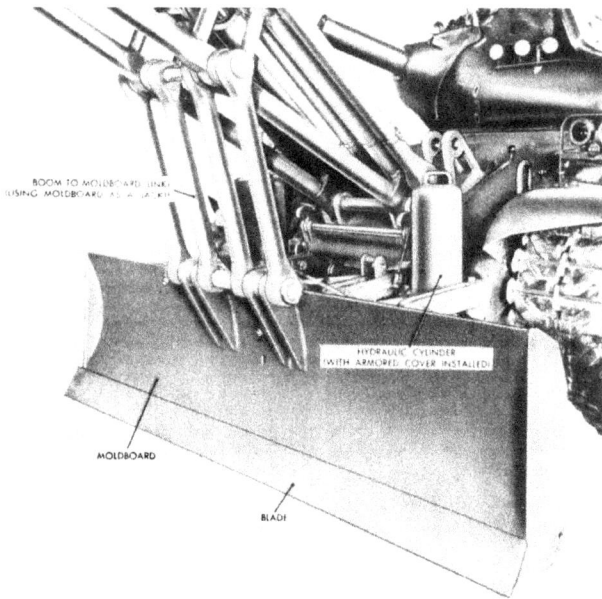

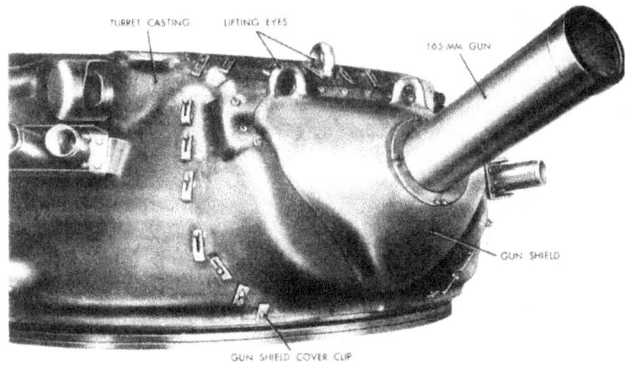

The modified M78 gun mount is installed above in the T39E2 turret with the shield cover removed. Below are two views of the 165mm gun T156. The .30 caliber coaxial machine gun has not been mounted in these photographs. At the left, details can be seen of the modified M6 bulldozer installation. Note the use of the dozer moldboard to support the boom.

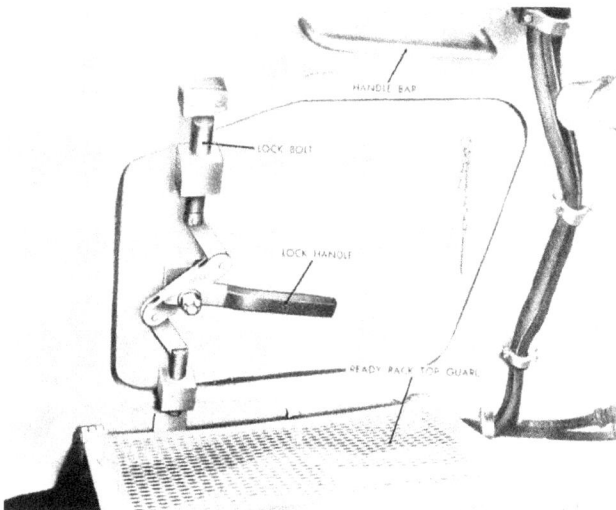

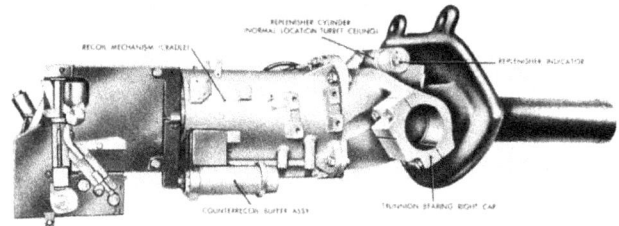

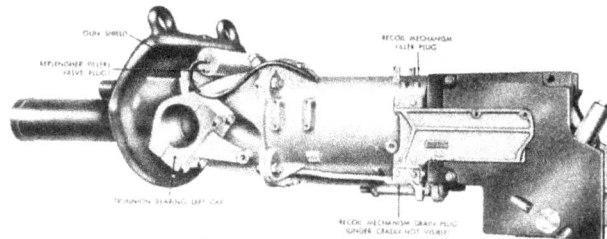

An interior view of the door in the left turret wall appears at the left above. Below, the T39E2 hull ammunition stowage can be seen at the left and the 165mm ready racks are at the right.

The two proposals for a combined recovery and combat engineer vehicle based on the T95 tank are sketched at the left. Proposal A (upper) has a fixed cab with limited traverse for the demolition gun. Proposal B (lower) mounts the weapon in a fully traversing turret.

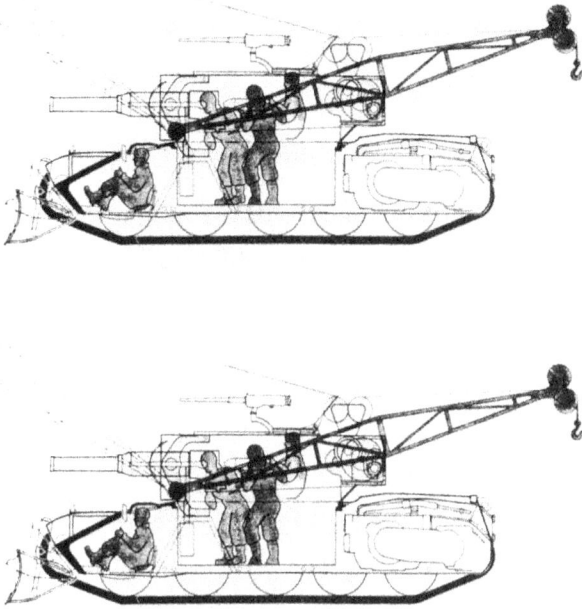

During the Fall of 1956, the T95 tank development program showed great promise and there was every indication that it would be the Army's future standard tank. An obvious solution to the engineer vehicle requirement was to develop a CEV based on the T95 so that it would be ready for service at the same time as the tank. On 11 October 1956, a concept of a CEV based on the T95 chassis was presented at the Interim Tank Program Reorientation Conference held at Detroit Arsenal. This concept was armed with the M57 demolition gun and equipped with the usual "A" frame boom and winch arrangement. A new feature of this proposal was the provision for a two station Dart missile launcher and guidance equipment.

Before any action was taken on the development of a new CEV, the entire question was reviewed again at Detroit in January 1957. This was the meeting that considered the feasibility of combining the functions of the CEV with those of the tank recovery vehicle. Two proposals for such a combination were presented, both based on the T95. One vehicle mounted the M57 demolition gun in a fixed cab with limited traverse and the other carried the same weapon in the fully rotating modified tank turret. Further studies of the combination vehicle more clearly revealed the numerous compromises such a design required. For example, the recovery vehicle did not need the demolition cannon with its bulky ammunition which would greatly reduce stowage space for equipment useful to the recovery mission. In view of these conflicting requirements, the decision was made to pursue separate development programs. In May 1957, the Chief of Staff approved a set of military characteristics for a new CEV based on the T95 tank. Design studies were undertaken by the Army Tank Automotive Command (ATAC) which was the renamed Ordnance Tank Automotive Command (OTAC). The project was officially established in January 1958 for the vehicle now designated as the CEV T118 and the manufacture of three pilots was authorized. A mock-up of the T118 turret was demonstrated at Detroit Arsenal in May 1958 and the fabrication of the three pilot vehicles began shortly thereafter. The first pilot was delivered to Aberdeen Proving Ground in December 1959. Subsequently, the second and third vehicles were shipped to the Armor Board at Fort Knox and the Corps of Engineers at Fort Belvoir respectively.

The dimensions of the combat engineer vehicle T118 are sketched at the left below. At the right is a view of the vehicle itself. Note the folding section in the dozer blade to provide a better view for the driver when the blade was raised to the travel position.

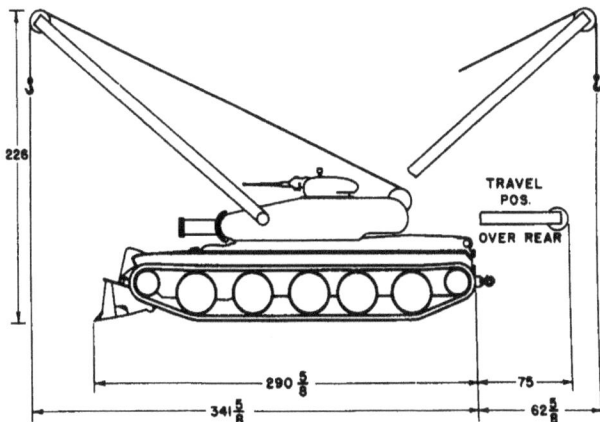

BOOM LIFTING OVER SIDE

C. E. V. T-118EI

SIDE ELEVATION

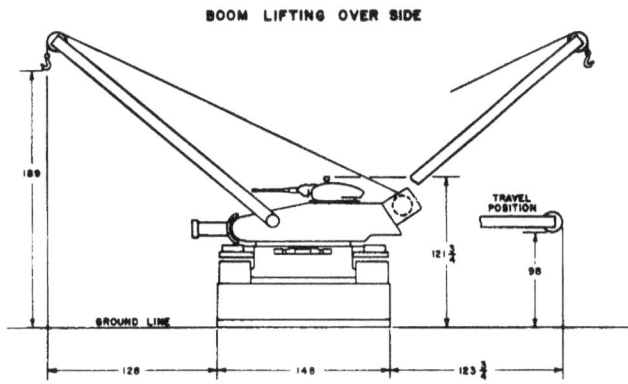

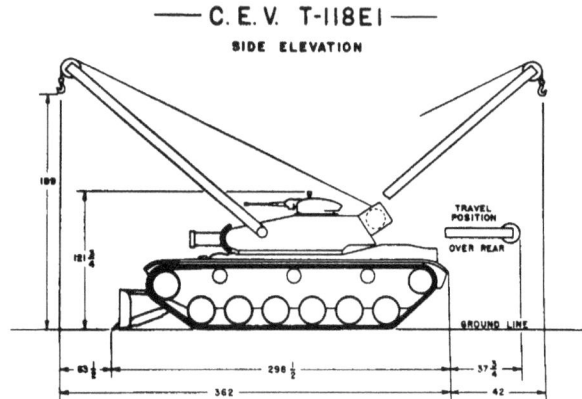

The dimensions of the combat engineer vehicle as originally proposed based on the M60 tank chassis are shown in the sketches above.

Unfortunately, luck was still against the CEV program. By the time the T118 tests began, the T95 had been replaced by the M60 as the Army's new main battle tank. At this point, the decision was made to use the T118 pilots in a test program to evaluate the turret components, the hydraulic system, and the armament. However, all production vehicles would be based on the M60 chassis and designated as the combat engineer vehicle T118E1.

The tests of the T118 at Aberdeen lasted for about a year and revealed the need for modification of the hydraulic system. Another test program also had noted some problems with the M57 demolition gun as well as its mount and although the weapon was installed in the T118 pilots, an improved version was recommended for the production vehicles. In August 1961, the test work using the T118 was

complete and there was no further requirement for the vehicle based on the T95 chassis.

Although the T118E1 was approved for limited production in October 1962, a review of the program indicated that extensive modifications would be required to adapt the T118 turret to the M60 chassis. This resulted in the decision to shift the engineer equipment to the M60A1 turret thus using both the turret and chassis of the new tank. However, the designation CEV T118E1 was retained for this modified version.

To correct the problems with the M57 demolition gun, it was replaced by a new weapon designated as the 165mm gun cannon XM135. This was a redesigned version of the British L9A1 cannon with the bore evacuator deleted, the breech bushing integrated with the tube, and the semiautomatic breech

Below, the combat engineer vehicle T118E1 has the boom stowed and the bulldozer blade raised to the travel position. The turret has been converted from a standard M60A1 casting. Note the plates covering the range finder ports of the tank turret.

The CEV T118E1 is shown here with the boom raised to the operating position.

mechanism replaced by a manual version. The obturator, breechblock latch, and firing mechanism also were redesigned. The XM135 fired the same HEP M123 ammunition as the M57. Stowage space was available for 30 of the M123 rounds. The XM135 was installed in the XM150 mount which was a slightly modified version of the M140 used with the 105mm gun in the M60A1. A coaxial 7.62mm M73 machine gun was fitted to the left of the cannon with the gunner's M105F telescope to the right. The normal four man crew occupied the same positions as in the M60A1 and the M19 commander's cupola was retained with its .50 caliber machine gun.

The engineer equipment on both the T118 and T118E1 differed from that on the earlier M102. A single "A" frame boom pivoted near the front of the turret replaced the high and low "A" frames mounted at each end of the earlier vehicle. The single boom could be raised or lowered from within the turret and it had a lifting capacity of 17,500 pounds with a single-part line. The T118E1 also was equipped with a 25,000 pound capacity winch installed in the turret bustle and the M9 bulldozer designed for the M60 series tanks. Both winch and dozer were hydraulically operated. Combat loaded, the new CEV weighed about 57½ tons.

Here, the CEV T118E1 has the turret traversed to the rear. Again, note the elevated headlights required for use with the dozer blade in the travel position.

The combat engineer vehicle M728 is shown here in travel order. This vehicle is fitted with the 2.2kw xenon searchlight.

In October 1963, three M60A1 tanks were diverted from production for conversion to the T118E1 configuration. One each was delivered in March 1964 to Aberdeen Proving Ground, Fort Knox, and ATAC for tests which were completed by July. The T118E1 was initially released for production that same month with a supplemental drawing package released in September 1965. On 5 November 1965, AMCTC 3942 classified the T118E1 as Standard A and designated it as the full tracked combat engineer vehicle M728. At this same time, the gun and mount dropped the X prefix becoming the M135 and M150 respectively.

In early 1966, an improved round of ammunition was standardized for the 165mm demolition gun. It differed from the earlier M123 in several respects. These included a quick release handle assembly to speed up the loading operation, a two piece welded cartridge case replacing the three piece design, a polyethylene sealed propellant charge assembly, a castable explosive filler in place of a pressed filler, and an improved T50 electric primer igniter.

Chrysler produced the M728 at the Detroit Tank Arsenal. A total of 243 CEVs were completed during the period 1966–1972.

Another view of CEV M728, registration number 9B 7112, appears at the bottom of the page. All of these photographs were dated 17 June 1966.

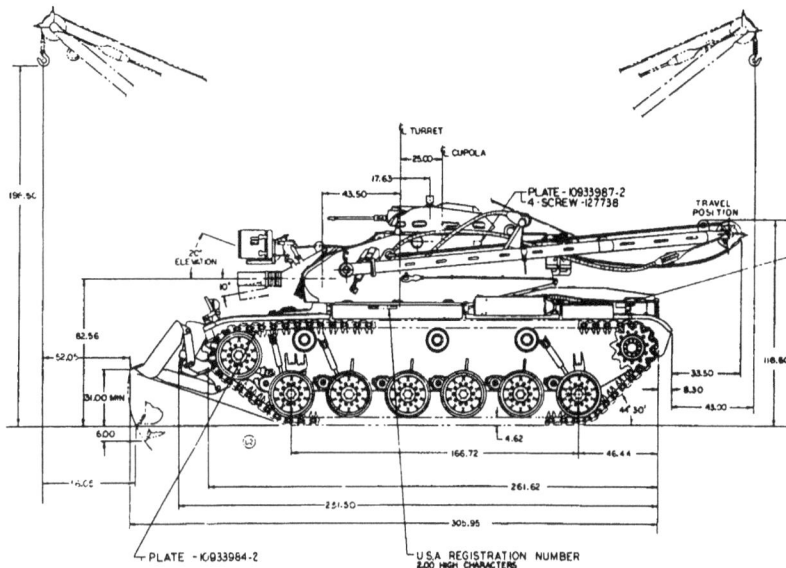

The M728 above is lifting a load with the boom winch during test operations. At the right, the front and side view layout drawings of the M728 are reproduced.

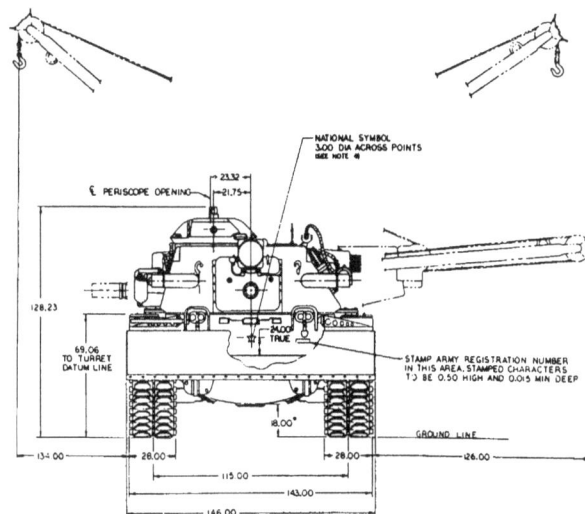

Below, an M728 combat engineer vehicle of the 7th Engineer Battalion is at Fort Carson, Colorado on 25 April 1967. This was the first operational battalion in the United States to be equipped with the M728.

Above a CEV M728 from the 7th Engineer Battalion fires a 165mm HEP round during a demonstration at Fort Carson on 25 April 1967. Below, an M728 lifts a smashed jeep out of the roadway in a simulated combat situation at Fort Belvoir, Virginia on 28 September 1967.

The view above shows the rapid railway destructor assembled and ready for operation. The ammunition chutes feeding the six .50 caliber machine guns used to notch the rails can be seen below.

The rapid destruction of railway tracks during retrograde operations was the subject of a Corps of Engineers development project. One experimental solution to the problem was the use of a machine based on the M46 chassis. Referred to as the rapid railway destructor, it lifted the complete track from the road bed. Moving at a speed of 6 miles/hour, the machine broke both ties and rails, the latter after notching by armor piercing bullets fired from six .50 caliber machine guns. Built under contract by Sperry Products Company, the rapid railway destructor was successfully tested in March 1952.

The front (left) and rear (right) of the rapid railway destructor are illustrated below. Note the broken rails and ties behind the machine.

Above, the 60 foot span armored vehicle launched bridge is mounted on the M46 tank chassis. Below, the bridge is unfolding (left) and is fully emplaced (right). This demonstration was at Fort Belvoir, Virginia on 7 December 1953.

ARMORED VEHICLE LAUNCHED BRIDGES

The capability of launching a 40 foot span assault bridge was dropped from the specifications for the pioneer tank T39 when it became obvious that too many requirements were complicating the design. However, the need for such an armored assault bridge launcher remained and in 1949, work began on an armored vehicle launched bridge (AVLB) based on the M46 tank chassis. This prototype featured a 60 foot span scissoring type bridge fitted with an hydraulic launching system. It was designed to permit both the launching and retrieval of the bridge by an operator inside the tank hull. Extensive tests showed the superiority of the AVLB compared to other portable bridge designs, but by the time the program was complete, the M46 was considered obsolete and interest shifted to the use of a newer vehicle.

Above is a 43 foot portable assault bridge pushed into place by an M46 tank. It also was demonstrated at Fort Belvoir on 7 December 1953. The M46 AVLB can be seen in the background of the right-hand photograph.

Above, an M48 tank pushes a 40 foot span folding bridge into place at Fort Belvoir during a demonstration in the Summer of 1955. Below, the bridge is in the process of unfolding.

Above, an AVLB based on the M48 tank chassis launches its 60 foot span bridge during a demonstration by the 7th Engineer Battalion at Fort Carson, Colorado on 20 April 1962.

Two scissoring type hydraulically operated bridges were designed to span gaps of 40 and 60 feet. Installed on the M48 tank chassis, they were tested during 1956 and 1957 confirming the effectiveness of the AVLB concept. Some of the M48C tanks were converted to this application.

Below, an M48A1 tank of C Company, 77th Armor crosses an armored vehicle launched bridge during exercises at Fort Carson on 21 June 1962.

The AVLB based on the M48A2 tank chassis is shown above. Below, the overhead cylinder hose armor can be seen between the cupolas for the two crew members.

Since by then the production tank was the M48A2, this was the chassis selected as the basis for the standard AVLB. Produced by Unit Rig and Equipment Company of Tulsa, Oklahoma, the new AVLB carried either a 40 foot span or a 60 foot span aluminum bridge. Designed to support class 60 loads, the actual lengths of these two bridges were 43 and 63 feet with a roadway width of 12½ feet. They were identical except for the addition of the two 10 foot center sections in the longer model. The launcher chassis was the M48A2C modified by the removal of the turret and the installation of a boom assembly, a tongue assembly, and the bridge seat and hydraulic system. The driver's controls were relocated to the left side of the hull below the turret ring. The two man crew was seated side by side below the turret ring with the driver on the left and the commander on the right. A vision cupola was installed above each crew station. The controls for launching and retrieving the bridge were located just in front of the driver's seat. When fitted with the 60 foot bridge, the total weight of the vehicle was about 63 tons.

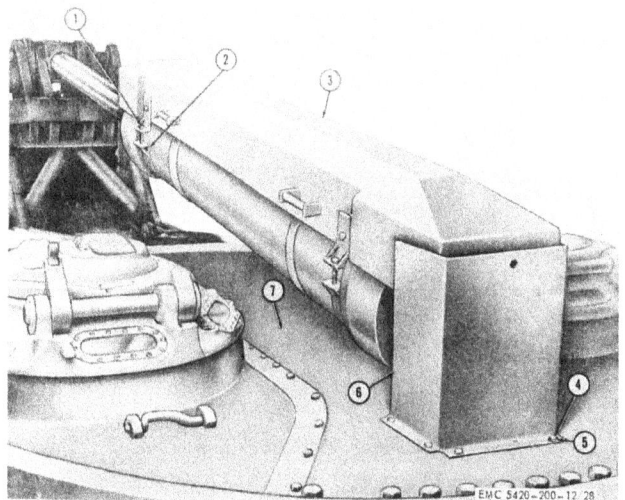

EMC 5420–200–12 26

1 Screw, mach., hex. hd., 3/8-24 UNC (4 rqr)	5 Washer, lock, 3/8 in. (6 rqr)
2 Strap (4 rqr)	6 Rear hose armor
3 Top hose armor	7 Cover plate center section
4 Screw, mach., hex. hd., 3/8-16 UNC-2A x 3/4 in. (6 rqr)	

Below is a drawing of the 60 foot span scissors bridge which was installed on the M48A2 chassis.

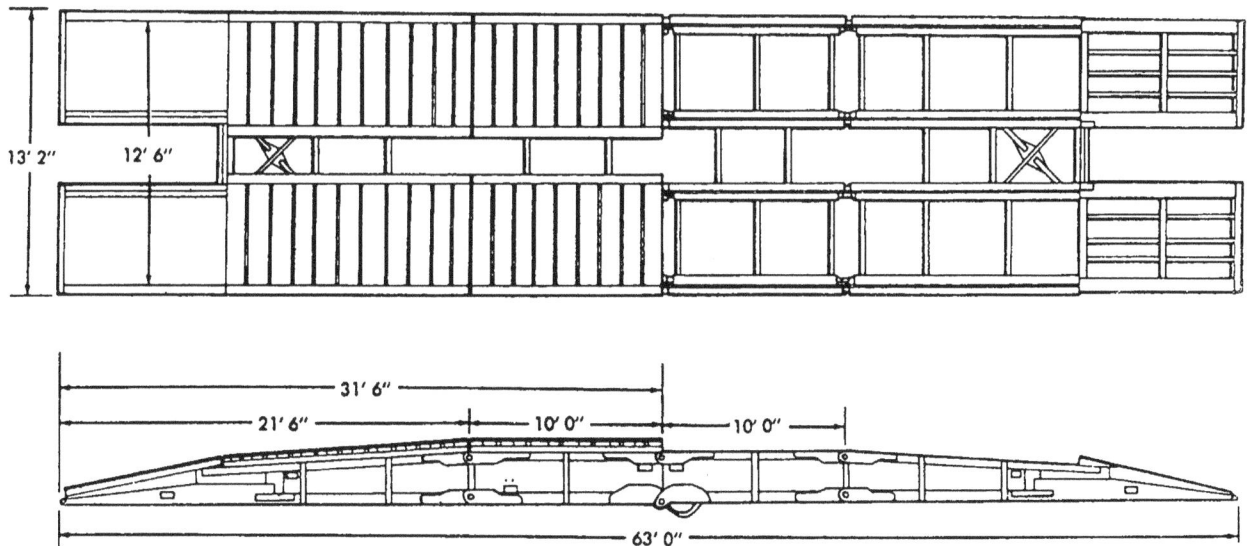

13' 2"   12' 6"

31' 6"
21' 6"   10' 0"   10' 0"
63' 0"

307

Above is the AVLB based on the M60A1 tank chassis with the bridge mounted. Below at the right, details of the commander's cupola can be seen in this photograph illustrating the use of the hydraulic slave system to power a disabled launcher.

After the M60 series tanks entered production, the chassis of the new vehicle was modified for use as an AVLB utilizing the same two sizes of aluminum bridges as on the earlier launcher. During the period 1964 to 1973, Chrysler produced 373 AVLBs based on the M60A1 chassis. An additional 14 were built during fiscal year 1978 followed by six in 1980–1981. Some of the M60A1 AVLBs were deployed to Vietnam where they were highly successful in operation.

The conversion of the M60A1 to the AVLB configuration was similar to that for the M48A2C, with the two man crew relocated beneath the turret ring. The vision cupolas over each crew station were somewhat different in design from those on the earlier vehicle. The most obvious difference was the higher silhouette pivoting hatch cover. The M60A1

AVLB weighed a little over 61 tons with the 60 foot bridge on board. With its diesel engine, it had much greater cruising range than the gasoline powered models.

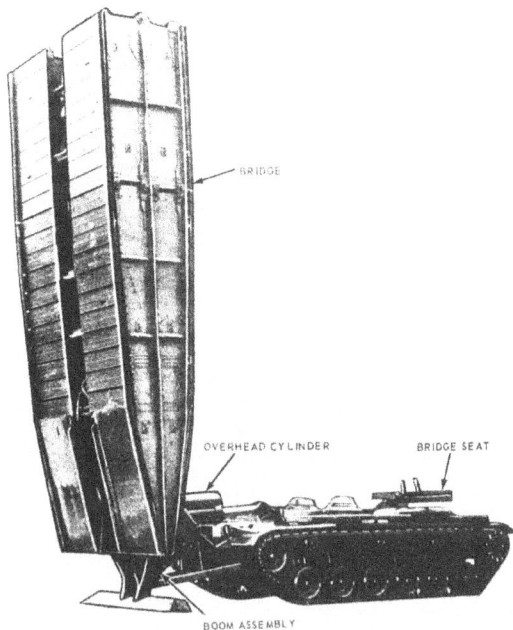

Two stages in the launch sequence are shown at the left and below. The M60A1 AVLB is launching the 60 foot span aluminum bridge in these views.

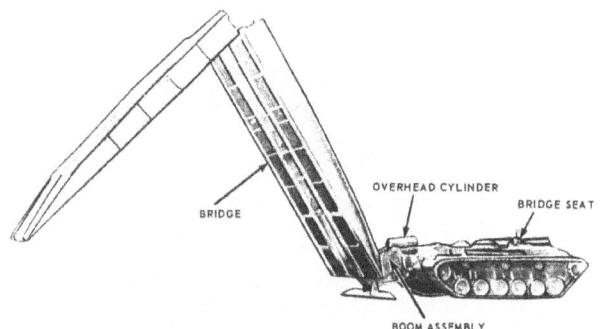

Above is the AVLB chassis adapted from the M48A5 tank without the bridge installed.

In the late 1970s, a project was started to upgrade the M48 series AVLBs to the diesel powered M48A5 configuration. This made use of the earlier vehicles in the inventory and was considerably less expensive than procuring new M60A1 chassis. Such a conversion gave the earlier AVLB performance essentially equivalent to the new vehicle based on the M60A1. An improved vision cupola was introduced with the M48A5 conversion and it was planned to retrofit this item on AVLBs already in the field.

Further development work using high strength aluminum alloys resulted in a new bridge weighing about 14,000 pounds compared to 29,000 pounds for the standard 60 foot span. Designed to fit the M60A1 launcher, the two sections of the lightweight bridge also were intended to become the end ramp portions of a new 90 foot folding bridge.

Below, the M60A1 AVLB launches the lightweight 60 foot span aluminum bridge.

Above is the medium tank M26 equipped for deep water fording and fitted with the T8 flotation device. This equipment was considered to be adaptable to the M46, if required.

FLOTATION DEVICES AND FORDING EQUIPMENT

Battle experience during World War II repeatedly proved the great value of armor during the initial stages of amphibious landings or river crossing operations. To permit the participation of tanks and other armored vehicles in such actions, deep water fording equipment and some special flotation devices were developed. Many of these designs were carried over for use during the postwar period. The fording and flotation equipment developed for the M26 medium tank was considered suitable for adaptation to the later M46 and M46A1. Following wartime design practice, this equipment sealed and excluded water from the entire engine compartment and required large stacks for the intake and exhaust of cooling air. In addition, a separate stack was provided for the engine exhaust. This was a heavy, bulky arrangement and the preparation and installation were complicated and time consuming.

To provide complete flotation of the tank during ship to shore operations, the T8 device was developed for the M26. Used with the deep water fording kit, it consisted of metal floats in four jettisonable assemblies with two rudders mounted at the stern. Propulsion was provided by the action of the vehicle tracks in the water. This combination was 65

Below, the T15 flotation device and the deep water fording kit are installed on the 90mm gun tank M47. This photograph was taken at Detroit Arsenal on 3 June 1953.

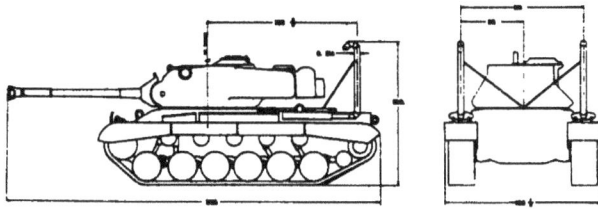

These sketches show the dimensions of the deep water fording kit (above) and the T15 flotation device (at right) for the 90mm gun tank M47.

feet long, 16 feet wide, and had a maximum speed in the water of 5.5 miles/hour.

With the introduction of the M47 tank, a new deep water fording kit was developed to take advantage of the ability of the AV-1790 air-cooled engine to operate fully submerged in water. This new design permitted complete flooding of the engine compartment and only required stacks attached to the engine exhausts. The engine air intake was through the crew compartment. Seals for the turret ring and gun shield, as well as waterproofing for the auxiliary generator engine and other minor components, increased the fording depth from the normal 48 inches to 101 inches. At this point, the water reached the top of the turret bustle. A bilge pump was installed in the crew compartment to take care of any leakage. With far fewer parts, the installation of the new deep water fording kit was greatly simplified over the earlier version with the dry engine compartment.

During October 1950, the Research Designing Service of Center Line, Michigan began a study project for Detroit Arsenal of several tank flotation devices. Based on this work, the design of the flotation device T15 was selected for application to the M47 tank and a pilot model was constructed. Shorter and wider than the T8 device, it had a length of over 41 feet and a width of about 24 feet. The angle of approach was increased from 14 to 22 degrees allowing the vehicle to climb out on banks with a greater slope. This was a major problem with all of these large flotation devices. With a highly sloped shore line, the floats would strike the shore before the tracks contacted the bottom preventing the vehicle from climbing out.

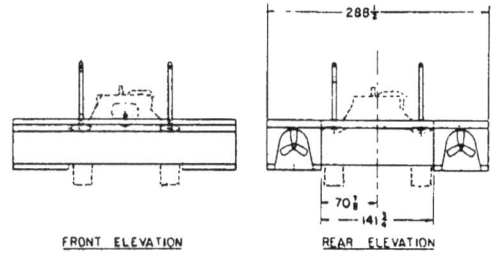

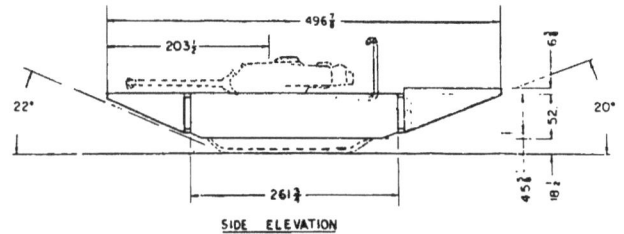

The floats on the T15 consisted of plastic blocks secured by sectional steel framing to brackets on the tank. These floats could be jettisoned by electrically fired explosive cartridges. Unlike the earlier T8, propulsion of the T15 was by two 36 inch diameter propellers driven by the tank drive sprockets. Twin rudders controlled from the tank commander's position provided steering. Tested in the water at Detroit in June 1953, the T15 attained a maximum speed of 7½ miles/hour and had an estimated cruising range of 30 miles. The deep water fording kit previously described was required with the T15.

Below, the M47 is afloat using the T15 device during tests by Detroit Arsenal on 3 June 1953.

Above, the deep water fording kit is installed on the 90mm gun tank M48. Note the outlet for the bilge pump on the right front top of the hull.

The appearance of the medium tank T48 required some modification of the deep water fording equipment. Like the M47, its air-cooled power plant was capable of fully submerged operation using only engine exhaust stacks. With the air intake through the crew compartment and the sealing of the turret ring and other components, the fording depth was increased to a maximum of 100 inches.

An attempt was made to develop a flotation device for the T48. Designated as the flotation device T44, it was a slightly enlarged version of the T15 used with M47. The length and width were increased to a little over 42 feet and 26 feet respectively. This increased the freeboard to approximately 16 inches in front and 7 inches in the rear compared to 11 inches and 5 inches for the T15. The two 36 inch diameter propellers were driven from the tank sprockets with a ratio between propeller rpm and sprocket rpm of 6.25:1. The two rudders were operated electrically by either the commander or the driver. Development of the T44 was contracted by Detroit Arsenal to Sparkman & Stephens, Inc. of New York City. However, the limitations of these large flotation devices had long been apparent. Unlike the wartime DD tanks where flotation was obtained by extending collapsible canvas side walls above the tank hull with little or no increase in the deck area occupied by the tank, the large size of the T44 and similar flotation devices greatly reduced the number of tanks which could be carried on shipboard. Also, as previously mentioned, the large overhang limited their operations to shore lines with slopes not exceeding their approach angles. The bulky design of the devices resulted from the requirement that the tank be capable of using its main armament while afloat unlike the DD, where the tank was suspended below the water line. In any case, the many limitations of the large flotation devices outweighed their advantages and further development was dropped.

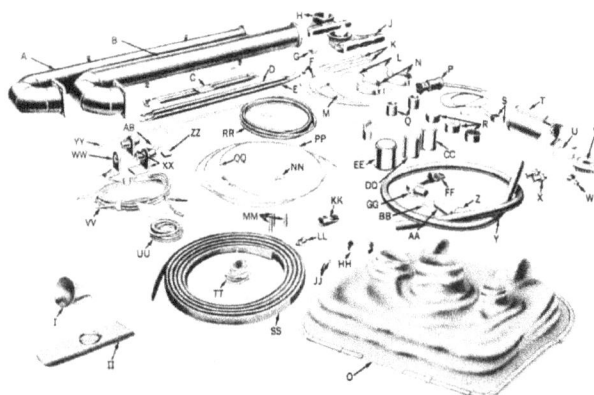

Components of the deep water fording kit for the M48 are illustrated at the left. The sketch below shows the installation of this equipment on the tank.

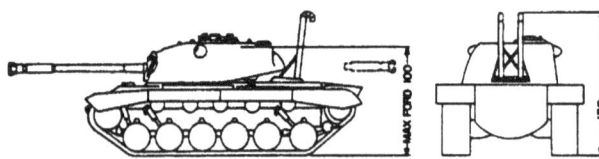

A	Right stack assembly	I	Collar cap assembly	R	Fire extinguisher horn cap
AA	Nipple	II	Machine gun cover assembly	RR	Fuel tank vent hose
AB	Right stack outer brace anchor	J	Right duct assembly	S	Switch assembly
B	Left stack assembly	JJ	Oil fan drive brush box plug	SS	Turret-to-hull seal
BB	Bilge pump housing outlet jacket	K	Gasket	T	Bilge pump and motor assembly
C	Stack cross brace	KK	Auxiliary engine exhaust outlet seal	TT	Elastic shock absorber cord
CC	Water pump grease	L	Exhaust stack gasket	U	Pump and base gasket
D	Left stack outer brace	LL	Hot spot control shut-off valve	UU	Rangefinder blister cord
DD	Asbestos grease	M	Clamp	V	Bilge pump base
E	Right stack outer brace	MM	Turret race drain plug	VV	Stack release cord assembly
EE	Nonhygroscopic adhesive tape	N	Clamp	WW	Fording cord guide
F	Clamp	NN	Final drive vent line tee assembly	X	Bilge pump hose nipple
FF	Bilge pump housing outlet seal	O	Gun shield cover	XX	Cord guide assembly
G	Exhaust duct support	P	Bilge pump harness	Y	Bilge pump hose
GG	Bilge pump outlet housing	PP	Final drive breather tubing	YY	Fuel tank vent tube
H	Left duct assembly	Q	Fuel tank cap seal	Z	Elbow
HH	Auxiliary engine bellows gasket			ZZ	Left stack outer brace anchor

REAR

FRONT

A BILGE PUMP INSTALLATION   D FIRE EXTINGUISHER HORN CAP—8671589   E FUEL TANK VENT HOSE—8744454
C CORD ASSY—8744484   H FINAL DRIVE BREATHER HOSE—8671173   F STACK ASSY—8744476
B SUPPORT—8744473   I SUPPORT—8744473   J TUBE ASSY—8381882

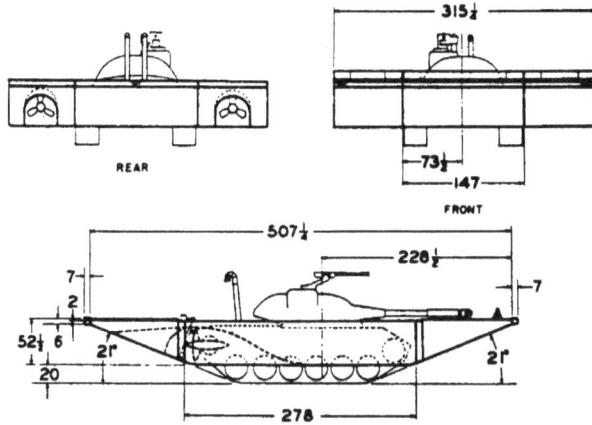

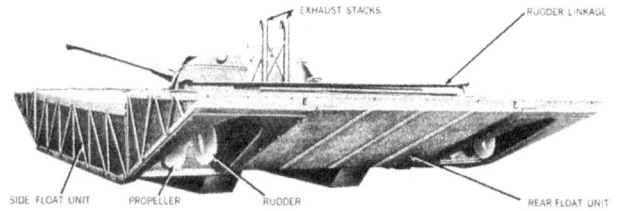

The installation of the T44 flotation device on the M48 tank is sketched at the left and illustrated above. Once again, the great bulk of this unwieldy arrangement is obvious. Below are two drawings showing the deep water fording equipment developed for the M48A2 tank. Note the complications resulting from the new rear deck and exhaust system.

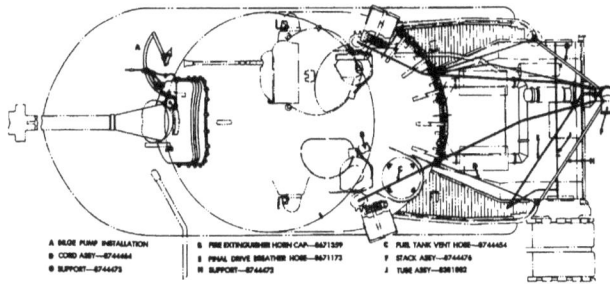

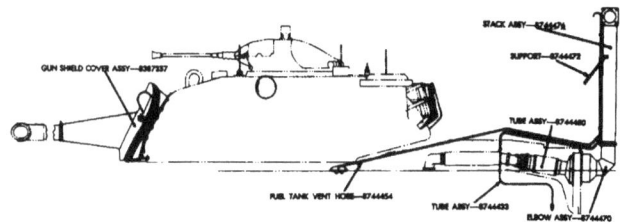

The new design of the rear deck and exhaust system of the M48A2 complicated the application of the deep water fording equipment. The kit was modified to include exhaust pipe extensions for use between the mufflers and a mounting pad on the inside of the right rear grill door. A single exhaust stack was then attached to the pad on the outside of the same door. After the introduction of the supercharged diesel engine in the M60 and the M48A3, further modifications were required. These included a new extension pipe connecting the engine exhaust from the turbochargers to the mounting pad on the right rear grill door. The same single exterior exhaust stack was retained. A square mounting pad on the grill door was used for the early production tanks, but this was changed to one of irregular shape on the later vehicles. These tanks also were fitted with an inflatable seal for the turret ring.

The photograph above shows the connections required to the exhaust system on the M48A2 when prepared for deep water fording. Below are views of the M48A2 with the complete kit installed.

Here is the 105mm gun tank M60, pilot number 4, equipped with the deep water fording kit at Chrysler on 7 April 1960. In the view above, the bilge pump outlet can be seen on top of the right front hull near the personnel heater exhaust pipe.

Above, M60 pilot number 2 is in the fording pond at Detroit Arsenal. The effects of the cold Michigan winter are obvious in this photograph dated 24 February 1960. Below are views of the exhaust system ducting (left) and the right-hand grill door (right) with the mounting pad for the exhaust stack connection.

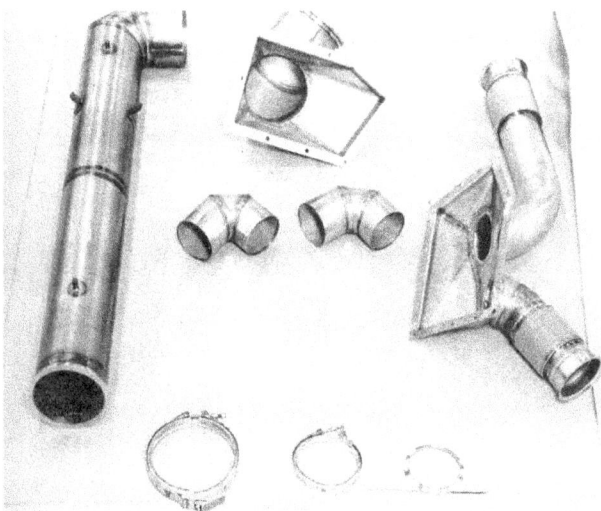

The underwater fording kit improvised at Fort Knox is illustrated above. The conning tower installed over the loader's hatch and the flexible exhaust stack can be seen in these views.

Field tests of the M60 with the deep water fording kit at Fort Knox were generally satisfactory, although there were some objections to the time required to install the equipment, particularly the extensions to the engine exhaust system. Modification of the kit at the Armor School during 1961 replaced the fixed exhaust stack with a pair of flexible tubes which could be installed more rapidly. This modified kit also sealed the M19 commander's cupola with a cover and fitted a conning tower over the loader's hatch. With this arrangement, the tank could submerge below the turret top. If some permanent modifications were made, the tank could be prepared for deep water fording in less than 30 minutes.

Further work at Detroit Arsenal during 1962 developed an underwater fording kit to provide the M60 and M60A1 the capability of fording rivers and waterways to a depth of 15 feet. The engineering and service tests were completed during 1963 and the equipment was evaluated by the Armor and Engineer Board at Fort Knox during November and December 1964.

The new underwater fording kit (note the difference in designation from the deep water fording kit) consisted of two major groups. The first was a base group installed in the shop on a vehicle intended for use in fording operations. The second was a shore installation group which was stowed on the vehicle and was installed by the crew just prior

Below is the complete underwater fording kit for the M60 series tanks. Note the different gun shield covers for the M60 and the M60A1.

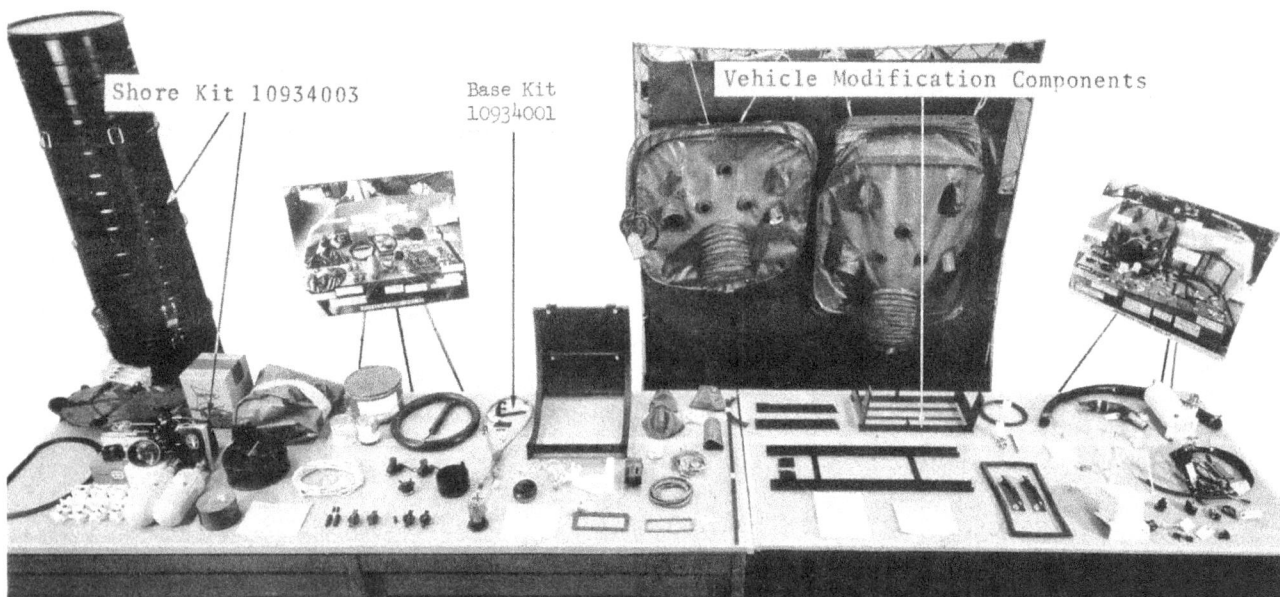

The M60A1 above at Fort Knox is equipped for underwater fording. A close-up view of the four section conning tower (below left) is compared with an experimental five section tower (below right).

to entering the water. The base group consisted of the gun shield cover, bilge pump, conning tower stowage bracket, and numerous seals, covers, and caps. This part of the kit could be applied to the M60 or M60A1 by two men during an eight hour day

and the installation did not reduce the combat effectiveness of the tank. The shore kit consisted of the four aluminum conning tower sections which were stowed nested on the rear of the turret, providing a container for the remaining items. After assembly,

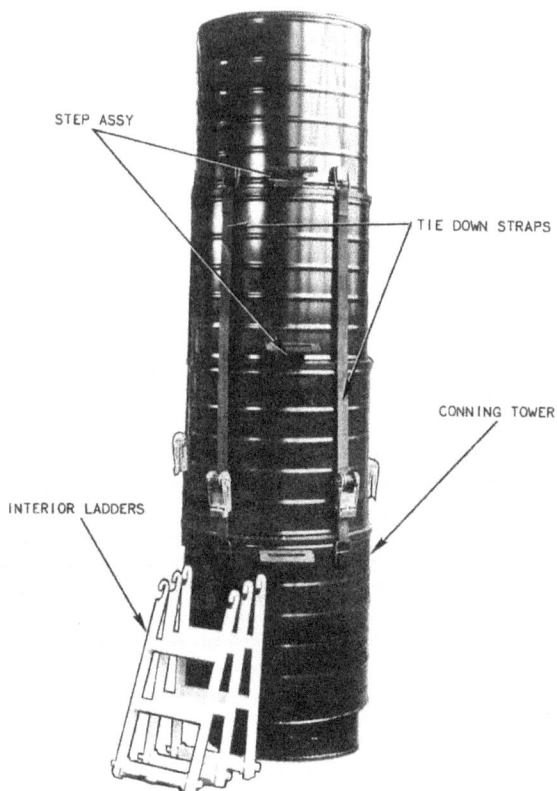

STEP ASSY

TIE DOWN STRAPS

CONNING TOWER

INTERIOR LADDERS

Conning Tower 5th Section TD 111888

Conning Tower 4th Section LO 6086-1

Conning Tower Retaining Strap Assemblies TD111881

Conning Tower 3rd Section LO6086-8

Conning Tower 2nd Section 10933831

Commanders Cupola Cover LO5869

Conning Tower Base Assy. 10933811

On the original underwater fording kit (above), the exterior steps were installed on the left rear of the conning tower. The close proximity of the radio antenna made it difficult for the crew members to use the steps in this position. The steps were then relocated on the modified kit (below) to the front of the conning tower.

The M60A1 above is fitted with the modified underwater fording kit at Fort Knox on 14 January 1964. Note the guide rod installed on the right front fender to orient the crew with respect to direction and water depth during fording operations.

the conning tower was fitted over the loader's hatch, projecting almost 8½ feet above the turret. It was equipped with a ladder and the open top served as an observation post for the tank commander. Underwater breathing apparatus (SCUBA) was provided for emergency use by the three remaining crew members. The commander's cupola was sealed with a cover after removal of the .50 caliber machine gun. The remaining components consisted of the turret ventilator blower air inlet seal and pump, various plugs and seals, as well as a sealing compound and a repair kit. In addition, four inflatable life preservers were included for the crew. The shore group could be installed by a trained crew in about 20 minutes.

Unlike earlier fording kits, no engine exhaust stacks were provided. It was noted that the inherently high exhaust pressure of the diesel engine permitted it to run satisfactorily submerged. An additional benefit was the almost silent operation of the tank with the underwater exhaust. Flapper valves were installed on the exhaust ports to prevent water from flooding into the engine if it stalled. However, these valves did not seal very well, particularly after long operation of the high temperature exhaust system. A restart within 10 to 15 seconds was required to prevent excessive leakage.

Tests showed that the M60 and M60A1 could operate in water up to a depth of 15 feet at a maximum speed of 9 miles/hour. The tank was capable of climbing out on slopes of 40 to 60 per cent, depending upon the soil conditions. It was noted that radio communications with the AN/VRC-12 radio was possible up to three miles with the tank under water. Seventy-one of the underwater fording kits were manufactured and placed in depot storage.

These additional views of the M60A1 at Fort Knox on 14 January 1964 show the modified underwater fording kit stowed on the turret bustle. The conning tower sections fitted inside each other and provided a container for the other components.

Above and below is the medium recovery vehicle M88 fitted with deep water fording equipment. Twin exhaust stacks are installed at the rear and large rectangular air inlet ducts are located midway on each side of the vehicle. The small circular stacks near the center provide cooling air for the main generator and serve as inlet air and exhaust pipes for the auxiliary generator and personnel heater.

Above, the combat engineer vehicle M728 equipped with the underwater fording kit is dozing a path onto the shore. Below, another M728 is lifting a roll of mat with the boom winch. Note the underwater fording kit stowed on the boom.

The M26 tank above is fitted with the jet mine clearing device using solid fuel rockets mounted on the front of the vehicle. In the right-hand view, the device is operating during a demonstration at Fort Belvoir, Virginia in November 1947.

MINE CLEARANCE EQUIPMENT

The detection and clearance of antitank mines was the subject of intensive research and development programs throughout World War II. Despite the high level of effort, no completely satisfactory solution to the problem had emerged by the end of the war. The development proceeded along two basic directions. The first utilized mechanical equipment to explode or remove the mine ahead of the armored vehicles and the second made use of explosive charges to detonate the mines and clear a safe passageway. Several types of both classes of mine clearance equipment were developed, each with its own advantages and limitations.

The most widely used of the mechanical mine exploders was the flail type developed in Great Britain. An advantage of the flail tank was that it was not extremely heavy and the tank retained most of its speed and mobility, particularly when the device was not operating. Unfortunately, the flail was far from 100 per cent effective, although it undoubtedly reduced casualties by eliminating many mines.

A few of the large American roller type mine exploders were tested in Europe. When the terrain was suitable for their use, they were extremely effective in detonating mines. However, the bulk and weight of this unwieldy equipment drastically reduced the tank's mobility and limited the kind of terrain on which they could be used.

A special problem existed with mines on the sandy beaches of the Pacific. To simplify their removal, a number of bulldozer type mine excavators were procured during the closing months of the war.

Among the wartime explosive mine clearance devices, the demolition snake emerged as the most

useful. Successfully employed at Anzio and in later operations, it was recognized at the end of the war as a standard method of breaching minefields. However, by its very nature, the main use of the demolition snake was in breaking through prepared positions protected by transverse minefields which were relatively shallow in depth. It was recognized that future operations might require the clearance of longitudinal or random minefields extending for miles along roads or other narrow defiles. Such routes would have to be rapidly cleared to permit the maneuver of fast moving columns and for this type of mine clearance, the demolition snake was clearly inadequate.

Following the end of the war, the work on mine clearance equipment was drastically reduced along with other military development programs. However, the continuing need for such devices was recognized in the requirements outlined by the Stilwell Board and the later Army Equipment Development Guide. Although the level of effort decreased, work continued to develop new mine clearance equipment based on the wartime experience as well as several new concepts.

One new approach to the problem utilized the erosive action of high velocity gas jets to remove both the soil and the mines. Three solid rockets were mounted on the front plate of an M26 tank, angled so that the jet would strike the ground in front of the vehicle. Tested at Fort Belvoir in late 1947, the device did not work well against hard dry clay soils. An even greater limitation was the short duration of firing with the solid rockets. In an effort to solve the duration problem, a tank was equipped

Above, four liquid rocket motors (two in front and one on each side) are installed on the front of an M26 tank as an experimental mine clearing device. Note the large tanks required for fuel and oxidizer.

with four liquid fuel rocket motors. Referred to as Squirt, the vehicle was modified to carry a large load of fuel and oxidizer to extend the operating time. However, the liquid rocket also suffered from poor performance in hard dry clay. A small scale experiment attempted to improve the excavation in clay by utilizing sand in a fuel oil-air jet at a velocity of about 800 feet/second. Unfortunately, there was no significant improvement in performance. All of these jet type excavators created large dust clouds during operation and would have undoubtedly attracted the unwelcome attention of the enemy.

Below is an artist's concept of a jet type mine excavator dated 8 August 1949. Like all such equipment, it raised very large clouds of dust and debris and tests showed that its effectiveness was limited on hard ground.

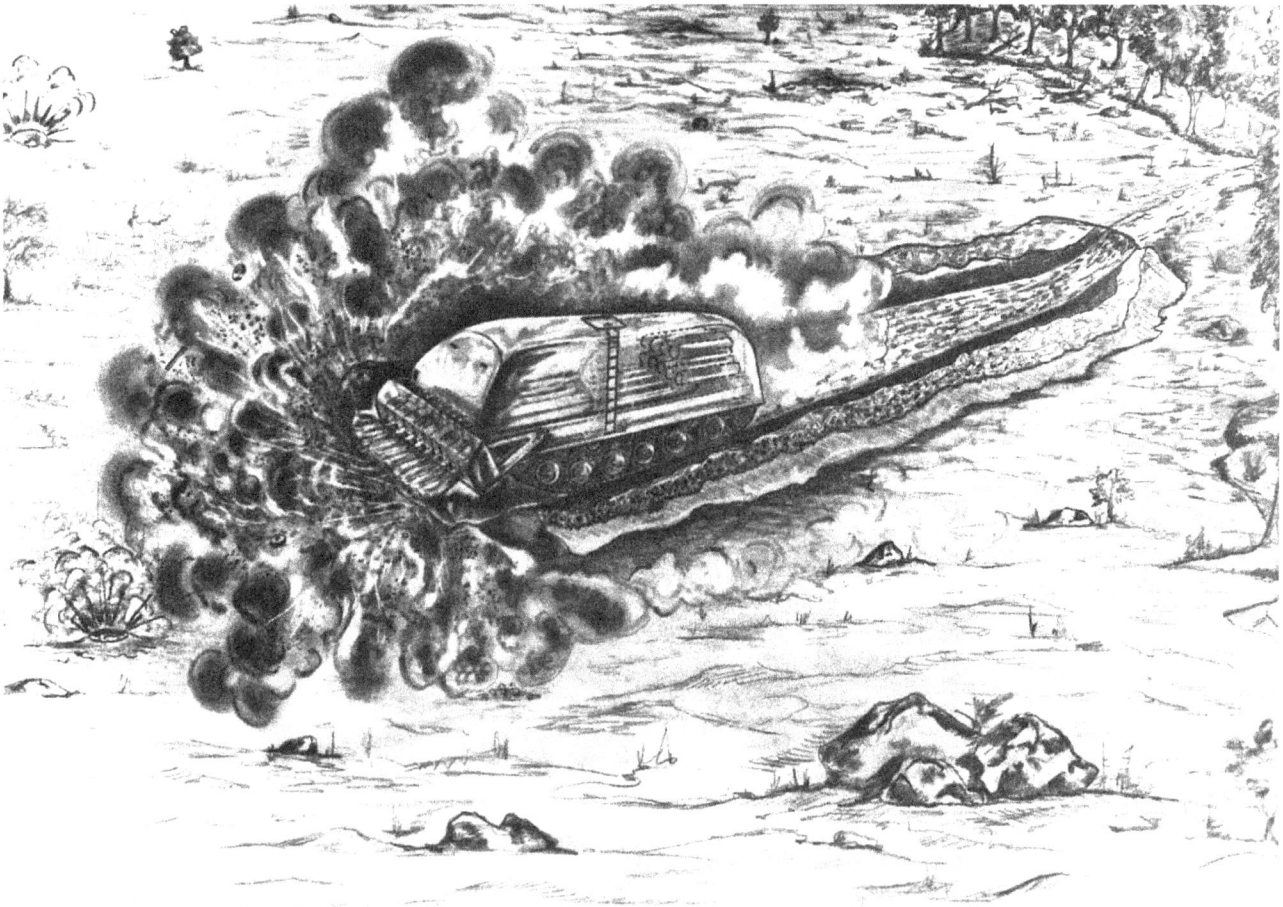

Above, the model I heavy mine excavator is at the left and at the right is the model II intended for engineering and service tests. Below, the model II appears on the left with four pusher tanks and at the right, it is shown during excavation tests using five pusher tanks.

Although the T5E3 mine excavator was procured in 1945, no further work was done on plow type devices until 1950. At that time, a new project was approved which resulted in the construction of two large excavators by R. G. LeTourneau, Inc. Designated as the heavy pusher type mine clearing excavator, this device was intended to excavate a path 14 feet wide and 3 feet deep. The first item, model I, was tested at Longview, Texas in 1951 to determine its plowing ability and its resistance to mine explosions. Lessons learned during these preliminary tests were applied to model II which was tested at Longview in January 1952 and at Fort Knox from April through July. Both models I and II were driven by an Allison V-1710 engine derated to 500 horsepower which generated power for the electrically driven wheels. This was, of course, insufficient for excavating, but it allowed the device to move under its own power. Tests showed that to clear a path 14 feet wide and 3 feet deep, four or preferably five M46 pusher tanks were required. The speed of excavation was 2 to 3 miles/hour.

Although the heavy excavator, also called Pistol Pete, approached its design specifications, the size and complexity of the device greatly limited its tactical use. On 13 August 1953, the project was cancelled before the evaluation was fully completed.

A less ambitious mine excavator was the Peter Pan tank mounted plow. A V-blade plow was installed on the M47 tank and it could be jettisoned by firing explosive bolts from inside the vehicle. Although it plowed a path about a foot deep, it was difficult to maintain a level or equal cut on both sides and further work was discontinued.

At the right is the light mine excavating plow, Peter Pan, installed on an M47 tank.

Another view of the Peter Pan light mine excavating plow appears above on the M47 tank. At the right below, is the Fargo Express rotating hoe type mine excavator.

Another mine excavator was the rotating hoe device dubbed the Fargo Express. It was fitted with 16 arms which rotated in the reverse direction to cut slots in the road and throw mines and debris forward. Unfortunately, some debris and live mines frequently landed on top of the vehicle. Susceptibility to blast damage combined with high power requirements and slow speed also served to kill the project.

Later mine plows developed for the M60 series tanks were more successful. They were designed to scoop up mines in front of each track and deposit them to the side. This equipment consisted of two separate plows connected by a chain or bar assembly to detonate tilt rod mines in front of the tank.

Below is the track width mine plow designed for the M60 tank series. This side view shows the plow lowered to the operating position.

These photographs show the track width mine plow lowered to the operating position (above) and raised for travel (below). Installed here on the M60 tank, the equipment also was intended for use on the later tanks of the M60 series and the combat engineer vehicle M728.

High Herman, the heavy roller type mine exploder appears above at the left. At the right, the original 16 disc version (above) of the light roller type mine exploder, Larruping Lou, can be compared with the later model (below) with 12 discs.

The development of the roller type mine exploder stopped at the end of World War II. However, after the invasion of South Korea, interest was revived in such equipment. A development project was launched to adapt the best features of the World War II apparatus to a roller suitable for use with the new wider tanks then in service. This program produced a heavy tank mounted roller consisting of 25 one ton discs. Named High Herman, this device successfully withstood the repeated blasts of antitank mines. However, the 36 ton weight of the installed roller placed an excessive load on the tank's final drive. In addition, the boom assembly which attached the device to the tank prevented the main gun from being used in the forward direction.

To correct the problems with High Herman, a new program was initiated in 1953 to develop a light tank mounted mine clearing roller. Named Larruping Lou, the new device utilized the same articulated discs as High Herman. The experimental roller

used at first eight and later six discs in front of each track. Thus it cleared two paths, one for each track, and the weight of the device was reduced to 20 tons. The attachment to the tank was simplified and did not interfere with the main gun. Tests showed that the roller had excellent resistance to mine blast and the lower weight improved the mobility and reduced the wear on the tank final drive.

Below, the late version of Larruping Lou is installed on an M48 for tests on 20 November 1957. Note that this M48 has the T-shape blast deflector on the 90mm gun.

The view above from the tank turret shows the arrangement of Larruping Lou's six rollers in front of each track. Below, the device, installed on and M48A1, is exploding mines. Normally, the tank cannon would be aimed to the rear during such operations to prevent damage.

Above, Larruping Lou has been adapted to the M60 tank. Its appearance after the explosion of a statically fired M15 mine is shown in the right-hand photograph. Despite its appearance, there was no significant damage.

Another attempt to solve the mine clearance problem was the self-propelled exploder called the Galloping Ghost. Constructed using many parts from the late model Sherman tank, this device had a roller consisting of 12 cast steel discs. The 500 horsepower engine drove the vehicle tracks and also applied power to the perimeter of each roller disc through a system of gears. The self-propelled exploder could be driven by a single crewman or it could be operated by remote control.

The Army Equipment Development Guide of May 1954 included the requirement for light, low cost, expendable mine clearing devices for use with tanks. Neither High Herman nor Larruping Lou was designed to fit this job. In the past, several attempts had been made to improvise such an expendable roller in the field. During the advance of the United Nations forces into North Korea in late 1950, tanks were frequently immobilized by mines planted in the roadway. To minimize these losses, Captain Chester F. Kopicki designed an expendable roller which was fabricated from parts salvaged in the field. Intended as a one-shot device, it was expected that it would

be destroyed by the first mine that it detonated, but it would save the tank's tracks and suspension. Unfortunately, the Kopicki roller was never tested as it was abandoned in North Korea during the retreat from the first Chinese offensive.

Studies by the U.S. Army Engineer Research and Development Laboratories (ERDL) at Fort Belvoir, Virginia indicated early in 1953 that the weight of a light roller alone would be insufficient to detonate antitank mines. To obtain satisfactory performance, the ground pressure had to be increased by transferring part of the tank's weight to the exploder. A development program was initiated which evaluated four expendable roller designs. These were designated as the ERDL Model I, the ERDL Model II, the Birmingham, and the Lockley.

The ERDL Model I consisted of two separate drum type rollers, one mounted in front of each track. The effective width of each drum was 28 inches, the same as the track width of the M48. Each roller assembly consisted of a drum type exploder, a break-away boom, and a weight transfer mechanism. The latter used torsion elements to load

Below, the self-propelled mine exploder Galloping Ghost appears at the left. At the right is the expendable exploder improvised in Korea by Captain Kopicki.

Above, the ERDL Model I expendable roller is shown at the left and the ERDL Model II is at the right. Note the use of the standard tank road wheels on the latter.

the drum with part of the tank's weight. However, tests showed that the weight transfer mechanism and the tank hull adapter were unsatisfactory, although the drum type exploder and yoke system proved to be the best of the four devices tested.

The ERDL Model II was similar to the Model I, except that its two rollers each consisted of four individually articulated tank road wheels. These wheels were selected since they were already in the supply system and would be readily available in the field. The effective width of each roller assembly was 30 inches. Tests showed that the road wheel exploders were somewhat less effective than the drum type used on Model I.

After test and modification of the two ERDL exploders, a contract was awarded to the Birmingham Fabricating Company of Birmingham, Alabama to construct an improved version. Like the ERDL Model II, this device used two rollers assembled from tank road wheels. However, the number of road wheels per roller was increased to six giving an effective width of about 50 inches. Torsion elements were again used to transfer load to the rollers. Pressure was applied to the rollers by a screw thread system and the load was directly proportional to the tightness of the applicator nuts. Each wheel in the roller could articulate individually plus or minus six inches from its normal position while applying a force of about 2000 pounds to the ground. The two roller assemblies and their booms could be jettisoned leaving the adapter attached to the tank. Tests indicated that the weight transfer mechanism and the tank hull adapter were the best of any evaluated. However, there were still problems with the articulated road wheel assemblies.

Another contract was let to the Lockley Machine Company of Newcastle, Pennsylvania to design and build an expendable roller mine exploder.

The views below show the improved mine clearing roller built by the Birmingham Fabricating Company (left) and the independent design from the Lockley Machine Company (right).

The additional views of the Lockley mine clearing roller above show the overload condition on the fifth and sixth road wheels (left) and the low ground clearance under the power pack at the rear. Both conditions existed when the roller was loaded for operation.

This design featured an hydraulically operated weight transfer system and two rollers, each consisting of four cast steel discs. Each roller assembly covered a width of approximately 42 inches in front of its track. The individual discs were designed to articulate plus or minus four inches from their normal positions while the hydraulic system applied a load of 3000 pounds to the ground. The hydraulic pressure was provided by a power pack mounted on the rear of the tank and driven by a power take-off shaft from the engine. The complete exploder, except for the power pack, could be jettisoned in an emergency. The Lockley device weighed about 32,000 pounds and tests revealed that it imposed excessive loads on the tank's rear suspension when it was clearing mines and on the front suspension when it was being transported in the carry position. In addition, its complexity and expensive price did not meet the criteria for a low cost, expendable mine exploder.

Below is the modified Birmingham type exploder during the development program for a new expendable mine clearing roller. Note the changes in the frame used to attach the roller assemblies.

332

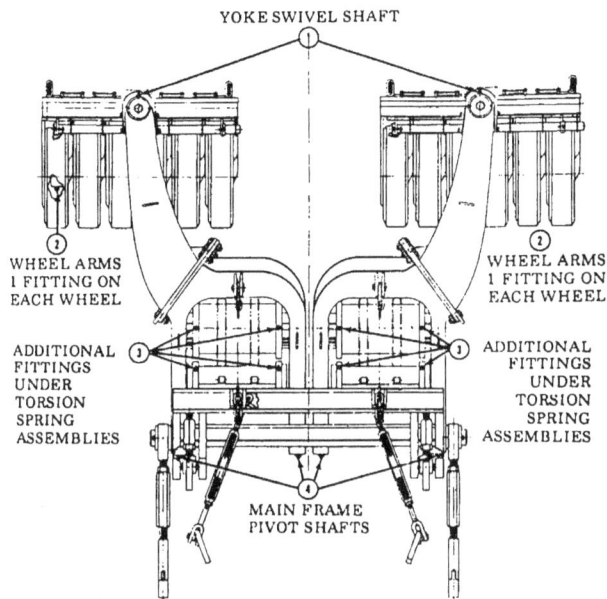

YOKE SWIVEL SHAFT
①

WHEEL ARMS
I FITTING ON
EACH WHEEL

WHEEL ARMS
I FITTING ON
EACH WHEEL

ADDITIONAL
FITTINGS
UNDER
TORSION
SPRING
ASSEMBLIES

ADDITIONAL
FITTINGS
UNDER
TORSION
SPRING
ASSEMBLIES

MAIN FRAME
PIVOT SHAFTS

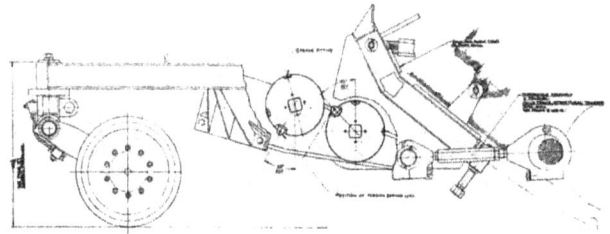

The configuration of the ENSURE 202 expendable mine clearing roller is illustrated in the drawings above and at the left.

After the commitment of United States forces to active service in Vietnam, a means was required to clear the mines frequently planted on roadways throughout the country. With a high level of guerrilla activity, roads which were clear one day might be mined the next. To meet this requirement, ERDL drew on the experience of the earlier research and designed an expendable mine exploder for use with the M48A3 tank. The work was performed under the Expedited, Non-Standard, Urgent Requirement for Equipment (ENSURE) and as the 202nd item under the program, it was referred to as the EN-SURE 202 tank mounted expendable mine roller. Like the earlier Birmingham model, it consisted of two rollers, one in front of each track. Six standard road wheels in each roller applied a ground pressure similar to that of the tank's road wheels. Each roller assembly was 47½ inches wide and the point of contact was about 9½ feet forward of the front tank road wheel. The two roller assemblies were attached to frame extensions which, in turn, were bolted to the main frame. These bolts were intended to break away upon the detonation of a mine, thus protecting the main frame. The latter was attached to a tank mounted structural frame by means of a pivot shaft. This structural frame was installed on the tank by the use of turnbuckles to the lifting eyes, turnbuckles to two nesting blocks welded to the tank hull, and pins through the towing eyes on the hull.

Part of the tank's weight was transferred to the roller assemblies by means of rubber torsion springs. One torsion spring assembly was fixed to the structural frame and another was mounted in the main frame. The force applied to the roller was proportional to the weight transferred from the tank through the springs by lengthening or shortening the chains on the spring assemblies. The expendable roller weighed about 10,000 pounds including the mounting frame.

The ENSURE 202 expendable mine clearing roller is installed on an M48 tank in these photographs. Below, the ability of this equipment to operate on a slope is illustrated. The view at the left shows the blocking of the tracks necessary for the installation of the torsion spring chains.

Above, a crew from the 1st Squadron, 10 Cavalry install an ENSURE 202 mine clearing roller on their M48A3 prior to operations in Vietnam. Below, another tank from the same squadron moves out with the mine exploder on Highway 19 near An Khe, Vietnam on 15 October 1970.

Twenty-seven of the ENSURE 202 mine clearing rollers were shipped to Vietnam where they were used with varying degrees of success. However, they were not considered a solution to the mine clearing problem. For example, the 11th Armored Cavalry Regiment frequently used their tanks alone to clear the mines. It appeared to be as easy to repair the tank after a mine detonation as to replace the roller. In view of this, considerable research was devoted to mine resistant tank tracks and suspensions. The objective of later work in this area has been to develop a track and suspension system to withstand the blast of a 22 pound mine. These experimental components reduced the area exposed to the blast and used some components fabricated from high strength composite materials.

Above, the M48A3 named ''Butch'' from the 1st Squadron, 10 Cavalry moves along Highway 19 with its mine exploder on 6 August 1970. Below, the layout of the ENSURE 202 can be seen in the top view of the device in use by the 11th Armored Cavalry Regiment in Vietnam.

The tank mounted mine clearing roller (TMMCR) developed from the tests of the Soviet equipment can be seen in the views above. A close-up of the five roller exploder assembly is at the right.

Tests of the Soviet built PT54 and KMT5 tank mounted mine clearing rollers (TMMCR) resulted in a program to develop a similar device utilizing the best features of the Soviet equipment. Initiated in August 1976, this project tested two prototype versions of the TMMCR. Each was fitted with two roller assemblies, one in front of each track. One prototype used four-wheel roller assemblies and on

the other, the number of wheels per roller was increased to five. A weighted chain was suspended between the rollers to detonate any tilt rod mines. Tests indicated that the rollers were about 85 per cent effective against pressure fuzed mines buried up to four inches deep at speeds of 10 miles/hour. Weighing under ten tons, these rollers could be installed in the field in less than 15 minutes and an hydraulic disconnect system released the device in under 30 seconds.

The roller assembly from the General Dynamics built TMMCR is sketched at the left. Below are attachment diagrams for this equipment to the M60 series tanks (left) and the M48 series tanks (right).

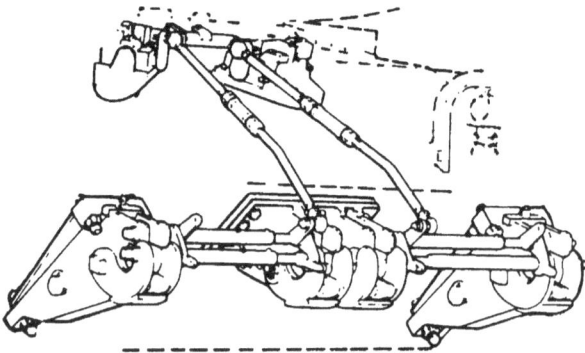

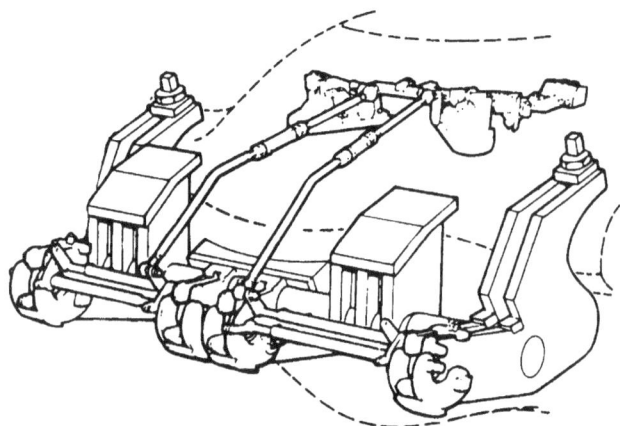

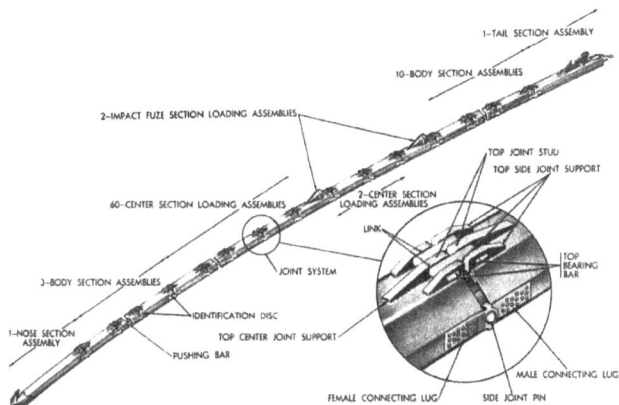

The assembly of the M157 projected charge demolition kit is illustrated above. At the right, the snake is being pushed into position by a tank.

The demolition snakes M2 and M3 developed during World War II were the standard equipment for breaching minefields in the early postwar period. These semirigid line charges were 400 feet long constructed from steel (M2) or aluminum (M3) plates. A 320 foot length of each snake was loaded with 14 pounds/foot of Amatol explosive. An improved snake, also constructed of aluminum, was introduced during the postwar period. Standardized as the projected charge demolition kit M157, it also was 400 feet long with an explosive loaded length of 320 feet. However the 10 pounds/foot Composition B and C4 explosive was in the form of two parallel shaped charges angled to direct the blast down and outward at 45 degrees. In most soils, the new snake blasted a crater 320 feet long, 12 to 16 feet wide, with a maximum depth of 3 to 5 feet providing an obvious path for tanks or other vehicles. The M157 was towed forward and pushed into the minefield by a tank as with the earlier demolition snakes. It was detonated by firing the tank's machine gun at one of the two impact fuzes on the top rear of the snake.

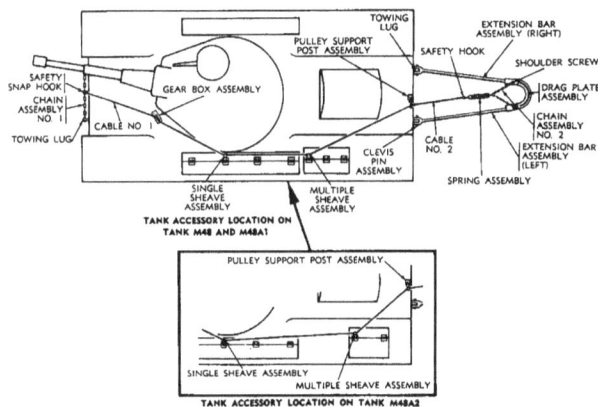

The rigging required on the M48 series tanks to push and tow the M157 projected charge demolition kit is sketched at the right above. Below are views of the rigging on the rear (left) and the front (right) of an M48A1.

At the right is the point charge snake Daisey Mae. In this view, the arms holding the 20 pound explosive charges are extended in the firing position.

Another type of semirigid demolition snake utilized point charges in place of the linear explosive. This experimental device, dubbed Daisey Mae, also was 400 feet long with the explosive carried in 32 center sections, each of which was 10 feet in length. Each center section contained two 20 pound Composition B charges mounted on extension arms which folded into the structure. After the snake was pushed into place, these arms were released and forced outward to hold the charges above the ground prior to detonation. Because of its complex design, this device was more difficult to assemble and use than the M157 and it did not progress beyond the experimental stage.

The wartime experiments with flexible line charges evolved into the projected charge demolition kit M173. This consisted of a 300 foot flexible line charge with the appearance of a giant string of sausages. It was stowed inside a boat shaped carrier which could be towed over land or water. The line charge was connected by a cable to a solid rocket in the launcher tube located in the nose of the carrier. The demolition kit was towed by a tank or other vehicle to the desired launch area. The tow coupling was released and the vehicle moved away, still connected by a 250 foot electric cable. A signal through this cable removed the carrier cover and elevated the launch tube to the required angle. Firing the rocket carried the flexible charge across the minefield. After coming to rest, the charge was detonated electrically still using the same connecting cable. The flexible charge contained 1500 pounds of Composition C4 for a total weight of 1720 pounds. A similar practice projected charge demolition kit M174 was provided for training purposes. It was identical to the M173, except that the charge was inert and it was painted blue instead of olive drab.

Other experiments with flexible line charges included a triple system where three parallel line charges, spaced two meters apart, were propelled by a rocket and controlled by nylon ropes. Each line contained seven pounds of explosive per foot or a total of 21 pounds/foot to clear a wide path. Named Mountain Lion, this project was dropped because a suitable rocket was not available at the time to propel the triple charge.

The U.S. Army also evaluated the British Giant Viper flexible line charge. This device was 750 feet long, loaded with plastic explosive, and stowed in a trailer. It was towed to the edge of the minefield and a cluster of eight rocket motors carried the explosive filled line into position. An arrester gear

Below, the projected charge demolition kit M173 is shown closed (right) and with the cover removed (left). Above, the M173 is in firing position with the launch tube erected.

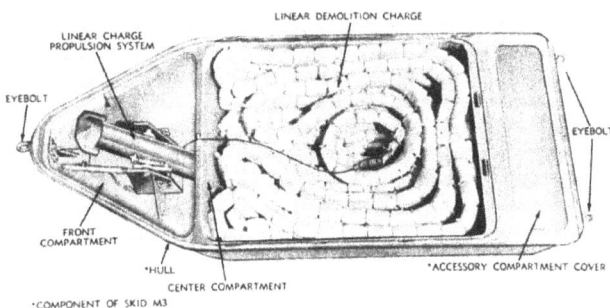

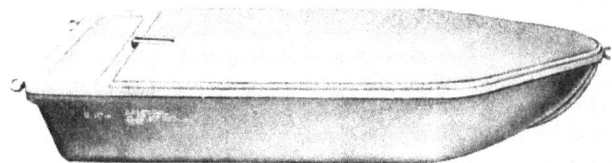

Above is the triple projected line charge system named Mountain Lion set up in firing position. At the right (above) is the British Giant Viper being towed behind an M60 tank. Below is an artist's concept of the Wild Bill point charge projector. This device was proposed for installation on several vehicle chassis.

consisting of three parachutes straightened the charge in flight. It was then detonated to clear a safe path about 600 feet long for tanks and other vehicles.

A vehicle mounted projector for small point charges was the subject of another experimental program. Intended for installation on a variety of chassis, this device consisted of an array of magazine fed, spring loaded launchers which projected charges weighing about one kilogram each onto a two-foot grid pattern. Because of the small pattern and the substantial charge, the blast action was extremely effective against most mines. Referred to as both the Peter Rabbit mine clearing system and the Wild Bill point charge projector, the major difficulty with this device was the lack of a low cost, reliable, safe, omnidirectional impact fuze.

Below are two views of the ROBAT (Robotic Obstacle Breaching Assault Tank) under development by the Corps of Engineers. One of the early prototypes appears at the left and a model of a later concept is at the right. This equipment was designed to use both the roller type mine exploder as well as linear explosive charges.

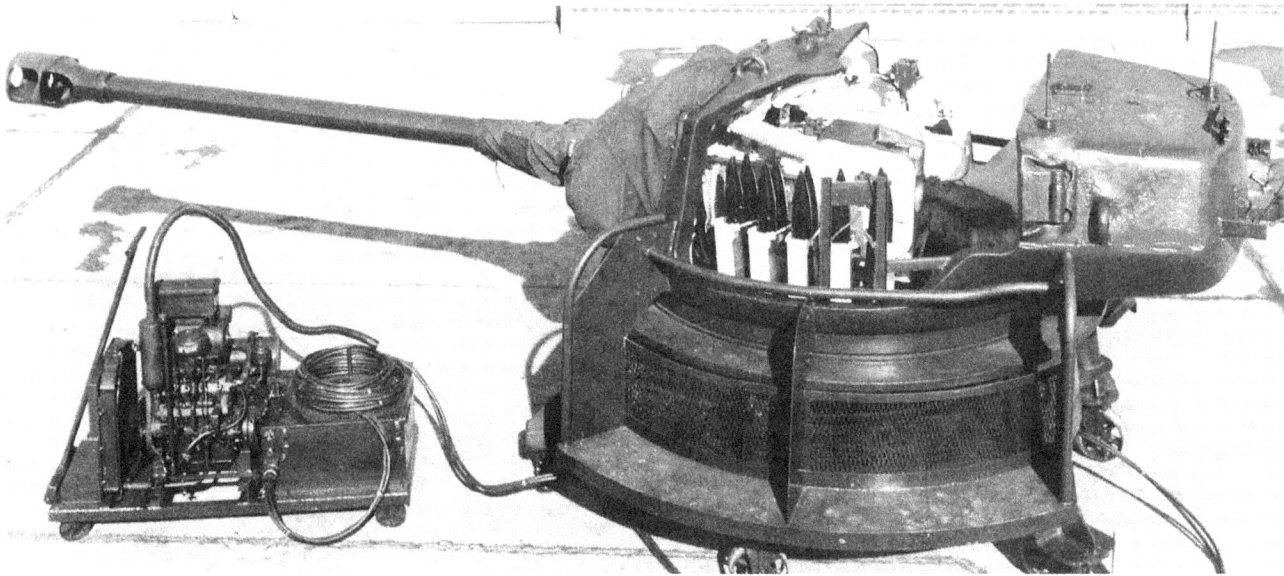

Above is the T13 tank gunnery trainer developed for use with the M26 tank. It could, of course, also serve as a turret trainer for the M46.

TRAINING DEVICES AND SPECIAL TEST EQUIPMENT

Experience dating from World War II had shown that troops could be trained much more rapidly and at less cost by using training aids such as cutaway or simulated tank turrets. Training turrets of this type were developed for all of the major postwar tank models and they continue in use to the present time. Similar equipment was developed to instruct drivers and later, more complex devices simulated a wide variety of combat situations for the training of both the gunner and the tank commander.

The tank chassis also provided the basis for other specialized training equipment such as the

target tank. Several of these were constructed using the M48 series tank.

Below are two photographs of the tank gunnery trainer T18 designed to simulate the M47 tank. A lightweight turret trainer, also intended for use with the M47, appears above at the right. This was a low cost unit particularly suitable for use by the Reserves and the National Guard.

The tank gunnery trainer T20 is shown above. This turret trainer simulated the 90mm gun tanks M48 and M48A1. It was fully equipped with a complete fire control system, ammunition stowage, and radio equipment.

The two views above show the tank gunnery trainer XM30 which duplicated the turret of the 105mm gun tank M60. At the left, the standardized version, the M30, is being used to instruct officers at Fort Knox. The turret trainer for use with the M60A1 tanks can be seen below. Designated as the tank gunnery trainer M30A1, it is shown after completion at Chrysler on 2 December 1963. In this photograph, it is still mounted on the build-up stand.

Two photographs of the tank gunnery trainer M37 appear above and at the left. Armed with the 152mm gun-launcher, it served to train crews for the 152mm gun tank M60A1E1.

Below at the left is the driver trainer M34 designed to train drivers for the M60A1 tank. Note the T-bar steering in this device. Directly below is an earlier driver trainer which simulated the M48A2 tank with the steering wheel.

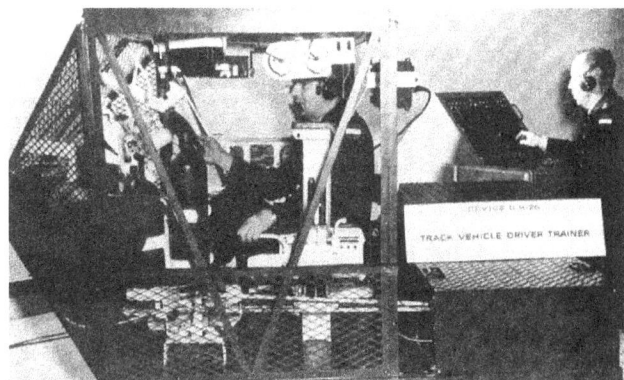

Above is the Telfare device installed on an M60A1 tank. Originated by Sergeant First Class Nathaniel Telfare at the Armor School, it used the .50 caliber machine gun to simulate the firing of the 105mm gun thus greatly reducing costs. At the left is a laser gunnery trainer which replaced the coaxial machine gun and simulated the main weapon during the training programs.

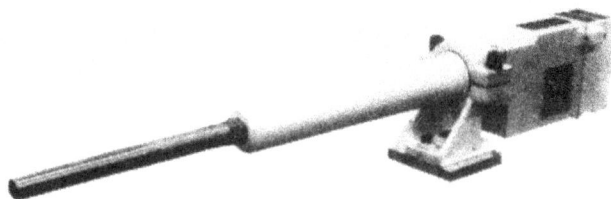

The M60A1 tank at the right is equipped with the SIMFIRE hit/kill indicator. When this device is fitted to a tank, it produces a flash and a bang to simulate the firing of the cannon. If the tank is hit by the laser beam from an opposing tank, a smoke generator is activated indicating a kill as at the right below. The various components of the SIMFIRE apparatus are sketched below.

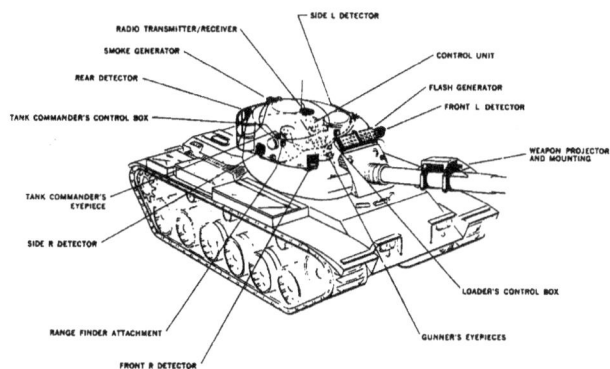

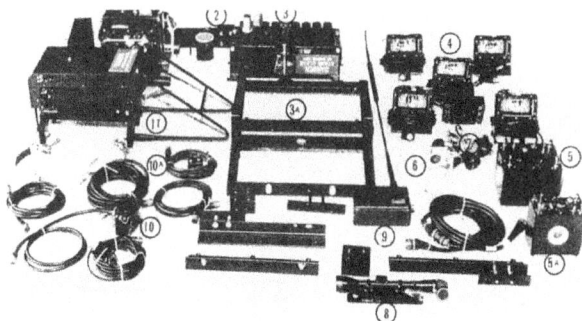

1 Control Unit 2 Smoke Generator 3 Flash Bang Generator 3a Mounting Brackets. 4 Test Box Detector, 5a Test Box Laser Projector 6 Rangefinder Attachment 7 Three Eyepiece Adapters 8 Alignment Aid 9 Radio Transmitter/Receiver 10 Tank Commander's Remote Box, 10a Loader's Remote Box 11 Laser Projector with tree guard

Above is the M48A3 evasive target tank (ETT) early (left) and late (right) in its career. Manned by a crew of two, it provided a moving target for TOW missiles during tests in late 1973. Below are two additional views of the ETT during the test program.

The M60A1 directly above has been modified to suppress infrared radiation. At the top right is an M60 with a breadboard automatic defense system designed to detect antitank missiles and to launch smoke grenades or thermoflares as a countermeasure. At the right, the tank has launched a thermoflare.

Other tanks served as test beds to evaluate new equipment prior to its adoption for general service. Such programs studied methods of reducing the vehicle's infrared or radar signature and tested techniques for providing a screen from enemy observation.

In 1961, an M48 chassis without a turret was fitted with television equipment to determine the feasibility of using such apparatus to operate the tank when completely closed up. This vehicle was equipped with television screens for the driver and other crew members to determine their performance when operating under these conditions.

The M48 test bed with the television equipment installed appears above. Below are the TV screens for the gunner and the tank commander. The driver's screen is at the right.

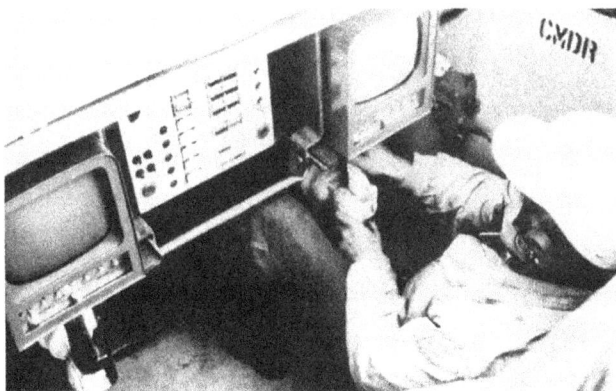

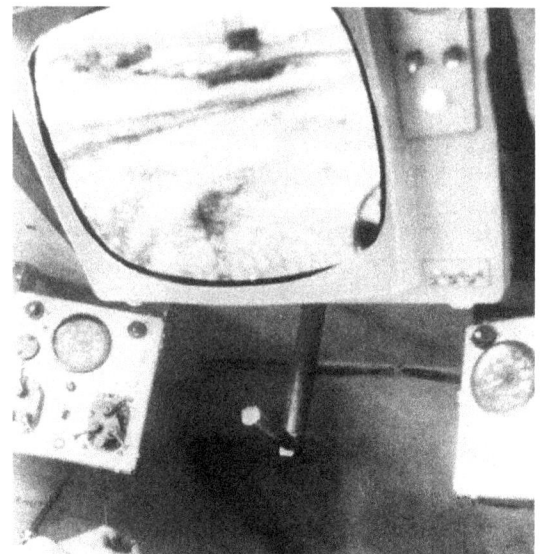

The M48 chassis modified to test the AGT-1500 gas turbine is shown in the view above and below at the left. A cab for the test crew replaced the turret, but the driver remained in his usual position.

Above, the AGT-1500 gas turbine is removed from the vehicle. The top view at the left identifies the various components. The major units of this power plant also are shown in the sketch at the bottom left.

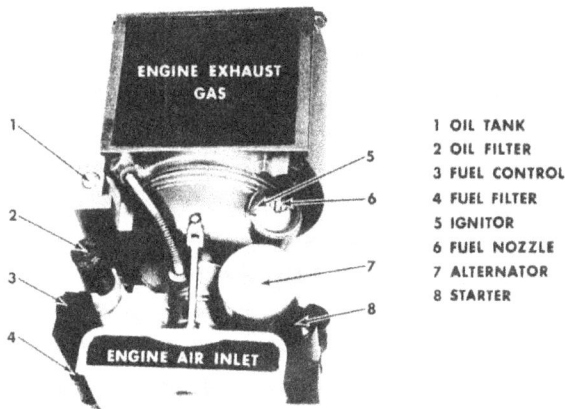

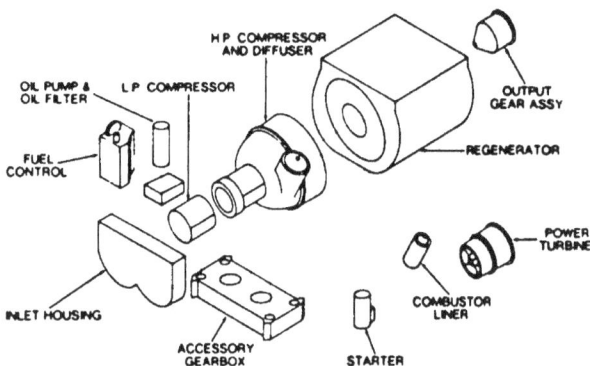

ENGINE EXHAUST GAS

ENGINE AIR INLET

1 OIL TANK
2 OIL FILTER
3 FUEL CONTROL
4 FUEL FILTER
5 IGNITOR
6 FUEL NOZZLE
7 ALTERNATOR
8 STARTER

OIL PUMP & OIL FILTER
LP COMPRESSOR
H P COMPRESSOR AND DIFFUSER
OUTPUT GEAR ASSY
REGENERATOR
FUEL CONTROL
POWER TURBINE
INLET HOUSING
ACCESSORY GEARBOX
STARTER
COMBUSTOR LINER

The Avco-Lycoming AGT-1500 gas turbine was installed in an M48 as a test bed to evaluate the new power plant which was used eventually in the M1 tank. Later in 1977, the M48A1 was fitted with the Avco-Lycoming Six Fifty vehicular turbine. Rated at 650 horsepower, it weighed only 490 pounds. The small size of this unit permitted the use of larger fuel tanks to compensate for the turbine's more rapid rate of consumption. The cruising range was estimated as 300 miles. A diesel auxiliary power unit was installed for use during silent watch conditions. Tests at Fort Knox showed that the performance of the turbine powered tank was superior to the M48A5 when the temperature was below 65 degrees Fahrenheit. However, its advantage was lost when the temperature rose to 70 degrees F or higher. Further work was in progress to correct this problem. Another gas turbine candidate for use as a tank power plant was the Garrett GT601. It also utilized the M48 series chassis as a test bed.

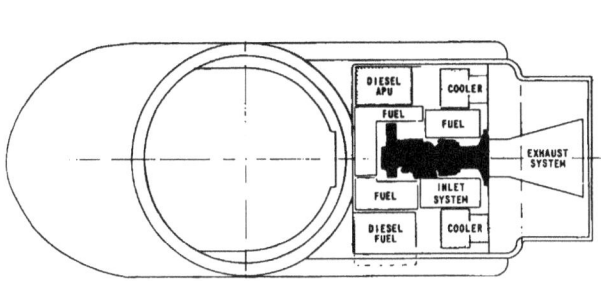

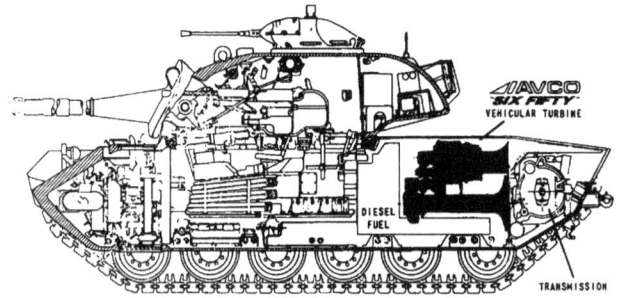

The drawings above depict the Avco-Lycoming Six Fifty vehicular turbine installed in an M48A1 tank. Note the small size of this power plant. The sketches below show the modernized German M48A2G powered by the Garrett GT601-750 gas turbine.

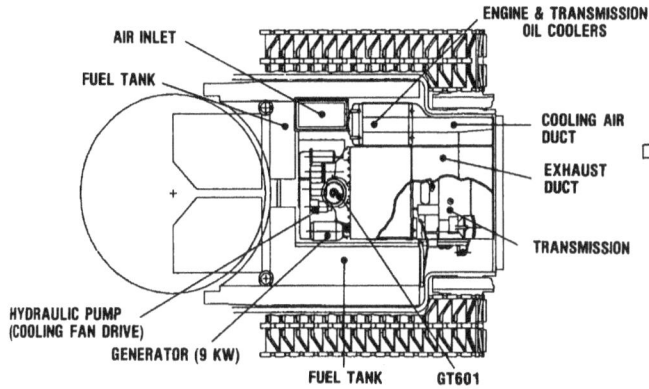

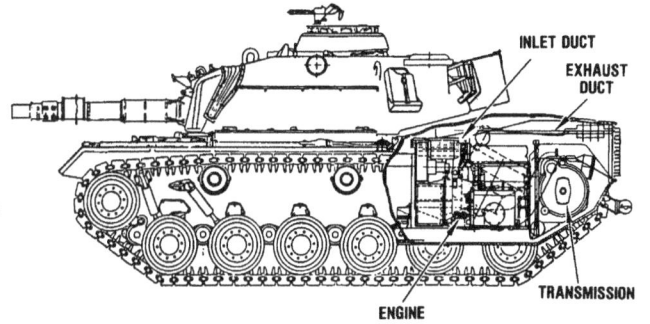

The M48A2G test vehicle powered by the Garrett GT601 gas turbine is shown below during the test program in Germany.

Above the Garrett GT601 gas turbine is being lowered into the engine compartment of the modified M48A2G. Below, the tank can be seen with the new engine installed.

PART V

THE PATTON ON ACTIVE SERVICE

Above and below, M47 registration number 30164453 is being used to conduct a fording school for the 1st Battalion, 14th Armored Command. The tanks are operating in the Fulda River near Fulda, Germany on 11 December 1952.

U.S. FORCES IN THE UNITED STATES AND OVERSEAS

Although the hot war was being waged in Korea in the early 1950s, the major threat to the free world was considered to be in Europe. Thus the armor role in Korea was borne by the M46 as well as the Shermans and Pershings of World War II

vintage. As soon as the newly produced M47 and M48 tanks were accepted for troop use, they were rushed to Europe to strengthen U.S. Army and NATO units. Although never used in battle by American troops, these early Pattons played a vital role in maintaining the peace during this critical period.

In addition to the Army, the U.S. Marine Corps also received a quota of the new vehicles and when troops were ordered into Lebanon in July 1958, Marine Corps M48A1s were the first tanks to roll into Beirut. These early Pattons continued to serve in the training role both in the United States and Europe until they were supplemented or replaced by the later models such as the M48A2 and M48A2C.

In December 1960, the first of the new diesel powered M60 tanks appeared in Europe and rapidly took over the role of the first line main battle tank. When it in turn was superseded by the M60A1, the U.S. main battle tank had assumed the form that it would retain with numerous modifications for more than 20 years.

The crew from an M47 of A Company, 29th Tank Battalion takes a roadside break (above) during Exercise Sledgehammer in Germany, September 1953. Below, Chief Wahoo, an M47 from the 57th Tank Battalion, waits for orders during Exercise Monte Carlo. This photograph also was dated September 1953 in Germany.

The M47 above from A Company, 57th Tank Battalion, 2nd Armored Division is engaged in training at Coleman Barracks outside of Mannheim, Germany on 5 April 1955. Below, an M47 from the 759th Tank Battalion prepares for a fire mission on the range at Grafenwoehr, Germany during July 1955.

Reservists above are training with M48A1 tanks in the desert at Camp Irwin, California on 17 June 1957. Note the white paint on the upper surface of the tanks. Below, an M48 and an M48A1 from the 710th Tank Battalion fire their machine guns during a training period at Fort Stewart, Georgia on 7 September 1957.

Above, an M48A1 of the 3rd Battalion, 35th Armor maneuvers into position on the airport at Beirut, Lebanon during the operations in July 1958. Note the flash hiders on the .50 caliber machine guns and the jettisonable fuel drums on the rear of each tank. Below, the first U.S. Marine Corps tank enters Beirut on 16 July 1958.

M48A1s from the 4th Armored Division are lined up for firing exercises near Grafenwoehr and Vilseck, Germany on 23 July 1959. Below, M48A1s of the 3rd Armored Division cross a Bailey Bridge spanning the Vils River near Amberg, Germany during Exercise Winter Shield on 7 February 1960.

Above, a camouflaged M48A2C moves along a mountain road in Korea during training exercises in the Fall of 1962. Note the 18 inch diameter searchlight on this tank. At the left, the tanker in the turret of this M48A2C is bundled up against the cold of the Korean winter. Below is another M48A2C from the 1st Cavalry Division in Korea during the Winter of 1963.

M48A2Cs from the 3rd Tank Battalion, 40th Armor of the 1st Cavalry Division maneuver over a frozen rice paddy near Wangchon, Korea on 18 January 1963. Below, one of the tanks is immobilized while trying to cross a dike at an angle. Both tracks are off the ground.

Above, details of the M1 commander's cupola can be seen at the left. At the right, an M48A1 patrols along the border between East and West Germany.

Below, armored vehicles for use by the 2nd Armored Division are prepositioned at the Germersheim Ordnance Depot on 23 October 1963 in preparation for Operation Big Lift. Note that all of the M48A1s are fitted with racks for jettisonable fuel drums.

Above, an M60 from the 3rd Armored Division operates near Wickershofe, Germany on 30 October 1963 during Operation Big Lift. This tank has been fitted with the friction snubbers on the first and last road wheel arms. Below, an M48A1 from the 4th Battalion, 69th Armor engages a target with its .50 caliber machine gun while training at Fort Stewart, Georgia on 10 January 1964.

Above, the .50 caliber machine gun on this M48A1 from A Company, 4th Battalion, 69th Armor is firing tracer ammunition during qualification tests at Fort Stewart, Georgia. Below, M48A1s of C Company, 1st Battalion, 69th Armor are training with the 4th Marines in Hawaii on 30 January 1964.

The M60 tanks above are operating with infantry during training exercises at Grafenwoehr, Germany on 7 March 1964. These M60s are still armed with the .50 caliber M2 machine guns on the external cupola mount. Below, an M60 is showing the firepower of the 105mm gun during a demonstration in Germany.

The belly of this 4th Armored Division M60 (above) is exposed while climbing at Grafenwoehr, Germany on 18 September 1970. Below, M60s are operating north of the Amper River near Palzing, Germany during Exercise Certain Forge on 11 October 1971.

Above, an M60A1 from the 4th Battalion, 69th Armor moves out for a training exercise in Germany on 6 April 1974. At the right, a crew member loads APDS ammunition into the turret of an M60. Below are M60A1s from the platoon commanded by 1st Lieutenant Dwight McLemore of the 40th Armor in Berlin.

Above, a refurbished M60A1 runs through the test course at the U.S. Army Maintenance Plant in Mainz, Germany on 25 April 1975. A rear view of the tank (below) climbing the same slope shows the details of the rear deck and turret roof.

These Marine Corps M60A1s are participating in Operation Solid Shield-75 at Camp Lejeune, North Carolina on 30 May 1975. The tanks have been prepared for deep water fording with the exhaust stacks and various seals installed.

The M60A1s of the 2nd Marine Tank Battalion operate above with an LVTP-7 during exercises on Sardinia. Below, M60A1s, also from the 2nd Marine Tank Battalion, move across terrain during Phase I of Exercise Bonded Item in Denmark during October 1976.

Above, an M60A1 from the 2nd Marine Division passes through a German town as part of Task Force Gary on 22 October 1976. The bilge pump outlet on the Marine Corps tanks is visible on the hull roof next to the personnel heater exhaust. Below, an M60A1 of the 1st Marine Tank Battalion is shown during training at Camp Pendleton, California on 9 December 1976. Both of these tanks have the new AN/VSS-3 searchlight.

The M60A1 above from B Company, 1st Battalion, 72nd Armor in Korea proved that it was not amphibious in January 1977. As the tank, commanded by 2nd Lieutenant Gary Kiljan, rounded a corner, it slid out onto the river ice and broke through. As can be seen in the right-hand photograph, the efforts of two M88 recovery vehicles were required for the rescue. The tank was subsequently dried out and returned to service. These photographs were taken by Colonel Charles D. McGaw who also supplied the background information.

Below, an M60A1 waits in firing position at the Yakima Firing Center during gunnery training at Fort Lewis, Washington on 29 April 1977.

The M60A1 above is taking part in Operation Fortress Lightning in the Philippines on 15 October 1977. The tanks were from the 3rd Marine Division based on Okinawa. Note that the tank is sealed for deep water fording. Below are the M48A5s delivered to the 72nd Armor in Korea. This photograph, taken by Russell P. Vaughan, shows the vehicles in the motor park at Camp Casey. The mount for the commander's .50 caliber machine gun can be seen on the low silhouette cupola.

Above is an M48A5 photographed by Major Fred Crismon at Fort Knox during February 1984. This tank was being used to train drivers for the DIVAD Sergeant York which utilized the same chassis. The top loading air cleaners can be seen on the fenders. Below is an M60A3 from the 2nd Armored Cavalry Regiment in Germany. It was photographed by Russell P. Vaughan in October 1983. Note the use of artillery ammunition canisters to stow maps on the outside of the turret.

This M60A3 also was photographed by Major Fred Crismon at Fort Knox in February 1984. The markings indicated that it belonged to the 194th Armored Brigade. Note that the wind sensor head is not mounted on the turret top mast.

The M48A3s of B Company, 3rd Marine Tank Battalion are shown (above and below) in their landing craft off Da Nang, Vietnam on 9 March 1965.

VIETNAM

The combined-pressure of the Viet Cong guerrilla forces and the regular North Vietnamese Army had increased by early 1965 to the point that the survival of South Vietnam depended upon the active intervention by U.S. forces. At that time, U.S. air and naval units began a limited air and sea bombardment of North Vietnam operating from bases in the South. Initially, the local security for these bases was provided by South Vietnamese troops. In February, General William Westmoreland requested two U.S. Marine battalion landing teams to improve the security of the airfield at Da Nang. Although the planners had envisaged an infantry unit, the battalion landing team was a combined arms force and the marines saw no reason to leave their tanks behind. Thus on 9 March 1965, the 3rd Platoon, Company B, 3rd Marine Tank Battalion rolled off its landing craft onto the beach at Da Nang becoming the first U.S. tank unit to arrive in Vietnam. The second battalion landing team arrived later in March accompanied by the 1st Platoon, Company A, 3rd Marine Tank Battalion. Equipped with M48A3 tanks, the two platoons were followed on 8 July by the remainder of the 3rd Marine Tank Battalion. In mid-August, they were the first Americans to take their tanks into action against the Viet Cong during Operation Starlight. Three marine battalion teams, each including a

Above, an M48A3 from A Company, 3rd Marine Tank Battalion prepares to move out for the firing exercise shown below. As can be seen by the turret markings, tanks from both A and B Companies are participating in this exercise near Da Nang, Vietnam on 1 May 1965.

tank platoon, were engaged during the two day operation which killed approximately 700 men from a large Viet Cong force. Seven tanks were damaged in the fighting and one of these had to be destroyed by a demolition team. This was just the first of a long series of operations. Until their withdrawal in late 1969, these tankers participated in fighting throughout Vietnam in support of marine infantry.

The Marine Corps tank above is coming ashore at Quing Tin, Vietnam on 7 May 1965. Below, another M48A3 lands at Chu Lai, about 50 miles south of Da Nang, also in May. The tanks are sealed for deep water fording. Note the diesel engine smoke from the exhaust stack.

The tanks above support the 2nd Battalion, 9th Marines in a sweep south of Da Nang Air Base on 3 August 1965. Below, a 3rd Marine Division M48A3 comes ashore during Operation Piranha on 7 September 1965. The mounting of the deep water fording exhaust stack is easily seen in this photograph.

Above, an M48A3 of A Company, 3rd Marine Tank Battalion is operating about eight miles south of Phu Bai, Vietnam on 7 October 1965. Below, an M67A2 flame thrower tank from B Company, 3rd Marine Tank Battalion guards the perimeter at Hoa Long village. Note the .30 caliber machine gun mounted on top of the commander's cupola.

The M67A2 flame thrower tank in these two photographs is burning off the fields near the 1st Battalion, 3rd Marines command post. Such operations reduced the concealment available to the enemy.

Above, a 3rd Marine Division M48A3 moves out of Da Nang on 20 April 1966 during Operation Sitting Duck. Below are 1st Marine Division tanks participating in Operation Incinerator on 1 June 1966. Note the usual Marine Corps stowage of the deep water fording exhaust stack on the side of the turret in the left photograph. At the right, the crew cleans the 90mm gun after operations.

The M48A3 above moves into action supporting the 2nd Battalion, 7th Marines five miles west of Dong Ha on 30 August 1966. The M67A2 flame thrower tank below is burning a suspected enemy area near Binh Son during Operation Dozer on 5 October 1966.

M48A3 tanks of the 3rd Squadron, 11th Armored Cavalry Regiment move into position after reaching their objective east of Bien Hoa during Operation Hickory on 7 October 1966.

When the U.S. 1st Infantry Division was alerted for deployment to Vietnam in 1965, it was directed to reorganize by eliminating its two tank battalions and converting its mechanized infantry to dismounted units. The divisional cavalry, the 1st Squadron, 4th Cavalry, was permitted to keep its M48A3 tanks to allow their evaluation in the Vietnamese jungle terrain. At that time, the prevailing opinion of both the Chief of Staff and the senior command in Vietnam was that tanks and mechanized infantry were unsuitable for use in that country. In fact, when the 1st Squadron, 4th Cavalry arrived, its M48A3 tanks were withdrawn and the squadron was broken up with each troop assigned to a brigade in different locations. It was six months before the division and squadron commanders could convince the Military Assistance Command, Vietnam (MACV) that the tanks could be effectively used in operations. Thus in its initial combat, the squadron, deprived of its heavy weapons, was mounted in armored personnel carriers (APCs) which actually were employed in the light tank role. However, its operations did prove the value of armor in Vietnam and opened the way for the deployment of additional units in the future.

Although the tanks could not traverse all of the areas covered by the amphibious APCs, their heavy armor and firepower were a great advantage in many situations. The powerful vehicles were extremely effective in breaking through the heavy jungle growth (jungle-busting) and the thick armor offered greater protection from enemy fire and mine blasts. The 90mm gun had a devastating effect on both enemy bunkers and personnel using high explosive and canister ammunition. As usual, the troops in the field improvised a number of additions and modifications to the tanks. The turrets were frequently covered with sand bags to reduce the effect of the shaped charge warheads of the enemy rocket propelled grenades (RPGs). Wide use also was made of chain-link or Cyclone fencing to protect the tanks from the RPGs. Rolls of this fencing were carried on the vehicles and each crew set up a section of fence in front of its position to detonate the shaped charge warheads before they struck the vehicles. As mentioned earlier, the .50 caliber machine gun was removed from the cramped M1 cupola and remounted on a pedestal welded to the cupola roof. It was then fired by the tank commander standing in the open cupola hatch.

In addition to offensive operations, a major task for armored forces in Vietnam was to insure safe routes of communication and supply by clearing the roads and providing escort for troop and supply convoys. The fighting which occurred during the execution of this mission required the development of new tactics and technique to defeat the frequent attempts of the enemy to mine the roads and ambush the convoys. The counterambush tactics which evolved from these engagements made use of the firepower and mobility of the armored troops to reinforce the escorts, beat off the main attack, and pin down the attacking force. This enabled other elements of the combined arms team, such as air mobile infantry, to block their escape routes. The enemy units were then destroyed by the combined efforts of the ground and air forces. When properly carried out, these tactics were extremely effective. However, carelessness in planning or execution could, and did, result in disaster and many painful lessons were learned, particularly by inexperienced troops.

Above, an M48A3 from the 2nd Battalion, 34th Armor rumbles past the main gate of the Cu Chi base camp. Note that the .50 caliber machine gun has been removed from its mount and relocated onto the cupola roof. Below, the 34th Armor lines up its tanks on a road near Cu Chi in support of the 25th Infantry Division.

The M48A3s above of the 1st Squadron, 11th Armored Cavalry Regiment deploy in the herringbone formation to give fire cover at Ben Dong on 15 November 1966. The battered condition of the tank's fenders shows the hazards of operating in the rough jungle terrain.

The escort techniques developed during the early operations of the 1st Squadron, 4th Cavalry called for the convoy to be led by an armored vehicle, frequently a tank. Other armored vehicles were dispersed throughout the column. When ambushed, the armored vehicles pivoted outward at an angle from the road with alternate vehicles facing in opposite directions. They then opened fire with all weapons. This herringbone formation provided good fields of fire and allowed the soft-skinned vehicles to seek shelter behind the tanks or APCs. These tactics were adopted and further developed by other units engaged in convoy escort. Their effectiveness is illustrated by the following description of an action of the 1st Squadron, 11th Armored Cavalry Regiment on 2 December 1966 during Operation Atlanta. It is quoted from the Department of the Army publication "Mounted Combat in Vietnam" by General Donn Starry. On that day, a resupply convoy consisting of two tanks, three ACAVs (armored cavalry assault vehicles which were modified M113 APCs), and two 2½ ton trucks was returning from Blackhorse base camp to Gia Ray through Suoi Cat about 50 kilometers east of Saigon.

"At 1600 the convoy commander, Lieutenant Wilbert Radosevich, readied his convoy for the return trip to Gia Ray. The column had a tank in the lead, followed by two ACAVs, two trucks, another ACAV,

and, finally, the remaining tank. Lieutenant Radosevich was in the lead tank, and after making sure that he had contact with the forward air controller in an armed helicopter overhead, moved his convoy out toward Suoi Cat. As the convoy passed through Suoi Cat, the men in the column noticed an absence of children and an unusual stillness. Sensing danger, Lieutenant Radosevich was turning in the tank commander's hatch to observe closely both sides of the road when he accidently tripped the turret control handle. The turret moved suddenly to the right, evidently scaring the enemy into prematurely firing a command detonated mine approximately ten meters in front of the tank. Lieutenant Radosevich immediately shouted "Ambush! Ambush! Claymore Corner!" over the troop frequency (This was the nickname for a landmark known to everyone in the squadron) and led his convoy in a charge through what had become a hail of enemy fire while he blasted both sides of the road. Even as Lieutenant Radosevich charged, help was on the way. Troop B, nearest the scene, immediately headed toward the action. At squadron headquarters, Company D, a tank company, Troop C, and the howitzer battery hastened toward the ambush. Troop A, on perimeter security at the regimental base camp, followed as soon as it was released. The gunship on station immediately began delivering fire and called for additional assistance, while the forward air controller radioed for air support.

The tank at the right from the 1st Battalion, 69th Armor uses its xenon searchlight to pick out targets for night fire during Operation Paul Revere IV. This action was 25 miles west of Pleiku on 25 November 1966.

When the convoy reached the eastern edge of the ambush, one of the ACAVs, already hit three times, was struck again and caught fire. At this point Troop B arrived, moved into the ambush from the east, and immediately came under intense fire as the enemy maneuvered toward the burning ACAV. Troop B fought its way through the ambush, alternately employing the herringbone formation and moving west, and encountering the enemy in sizable groups.

Lieutenant Colonel Martin D. Howell, the squadron commander, arrived over the scene by helicopter ten minutes after the first fire. He immediately designated National Highway 1 a fire coordination line, and directed tactical aircraft to strike to the east and south while artillery fired to the north and west. As Company D and Troop C reached Suoi Cat, he ordered them to begin firing as they left the east side of the village. The howitzer battery went into position in Suoi Cat. By this time, Troop B had traversed the entire ambush area, turned around, and was making a second trip back toward the east. Company D and Troop C followed close behind, raking both sides of the road with fire as they moved. The tanks fired 90mm canister, mowing down the charging Viet Cong and destroying a 57mm recoilless rifle. Midway through the ambush zone, Troop B halted in a herringbone formation, while Company D and Troop C continued to the east toward the junction of Route 333 and Route 1. Troop A, now to the west of the ambush, entered the area, surprised a scavenging party, and killed fifteen Viet Cong.

The squadron commander halted Troop A to the west of Troop B. Company D was turned around at the eastern side of the ambush and positioned to the east of Troop B. Troop C was sent southeast on Route 1 to trap enemy forces if they moved in that direction. As Troops A and B and Company D consolidated at the ambush site, enemy fire became intense around Troop B. The Viet Cong forces were soon caught in a deadly crossfire when the cavalry units converged. As darkness approached, the American troops prepared night defensive positions and artillery fire was shifted to the south to seal off enemy escape routes. A search of the battlefield next morning revealed over 100 enemy dead. The

Below, a helicopter lands among the tanks and personnel carriers of the 11th Armored Cavalry Regiment near Bien Hoa.

Above, an M48A3 from the 1st Squadron, 11th Armored Cavalry Regiment moves through a rubber plantation to a new command post west of Ben Dong on 15 November 1966.

toll, however, was heavier than that. Enemy documents captured in May 1967 recorded the loss of three Viet Cong battalion commanders and four company commanders in the Suoi Cat action."

The lessons learned in this and other engagements provided the basis for the counterambush tactics employed for the remainder of the war in Vietnam. Basically, they called for the ambushed unit to fight its way clear of the enemy killing zone using its firepower to protect the escorted vehicles. Once clear, the armored vehicles would regroup and counterattack. All available reinforcements would be used to attack the flanks of the ambush and the maximum available artillery and air support would be provided.

The Marine Corps M48A3 at the right is still equipped with the early 18 inch diameter searchlight. This tank was photographed guarding the perimeter of the Ky Ha Air Base on 18 November 1966.

Above, an M48A3 from C Company, 2nd Battalion, 34th armor has a damaged track after hitting a mine during Operation Cedar Falls on 17 January 1967. Note that the track has been thrown, but it is not broken and the damage to the suspension appears to be slight. Below, other tanks of C Company move on line during a search and destroy mission on 18 January 1967.

Above, a tank from L Troop, 3rd Squadron, 11th Armored Cavalry Regiment moves up a hill during Operation Junction City northeast of Lai Khe on 13 April 1967. At the right, an M88 from the same squadron crosses a river bed on 7 April 1967. Below, an M48A3 of the 1st Battalion, 7th Cavalry prepares to climb out of the An Lao River during a sweep on 12 April 1967.

Above a tank supporting the 25th Infantry Division comes ashore at Bong Son. Note the spare road wheels stowed on the turret. Below, the crew of an M48A3 from the 3rd Platoon, B Company, 3rd Marine Tank Battalion check fires their 90mm gun near Khe Sanh on 9 July 1967. Spare track links are attached to the turret front and a lot of unofficial stowage is on the rear of the tank.

Above, a 3rd Marine Division tank is fueled at Camp Carrol, Vietnam on 18 July 1967. This tank still retains the early searchlight. Below at the left, the loader of a 3rd Marine Division tank scans the area south of the Demilitarized Zone for enemy troops. At the right below, a disabled M48A3 is towed by another tank. Note the .30 caliber machine gun on top of the cupola and the .50 caliber coaxial weapon. Both of the bottom photographs were dated September 1967.

Above, an M48A3 from A Company, 69th Armor moves into position with the 1st Cavalry Division in Operation Pershing. Below, a Marine Corps tank waits to fire during Operation Badger Tooth in Quang Tri Province on 11 January 1968. Compare the xenon searchlight in this photograph with the early model in the upper view.

The tanks shown here are supporting the 1st Battalion, 5th Marines during the battle for Hue in February 1968. The effect of hits on the laminated vision blocks can be seen in the upper photograph. The M48A3 below is alongside the walls of the Imperial Citadel. Its .50 caliber machine gun has been remounted on top of the cupola.

This Marine Corps tank is dug in (above) during the defense of Khe Sanh on 2 March 1968. A .50 caliber coaxial machine gun is installed and a .30 caliber machine gun is mounted on top of the cupola. Below, a tank from the 3rd Squadron, 4th Cavalry has its turret covered with sand bags to increase the protection from rocket propelled grenades.

Above is one of the few M48A2 tanks to serve in Vietnam. Note the three track return rollers and the absence of the fender mounted air cleaners. This tank was with the 1st Battalion, 5th Mechanized Division during Operation Fisher on 6 January 1969. Below is an M728 combat engineer vehicle from the 11th Armored Cavalry photographed by James Loop in January 1969.

The M88 medium recovery vehicle above from the 11th Armored Cavalry has the gun shield frequently installed in Vietnam on the .50 caliber machine gun mount. Below, the crew of another M88 repairs the track and suspension of an M48A3 in the Khe Sanh area on 23 June 1969.

Above, an M48A3 tank commander from C Company, 1st Battalion, 69th Armor watches a convoy pass on Highway 14 west of An Khe on 17 June 1969. This tank seems to be missing its .50 caliber machine gun entirely.

Only one tank versus tank action occurred between American and enemy forces during the war. This encounter was hardly a test of the M48A3 since the adversary was the Soviet built PT76 amphibious tank. With armor 11mm to 14mm thick and a 76.2mm gun, it was a reconnaissance vehicle and not in the same class as the Patton. The action occurred at the Ben Het special forces camp in March 1969. Located in the central highlands overlooking the Ho Chi Minh trail, the camp had been frequently under fire during February. To strengthen the defenses, Company B, 1st Battalion, 69th Armor occupied positions along the road between Ben Het and Dak To with one platoon stationed in the camp itself. Late in the evening of 3 March, sounds of engines and tracks were heard to the west where no friendly armor was operating. The area was scanned by the tank's infrared searchlight, but fog prevented the identification of any targets. The action started when one of the enemy vehicles was immobilized by an antitank mine at a range of 1100 meters. Otherwise undamaged, it opened fire on the camp and gun flashes were counted from seven other enemy tanks. The M48A3s returned the fire aiming at the gun

flashes and mortar rounds illuminated the area with flares revealing the enemy vehicles as PT76 tanks. During the exchange, two of the PT76s were knocked out and burned. One M48A3 was slightly damaged when an enemy high explosive round struck the loader's hatch killing the loader and the driver.

At the right, a tank from the 3rd Platoon, C Company, 1st Marine Tank Battalion climbs out of a big ditch during Operation Pipestone Canyon. This photograph was taken on 2 June 1969 eight miles south of Da Nang.

Above, an armored vehicle launched bridge, based on the M60A1 tank, moves into position during Operation Utah Mesa in the A Shau Valley. This AVLB was assigned to the 1st Brigade of the 5th Mechanized Division. Crew members of the tanks below from the 11th Armored Cavalry load 90mm ammunition prior to moving out of the command post area north of Hung Nghia in January 1971.

This M48A3 from C Troop, 1st Squadron, 77th Cavalry is in a night defensive position near the Rockpile on 13 March 1971. Note the use of the chain-link fence as protection against rocket propelled grenades.

No further tank versus tank engagements occurred while the U.S. forces were active in Vietnam. However, the South Vietnamese tankers fought several significant battles after the American departure. Equipped with M48A3s, the South Vietnamese 20th Tank Regiment distinguished itself during the enemy Spring offensive in 1972. By the end of March that year, most of the South Vietnamese positions along the Demilitarized Zone and in western Quang Tri Province had been overrun or evacuated. To stem the retreat, the 20th Tank Regiment with its 44 operational tanks took up a defensive position southeast of Can Lo at nightfall on 1 April. On the following morning, the regiment moved toward Dong Ha and deployed around the town with the 1st Squadron on the high ground approximately three kilometers to the west. About noon, a column of North Vietnamese tanks and infantry was observed moving south along National Highway 1 toward Dong Ha. The M48A3s opened fire at a range of 2500 to 3000 meters destroying nine

PT76s as well as two T54 tanks. The enemy column retreated in confusion without seeing their adversary or firing a single shot.

Later on 9 April, all three squadrons of the 20th Tank Regiment successfully fought enemy armor. This action occurred along National Highway 9 when the tankers engaged North Vietnamese tanks and infantry at ranges up to 2800 meters. In all, the

At the right is an M48A3 manned by the Army of the Republic of Vietnam (ARVN) in Quang Tri Province on 6 April 1972.

Above, an ARVN M48A3 guards the approach along National Highway 9 on 10-13 April 1972. This tank retained its .50 caliber machine gun in the original cupola mount.

regiment destroyed 16 T54 tanks that day and captured one Chinese built T59. Unlike the PT76, these were main battle tanks fully capable of engaging the M48A3 on equal terms. However, the M48A3s suffered only superficial damage in the battle. These actions as well as others that followed were the dividends for the long hours invested in training the South Vietnamese tankers. Originally they were skeptical of the value of the M48A3's range finder because of their earlier experience using the M41 light tank which was not so equipped. However, their success in these long range engagements showed that the reluctance to use the range finder was overcome and it was effectively utilized.

The ARVN tanks below are in position near the Dong Ha River overlooking National Highway 9 during the period 10-13 April 1972.

Above are two Israeli M48A2s in their original form as used during the Six-Day War in 1967.

SERVICE IN FOREIGN ARMIES

Neither the M47 nor the early M48 series tanks saw action with United States forces, however, they were taken into battle by foreign troops. During the war between India and Pakistan in the Fall of 1965, both the M47 and M48 were employed by the Pakistani Army. Their showing in this conflict was poor with Pakistan's losses in armor far exceeding those of India.

In the Middle East, both Israel and Jordan had obtained M48 series tanks prior to the Six-Day War in June 1967. Needless to say, this resulted in some confusion during the fighting. Although the use of the American tanks by most foreign countries had little or no effect on their development or employment in U.S. service, the lessons learned from Israel's battle experience were extremely valuable.

The first M48A2 tanks in Israel were obtained from Germany in 1965. These vehicles were used not only in their original form, but they also were

the subject of a continuing improvement program. As usual, the first objective was to increase the cruising range by installing a diesel engine. This was followed by upgunning the tank by replacing the 90mm weapon with the more powerful 105mm gun. However, only enough tanks to equip one company

At the right above is an Israeli M48A2 that has been modernized by rearming with the 105mm gun and installing the diesel engine. Note the fender mounted air cleaners. The Israeli Pattons below have been upgraded from M48A1s by installation of the new armament and the diesel engine. However, they still retain the five track return rollers.

Above, an Israeli M60 can be seen at the left and several M60A1s are at the right. Both types carry a .50 caliber machine gun on the gun mount above the cannon. A similar gun was installed on one of the Pattons at the bottom of page 399.

had been converted in time for the Six-Day war. During this fighting, over 100 Jordanian M48s and M48A1s were captured. After the war, these, along with numerous other M48 series tanks, were converted to the new standard with the diesel engine and the 105mm gun. To avoid confusion in the U.S. Army supply system, these upgraded Israeli tanks were listed as the M48A4, however, they had no connection with the M48A4 previously described.

Another feature of the Israeli modified Patton was the low silhouette cupola for the tank commander. This was the configuration subsequently adopted for the M48A5 in the U.S. Army. After upgrading, the Israeli Pattons were almost equivalent to the later M60 series tanks which Israel obtained in increasing numbers following the Six-Day war. The two types provided a major portion of Israel's armored strength during and after the Yom Kippur war of 1973, particularly when many losses of that conflict were replaced with M60A1 tanks.

Israeli battle experience also revealed the vulnerability of the M48 and M60 series tanks in several areas. One problem was with the constant pressure hydraulic system used to operate the turret and the elevation mechanism. The hydraulic fluid was flammable and a hit could rupture the system and ignite the high pressure fluid spraying all over the crew compartment. A change to a fire resistant hydraulic fluid minimized the danger of fire from this source, but the ammunition stowed in the turret bustle still presented a hazard. A hit on the turret could

These views show Israeli tanks fitted with the explosive armor packs dubbed Blazer, which were intended to defeat HEAT projectiles. The jet from the shaped charge round would detonate the explosive pack disrupting the jet. The view at the right shows one M60A1 without the explosive packs, but with the fittings for their attachment. Below at the right is a modernized M48A1 fitted with Blazer.

The arrangement of the Blazer explosive packs on the front hull and turret of the Israeli M60A1 can be seen above. This view also shows the .50 caliber machine gun above the main gun mount.

easily cause an ammunition fire resulting in the destruction of the tank. This danger was confirmed by similar experience in Vietnam which showed the vulnerability of the turret ammunition rack to hits from the shaped charge warhead of a rocket propelled grenade. New stowage arrangements were proposed to eliminate all ammunition stowage above the turret ring.

At the right, a track width mine plow is installed on this Israeli M60A1. Below, this M60A1 is protected by the Blazer packs and is fitted with the roller type mine exploder.

At the left is the Teledyne Continental Motors installation of the AVDS-1790-2C engine in the M48 hull without the extensive rework of the M48A3 type conversion. This design eliminated the infrared suppression rear deck, but it was far less expensive.

Numerous M47s as well as the later M48 and M60 series tanks were supplied to Iran prior to the fall of the Shah. Many of these early tanks were upgraded to the later standard with the diesel engine. These tanks, along with the newer M60A1s, were engaged in the fighting with Iraq which began in 1980.

Although many were not used in battle, the large numbers of M47, M48, and M60 series tanks distributed throughout the world provided a large market for modernization kits to bring them up to the latest combat vehicle standards. A few of the M47 modernization programs have been mentioned earlier and an even greater effort was expended to develop improvements for the much larger fleet of M48 series tanks. As might be expected, the prime modifications were the installation of some type of diesel engine to improve the cruising range and the replacement of the 90mm gun with a more powerful weapon.

A wide variety of diesel engines were proposed for use in the gasoline powered M48 series tanks. Teledyne Continental Motors developed a modified power pack using the AVDS-1790 engine and the CD-850 transmission for installation in the M48/M48A1 hull without the major cutting and welding required for the M48A3 type conversion. Such a simplified modification eliminated the distortion effects from the welding and greatly reduced the cost.

A joint effort by Napco Industries, Inc., the Detroit Diesel Allison Division of General Motors, and Airscrew Howden, Ltd. produced a power pack consisting of the water-cooled Detroit Diesel 12V71TA engine with the Allison CD-850-6 transmission. The Airscrew Howden cooling system was designed to permit fitting the complete unit into the M48 or M48A1 hull with only minor modifications. The 12V71TA engine developed 800 gross horsepower at 2500 rpm. The turbocharged, two-stroke cycle diesel had a displacement of 852 cubic inches. The increase in power over the AVDS-1790 gave the modified M48 performance slightly better in some respects than the M60. New 300 gallon fuel tanks were required to fit the shape of the new engine and the bulkhead between the engine and crew compartments was redesigned.

Below at the left is a view of the Detroit Diesel Allison 12V71TA engine and CD-850-6 transmission installed in an M48 hull with the rear deck grills open. The general arrangement of this installation is depicted in the sketch at the right below.

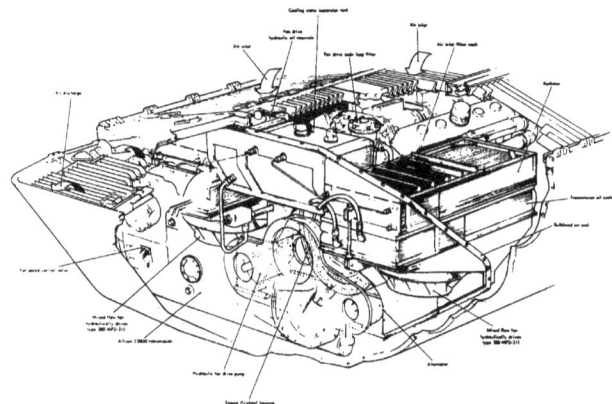

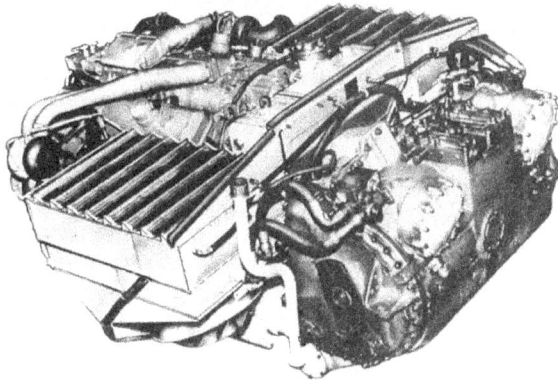

Additional views of the 12V71TA engine and the CD-850-6 transmission can be seen above. It can be compared with the MTU power pack below consisting of the MTU engine with the modified CD-850-4 transmission.

In West Germany, Motoren-und-Turbinen Union (MTU) adapted their MB-837 Ea-500 diesel engine for use in the gasoline powered M48 series tanks. The MTU engine was an eight cylinder, four-stroke cycle, water-cooled diesel with exhaust gas turbochargers. This was the same engine installed in the upgraded M47 tank and it was rated at 750 gross horsepower. The original Allison CD-850-4 transmission was retained after modification to −5 or −6 standards.

In regard to more powerful armament, the most popular candidate was the British 105mm gun, although new developments in ammunition, such as the long rod penetrator, greatly improved the performance of the original 90mm weapon. The French

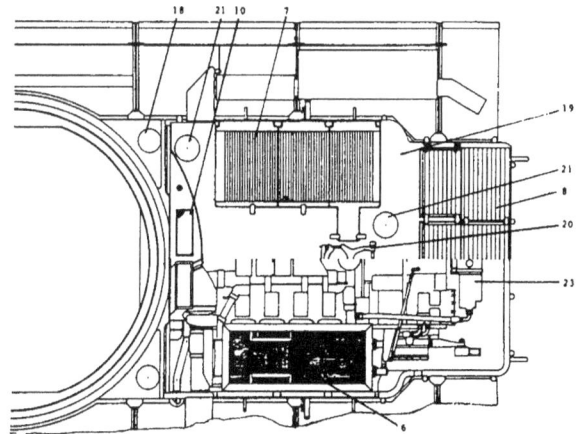

Drawings showing top (above) as well as transverse and side (below) views of the MTU power pack installation are reproduced here. The numbers indicate the following: 1. Engine MB-837Ea, 2. Transmission CD-850-4/5. 3. Cooling fans, 4. Engine radiators, 5. Transmission radiators, 6. Fuel cooler, 7,8. Air inlet grill, 9. Mufflers, 10. Engine air intake, 11. Air filter, 12. Dust ejection blower, 13. Fuze box, 14. Transmission mount, 15. Engine mount, 16,17. Fuel tanks, 18. Fuel caps, 19. Cover plate, 20. Gun travel lock, 21,22. Service doors, 23. Preheater.

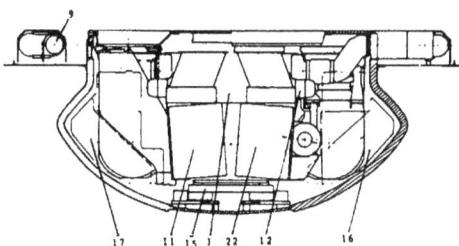

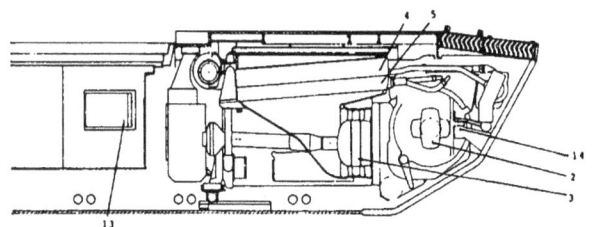

The photographs above show the original installation of the MB-837 Ea-500 engine with the modified CD-850-4 transmission in an M48A1 for Turkey. Below at the right, the MTU power pack is installed in an M48 which has been rearmed with the 105mm gun. This conversion also was for Turkey.

105mm gun from the AMX 30 tank also was proposed for installation as well as the Rheinmetall 105mm and 120mm smooth bore cannon.

A variety of new fire control systems are now available to upgrade the earlier tanks to the latest main battle tank standard. Including laser range finders, night vision equipment, solid state computers, and special sensors as well as stabilization systems and reinforced armor, these modifications give the modernized vehicles a combat potential essentially equivalent to many of the latest production tanks.

At the left and below is a German M48A2G in the modernization program at Wegmann and Company. Rearmed with the British L7A3 105mm gun, it was designated as the M48A2G (105mm Gun).

404

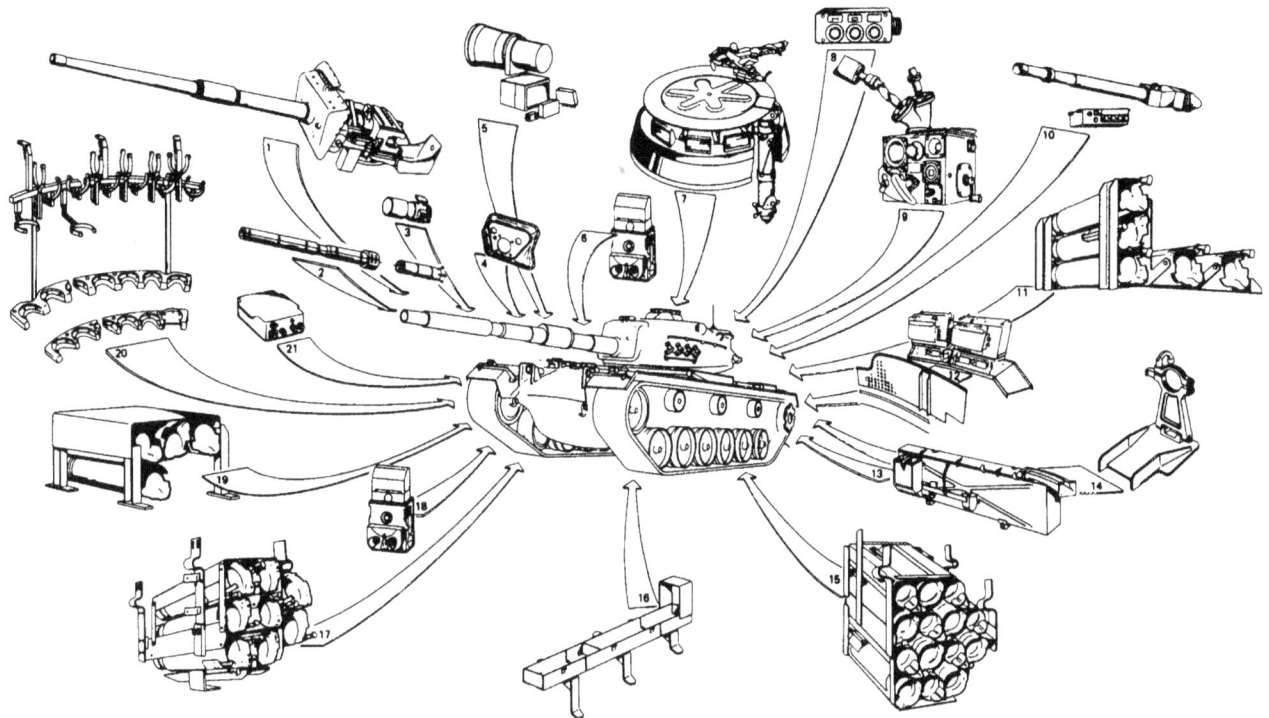

The items covered in the M48A2G modernization program at Wegmann are sketched above. The components indicated by the various numbers are as follows: 1. Gun and mount, 2. Gun thermal shroud, 3. Shock generator, 4. Gun mount joint, 5. PZB 200 passive sight and observation device, 5. BiV periscope for the tank commander, 7. Commander's hatch with AA machine gun mount, 8. Breech switch box for LS, 9. Computer with superelevator, 10. Telescope and transformer, 11. Turret bustle 105mm ammunition racks, 12. Turret bustle racks and components, 13. Coaxial machine gun ammunition box, 14. Gun travel lock, 15. Hull left front 105mm ammunition racks, 16. Machine gun ammunition box rack, 17. Hull right front 105mm ammunition rack, 18. BiV driver's periscope, 19. Turret platform 105mm ammunition rack, 20. 105mm ammunition ready rack, 21. Relay box.

The photographs at the right and below show the M48A2G (105mm Gun) after the modernization work was completed at Wegmann.

The photograph above shows a modernized M48A2G on its transporter near Nuernberg, Germany. Taken by Russell P. Vaughan in September 1983, it provides an excellent view of the rear of the tank and the gun travel lock.

The German M48A1 below and at the right has been fitted with applique armor to improve the protection on the front and sides of the turret. Note that the track tension idler also has been removed from this tank.

The Korean M48A3 above has some unusual features. Note what appears to be some sort of sighting device mounted on the gun tube just behind the blast deflector. In the turret close-up at the left, the special mount with a wire protective guard can be seen attached to the .50 caliber gun shield on the cupola. This tank was photographed near Munsan, Korea by Russell P. Vaughan during late April or early May 1979.

Below, the extensive modernization of the M48 series Patton in Korea can be seen in these two views. This tank is powered by the diesel engine and has been rearmed with the 105mm gun. It is fitted with the low silhouette cupola and hatch for the tank commander and the new AN/VSS-3 searchlight is installed on the gun mount. Also, note the side skirts added to protect the suspension.

HULL WELDMENT & MACHINING
COMMANDER'S CUPOLA
TOP DECK GRILLE
ENGINE & TRANS. MOUNTS
ELBOWS OF AIR CLEANER
PROTECTION PLATE

Although the Patton served the U.S. Army in all parts of the world, the greatest threat always was considered to be in Europe. Thus the units in that area had the highest priority for tanks. Above (left) an M47 with the 28th Infantry Division participates in a field exercise near Neu Ulm, Germany on 26 November 1953. Over seven years later, an M48A1 (above right) of the 4th Armored Division stands guard during Exercise Wintershield II on 3 February 1961.

CONCLUSION

It is difficult to draw conclusions from a story that is not yet complete and the life of the Patton tank family is far from over. It continues to serve as a main battle tank in the U.S. Army as the M48A5 and in other versions all over the world. The chassis is the basis of the new DIVAD gun, the Sergeant York, just entering service. Also, although it was never officially named as a Patton, the M60 series was only a product improvement of the earlier tank and thus a member of the same family. At the present time, the latest M60A3 remains, alongside the new M1, a first line main battle tank.

Many of the reasons for the longevity of the Patton may be found in the basic design. Intended from the first to be capable of mounting a gun larger than the 90, it incorporated an 85 inch diameter turret ring. Such a large ring diameter gave it great flexibility in adapting to new weapon systems. However, one of the most criticized features of the M48 was its size. The height and width exceeded the railway limitations requiring special preparation for shipping. All of the OTCMs covering the early development carried an attachment from the Transportation Corps objecting to the size and weight of the M48 and other vehicles based on the same chassis. Other sources criticized the relatively high silhouette, pointing out the greater probability of a hit compared to many foreign tanks. Despite this criticism, the very size of the tank contributed to its longevity. The large hull

and turret provided ample space for the installation of new equipment and stowage items never envisaged when the tank was designed. More room was allowed for the crew to operate efficiently and a larger supply of ammunition was carried than in some of its potential adversaries. As mentioned frequently, a major defect in the early Pattons was the short cruising range resulting from the high fuel consumption of the gasoline engine. When this was corrected by the introduction of the diesel, the basic design could be developed to its full potential.

In the late 1950s, the guided missile appeared to be the weapon of the future and numerous study programs considered its application as the main armament for tanks. The versatility of the Patton was again demonstrated as it provided test beds for several experimental programs and finally evolved into the standardized M60A2. This vehicle, like the original M60, was considered to be an interim tank until the ultimate fighting vehicle could be developed. However, with the demise of the MBT70 and the XM803, it became obvious that the M60 series would continue to serve for many years.

To permit the M60A1 to compete with the latest main battle tanks, a modernization program was launched to upgrade the basic vehicle by adapting many new components developed since the original design. A high reliability engine and power train reduced maintenance requirements and improved the availability of the tanks in the fleet. New developments in fire control equipment increased the ability to engage targets rapidly and at long range. These included the laser range finder and a new solid state computer. Fin stabilized armor piercing ammunition using long rod penetrators improved the penetration performance of the 105mm gun against heavy armor. With these developments, the trend in main armament shifted away from the missile and back toward the high velocity gun. Many other modifications also were considered. These included a higher power engine, new suspension systems, and improved armor for the hull and turret. These later changes were held in abeyance with the introduction of the new M1 tank. However, many of these may be revived in future modernization programs.

If all of the proposed modifications were applied to the M48 or M60 series tanks, about all that would remain of the original vehicle would be the hull and in some cases the turret. At the time of their introduction, the single piece cast hull and turret represented the most modern concept of tank design. It also was assumed that adequate foundry capacity would be available for quantity production of those heavy castings in time of war. Unfortunately, such a situation does not exist today with the greatly reduced heavy steel foundry capacity in the United States. As a result, the hull and turret of the new M1 are welded assemblies of rolled armor plate. However, although the large castings would present a production problem today, they were readily modified to accept the many changes necessary in the various modernization programs.

With the high cost of modern battle tanks, it is essential that they be capable of absorbing any modifications necessary to maintain their combat effectiveness. This ability was characteristic of the Patton and, perhaps, its greatest strength. More than thirty years after its original design, the modernized version is still rated as a first line main battle tank and will continue to be so in the foreseeable future. Considering the large numbers of these tanks in existence and that the potential for future modification has not been exhausted, the Patton must be rated as one of the most important tank designs of all time.

Below, the product improved Patton, the M60A1, carries on the tradition of providing the main offensive power in the U.S. Army late in 1978.

PART VI

REFERENCE DATA

The markings on this M48A3 indicate that it belongs to H Company, 2nd Squadron, 11th Armored Cavalry Regiment. The two rings around the bore evacuator identify a 2nd platoon tank. Although the drawing shows the .50 caliber machine gun in its standard position, the company remounted the weapon on the cupola roof during their service in Vietnam.

The 2.2 kilowatt searchlight is being checked on this M60A1 assigned to F Company, 40th Armor of the Berlin Brigade during 1969.

U.S. Army M60A1s move through a German town during Reforger exercises in the Fall of 1976. The extra stowage on top of the first tank's gun mount would certainly restrict the vision through the gunner's periscope.

Above, an M48A3 from B Company, 69th Armor moves up with its load of infantry during Operation Lincoln south of Pleiku on 27 March 1966. Below, ''COOL BUT CRUEL'', a 4th Cavalry Patton, has a track replaced 12 kilometers northwest of Saigon, Vietnam during August 1968. Note the liberal use of track links on the turret to improve the protection against rocket propelled grenades.

Above, tanks from H Company, 2nd Squadron, 11 Armored Cavalry Regiment are lined up after returning from operations in Vietnam. The officer in the center is the company commander Captain Robert D. Hurt III. ''Hell's Henchmen'' was the nickname of H Company. Below, the red dust and mud which frequently coated the tanks in Vietnam can be seen on this 1st Marine Division M48A3 named ''PUFF JR''.

Mud covered tanks were not limited to Vietnam as illustrated by the three M60s above during exercises at Fort Knox, Kentucky on 27 February 1964. Below, the 1st Battalion, 110th Armor have their M48A5s on the firing range at Fort Drum, New York during the Summer of 1977.

The M60A1E2 above undergoes winter road testing in the snow. Below, a Shillelagh missile is launched during a firepower demonstration. Both photographs were taken at Fort Knox and dated 1971.

Above, an M67A2 flame thrower tank from the 1st Marine Division fires napalm during operations in Vietnam. This photograph was dated 1968. Below, an AVLB based on the M60A1 tank launches its bridge during tests at Aberdeen Proving Ground on 30 December 1974. The cupolas and hatches for the two man crew are clearly visible in this photograph.

VEHICLE DATA SHEETS

Data sheets are included in this section for each of the major versions of the Patton and M60 series tanks as well as a number of other vehicles based on these chassis. These sheets cover both the original production models and the later vehicles such as the M48A3 and M48A5 which were converted from the earlier tanks. The data follow the format established in the earlier volumes in this series to permit easy comparison with those vehicles.

Where possible, the source documents for the data were the tank arsenal drawings for each production vehicle and the official characteristic sheets followed by test reports from Fort Knox or Aberdeen Proving Ground and the appropriate technical manuals. Needless to say, much of the data from the various sources frequently did not agree. In that case, the dimensions from the drawings were used unless obvious changes had been made. Some of the drawing dimensions were for reference only and would vary on the actual tank. For example, the fire height, defined as the distance from the ground to the centerline of the main weapon bore at zero elevation, would certainly vary from the drawing dimension which was given to one tenth of an inch. The same would apply to the ground clearance which also would change with the load on the tank. However, these design values are quoted to permit comparison between vehicles.

Although the meaning of most terms used in the data sheets is obvious, a few may need some explanation. The ground contact length at zero penetration is the distance between the centers of the front and rear road wheels. This value is used to calculate the ground contact area and then the ground pressure using the combat weight of the vehicle. The tread is the distance between the centerline of each track. Gross engine power and torque reflect the shaft output with only those accessories essential to engine operation, neglecting the effect of such items as air cleaners or generators. The net power and torque are the values actually obtained with the engine installed in the vehicle using all of its normal accessories. The weights are given for each vehicle unstowed and combat loaded. The latter includes the crew and a full load of fuel and ammunition. The power to weight ratios were calculated using the combat weight.

The actual weights of individual vehicles were obtained when possible. If it was a single experimental item, the exact weight is quoted. However, if it was intended to be representative of a production run, the weight is rounded off to the nearest 100 pounds. In many cases average values are used for production vehicles and these are given to the nearest 1000 pounds. For a few experimental vehicles, only approximate weights could be obtained.

Stowage arrangements frequently changed during the long operational life of much of this equipment. In that case, the stowage appropriate when the vehicle was new or during its period of greatest use is listed in the data sheets. Also, because of security restrictions, some items have been omitted.

MEDIUM TANKS M46 and M46A1

GENERAL DATA

Crew:	5	men
Length: Gun forward	333.6	inches
Length: Gun in travel position	277.6	inches
Length: without gun	250.3	inches
Gun Overhang: Gun forward	83.3	inches
Width: Over sandshields	138.3	inches
Height: Over AA MG	125.1	inches
Tread:	110.0	inches
Ground Clearance:	18.8	inches
Fire Height:	approx. 78	inches
Turret Ring Diameter: (inside)	69.0	inches
Weight, Combat Loaded:	approx. 97,000	pounds
Weight, Unstowed:	approx. 92,800	pounds
Power to Weight Ratio: Net	14.5	hp/ton
Gross	16.7	hp/ton
Ground Pressure: Zero penetration	14.0	psi

ARMOR

Type: Turret, cast homogeneous steel; Hull, rolled and cast homogeneous steel; Welded assembly

Hull Thickness:	Actual	Angle w/ Vertical
Front, Upper	4.0 inches (102mm)	46 degrees
Lower	3.0 inches (76mm)	53 degrees
Sides, Front	3.0 inches (76mm)	0 degrees
Rear	2.0 inches (51mm)	0 degrees
Rear, Upper	2.0 inches (51mm)	0 degrees
Lower	0.75 inches (20mm)	62 degrees
Top	0.88 inches(22mm)	90 degrees
Floor, Front	1.0 inches (25mm)	90 degrees
Rear	0.5 inches (13mm)	90 degrees
Turret Thickness:		
Gun Shield	4.5 inches (114mm)	0 degrees
Front	4.0 inches (102mm)	0 degrees
Sides	3.0 inches (76mm)	0 to 8 degrees
Rear	3.0 inches (76mm)	0 to 5 degrees
Top	1.0 inches (25mm)	90 degrees

ARMAMENT

Primary: 90mm Gun M3A1 in Mount M73 in turret

Traverse: Electric-hydraulic and manual	360 degrees
Traverse Rate: (max)	15 seconds/360 degrees
Elevation: Manual	+ 20 to − 10 degrees
Firing Rate: (max)	8 rounds/minute
Loading System:	Manual
Stabilizer System:	None

Secondary:
(1) .50 caliber MG HB M2 flexible AA mount on turret
(1) .30 caliber MG M1919A4 coaxial w/90mm gun in turret
(1) .30 caliber MG M1919A4 in bow mount
Provision for (1) .45 caliber SMG M3A1
Provision for (1) .30 caliber Carbine M2 w/grenade launcher

AMMUNITION

70 rounds 90mm	12 hand grenades
550 rounds .50 caliber	
180 rounds .45 caliber	
5500 rounds .30 caliber	
90 rounds .30 caliber (carbine)	

FIRE CONTROL AND VISION EQUIPMENT

Primary Weapon:	Direct	Indirect
	Telescope M46 (T152)	Azimuth Indicator M20
	or Telescope T40	Elevation Quadrant M9
	Periscope M10F	Gunner's Quadrant M1
Vision Devices	Direct	Indirect
Driver	Hatch	Periscope M13 (1)
Asst. Driver	Hatch	Periscope M13 (1)
Commander	Vision blocks (6) in cupola, hatch	Periscope M15A1 (1)
Gunner	None	Periscope M10F (1)
Loader	Hatch and pistol port	Periscope M13 (1)

Total Periscopes: M10F (1), M13 (3), M15A1 (1)
Total Pistol Ports: Hull (0), Turret (1)
Total Vision Blocks: (6) in cupola on turret top

ENGINE

Make and Model: Continental AV-1790-5A (M46) or AV-1790-5B (M46A1)	
Type: 12 cylinder, 4 cycle, 90 degree vee	
Cooling System: Air Ignition: Magneto	
Displacement:	1791.7 cubic inches
Bore and Stroke:	5.75 x 5.75 inches
Compression Ratio:	6.5:1
Net Horsepower (max):	704 hp at 2800 rpm
Gross Horsepower (max):	810 hp at 2800 rpm
Net Torque (max):	1440 ft-lb at 2000 rpm
Gross Torque (max):	1610 ft-lb at 2200 rpm
Weight:	2505 pounds dry
Fuel: 80 octane gasoline	232 gallons
Engine Oil: AV1790-5A (M46)	72 quarts
AV1790-5B (M46A1)	64 quarts

POWER TRAIN

Transmission: Cross-drive CD-850-3 (M46) or CD-850-4 (M46 or M46A1), 2 ranges forward, 1 reverse
Single stage multiphase hydraulic torque converter
Stall multiplication: 4.3:1

Overall Usable Ratios: low 13.0:1	reverse 17.8:1
high 4.5:1	

Steering Control: Mechanical, wobble stick
Steering Rate: 5.7 rpm
Brakes: Multiple disc
Final Drive: Spur gear   Gear Ratio: 3.95:1
Drive Sprocket: At rear of vehicle with 13 teeth
Pitch Diameter: 25.038 inches

RUNNING GEAR

Suspension: Torsion bar
12 individually sprung dual road wheels (6/track)
Tire Size: 26 x 6 inches
10 dual track return rollers (5/track)
Dual compensating idler at front of each track
Idler Tire Size: 26 x 6 inches
Dual track tension idler in front of each sprocket
Shock absorbers fitted on first 2 and last 2 road wheels on each side
Tracks: Center guide, T80E1, T80E4, and T84E1
Type: (T80E1) Double pin, 23 inch width, rubber backed steel
(T80E4) Double pin, 23 inch width, rubber backed steel
(T84E1) Double pin, 23 inch width, rubber chevron
Pitch: 6 inches
Shoes per Vehicle: 172 (86/track)
Ground Contact Length: 148.8 inches, left side
152.6 inches, right side

ELECTRICAL SYSTEM

Nominal Voltage: 24 volts DC
Main Generator: (1) 28.5 volts, 150 amperes, gear driven by main engine
Auxiliary Generator: (1) 28.5 volts, 150 amperes, driven by auxiliary engine
Battery: (4) 12 volts, 2 sets of 2 in series connected in parallel

COMMUNICATIONS

Radio: AN/GRC-3 thru 8 series or SCR 508 or SCR 528 in turret bustle
AN/VRC-3 or SCR 608B in turret bustle (infantry communication)
Interphone: RC-99, 5 stations plus external extension kit AN/VIA-1

FIRE PROTECTION

(3) 10 pound carbon dioxide, fixed
(2) 4 pound carbon dioxide, portable

PERFORMANCE

Maximum Speed: Sustained, level road	30 miles/hour
Maximum Tractive Effort: TE at stall	66,800 pounds
Per Cent of Vehicle Weight: TE/W	64 per cent
Maximum Grade:	60 per cent
Maximum Trench:	8.5 feet
Maximum Vertical Wall:	36 inches
Maximum Fording Depth:	48 inches
Minimum Turning Circle: (diameter)	pivot
Cruising Range: Roads	approx. 80 miles

90mm GUN TANK T42

GENERAL DATA

Crew:	4	men
Length: Gun forward	314.7	inches
Length: Gun in travel position	273.4	inches
Length: Without gun	232.1	inches
Gun Overhang: Gun forward	82.6	inches
Width: Over fenders	140.8	inches
Height: Over AA MG	126.3	inches
Tread:	111.0	inches
Ground Clearance:	17.1	inches
Fire Height: approx.	76	inches
Turret Ring Diameter: (inside)	73.0	inches
Weight, Combat Loaded:	74,500	pounds
Weight, Unstowed:	68,080	pounds
Power to Weight Ratio: Net	9.9	hp/ton
Gross	13.4	hp/ton
Ground Pressure: Zero penetration, T95 tracks	12.5	psi
T80E6 & T84E1 tracks	13.0	psi

ARMOR

Type: Turret, cast homogeneous steel; Hull, rolled and cast homogeneous
steel; Welded assembly

Hull Thickness:	Actual	Angle w/Vertical
Front, Upper	4.0 inches (102mm)	60 degrees
Lower	2.5 to 4.0 inches (64-102mm)	54 degrees
Sides, Front	2.5 to 3.0 inches (64-76mm)	0 degrees
Rear	1.5 inches (38mm)	0 degrees
Rear, Upper	1.0 inches (25mm)	60 degrees
Lower	1.0 inches (25mm)	50 degrees
Top	2.0 inches (51mm)	90 degrees
Floor, Front	1.0 inches (25mm)	90 degrees
Rear	0.5 inches (13mm)	90 degrees
Turret Thickness:		
Gun Shield	4.0 inches (102mm)	60 degrees
Front	4.0 inches (102mm)	45 degrees
Sides	2.5 to 3.5 inches (64-89mm)	40 degrees
Rear	3.6 inches (91mm)	7 degrees
Top	1.0 inches (25mm)	90 degrees

ARMAMENT

Primary: 90mm Gun T119 in Mount T139 in turret

Traverse: Electric-hydraulic and manual	360 degrees
Traverse Rate: (max)	10 seconds/360 degrees
Elevation: Electric-hydraulic and manual	+ 20 to − 10 degrees
Elevation Rate: (max)	4 degrees/second
Firing Rate: (max)	8 rounds/minute
Loading System:	Manual
Stabilizer System:	IBM proposed

Secondary:
(1) .50 caliber MG HB M2 flexible AA mount on turret
(1) .50 caliber MG HB M2E1 coaxial w/90mm gun in turret (This
weapon is interchangeable with a .30 caliber coaxial MG)
Provision for (1) .45 caliber SMG M3A1
Provision for (1) .30 caliber Carbine M2 w/grenade launcher

AMMUNITION

60 rounds 90mm	4000 rounds .30 caliber (If
2350 rounds .50 caliber (coax)	optional .30 cal. coax is used)
700 rounds .50 caliber (AA)	180 rounds .30 caliber (carbine)
180 rounds .45 caliber	8 hand grenades

FIRE CONTROL AND VISION EQUIPMENT

Primary Weapon:	Direct	Indirect
	Range Finder T41	Azimuth Indicator T24
	Periscope T35	Elevation Quadrant T21
	Ballistic Drive T23E1	Gunner's Quadrant M1

Vision Devices	Direct	Indirect
Driver	Hatch	Periscope T36 (4)
Commander	Vision blocks (6) in cupola, hatch	Periscope T35 (1)
Gunner	None	Periscope T35 (1)
Loader	Hatch and pistol port	Periscope M13 (1)

Total Periscopes: M13 (1), T35 (2), T36 (4)
Total Pistol Ports: Hull (0), Turret (1) none on later pilots
Total Vision Blocks: (6) in cupola on turret top

ENGINE

Make and Model: Continental AOS-895-3		
Type: 6 cylinder, 4 cycle, opposed, supercharged		
Cooling System: Air	Ignition: Magneto	
Displacement:		895.9 cubic inches
Bore and Stroke:		5.75 x 5.75 inches
Compression Ratio:		5.5:1
Net Horsepower (max):		370 hp at 2800 rpm
Gross Horsepower (max):		500 hp at 2800 rpm
Net Torque (max):		820 ft-lb at 2400 rpm
Gross Torque (max):		985 ft-lb at 2400 rpm
Weight:		1700 pounds, dry
Fuel: 80 octane gasoline		145 gallons
Engine Oil:		52 quarts

POWER TRAIN

Transmission: Cross-drive CD-500-3, 2 ranges forward, 1 reverse
w/automatic lock-up in high
Single stage hydraulic torque converter
Stall multiplication: 4:1
Overall Usable Ratios: low 14.7:1   direct: 1:1
high 3.9:1   reverse 14.7:1
Steering Control: Mechanical, wobble stick
Steering Rate: 6.8 rpm
Brakes: Multiple disc
Final Drive: Spur gear   Gear Ratio: 5.917:1
Drive Sprocket: At rear of vehicle with 12 teeth
Pitch Diameter: 23.182 inches

RUNNING GEAR

Suspension: Torsion bar
10 individually sprung dual road wheels (5/track)
Tire Size: 26 x 6 inches
6 dual track return rollers (3/track)
Dual compensating idler at front of each track
Idler Tire Size: 22.5 x 5 inches, steel
Shock absorbers fitted on first 2 and last road wheels on each side
Tracks: Center guide, T80E6, T84E1, and T95
Type: (T80E6) Double pin, 23 inch width, rubber backed steel
(T84E1) Double pin, 23 inch width, rubber chevron
(T95)   Single pin, 24 inch width, steel w/rubber pad
Pitch: 6 inches
Shoes per Vehicle: 152 (76/track)
Ground Contact Length: 127 inches

ELECTRICAL SYSTEM

Nominal Voltage: 24 volts DC
Main Generator: (1) 24 volts, 200 amperes driven by main engine
Auxiliary Generator: (1) 24 volts, 300 amperes driven by auxiliary
engine
Battery: (4) 12 volts, 2 sets of 2 in series connected in parallel

COMMUNICATIONS

Radio: AN/GRC-3 thru 8 series in turret bustle
Interphone: 4 stations plus external extension kit AN/VIA-1

FIRE PROTECTION

(2) 10 pound carbon dioxide, fixed
(1) 5 pound carbon dioxide, portable

PERFORMANCE

Maximum Speed: Sustained, level road	32 miles/hour
Maximum Tractive Effort: TE at stall	56,000 pounds
Per Cent of Vehicle Weight: TE/W	75 per cent
Maximum Grade:	60 per cent
Maximum Trench:	6 feet
Maximum Vertical Wall:	30 inches
Maximum Fording Depth:	48 inches
Minimum Turning Circle: (diameter)	pivot
Cruising Range: Roads approx.	70 miles

90mm GUN TANK T69

GENERAL DATA

Crew:	4	men
Length: Gun forward	323.8	inches
Length: Gun in travel position	290.4	inches
Length: Without gun	232.1	inches
Gun Overhang: Gun forward	91.7	inches
Width: Over fenders	140.8	inches
Height: Over AA MG mount w/o MG	112.9	inches
Tread:	111.0	inches
Ground Clearance:	17.1	inches
Fire Height:	approx. 81	inches
Turret Ring Diameter: (inside)	73.0	inches
Weight, Combat Loaded:	approx. 76,000	pounds
Weight, Unstowed:	approx. 72,000	pounds
Power to Weight Ratio: Net	9.7	hp/ton
Gross	13.2	hp/ton
Ground Pressure: Zero penetration, T95 tracks	12.5	psi
T80E6 & T84E1 tracks	13.0	psi

ARMOR

Type: Turret, cast homogeneous steel; Hull, rolled and cast homogeneous steel; Welded assembly

Hull Thickness:	Actual	Angle w/Vertical
Front, Upper	4.0 inches (102mm)	60 degrees
Lower	2.5 to 4.0 inches (64-102mm)	54 degrees
Sides, Front	2.5 to 3.0 inches (64-76mm)	0 degrees
Rear	1.5 inches (38mm)	0 degrees
*Rear, Upper	1.0 inches (25mm)	60 degrees
Lower	1.0 inches (25mm)	50 degrees
Top	2.0 inches (51mm)	90 degrees
Floor, Front	1.0 inches (25mm)	90 degrees
Rear	0.5 inches (13mm)	90 degrees
Turret Thickness:		
Front	4.0 inches (102mm)	60 degrees
Sides	equals 3.0 inches (76mm)	0 degrees
Rear	5.8 inches (147mm)	40 degrees
Top	1.0 inches (25mm)	90 degrees

ARMAMENT

Primary: 90mm Gun T178 in combination mount in turret

Traverse: Hydraulic and manual	360 degrees
Traverse Rate: (max)	15 seconds/360 degrees
Elevation: Hydraulic and manual	+ 15 to − 9 degrees
Elevation Rate: (max)	4 degrees/second
Firing Rate: (max w/auto load)	33 rounds/minute (1 type ammo)
	18 rounds/minute (3 types ammo)
Loading System:	Automatic w/8 round magazine
Stabilizer System:	None

Secondary:
(1) .50 caliber MG HB M2 flexible AA on turret hatch
(1) .30 caliber MG M1919A4E1 coaxial w/90mm gun in turret
Provision for (1) .45 caliber SMG M3A1
Provision for (1) .30 caliber Carbine M2 w/grenade launcher

AMMUNITION

40 rounds 90mm	8 hand grenades
1000 rounds .50 caliber	
180 rounds .45 caliber	
3500 rounds .30 caliber	
180 rounds .30 caliber (carbine)	

FIRE CONTROL AND VISION EQUIPMENT

Primary Weapon:	Direct	Indirect
	Range Finder T46E2	Azimuth Indicator M31
	Periscope T35	Elevation Quadrant M13 (T21)
	Range Drives T32 & T33	Gunner's Quadrant M1 or M1A1

Vision Devices	Direct	Indirect
Driver	Hatch	Periscope T36 (4)
Commander	Hatch	Periscope T36 (6)
Gunner	None	Periscope T35
Loader	Hatch	Periscope M13

Total Periscopes: T36 (10), T35 (1), M13 (1)
Total Vision Blocks: None

ENGINE

Make and Model: Continental AOS-895-3	
Type: 6 cylinder, 4 cycle, opposed, supercharged	
Cooling System: Air Ignition: Magneto	
Displacement:	895.9 cubic inches
Bore and Stroke:	5.75 x 5.75 inches
Compression Ratio:	5.5:1
Net Horsepower (max):	370 hp at 2800 rpm
Gross Horsepower (max):	500 hp at 2800 rpm
Net Torque (max):	820 ft-lb at 2400 rpm
Gross Torque (max):	985 ft-lb at 2400 rpm
Weight:	1700 pounds, dry
Fuel: 80 octane gasoline	145 gallons
Engine Oil:	52 quarts

POWER TRAIN

Transmission: Cross-drive CD-500-3, 2 ranges forward, 1 reverse w/automatic lock-up in high
Single stage hydraulic torque converter
Stall multiplication: 4:1

Overall Usable Ratios:	low 14.7:1	direct: 1:1
	high 3.9:1	reverse 14.7:1

Steering Control: Mechanical, wobble stick
Steering Rate: 6.8 rpm
Brakes: Multiple disc
Final Drive: Spur gear   Gear Ratio: 5.917:1
Drive Sprocket: At rear of vehicle with 12 teeth
Pitch Diameter: 23.182 inches

RUNNING GEAR

Suspension: Torsion bar
10 individually sprung dual road wheels (5/track)
Tire Size: 26 x 6 inches
6 dual track return rollers (3/track)
Dual compensating idler at front of each track
Idler Tire Size: 22.5 x 5 inches, steel
Shock absorbers fitted on first 2 and last road wheels on each side
Tracks: Center guide, T80E6, T84E1, and T95
Type: (T80E6) Double pin, 23 inch width, rubber backed steel
(T84E1) Double pin, 23 inch width, rubber chevron
(T95) Single pin, 24 inch width, steel w/rubber pad
Pitch: 6 inches
Shoes per Vehicle: 152 (76/track)
Ground Contact Length: 127 inches

ELECTRICAL SYSTEM

Nominal Voltage: 24 volts DC
Main Generator: (1) 24 volts, 200 amperes driven by main engine
Auxiliary Generator: (1) 24 volts, 300 amperes driven by auxiliary engine
Battery: (4) 12 volts, 2 sets of 2 in series connected in parallel

COMMUNICATIONS

Radio: Interphone only installed because of limited space in turret bustle
Interphone: 4 stations, AN/VIC-1

FIRE PROTECTION

(2) 10 pound carbon dioxide, fixed
(1) 5 pound carbon dioxide, portable

PERFORMANCE

Maximum Speed: Sustained, level road	32 miles/hour
Maximum Tractive Effort: TE at stall	56,000 pounds
Per Cent of Vehicle Weight: TE/W	74 per cent
Maximum Grade:	60 per cent
Maximum Trench:	6 feet
Maximum Vertical Wall:	30 inches
Maximum Fording Depth:	48 inches
Minimum Turning Circle: (diameter)	pivot
Cruising Range: Roads	approx. 70 miles

*Rear armor is shown for CD-500 transmission. With the XT-500, the rear plate was 1.0 inches at 0 degrees from the vertical.
#The data shown are for the CD-500 transmission proposed for the T69. However, the pilot turret was installed on T42 chassis number 3 in which the original transmission had been replaced by the later XT-500. No performance data are available for the T69 with the latter transmission.

90mm GUN TANK M47

GENERAL DATA

Crew:	5	men
Length: Gun forward	335.0	inches
Length: Gun in travel position	279.2	inches
Length: Without gun	250.3	inches
Gun Overhang: Gun forward	84.7	inches
Width: Over fenders	138.3	inches
Height: Over AA MG	131.0	inches
Tread:	110.0	inches
Ground Clearance:	18.5	inches
Fire Height: approx.	81	inches
Turret Ring Diameter: (inside)	73.0	inches
Weight, Combat Loaded:	101,800	pounds
Weight, Unstowed:	92,900	pounds
Power to Weight Ratio: Net	13.8	hp/ton
Gross	15.9	hp/ton
Ground Pressure: Zero penetration	14.7	psi

ARMOR
Type: Turret, cast homogeneous steel; Hull, rolled and cast homogeneous steel; Welded assembly

Hull Thickness:

	Actual	Angle w/Vertical
Front, Upper	4.0 inches (102mm)	60 degrees
Lower	3.0 to 3.5 inches (76-89mm)	53 degrees
Sides, Front	3.0 inches (76mm)	0 degrees
Rear	2.0 inches (51mm)	0 degrees
Rear, Upper	2.0 inches (51mm)	0 degrees
Lower	0.75 inches (20mm)	62 degrees
Top	0.88 inches(22mm)	90 degrees
Floor, Front	1.0 inches (25mm)	90 degrees
Rear	0.5 inches (13mm)	90 degrees

Turret Thickness:

Gun Shield	4.5 inches (114mm)	60 degrees
Front	4.0 inches (102mm)	40 degrees
Sides	2.5 inches (64mm)	30 degrees
Rear	3.0 inches (76mm)	3 degrees
Top	1.0 inches (25mm)	90 degrees

ARMAMENT
Primary: 90mm Gun M36 (T119E1) in Mount M78 in turret

Traverse: Electric-hydraulic and manual	360 degrees
Traverse Rate: (max)	15 seconds/360 degrees
Elevation: Electric-hydraulic and manual	+ 19 to − 10 degrees
Elevation Rate: (max)	4 degrees/second
Firing Rate: (max)	8 rounds/minute
Loading System:	Manual
Stabilizer System:	None

Secondary:
(1) .50 caliber MG HB M2 flexible AA mount on turret
(1) .30 caliber MG M1919A4E1 coaxial w/90mm gun in turret
(1) .30 caliber MG M1919A4E1 in bow mount
Provision for (1) .45 caliber SMG M3A1

AMMUNITION
71 rounds 90mm   8 hand grenades
1700 rounds .50 caliber
180 rounds .45 caliber
11,150 rounds .30 caliber

FIRE CONTROL AND VISION EQUIPMENT

Primary Weapon:	Direct	Indirect
	Range Finder M12 (T41E3)	Azimuth Indicator M31 (T24)
	Ballistic Drive M3 (T23E1)	Elevation Quadrant M13 (T21)
	Periscope M20 or M20A1	Gunner's Quadrant M1 or M1A1
Vision Devices	Direct	Indirect
Driver	Hatch	Periscope M13 (1) or Periscope M19 (1) (infrared)
Asst. Driver	Hatch	Periscope M13 (1)
Commander	Vision blocks (5) in cupola, hatch	Periscope M20 or M20A1 (1)
Gunner	None	Periscope M20 or M20A1 (1)
Loader	Hatch	Periscope M13 (1)

Total Periscopes: M13 (3), M20 or M20A1 (2); In addition (1) M19 carried for night use by driver
Total Pistol Ports: None
Total Vision Blocks: (5) in cupola on turret top

ENGINE
Make and Model: Continental AV-1790-5B, AV-1790-7, or AV-1790-7B
(data for AV-1790-5B)
Type: 12 cylinder, 4 cycle, 90 degree vee
Cooling System: Air   Ignition: Magneto

Displacement:	1791.7 cubic inches
Bore and Stroke:	5.75 x 5.75 inches
Compression Ratio:	6.5:1
Net Horsepower (max):	704 hp at 2800 rpm
Gross Horsepower (max):	810 hp at 2800 rpm
Net Torque (max):	1440 ft-lb at 2000 rpm
Gross Torque (max):	1610 ft-lb at 2200 rpm
Weight:	2505 pounds, dry
Fuel: 80 octane gasoline	232 gallons
Engine Oil:	64 quarts

POWER TRAIN
Transmission: Cross-drive CD-850-4, 2 ranges forward, 1 reverse
Single stage multiphase hydraulic torque converter
Stall multiplication: 4.3:1
Overall Usable Ratios: low 13.0:1   direct: 17.8:1
high 4.5:1
Steering Control: Mechanical, wobble stick
Steering Rate: 5.7 rpm
Brakes: Multiple disc
Final Drive: Spur gear   Gear Ratio: 4.47:1
Drive Sprocket: At rear of vehicle with 13 teeth
Pitch Diameter: 25.038 inches

RUNNING GEAR
Suspension: Torsion bar
12 individually sprung dual road wheels (6/track)
Tire Size: 26 x 6 inches
6 dual track return rollers (3/track)
Dual compensating idler at front of each track
Idler Tire Size: 26 x 6 inches
Dual track tension idler in front of each sprocket
Shock absorbers fitted on first 2 and last 2 road wheels on each side
(2 shock absorbers installed on front road wheels)
Tracks: Center guide, T80E6, T84E1
Type: (T80E6) Double pin, 23 inch width, rubber backed steel
(T84E1) Double pin, 23 inch width, rubber chevron
Pitch: 6 inches
Shoes per Vehicle: 172 (86/track)
Ground Contact Length: 148.8 inches, left side
152.6 inches, right side

ELECTRICAL SYSTEM
Nominal Voltage: 24 volts DC
Main Generator: (1) 28.5 volts, 150 or 300 amperes, gear driven by main engine
Auxiliary Generator: (1) 28.5 volts, 150 amperes, driven by auxiliary engine
Battery: (4) 12 volts, 2 sets of 2 in series connected in parallel

COMMUNICATIONS
Radio: AN/GRC-3 thru 8 series turret bustle
Interphone: 5 stations plus external extension kit AN/VIA-1

FIRE PROTECTION
(3) 10 pound carbon dioxide, fixed
(1) 5 pound carbon dioxide, portable

PERFORMANCE

Maximum Speed: Sustained, level road	30 miles/hour
Maximum Tractive Effort: TE at stall	66,800 pounds
Per Cent of Vehicle Weight: TE/W	68 per cent
Maximum Grade:	60 per cent
Maximum Trench:	8.5 feet
Maximum Vertical Wall:	36 inches
Maximum Fording Depth:	48 inches
Minimum Turning Circle: (diameter)	pivot
Cruising Range: Roads approx.	80 miles

90mm GUN TANK M47-M

GENERAL DATA

Crew:	4	men
Length: Gun forward	336.8	inches
Length: Gun in travel position	276.5	inches
Length: Without gun	246.8	inches
Gun Overhang: Gun forward	90.0	inches
Width: Over fenders	133.5	inches
Height: Over AA MG	133.8	inches
Tread:	110.0	inches
Ground Clearance:	18.5	inches
Fire Height: approx.	81	inches
Turret Ring Diameter: (inside)	73.0	inches
Weight, Combat Loaded:	103,200	pounds
Weight, Unstowed:	93,300	pounds
Power to Weight Ratio: Net	12.5	hp/ton
Gross	14.5	hp/ton
Ground Pressure: Zero penetration	14.5	psi

ARMOR

Type: Turret, cast homogeneous steel; Hull, rolled and cast homogeneous steel; Welded assembly

Hull Thickness:

	Actual	Angle w/Vertical
Front, Upper	4.0 inches (102mm)	60 degrees
Lower	3.0 to 3.5 inches (76-89mm)	53 degrees
Sides, Front	3.0 inches (76mm)	0 degrees
Rear	2.0 inches (51mm)	0 degrees
Rear, Grill	equals 1.0 inches (25mm)	0 degrees
Lower	0.75 inches (20mm)	62 degrees
Top	0.88 inches (22mm)	90 degrees
Floor, Front	1.0 inches (25mm)	90 degrees
Rear	0.5 inches (13mm)	90 degrees

Turret Thickness:

Gun Shield	4.5 inches (114mm)	60 degrees
Front	4.0 inches (102mm)	40 degrees
Sides	2.5 inches (64mm)	30 degrees
Rear	3.0 inches (76mm)	3 degrees
Top	1.0 inches (25mm)	90 degrees

ARMAMENT

Primary: 90mm Gun M36 in Mount M78 in turret

Traverse: Hydraulic and manual	360 degrees
Traverse Rate: (max)	15 seconds/360 degrees
Elevation: Hydraulic and manual	+ 19 to − 10 degrees
Elevation Rate: (max)	4 degrees/second
Firing Rate: (max)	8 rounds/minute
Loading System:	Manual
Stabilizer System:	None

Secondary:
(1) .50 caliber MG HB M2 flexible AA mount on turret
(1) .30 caliber MG M1919A4E1 coaxial w/90mm gun in turret
Provision for (1) .45 caliber SMG M3A1

AMMUNITION

79 rounds 90mm	8 hand grenades
1700 rounds .50 caliber	
180 rounds .45 caliber	
11,150 rounds .30 caliber	

FIRE CONTROL AND VISION EQUIPMENT

Primary Weapon:	Direct	Indirect
	Range Finder M12	Azimuth Indicator M31
	Ballistic Drive M3	Elevation Quadrant M13
	Periscope M20 or M20A1	Gunner's Quadrant M1A1

Vision Devices	Direct	Indirect
Driver	Hatch	Periscope M13 (1) or Periscope M19 (1) infrared
Commander	Vision blocks (5) in cupola, hatch	Periscope M20 or M20A1 (1)
Gunner	None	Periscope M20 or M20A1 (1)
Loader	Hatch	Periscope M13 (1)

Total Periscopes: M13 (2), M20 or M20A1 (2); In addition (1) M19 carried for night use by driver
Total Pistol Ports: None
Total Vision Blocks: (5) in cupola on turret top

ENGINE

Make and Model: Continental AVDS-1790-2A
Type: 12 cylinder, 4 cycle, 90 degree vee, supercharged
Cooling System: Air   Ignition: Compression

Displacement:	1791.7 cubic inches
Bore and Stroke:	5.75 x 5.75 inches
Compression Ratio:	16:1
Net Horsepower (max):	643 hp at 2400 rpm
Gross Horsepower (max):	750 hp at 2400 rpm
Net Torque (max):	1575 ft-lb at 1750 rpm
Gross Torque (max):	1710 ft-lb at 1800 rpm
Weight:	4700 pounds, dry
Fuel: 40 cetane diesel	400 gallons
Engine Oil:	72 quarts

POWER TRAIN

Transmission: Cross-drive CD-850-6A, 2 ranges forward, 1 reverse
Single stage multiphase hydraulic torque converter
Stall multiplication: 4:1
Overall Usable Ratios: low 12.0:1   reverse: 16.9:1
high 4.3:1
Steering Control: Mechanical, wobble stick
Steering Rate: 5.7 rpm
Brakes: Multiple disc
Final Drive: Spur gear   Gear Ratio: 4.47:1
Drive Sprocket: At rear of vehicle with 13 teeth
Pitch Diameter: 25.038 inches

RUNNING GEAR

Suspension: Torsion bar
12 individually sprung dual road wheels (6/track)
Tire Size: 26 x 6 inches
6 dual track return rollers (3/track)
Dual compensating idler at front of each track
Idler Tire Size: 26 x 6 inches
Friction snubbers fitted on first 2 and last 2 road wheels on each side (2 friction snubbers on front road wheels)
Tracks: Center guide, T80E6, T84E1
Type: (T80E6) Double pin, 23 inch width, rubber backed steel
(T84E1) Double pin, 23 inch width, rubber chevron
Pitch: 6 inches
Shoes per Vehicle: 172 (86/track)
Ground Contact Length: 152.5 inches, left side
156.5 inches, right side

ELECTRICAL SYSTEM

Nominal Voltage: 24 volts DC
Main Generator: (1) 24 volts, 300 amperes, gear driven by main engine
Auxiliary Generator: None
Battery: (6) 12 volts, 3 sets of 2 in series connected in parallel

COMMUNICATIONS

Radio: AN/VRC-12, 46, 47, 53, or 64 in turret bustle
AN/VRC-24 (air to ground) also may be fitted
Interphone: 4 stations plus external box

FIRE PROTECTION

(3) 10 pound carbon dioxide, fixed
(1) 5 pound carbon dioxide, portable

PERFORMANCE

Maximum Speed: Sustained, level road	35 miles/hour
Maximum Tractive Effort: TE at stall	70,700 pounds
Per Cent of Vehicle Weight: TE/W	69 per cent
Maximum Grade:	60 per cent
Maximum Trench:	8.5 feet
Maximum Vertical Wall:	36 inches
Maximum Fording Depth:	48 inches
Minimum Turning Circle: (diameter)	pivot
Cruising Range: Roads approx.	370 miles

90mm GUN TANK T48

GENERAL DATA

Crew:	4	men
Length: Gun forward (muzzle brake) w/fenders	343.7	inches
Length: Gun in travel position	289.5	inches
Length: Without gun w/fenders	270.5	inches
Gun Overhang: Gun forward	73.2	inches
Width: Over tracks	143.0	inches
Height: Over AA MG	127.6	inches
Tread:	115.0	inches
Ground Clearance:	16.5	inches
Fire Height: approx.	79	inches
Turret Ring Diameter: (inside)	85.0	inches
Weight, Combat Loaded:	98,400	pounds
Weight, Unstowed:	92,550	pounds
Power to Weight Ratio: Net	14.3	hp/ton
Gross	16.5	hp/ton
Ground Pressure: Zero penetration	11.2	psi

*ARMOR

Type: Turret, cast homogeneous steel; Hull, cast homogeneous
steel; Welded assembly

Hull Thickness:	Actual	Angle w/Vertical
#Front, Upper	4.33 inches (110mm)	60 degrees
Lower	4.0 to 2.4 inches (102-61mm)	53 degrees
Sides, Front	equals 3.0 inches (76mm)	0 degrees
Rear	equals 2.0 inches (51mm)	0 degrees
Rear, Upper	1.38 inches (35mm)	30 degrees
Lower	1.0 inches (25mm)	60 degrees
Top	2.25 inches (57mm)	90 degrees
Floor, Front	1.5 inches (38mm)	90 degrees
Center	1.25 inches (32mm)	90 degrees
Rear	0.5 inches (13mm)	90 degrees
Turret Thickness:		
Gun Shield	4.5 inches (114mm)	30 degrees
Front	equals 7.0 inches (178mm)	0 degrees
Sides	equals 3.0 inches (76mm)	0 degrees
Rear	equals 2.0 inches (51mm)	0 degrees
Top	1.0 inches (25mm)	90 degrees

ARMAMENT

Primary: 90mm Gun T139 in Mount T148 in turret

Traverse: Electric-hydraulic and manual	360 degrees
Traverse Rate: (max)	10 seconds/360 degrees (design)
Elevation: Electric-hydraulic and manual	+20 to −9 degrees
Elevation Rate: (max)	4 degrees/second
Firing Rate: (max)	8 rounds/minute
Loading System:	Manual
Stabilizer System:	None

Secondary:
(1) .50 caliber MG HB M2 in remote control AA mount on turret
(1) .50 caliber MG HB M2E1 coaxial w/90mm gun in turret
(1) .30 caliber MG M1919A4E1 coaxial w/90 mm gun in turret
(2) .30 caliber MG T153 may replace coaxial MGs above
Provision for (1) .45 caliber SMG M3A1
Provision for (1) .30 caliber Carbine M2 w/grenade launcher

AMMUNITION

60 rounds 90mm		8 hand grenades
2500 rounds .50 caliber, approx.		
180 rounds .45 caliber		
2500 rounds .30 caliber, approx.		
180 rounds .30 caliber (carbine)		

FIRE CONTROL AND VISION EQUIPMENT

Primary Weapon:	Direct	Indirect
	Telescope T161	Azimuth Indicator M31
	Periscope M20 or T37	Elevation Quadrant M13
	Ballistic Drive T24	Gunner's Quadrant M1
	Range Drive T25	

Vision Devices	Direct	Indirect
Driver	Hatch	Periscope T25 (3)
Commander	Hatch	Periscope M17 (4)
Gunner	None	Periscope M20 or T37 (1)
Loader	Hatch	None

Total Periscopes: T25 (3), M17 (4), M20 or T37 (1)

ENGINE

Make and Model: Continental AV-1790-5B	
Type: 12 cylinder, 4 cycle, 90 degree vee	
Cooling System: Air Ignition: Magneto	
Displacement:	1791.7 cubic inches
Bore and Stroke:	5.75 x 5.75 inches
Compression Ratio:	6.5:1
Net Horsepower (max):	704 hp at 2800 rpm
Gross Horsepower (max):	810 hp at 2800 rpm
Net Torque (max):	1440 ft-lb at 2000 rpm
Gross Torque (max):	1610 ft-lb at 2200 rpm
Weight:	2505 pounds, dry
Fuel: 80 octane gasoline	200 gallons
Engine Oil:	64 quarts

POWER TRAIN

Transmission: Cross-drive CD-850-4, 2 ranges forward, 1 reverse
Single stage multiphase hydraulic torque converter
Stall multiplication: 4.3:1
Overall Usable Ratios: low 13.0:1   reverse: 17.8:1
high 4.5:1
Steering Control: Mechanical, steering wheel
Steering Rate: 5.7 rpm
Brakes: Multiple disc
Final Drive: Spur gear   Gear Ratio: 5.08:1
Drive Sprocket: At rear of vehicle with 11 teeth
Pitch Diameter: 24.504 inches

RUNNING GEAR

Suspension: Torsion bar
12 individually sprung dual road wheels (6/track)
Tire Size: 26 x 6 inches
10 dual track return rollers (5/track)
Dual compensating idler at front of each track
Idler Tire Size: 26 x 6 inches
Shock absorbers fitted on first 2 and last road wheels on each side
Tracks: Center guide, T96 and T97
Type: (T96) Double pin, 28 inch width, rubber backed steel
(T97) Double pin, 28 inch width, rubber chevron
Pitch: 6.94 inches
Shoes per Vehicle: 156 (78/track)
Ground Contact Length: 157.5 inches

ELECTRICAL SYSTEM

Nominal Voltage: 24 volts DC
Main Generator: (1) 24 volts, 150 amperes, gear driven by main engine
Auxiliary Generator: (1) 28 volts, 300 amperes driven by auxiliary engine
Battery: (4) 12 volts, 2 sets of 2 in series connected in parallel

COMMUNICATIONS

Radio: AN/GRC-3 thru 8 series in turret bustle
Interphone: 4 stations plus external extension kit AN/VIA-1

FIRE PROTECTION

(3) 10 pound carbon dioxide, fixed
(1) 5 pound carbon dioxide, portable

PERFORMANCE

Maximum Speed: Sustained, level road	30 miles/hour
Maximum Tractive Effort: TE at stall	78,000 pounds
Per Cent of Vehicle Weight: TE/W	79 per cent
Maximum Grade:	60 per cent
Maximum Trench:	8.5 feet
Maximum Vertical Wall:	36 inches
Maximum Fording Depth:	48 inches
Minimum Turning Circle: (diameter)	pivot
Cruising Range: Roads	approx. 70 miles

*The surface angles and wall thicknesses of the cast hull and turret varied widely from point to point. Therefore, where indicated the data are for an equivalent thickness at 0 degrees from the vertical.

#The hull casting did not meet this requirement near the driver's periscopes.

90mm GUN TANK M48

GENERAL DATA
Crew:	4	men
Length: Gun forward (cyl. blast defl.) w/fenders	346.9	inches
Length: Gun in travel position	292.9	inches
Length: without gun w/fenders	274.3	inches
Gun Overhang: Gun forward	72.6	inches
Width: Over tracks	143.0	inches
Height: Over AA MG	127.6	inches
Tread:	115.0	inches
Ground Clearance:	16.5	inches
Fire Height: approx.	79	inches
Turret Ring Diameter: (inside)	85.0	inches
Weight, Combat Loaded:	99,000	pounds
Weight, Unstowed:	93,100	pounds
Power to Weight Ratio: Net	14.2	hp/ton
Gross	16.4	hp/ton
Ground Pressure: Zero penetration	11.2	psi

*ARMOR
Type: Turret, cast homogeneous steel; Hull, cast homogeneous
steel; Welded assembly

Hull Thickness:	Actual	Angle w/Vertical
#Front, Upper	4.33 inches (110mm)	60 degrees
Lower	4.0 to 2.4 inches (102-61mm)	53 degrees
Sides, Front	equals 3.0 inches (76mm)	0 degrees
Rear	equals 2.0 inches (51mm)	0 degrees
Rear, Upper	1.38 inches (35mm)	30 degrees
Lower	1.0 inches (25mm)	60 degrees
Top	2.25 inches (57mm)	90 degrees
Floor, Front	1.5 inches (38mm)	90 degrees
Center	1.25 inches (32mm)	90 degrees
Rear	0.5 inches (13mm)	90 degrees
Turret Thickness:		
Gun Shield	4.5 inches (114mm)	30 degrees
Front	equals 7.0 inches (178mm)	0 degrees
Sides	equals 3.0 inches (76mm)	0 degrees
Rear	equals 2.0 inches (51mm)	0 degrees
Top	1.0 inches (25mm)	90 degrees

ARMAMENT
Primary: 90mm Gun M41 (T139) in Mount M87 (T148) in turret
Traverse: Electric-hydraulic and manual	360 degrees
Traverse Rate: (max)	15 seconds/360 degrees
Elevation: Electric-hydraulic and manual	+ 19 to − 9 degrees
Elevation Rate: (max)	4 degrees/second
Firing Rate: (max)	8 rounds/minute
Loading System:	Manual
Stabilizer System:	None

Secondary:
(1) .50 caliber MG HB M2 in remote control AA mount on turret
(1) .30 caliber MG M1919A4E1 coaxial w/90mm gun in turret
Provision for (1) .45 caliber SMG M3A1
Provision for (1) .30 caliber Carbine M2 w/grenade launcher

AMMUNITION
60 rounds 90mm	8 hand grenades
500 rounds .50 caliber	
180 rounds .45 caliber	
5900 rounds .30 caliber	
180 rounds .30 caliber (carbine)	

FIRE CONTROL AND VISION EQUIPMENT
Primary Weapon:	Direct	Indirect
	Range Finder T46E1	Azimuth Indicator T28
	Periscope M20,	Elevation Quadrant M13
	M20A1 or M20A2	Gunner's Quadrant M1
	Ballistic Drive T24E2	or M1A1
	Ballistic Computer T30	
	Telescope T156E1	

Vision Devices	Direct	Indirect
Driver	Hatch	Periscope T25 (3)
Commander	Hatch	Periscope M17 (4)
Gunner	None	Periscope M20,
		M20A1, or
		M20A2 (1)
Loader	Hatch	None

Total Periscopes: T25 (3), M17 (4), M20, M20A1, or M20A2 (1)

ENGINE
Make and Model: Continental AV-1790-5B, AV-1790-7, AV-1790-7B, or
AV-1790-7C (data for AV-1790-7B)
Type: 12 cylinder, 4 cycle, 90 degree vee
Cooling System: Air   Ignition: Magneto
Displacement:	1791.7 cubic inches
Bore and Stroke:	5.75 x 5.75 inches
Compression Ratio:	6.5:1
Net Horsepower (max):	704 hp at 2800 rpm
Gross Horsepower (max):	810 hp at 2800 rpm
Net Torque (max):	1440 ft-lb at 2000 rpm
Gross Torque (max):	1610 ft-lb at 2200 rpm
Weight:	2581 pounds, dry
Fuel: 80 octane gasoline	200 gallons
Engine Oil:	64 quarts

POWER TRAIN
Transmission: Cross-drive CD-850-4A or CD-850-4B (data for
CD-850-4A) 2 ranges forward, 1 reverse
Single stage multiphase hydraulic torque converter
Stall multiplication: 4.3:1
Overall Usable Ratios: low 13.0:1   reverse: 17.8:1
high 4.5:1
Steering Control: Mechanical, steering wheel
Steering Rate: 5.7 rpm
Brakes: Multiple disc
Final Drive: Spur gear   Gear Ratio: 5.08:1
Drive Sprocket: At rear of vehicle with 11 teeth
Pitch Diameter: 24.504 inches

RUNNING GEAR
Suspension: Torsion bar
12 individually sprung dual road wheels (6/track)
Tire Size: 26 x 6 inches
10 dual track return rollers (5/track)
Dual compensating idler at front of each track
Idler Tire Size: 26 x 6 inches
Dual track tension idler in front of each sprocket
Shock absorbers fitted on first 2 and last road wheels on each side
Tracks: Center guide, T96 and T97
Type: (T96) Double pin, 28 inch width, rubber backed steel
(T97) Double pin, 28 inch width, rubber chevron
Pitch: 6.94 inches
Shoes per Vehicle: 158 (79/track)
Ground Contact Length: 157.5 inches

ELECTRICAL SYSTEM
Nominal Voltage: 24 volts DC
Main Generator: (1) 24 volts, 300 amperes, gear driven by main engine
Auxiliary Generator: (1) 28 volts, 300 amperes, driven by auxiliary engine
Battery: (4) 12 volts, 2 sets of 2 in series connected in parallel

COMMUNICATIONS
Radio: AN/GRC-3 thru 8 series or AN/VRC-7 in turret bustle
AN/ARC-3 or AN/ARC-27 (air to ground) also may be fitted
Interphone: 4 stations plus external extension kit AN/VIA-1

FIRE PROTECTION
(3) 10 pound carbon dioxide, fixed
(1) 5 pound carbon dioxide, portable

PERFORMANCE
Maximum Speed: Sustained, level road	28 miles/hour
Maximum Tractive Effort: TE at stall	78,000 pounds
Per Cent of Vehicle Weight: TE/W	79 per cent
Maximum Grade:	60 per cent
Maximum Trench:	8.5 feet
Maximum Vertical Wall:	36 inches
Maximum Fording Depth:	48 inches
Minimum Turning Circle: (diameter)	pivot
Cruising Range: Roads approx.	70 miles

*The surface angles and wall thicknesses of the cast hull and turret varied widely from point to point. Therefore, where indicated the data are for an
equivalent thickness at 0 degrees from the vertical.
#120 early hull castings did not meet this requirement near the driver's periscopes. These tanks were redesignated as the M48C.

90mm GUN TANK M48A1

GENERAL DATA

Crew:	4	men
Length: Gun forward (cyl. blast defl.) w/fenders	346.9	inches
Length: Gun in travel position	292.9	inches
Length: Without gun, w/fenders	274.3	inches
Gun Overhang: Gun forward	72.6	inches
Width: Over tracks	143.0	inches
Height: Over cupola periscope	121.6	inches
Tread:	115.0	inches
Ground Clearance:	16.5	inches
Fire Height: approx.	79	inches
Turret Ring Diameter: (inside)	85.0	inches
Weight, Combat Loaded:	104,000	pounds
Weight, Unstowed:	97,000	pounds
Power to Weight Ratio: Net	13.3	hp/ton
Gross	15.6	hp/ton
Ground Pressure: Zero penetration	11.8	psi

*ARMOR

Type: Turret, cast homogeneous steel; Hull, cast homogeneous
steel; Welded assembly

Hull Thickness:

	Actual	Angle w/Vertical
Front, Upper	4.33 inches (110mm)	60 degrees
Lower	4.0 to 2.4 inches (102-61mm)	53 degrees
Sides, Front	equals 3.0 inches (76mm)	0 degrees
Rear	equals 2.0 inches (51mm)	0 degrees
Rear, Upper	1.38 inches (35mm)	30 degrees
Lower	1.0 inches (25mm)	60 degrees
Top	2.25 inches (57mm)	90 degrees
Floor, Front	1.5 inches (38mm)	90 degrees
Rear	1.0 inches (25mm)	90 degrees

Turret Thickness:

Gun Shield	4.5 inches (114mm)	30 degrees
Front	equals 7.0 inches (178mm)	0 degrees
Sides	equals 3.0 inches (76mm)	0 degrees
Rear	equals 2.0 inches (51mm)	0 degrees
Top	1.0 inches (25mm)	90 degrees

ARMAMENT

Primary: 90mm Gun M41 (T139) in Mount M87 (T148) in turret

Traverse: Electric-hydraulic and manual	360 degrees
Traverse Rate: (max)	15 seconds/360 degrees
Elevation: Electric-hydraulic and manual	+ 19 to − 9 degrees
Elevation Rate: (max)	4 degrees/second
Firing Rate: (max)	8 rounds/minute
Loading System:	Manual
Stabilizer System:	None

Secondary:
(1) .50 caliber MG HB M2 in cupola mount on turret
(1) .30 caliber MG M1919A4E1 coaxial w/90 mm gun in turret
Provision for (1) .45 caliber SMG M3A1
Provision for (1) .30 caliber Carbine M2 w/grenade launcher

AMMUNITION

60 rounds 90mm	8 hand grenades
500 rounds .50 caliber	
180 rounds .45 caliber	
5900 rounds .30 caliber	
180 rounds .30 caliber (carbine)	

FIRE CONTROL AND VISION EQUIPMENT

Primary Weapon:	Direct	Indirect
	Range Finder	Azimuth Indicator
	T46E1	T28
	Periscope M20, M20A1, or M20A2	Elevation Quadrant M13 Gunner's Quadrant M1
	Ballistic Drive T24E2	or M1A1
	Ballistic Computer T30	
	Telescope T156E1	
Vision Devices	Direct	Indirect
Driver	Hatch	Periscope M7 (T36) (3) and Periscope T41 infrared
Commander	Vision blocks (5) in cupola, hatch	Periscope T42 (AA MG sight)
Gunner	None	Periscope M20, M20A1, or M20A2 (1)
Loader	Hatch	None

Total Periscopes: M7 (T36) (3), T41 infrared (1), M20, M20A1,
or M20A2 (1), T42 (1)

Total Vision Blocks: (5) in M1 cupola on turret top

ENGINE

Make and Model: Continental AV-1790-5B, AV-1790-7, AV-1790-7B, or
AV-1790-7C (data for AV-1790-7C)

Type: 12 cylinder, 4 cycle, 90 degree vee

Cooling System: Air   Ignition: Magneto

Displacement:	1791.7 cubic inches
Bore and Stroke:	5.75 x 5.75 inches
Compression Ratio:	6.5:1
Net Horsepower (max):	690 hp at 2800 rpm
Gross Horsepower (max):	810 hp at 2800 rpm
Net Torque (max):	1410 ft-lb at 2200 rpm
Gross Torque (max):	1600 ft-lb at 2200 rpm
Weight:	2647 pounds, dry
Fuel: 80 octane gasoline	200 gallons
Engine Oil:	64 quarts

POWER TRAIN

Transmission: Cross-drive CD-850-4A or CD-850-4B (data for
CD-850-4B) 2 ranges forward, 1 reverse

Single stage multiphase hydraulic torque converter
Stall multiplication: 4:1

Overall Usable Ratios: low 13.0:1   reverse: 18.1:1
high 4.5:1

Steering Control: Mechanical, steering wheel
Steering Rate: 5.7 rpm
Brakes: Multiple disc
Final Drive: Spur gear   Gear Ratio: 5.08:1
Drive Sprocket: At rear of vehicle with 11 teeth
Pitch Diameter: 24.504 inches

RUNNING GEAR

Suspension: Torsion bar
12 individually sprung dual road wheels (6/track)
Tire Size: 26 x 6 inches
10 dual track return rollers (5/track)
Dual compensating idler at front of each track
Idler Tire Size: 26 x 6 inches
Dual track tension idler in front of each sprocket
Shock absorbers fitted on first 2 and last road wheels on each side

Tracks: Center guide, T97E2
Type: (T97E2) Double pin, 28 inch width, rubber chevron
Pitch: 6.94 inches
Shoes per Vehicle: 158 (79/track)
Ground Contact Length: 157.5 inches

ELECTRICAL SYSTEM

Nominal Voltage: 24 volts DC
Main Generator: (1) 24 volts, 300 amperes, gear driven by main engine
Auxiliary Generator: (1) 28 volts, 300 amperes driven by auxiliary
engine
Battery: (4) 12 volts, 2 sets of 2 in series connected in parallel

COMMUNICATIONS

Radio: AN/GRC-3 thru 8 series or AN/VRC-7 in turret bustle
AN/ARC-3 or AN/ARC-27 (air to ground) also may be fitted
Interphone: 4 stations plus external extension kit AN/VIA-1

FIRE PROTECTION

(3) 10 pound carbon dioxide, fixed
(1) 5 pound carbon dioxide, portable

PERFORMANCE

Maximum Speed: Sustained, level road		28 miles/hour
Maximum Tractive Effort: TE at stall		83,000 pounds
Per Cent of Vehicle Weight: TE/W		80 per cent
Maximum Grade:		60 per cent
Maximum Trench:		8.5 feet
Maximum Vertical Wall:		36 inches
Maximum Fording Depth:		48 inches
Minimum Turning Circle: (diameter)		pivot
Cruising Range: Roads	approx.	70 miles
	w/jettison fuel tanks approx.	135 miles

*The surface angles and wall thicknesses of the cast hull and turret varied widely from point to point. Therefore, where indicated the data are for an equivalent thickness at 0 degrees from the vertical.

105mm GUN TANK T54

GENERAL DATA

Crew:	4	men
Length: Gun forward w/fenders	450.2	inches
Length: Gun in travel position, w/fenders	395.9	inches
Length: Without gun, w/fenders	274.3	inches
Gun Overhang: Gun forward	175.9	inches
Width: Over tracks	143.0	inches
Height: Over cupola periscope	121.8	inches
Tread:	115.0	inches
Ground Clearance:	16.5	inches
Fire Height: approx.	79	inches
Turret Ring Diameter: (inside)	85.0	inches
Weight, Combat Loaded: approx.	110,000	pounds
Weight, Unstowed: approx.	103,000	pounds
Power to Weight Ratio: Net	11.7	hp/ton
Gross	13.6	hp/ton
Ground Pressure: Zero penetration	12.5	psi

*ARMOR

Type: Turret, cast homogeneous steel; Hull, cast homogeneous steel; Welded assembly

Hull Thickness:	Actual	Angle w/Vertical
Front, Upper	4.33 inches (110mm)	60 degrees
Lower	4.0 to 2.4 inches (102-61mm)	53 degrees
Sides, Front	equals 3.0 inches (76mm)	0 degrees
Rear	equals 2.0 inches (51mm)	0 degrees
Rear, Upper	1.38 inches (35mm)	30 degrees
Lower	1.0 inches (25mm)	60 degrees
Top	2.25 inches (57mm)	90 degrees
Floor, Front	1.5 inches (38mm)	90 degrees
Rear	1.0 inches (25mm)	90 degrees
Turret Thickness:		
Gun Shield	equals 4.3 inches (108mm)	50 degrees
Front	4.3 inches (108mm)	50 degrees
Sides	2.8 inches (70mm)	25 degrees
Rear	3.5 inches (89mm)	45 degrees
Top	1.3 inches (32mm)	90 degrees

ARMAMENT

Primary: 105mm Gun T140 in Mount T156 in turret

Traverse: Hydraulic and manual	360 degrees
Traverse Rate: (max)	10 seconds/360 degrees
Elevation: Hydraulic and manual	+ 15 to − 10 degrees
Elevation Rate: (max)	4 degrees/second
Firing Rate: (max w/auto load)	30 rounds/minute (design rate)
Loading System:	Automatic w/14 round magazine
Stabilizer System:	None

Secondary:

(1) .50 caliber MG HB M2 in cupola mount on turret

(1) .30 caliber MG M1919A4E1 coaxial/105mm gun in turret

Provision for (1) .45 caliber SMG M3A1

Provision for (1) .30 caliber Carbine M2 w/grenade launcher

AMMUNITION

47 rounds 105mm	8 hand grenades
1850 rounds .50 caliber	
180 rounds .45 caliber	
4500 rounds .30 caliber	
180 rounds .30 caliber (carbine)	

FIRE CONTROL AND VISION EQUIPMENT

Primary Weapon:	Direct	Indirect
	Range Finder T46E3	Azimuth Indicator T28
	Periscope M16E1	Elevation Quadrant M13 (T21)
	Ballistic Drive T24 mod.	Gunner's Quadrant M1 or M1A1
	Ballistic Computer T32	
	Telescope T174	

Vision Devices	Direct	Indirect
Driver	Hatch	Periscope M7 (T36) (3) and Periscope T41 infrared
Commander	Vision blocks (5) in cupola, hatch	Periscope T42 (AA MG sight)
Gunner	None	Periscope M16E1
Loader	Hatch	Periscope M13

Total Periscopes: M7 (T36) (3), T41 infrared (1), M16E1 (1), M13 (1), T42 (1)

Total Vision Blocks: (5) in Aircraft Armaments Mod. 108 cupola on turret top

ENGINE

Make and Model: Continental AV-1790-5B	
Type: 12 cylinder, 4 cycle, 90 degree vee	
Cooling System: Air Ignition: Magneto	
Displacement:	1791.7 cubic inches
Bore and Stroke:	5.75 x 5.75 inches
Compression Ratio:	6.5:1
Net Horsepower (max):	704 hp at 2800 rpm
Gross Horsepower (max):	810 hp at 2800 rpm
Net Torque (max):	1440 ft-lb at 2000 rpm
Gross Torque (max):	1610 ft-lb at 2200 rpm
Weight:	2505 pounds, dry
Fuel: 80 octane gasoline	200 gallons
Engine Oil:	64 quarts

POWER TRAIN

Transmission: Cross-drive CD-850-4, 2 ranges forward, 1 reverse

Single stage multiphase hydraulic torque converter

Stall multiplication: 4.3:1

Overall Usable Ratios: low 13.0:1   reverse: 17.8:1

high 4.5:1

Steering Control: Mechanical, steering wheel

Steering Rate: 5.7 rpm

Brakes: Multiple disc

Final Drive: Spur gear   Gear Ratio: 5.08:1

Drive Sprocket: At rear of vehicle with 11 teeth

Pitch Diameter: 24.504 inches

RUNNING GEAR

Suspension: Torsion bar

12 individually sprung dual road wheels (6/track)

Tire Size: 26 x 6 inches

10 dual track return rollers (5/track)

Dual compensating idler at front of each track

Idler Tire Size: 26 x 6 inches

Dual track tension idler in front of each sprocket

Shock absorbers fitted on first 2 and last road wheels on each side

Tracks: Center guide, T96 and T97

Type: (T96) Double pin, 28 inch width, rubber backed steel

(T97) Double pin, 28 inch width, rubber chevron

Pitch: 6.94 inches

Shoes per Vehicle: 158 (79/track)

Ground Contact Length: 157.5 inches

ELECTRICAL SYSTEM

Nominal Voltage: 24 volts DC

Main Generator: (1) 24 volts, 150 amperes, gear driven by main engine

Auxiliary Generator: (1) 28 volts, 300 amperes driven by auxiliary engine

Battery: (4) 12 volts, 2 sets of 2 in series connected in parallel

COMMUNICATIONS

Radio: AN/GRC-3 thru 8 series or AN/VRC-7 in turret bustle

AN/ARC-3 or AN/ARC-27 (air to ground) also may be fitted

Interphone: 4 stations plus external extension kit AN/VIA-1

FIRE PROTECTION

(3) 10 pound carbon dioxide, fixed

(1) 5 pound carbon dioxide, portable

PERFORMANCE

Maximum Speed: Sustained, level road	27 miles/hour
Maximum Tractive Effort: TE at stall	78,000 pounds
Per Cent of Vehicle Weight: TE/W	71 per cent
Maximum Grade:	60 per cent
Maximum Trench:	8.5 feet
Maximum Vertical Wall:	36 inches
Maximum Fording Depth:	48 inches
Minimum Turning Circle: (diameter)	pivot
Cruising Range: Roads approx.	70 miles

---

*The surface angles and wall thicknesses of the cast hull and turret varied widely from point to point. Therefore, where indicated the data are for an equivalent thickness at 0 degrees from the vertical.

105mm GUN TANK T54E1

GENERAL DATA

Crew:	4	men
Length: Gun forward, w/fenders	441.3	inches
Length: Gun in travel position, w/fenders	394.3	inches
Length: Without gun, w/fenders	274.3	inches
Gun Overhang: Gun forward	167.0	inches
Width: Over tracks	143.0	inches
Height: Over AA MG mount w/o MG	121.3	inches
Tread:	115.0	inches
Ground Clearance:	16.5	inches
Fire Height:	approx. 93	inches
Turret Ring Diameter: (inside)	85.0	inches
Weight, Combat Loaded:	approx. 120,000	pounds
Weight, Unstowed:	approx. 113,000	pounds
Power to Weight Ratio: Net	10.7	hp/ton
Gross	12.5	hp/ton
Ground Pressure: Zero penetration	13.6	psi

*ARMOR

Type: Turret, cast homogeneous steel; Hull, cast homogeneous
steel; Welded assembly

Hull Thickness:	Actual	Angle w/Vertical
Front, Upper	4.33 inches (110mm)	60 degrees
Lower	4.0 to 2.4 inches (102-61mm)	53 degrees
Sides, Front	equals 3.0 inches (76mm)	0 degrees
Rear	equals 2.0 inches (51mm)	0 degrees
Rear, Upper	1.38 inches (35mm)	30 degrees
Lower	1.0 inches (25mm)	60 degrees
Top	2.25 inches (57mm)	90 degrees
Floor, Front	1.5 inches (38mm)	90 degrees
Rear	1.0 inches (25mm)	90 degrees
Turret Thickness:		
Front	5.0 inches (127mm)	60 degrees
Sides	equals 2.5 inches (64mm)	30 degrees
Rear	2.0 inches (51mm)	30 degrees
Top	1.0 inches (25mm)	90 degrees

ARMAMENT

Primary: 105mm Gun T140E2 in Mount T157 in turret

Traverse: Amplidyne and manual	360 degrees
Traverse Rate: (max)	20 seconds/360 degrees
Elevation: Amplidyne and manual	+ 13 to − 8 degrees
Elevation Rate: (max)	4 degrees/second
Firing Rate: (max w/auto load)	35 rounds/minute (design rate)
Loading System:	Automatic w/9 round magazine
Stabilizer System:	None

Secondary:
(1) .50 caliber MG HB M2 flexible AA mount on cupola
(1) .30 caliber MG M1919A4E1 coaxial w/105mm gun in turret
Provision for (1) .45 caliber SMG M3A1
Provision for (1) .30 caliber Carbine M2 w/grenade launcher

AMMUNITION

36 rounds 105mm 8 hand grenades
900 rounds .50 caliber
180 rounds .45 caliber
3000 rounds .30 caliber
90 rounds .30 caliber (carbine)

FIRE CONTROL AND VISION EQUIPMENT

Primary Weapon:	Direct	Indirect
	Range Finder T50	Azimuth Indicator T28
	Periscope M20 (T35)	Elevation Quadrant M13
	Range Drives T32E1	(T21)
	& T33E1	Gunner's Quadrant
	Ballistic Computer T34	M1 or M1A1
	Telescope T170E1	

Vision Devices	Direct	Indirect
Driver	Hatch	Periscope M7 (T36) (3), and Periscope T41 infrared
Commander	Vision blocks (6) in cupola, hatch	Periscope M20 (T35)
Gunner	None	Periscope M20 (T35)
Loader	Hatch	Periscope M13

Total Periscopes: M27 (T36) (3), T41 infrared (1), M20 (T35) (2), M13 (1)
Total Vision Blocks: (6) in cupola on turret top

ENGINE

Make and Model: Continental AV-1790-5B	
Type: 12 cylinder, 4 cycle, 90 degree vee	
Cooling System: Air Ignition: Magneto	
Displacement:	1791.7 cubic inches
Bore and Stroke:	5.75 x 5.75 inches
Compression Ratio:	6.5:1
Net Horsepower (max):	704 hp at 2800 rpm
Gross Horsepower (max):	810 hp at 2800 rpm
Net Torque (max):	1440 ft-lb at 2000 rpm
Gross Torque (max):	1610 ft-lb at 2200 rpm
Weight:	2505 pounds, dry
Fuel: 80 octane gasoline	200 gallons
Engine Oil:	64 quarts

POWER TRAIN

Transmission: Cross-drive CD-850-4, 2 ranges forward, 1 reverse
Single stage multiphase hydraulic torque converter
Stall multiplication: 4.3:1
Overall Usable Ratios: low 13.0:1   reverse: 17.8:1
high 4.5:1
Steering Control: Mechanical, steering wheel
Steering Rate: 5.7 rpm
Brakes: Multiple disc
Final Drive: Spur gear   Gear Ratio: 5.08:1
Drive Sprocket: At rear of vehicle with 11 teeth
Pitch Diameter: 24.504 inches

RUNNING GEAR

Suspension: Torsion bar
12 individually sprung dual road wheels (6/track)
Tire Size: 26 x 6 inches
10 dual track return rollers (5/track)
Dual compensating idler at front of each track
Idler Tire Size: 26 x 6 inches
Shock absorbers fitted on first 2 and last road wheels on each side
Tracks: Center guide, T96 and T97
Type: (T96) Double pin, 28 inch width, rubber backed steel
(T97) Double pin, 28 inch width, rubber chevron
Pitch: 6.94 inches
Shoes per Vehicle: 156 (78/track)
Ground Contact Length: 157.5 inches

ELECTRICAL SYSTEM

Nominal Voltage: 24 volts DC
Main Generator: (1) 24 volts, 150 amperes, gear driven by main engine
Auxiliary Generator: (1) 28 volts, 300 amperes driven by auxiliary engine
Battery: (4) 12 volts, 2 sets of 2 in series connected in parallel

COMMUNICATIONS

Radio: AN/GRC-3 thru 8 series in turret bustle
Interphone: 4 stations plus external extension kit AN/VIA-1

FIRE PROTECTION

(3) 10 pound carbon dioxide, fixed
(1) 5 pound carbon dioxide, portable

PERFORMANCE

Maximum Speed: Sustained, level road	27 miles/hour
Maximum Tractive Effort: TE at stall	78,000 pounds
Per Cent of Vehicle Weight: TE/W	65 per cent
Maximum Grade:	60 per cent
Maximum Trench:	8.5 feet
Maximum Vertical Wall:	36 inches
Maximum Fording Depth:	48 inches
Minimum Turning Circle: (diameter)	pivot
Cruising Range: Roads	approx. 70 miles

*The surface angles and wall thicknesses of the cast hull and turret varied widely from point to point. Therefore, where indicated the data are for an equivalent thickness at 0 degrees from the vertical.

105mm GUN TANK T54E2

GENERAL DATA

Crew:		4	men
Length: Gun forward, w/fenders		452.1	inches
Length: Gun in travel position, w/fenders		397.9	inches
Length: Without gun, w/fenders		274.3	inches
Gun Overhang: Gun forward		177.8	inches
Width: Over tracks		143.0	inches
Height: Over cupola periscope		123.1	inches
Tread:		115.0	inches
Ground Clearance:		17.5	inches
Fire Height:	approx.	79	inches
Turret Ring Diameter: (inside)		85.0	inches
Weight, Combat Loaded:	approx.	109,000	pounds
Weight, Unstowed:	approx.	102,000	pounds
Power to Weight Ratio: Net		11.8	hp/ton
Gross		13.8	hp/ton
Ground Pressure: Zero penetration		12.4	psi

*ARMOR

Type: Turret, cast homogeneous steel; Hull, cast homogeneous
    steel; Welded assembly

Hull Thickness:	Actual	Angle w/Vertical
Front, Upper	4.33 inches (110mm)	60 degrees
Lower	4.0 to 2.4 inches (102-61mm)	53 degrees
Sides, Front	equals 3.0 inches (76mm)	0 degrees
Rear	equals 2.0 inches (51mm)	0 degrees
Rear, Upper	1.38 inches (35mm)	30 degrees
Lower	1.0 inches (25mm)	60 degrees
Top	2.25 inches (57mm)	90 degrees
Floor, Front	1.5 inches (38mm)	90 degrees
Rear	1.0 inches (25mm)	90 degrees
Turret Thickness:		
Gun Shield	equals 4.0 inches (102mm)	60 degrees
Front	4.0 inches (102mm)	60 degrees
Sides	2.0 to 3.5 inches (51-89mm)	10 degrees
Rear	2.5 inches (64mm)	0 degrees
Top	1.0 inches (25mm)	90 degrees

ARMAMENT

Primary: 105mm Gun T140E3 in Mount T174 in turret
    Traverse: Hydraulic or amplidyne and manual 360 degrees

Traverse Rate: (max)	17 seconds/360 degrees
Elevation: Hydraulic or amplidyne & manual	+ 20 to – 10 degrees
Elevation Rate: (max)	4 degrees/second
Firing Rate: (max)	6 rounds/minute
Loading System:	Manual
Stabilizer System:	None

Secondary:
    (1) .50 caliber MG HB M2 in cupola mount on turret
    (1) .30 caliber MG M1919A4E1 coaxial w/105 mm gun in turret
    Provision for (1) .45 caliber SMG M3A1
    Provision for (1) .30 caliber Carbine M2 w/grenade launcher

AMMUNITION

39 rounds 105mm	8 hand grenades
1850 rounds .50 caliber	
180 rounds .45 caliber	
4750 rounds .30 caliber	
180 rounds .30 caliber (carbine)	

FIRE CONTROL AND VISION EQUIPMENT

Primary Weapon:	Direct	Indirect
	Range Finder T46E3	Azimuth Indicator T28
	Periscope M16E1	Elevation Quadrant M13
	Ballistic Drive T37	(T21)
	Ballistic Computer T32	Gunner's Quadrant
	Telescope T156E2	M1 or M1A1
Vision Devices	Direct	Indirect
Driver	Hatch	Periscope M7 (T36) (3), and
		Periscope T41 infrared
Commander	Vision blocks (5)	Periscope T42
	in cupola, hatch	(AA MG sight)
Gunner	None	Periscope M16E1
Loader	Hatch	None

Total Periscopes: M27 (T36) (3), T41 infrared (1), M16E1 (1), T42 (1)
Total Vision Blocks: (5) in Aircraft Armaments Mod. 108 cupola on turret top

ENGINE

Make and Model: Continental AV-1790-5B	
Type: 12 cylinder, 4 cycle, 90 degree vee	
Cooling System: Air Ignition: Magneto	
Displacement:	1791.7 cubic inches
Bore and Stroke:	5.75 x 5.75 inches
Compression Ratio:	6.5:1
Net Horsepower (max):	704 hp at 2800 rpm
Gross Horsepower (max):	810 hp at 2800 rpm
Net Torque (max):	1440 ft-lb at 2000 rpm
Gross Torque (max):	1610 ft-lb at 2200 rpm
Weight:	2505 pounds, dry
Fuel: 80 octane gasoline	200 gallons
Engine Oil:	64 quarts

POWER TRAIN

Transmission: Cross-drive CD-850-4, 2 ranges forward, 1 reverse
    Single stage multiphase hydraulic torque converter
    Stall multiplication: 4.3:1

Overall Usable Ratios:	low 13.0:1	reverse:	17.8:1
	high 4.5:1		

Steering Control: Mechanical, steering wheel
    Steering Rate: 5.7 rpm
Brakes: Multiple disc
Final Drive: Spur gear   Gear Ratio: 5.08:1
Drive Sprocket: At rear of vehicle with 11 teeth
    Pitch Diameter: 24.504 inches

RUNNING GEAR

Suspension: Torsion bar
    12 individually sprung dual road wheels (6/track)
    Tire Size: 26 x 6 inches
    10 dual track return rollers (5/track)
    Dual compensating idler at front of each track
    Idler Tire Size: 26 x 6 inches
    Shock absorbers fitted on first 2 and last road wheels on each side
Tracks: Center guide, T96 and T97
    Type: (T96) Double pin, 28 inch width, rubber backed steel
    (T97) Double pin, 28 inch width, rubber chevron
    Pitch: 6.94 inches
    Shoes per Vehicle: 156 (78/track)
    Ground Contact Length: 157.5 inches

ELECTRICAL SYSTEM

Nominal Voltage: 24 volts DC
Main Generator: (1) 24 volts, 150 amperes, gear driven by main engine
Auxiliary Generator: (1) 28 volts, 300 amperes driven by auxiliary
    engine
Battery: (4) 12 volts, 2 sets of 2 in series connected in parallel

COMMUNICATIONS

Radio: AN/GRC-3 thru 8 series in turret bustle
Interphone: 4 stations plus external extension kit AN/VIA-1

FIRE PROTECTION

    (3) 10 pound carbon dioxide, fixed
    (1) 5 pound carbon dioxide, portable

PERFORMANCE

Maximum Speed: Sustained, level road	27 miles/hour
Maximum Tractive Effort: TE at stall	78,000 pounds
Per Cent of Vehicle Weight: TE/W	72 per cent
Maximum Grade:	60 per cent
Maximum Trench:	8.5 feet
Maximum Vertical Wall:	36 inches
Maximum Fording Depth:	48 inches
Minimum Turning Circle: (diameter)	pivot
Cruising Range: Roads	approx. 70 miles

---

*The surface angles and wall thicknesses of the cast hull and turret varied widely from point to point. Therefore, where indicated the data are for an
  equivalent thickness at 0 degrees from the vertical.

120mm GUN TANK T77

GENERAL DATA

Crew:	4	men
Length: Gun forward, w/fenders	449.3	inches
Length: Gun in travel position, w/fenders	395.0	inches
Length: Without gun, w/fenders	274.3	inches
Gun Overhang: Gun forward	175.0	inches
Width: Over tracks	143.0	inches
Height: Over AA MG w/o MG	119.1	inches
Tread:	115.0	inches
Ground Clearance:	16.5	inches
Fire Height:	approx. 89	inches
Turret Ring Diameter: (inside)	85.0	inches
Weight, Combat Loaded:	approx. 115,000	pounds
Weight, Unstowed:	approx. 108,000	pounds
Power to Weight Ratio: Net	11.2	hp/ton
Gross	13.0	hp/ton
Ground Pressure: Zero penetration	13.0	psi

*ARMOR

Type: Turret, cast homogeneous steel; Hull, cast homogeneous
    steel; Welded assembly

Hull Thickness:	Actual	Angle w/Vertical
Front, Upper	4.33 inches (110mm)	60 degrees
Lower	4.0 to 2.4 inches (102-61mm)	53 degrees
Sides, Front	equals 3.0 inches (76mm)	0 degrees
Rear	equals 2.0 inches (51mm)	0 degrees
Rear, Upper	1.38 inches (35mm)	30 degrees
Lower	1.0 inches (25mm)	60 degrees
Top	2.25 inches (57mm)	90 degrees
Floor, Front	1.5 inches (38mm)	90 degrees
Rear	1.0 inches (25mm)	90 degrees
Turret Thickness:		
Front	5.0 inches (127mm)	50 degrees
Sides	5.4 inches (137mm)	20 to 40 degrees
Rear	equals 2.0 inches (51mm)	40 degrees
Top	1.5 inches (38mm)	90 degrees

ARMAMENT

Primary: 120mm Gun T179 in Mount T169 (rigid) in turret

Traverse: Hydraulic and manual	360 degrees
Traverse Rate: (max)	15 seconds/360 degrees
Elevation: Hydraulic and manual	+15 to −8 degrees
Elevation Rate: (max)	4 degrees/second
Firing Rate: (max w/auto load)	30 rounds/minute (design rate)
Loading System:	Automatic w/8 round magazine
Stabilizer System:	None

Secondary:
    (1) .50 caliber MG HB M2 flexible AA on turret hatch
    (1) .30 caliber MG M1919A4E1 coaxial w/120 mm gun in turret
    Provision for (1) .45 caliber SMG M3A1
    Provision for (1) .30 caliber Carbine M2 w/grenade launcher

AMMUNITION

18 rounds 120mm	12 hand grenades
3425 rounds .50 caliber	
180 rounds .45 caliber	
3000 rounds .30 caliber	
180 rounds .30 caliber (carbine)	

FIRE CONTROL AND VISION EQUIPMENT

Primary Weapon:	Direct	Indirect
	Range Finder T50	Azimuth Indicator T28
	Periscope M16E1	Elevation Quadrant M13
	Range Drive T33E2	(T21)
	Ballistic Computer T30	Gunner's Quadrant
	Telescope T173	M1 or M1A1
Vision Devices	Direct	Indirect
Driver	Hatch	Periscope M7 (T36) (3), and Periscope T41 infrared
Commander	Hatch	Periscope M7 (T36) (6)
Gunner	None	Periscope M16E1
Loader	Hatch	Periscope M13

Total Periscopes: M27 (T36) (9), T41 infrared (1), M16E1 (1), M13 (1)
Total Vision Blocks: None

ENGINE

Make and Model: Continental AV-1790-5B	
Type: 12 cylinder, 4 cycle, 90 degree vee	
Cooling System: Air Ignition: Magneto	
Displacement:	1791.7 cubic inches
Bore and Stroke:	5.75 x 5.75 inches
Compression Ratio:	6.5:1
Net Horsepower (max):	704 hp at 2800 rpm
Gross Horsepower (max):	810 hp at 2800 rpm
Net Torque (max):	1440 ft-lb at 2000 rpm
Gross Torque (max):	1610 ft-lb at 2200 rpm
Weight:	2505 pounds, dry
Fuel: 80 octane gasoline	200 gallons
Engine Oil:	64 quarts

POWER TRAIN

Transmission: Cross-drive CD-850-4, 2 ranges forward, 1 reverse
    Single stage multiphase hydraulic torque converter
    Stall multiplication: 4.3:1
    Overall Usable Ratios: low 13.0:1   reverse: 17.8:1
        high 4.5:1
Steering Control: Mechanical, steering wheel
Steering Rate: 5.7 rpm
Brakes: Multiple disc
Final Drive: Spur gear   Gear Ratio: 5.08:1
Drive Sprocket: At rear of vehicle with 11 teeth
    Pitch Diameter: 24.504 inches

RUNNING GEAR

Suspension: Torsion bar
    12 individually sprung dual road wheels (6/track)
    Tire Size: 26 x 6 inches
    10 dual track return rollers (5/track)
    Dual compensating idler at front of each track
    Idler Tire Size: 26 x 6 inches
    Shock absorbers fitted on first 2 and last road wheels on each side
Tracks: Center guide, T96 and T97
    Type: (T96) Double pin, 28 inch width, rubber backed steel
    (T97) Double pin, 28 inch width, rubber chevron
    Pitch: 6.94 inches
    Shoes per Vehicle: 156 (78/track)
    Ground Contact Length: 157.5 inches

ELECTRICAL SYSTEM

Nominal Voltage: 24 volts DC
Main Generator: (1) 24 volts, 150 amperes, gear driven by main engine
Auxiliary Generator: (1) 28 volts, 300 amperes driven by auxiliary
    engine
Battery: (4) 12 volts, 2 sets of 2 in series connected in parallel

COMMUNICATIONS

Radio: AN/GRC-3 thru 8 series in turret bustle
Interphone: 4 stations plus external extension kit AN/VIA-1

FIRE PROTECTION

    (3) 10 pound carbon dioxide, fixed
    (1) 5 pound carbon dioxide, portable

PERFORMANCE

Maximum Speed: Sustained, level road	27 miles/hour
Maximum Tractive Effort: TE at stall	78,000 pounds
Per Cent of Vehicle Weight: TE/W	68 per cent
Maximum Grade:	60 per cent
Maximum Trench:	8.5 feet
Maximum Vertical Wall:	36 inches
Maximum Fording Depth:	48 inches
Minimum Turning Circle: (diameter)	pivot
Cruising Range: Roads	approx. 70 miles

*The surface angles and wall thicknesses of the cast hull and turret varied widely from point to point. Therefore, where indicated the data are for an equivalent thickness at 0 degrees from the vertical.

90mm GUN TANKS M48A2 and M48A2C

GENERAL DATA

Crew:	4	men
Length: Gun forward	341.8	inches
Length: Gun in travel position	292.8	inches
Length: Without gun	270.5	inches
Gun Overhang: Gun forward	71.3	inches
Width: Over tracks	143.0	inches
Height: Over cupola periscope	121.6	inches
Tread:	115.0	inches
Ground Clearance:	16.5	inches
Fire Height: approx.	79	inches
Turret Ring Diameter: (inside)	85.0	inches
Weight, Combat Loaded:	105,000	pounds
Weight, Unstowed:	99,200	pounds
Power to Weight Ratio: Net	13.1	hp/ton
Gross	15.7	hp/ton
Ground Pressure: Zero penetration	11.9	psi

*ARMOR

Type: Turret, cast homogeneous steel; Hull, cast homogeneous steel; Welded assembly

Hull Thickness:	Actual	Angle w/Vertical
Front, Upper	4.33 inches (110mm)	60 degrees
Lower	4.0 to 2.4 inches (102-61mm)	53 degrees
Sides, Front	equals 3.0 inches (76mm)	0 degrees
Rear	equals 2.0 inches (51mm)	0 degrees
Rear, Grill	equals 1.0 inches (25mm)	0 degrees
Lower	1.6 to 1.2 inches (41-30mm)	30 to 60 degrees
Top	2.25 inches (57mm)	90 degrees
Floor, Front	1.5 inches (38mm)	90 degrees
Rear	1.0 inches (25mm)	90 degrees
Turret Thickness:		
Gun Shield	4.5 inches (114mm)	30 degrees
Front	equals 7.0 inches (178mm)	0 degrees
Sides	equals 3.0 inches (76mm)	0 degrees
Rear	equals 2.0 inches (51mm)	0 degrees
Top	1.0 inches (25mm)	90 degrees

ARMAMENT

Primary: 90mm Gun M41 in Mount M87 in turret

Traverse: Hydraulic and manual	360 degrees
Traverse Rate: (max)	15 seconds/360 degrees
Elevation: Hydraulic and manual	+ 19 to − 9 degrees
Elevation Rate: (max)	4 degrees/second
Firing Rate: (max)	8 rounds/minute
Loading System:	Manual
Stabilizer System:	None

Secondary:
- (1) .50 caliber MG HB M2 in cupola mount on turret
- (1) .30 caliber MG M37 coaxial w/90mm gun in turret
- Provision for (1) .45 caliber SMG M3A1
- Provision for (1) .30 caliber Carbine M2

AMMUNITION

64 rounds 90mm	8 hand grenades
1360 rounds .50 caliber	
180 rounds .45 caliber	
5950 rounds .30 caliber	
180 rounds .30 caliber (carbine)	

FIRE CONTROL AND VISION EQUIPMENT

Primary Weapon:	Direct	Indirect
(M48A2)	Range Finder M13A1	Azimuth Indicator
(M48A2C)	Range Finder M17	M28A1
	(M13A1E1) (M48A2)	Elevation Quadrant M13
(M48A2)	Periscope M20A3 (M48A2C)	Elevation Quadrant
(M48A2C)	Periscope M32	M13B1
	(day & IR)	Gunner's Quadrant
	Ballistic Drive M5A1	M1A1
	Ballistic Computer	
	M13A1	
(M48A2)	Telescope M97C	
(M48A2C)	Telescope M105	

Vision Devices	Direct	Indirect
Driver	Hatch	Periscope M7 (T36) (3), and Periscope M24 (1) infrared
Commander	Vision blocks (5) in cupola, hatch	Periscope M28 (AA MG sight)
Gunner	None	(M48A2) Periscope M20A3 (M48A2C) Periscope M32 (day & IR)
Loader	Hatch	None

Total Periscopes: M27 (3), M24 infrared (1), M20A3 or M32 (1), M28 (1)
Total Vision Blocks: (5) in M1 cupola on turret top

ENGINE

Make and Model: Continental AVI-1790-8	
Type: 12 cylinder, 4 cycle, 90 degree vee, fuel injection	
Cooling System: Air Ignition: Magneto	
Displacement:	1791.7 cubic inches
Bore and Stroke:	5.75 x 5.75 inches
Compression Ratio:	6.35:1
Net Horsepower (max):	690 hp at 2800 rpm
Gross Horsepower (max):	825 hp at 2800 rpm
Net Torque (max):	1470 ft-lb at 2200 rpm
Gross Torque (max):	1670 ft-lb at 2200 rpm
Weight:	2975 pounds, dry
Fuel: 80 octane gasoline	335 gallons
Engine Oil:	64 quarts

POWER TRAIN

Transmission: Cross-drive CD-850-5, 2 ranges forward, 1 reverse
Single stage multiphase hydraulic torque converter
Stall multiplication: 4:1

Overall Usable Ratios: low 13.0:1   reverse: 18.1:1
high 4.5:1

Steering Control: Mechanical, steering wheel
Steering Rate: 5.7 rpm
Brakes: Multiple disc
Final Drive: Spur gear   Gear Ratio: 5.08:1
Drive Sprocket: At rear of vehicle with 11 teeth
Pitch Diameter: 24.504 inches

RUNNING GEAR

Suspension: Torsion bar
- 12 individually sprung dual road wheels (6/track)
- Tire Size: 26 x 6 inches
- 6 dual track return rollers (3/track)
- Dual compensating idler at front of each track
- Idler Tire Size: 26 x 6 inches
- Dual track tension idler in front of each sprocket (M48A2, none on M48A2C)
- Friction snubbers fitted on first 2 and last road wheels on each side

Tracks: Center guide, T97E2
- Type: (T97E2) Double pin, 28 inch width, rubber chevron
- Pitch: 6.94 inches
- Shoes per Vehicle: 158 (79/track)
- Ground Contact Length: 157.5 inches

ELECTRICAL SYSTEM

Nominal Voltage: 24 volts DC
Main Generator: (1) 24 volts, 300 amperes, gear driven by main engine
Auxiliary Generator: (1) 28 volts, 300 amperes driven by auxiliary engine
Battery: (4) 12 volts, 2 sets of 2 in series connected in parallel

COMMUNICATIONS

Radio: AN/GRC-3 thru 8 series or AN/VRC-7 in turret bustle
AN/ARC-3 or AN/ARC-27 (air to ground) also may be fitted
Interphone: AN/VIC-1 (3 stations) plus external extension kit AN/VIA-4

FIRE PROTECTION

(3) 10 pound carbon dioxide, fixed
(1) 5 pound carbon dioxide, portable

PERFORMANCE

Maximum Speed: Sustained, level road		30 miles/hour
Maximum Tractive Effort: TE at stall		84,000 pounds
Per Cent of Vehicle Weight: TE/W		80 per cent
Maximum Grade:		60 per cent
Maximum Trench:		8.5 feet
Maximum Vertical Wall:		36 inches
Maximum Fording Depth:		40 inches
Minimum Turning Circle: (diameter)		pivot
Cruising Range: Roads	approx.	160 miles
	w/jettison fuel tanks approx.	250 miles

*The surface angles and wall thicknesses of the cast hull and turret varied widely from point to point. Therefore, where indicated the data are for an equivalent thickness at 0 degrees from the vertical.

105mm GUN TANK M48A1E1

GENERAL DATA

Crew:	4	men
Length: Gun forward	366.5	inches
Length: Gun in travel position	317.0	inches
Length: Without gun	270.5	inches
Gun Overhang: Gun forward	96.0	inches
Width: Over fenders	143.0	inches
Height: Over cupola periscope	123.5	inches
Tread:	115.0	inches
Ground Clearance:	14.5	inches
Fire Height: approx.	79	inches
Turret Ring Diameter: (inside)	85.0	inches
Weight, Combat Loaded:	105,200	pounds
Weight, Unstowed:	95,500	pounds
Power to Weight Ratio: Net	12.2	hp/ton
Gross	14.3	hp/ton
Ground Pressure: Zero penetration	11.9	psi

*ARMOR

Type: Turret, cast homogeneous steel; Hull, cast homogeneous
steel; Welded assembly

Hull Thickness:	Actual	Angle w/Vertical
Front, Upper	4.33 inches (110mm)	60 degrees
Lower	4.0 to 2.4 inches (102-61mm)	53 degrees
Sides, Front	equals 3.0 inches (76mm)	0 degrees
Rear	equals 2.0 inches (51mm)	0 degrees
Rear, Grill	equals 1.0 inches (25mm)	0 degrees
Lower	1.6 to 1.2 inches (41-30mm)	30 to 60 degrees
Top	2.25 inches (57mm)	90 degrees
Floor, Front	1.5 inches (38mm)	90 degrees
Rear	1.0 inches (25mm)	90 degrees
Turret Thickness:		
Gun Shield	4.5 inches (114mm)	30 degrees
Front	equals 7.0 inches (178mm)	0 degrees
Sides	equals 3.0 inches (76mm)	0 degrees
Rear	equals 2.0 inches (51mm)	0 degrees
Top	1.0 inches (25mm)	90 degrees

ARMAMENT

Primary: 105mm Gun M68 in modified Mount M87 in turret

Traverse: Hydraulic and manual	360 degrees
Traverse Rate: (max)	15 seconds/360 degrees
Elevation: Hydraulic and manual	+ 19 to − 9 degrees
Elevation Rate: (max)	4 degrees/second
Firing Rate: (max)	7 rounds/minute
Loading System:	Manual
Stabilizer System:	None

Secondary:
(1) .50 caliber MG HB M2 in cupola mount on turret
(1) 7.62mm MG M73 coaxial w/105mm gun in turret
Provision for (2) .45 caliber SMG M3A1

AMMUNITION

57 rounds 105mm	8 hand grenades
1040 rounds .50 caliber	
360 rounds .45 caliber	
5000 rounds 7.62mm	

FIRE CONTROL AND VISION EQUIPMENT

Primary Weapon:	Direct	Indirect
	Range Finder M17C	Azimuth Indicator
	Periscope M31	M28A1
	Ballistic Drive M10	Elevation Quadrant
	Ballistic Computer	M13A1
	M13A1D	Gunner's Quadrant
	Telescope M105C	M1A1
Vision Devices	Direct	Indirect
Driver	Hatch	Periscope M27 (3) and
		Periscope M24 (1)
		infrared
Commander	Vision blocks (5)	Periscope M28C
	in cupola, hatch	(AA MG sight)
Gunner	None	Periscope M31
Loader	Hatch	None

Total Periscopes: M27 (3), M24 infrared (1), M31 (1), M28C (1)
Total Vision Blocks: (5) in M1 cupola on turret top

ENGINE

Make and Model: Continental AVDS-1790-2	
Type: 12 cylinder, 4 cycle, 90 degree vee, supercharged	
Cooling System: Air Ignition: Compression	
Displacement:	1791.7 cubic inches
Bore and Stroke:	5.75 x 5.75 inches
Compression Ratio:	16:1
Net Horsepower (max):	643 hp at 2400 rpm
Gross Horsepower (max):	750 hp at 2400 rpm
Net Torque (max):	1575 ft-lb at 1750 rpm
Gross Torque (max):	1710 ft-lb at 1800 rpm
Weight:	4527 pounds, dry
Fuel: 40 cetane diesel	385 gallons
Engine Oil:	64 quarts

POWER TRAIN

Transmission: Cross-drive CD-850-6A, 2 ranges forward, 1 reverse
Single stage multiphase hydraulic torque converter
Stall multiplication: 4:1
Overall Usable Ratios: low 12.0:1   reverse: 16.9:1
high 4.3:1

Steering Control: Mechanical, steering wheel	
Steering Rate: 5.7 rpm	
Brakes: Multiple disc	
Final Drive: Spur gear Gear Ratio: 5.08:1	
Drive Sprocket: At rear of vehicle with 11 teeth	
Pitch Diameter: 24.504 inches	

RUNNING GEAR

Suspension: Torsion bar
12 individually sprung dual road wheels (6/track)
Tire Size: 26 x 6 inches
6 dual track return rollers (3/track)
Dual compensating idler at front of each track
Idler Tire Size: 26 x 6 inches
Friction snubbers fitted on first 2 and last road wheels on each side
Tracks: Center guide, T97E2
Type: (T97E2) Double pin, 28 inch width, rubber chevron
Pitch: 6.94 inches
Shoes per Vehicle: 158 (79/track)
Ground Contact Length: 157.5 inches

ELECTRICAL SYSTEM

Nominal Voltage: 24 volts DC
Main Generator: (1) 24 volts, 300 amperes, gear driven by main engine
Auxiliary Generator: None
Battery: (6) 12 volts, 3 sets of 2 in series connected in parallel

COMMUNICATIONS

Radio: AN/VRC-44 in turret bustle
AN/VRC-24 (air to ground) also may be fitted
Interphone: 4 stations plus external box

FIRE PROTECTION

(1) 30 pound carbon dioxide, fixed (2 shot)
(1) 5 pound carbon dioxide, portable

PERFORMANCE

Maximum Speed: Sustained, level road		30 miles/hour
Maximum Tractive Effort: TE at stall		70,000 pounds
Per Cent of Vehicle Weight: TE/W		67 per cent
Maximum Grade:		60 per cent
Maximum Trench:		8.5 feet
Maximum Vertical Wall:		36 inches
Maximum Fording Depth: w/o kit		48 inches
Minimum Turning Circle: (diameter)		pivot
Cruising Range: Roads	approx.	300 miles

*The surface angles and wall thicknesses of the cast hull and turret varied widely from point to point. Therefore, where indicated the data are for an equivalent thickness at 0 degrees from the vertical.

90mm GUN TANK M48A3

GENERAL DATA

Crew:	4	men
Length: Gun forward	341.8	inches
Length: Gun in travel position	292.5	inches
Length: Without gun	270.5	inches
Gun Overhang: Gun forward	71.3	inches
Width: Over tracks	143.0	inches
Height: Over cupola periscope	129.3	inches
Tread:	115.0	inches
Ground Clearance:	16.5	inches
Fire Height:	approx. 79	inches
Turret Ring Diameter: (inside)	85.0	inches
Weight, Combat Loaded:	107,000	pounds
Weight, Unstowed:	101,000	pounds
Power to Weight Ratio: Net	12.0	hp/ton
Gross	14.0	hp/ton
Ground Pressure: Zero penetration	12.1	psi

*ARMOR

Type: Turret, cast homogeneous steel; Hull, cast homogeneous steel; Welded assembly

Hull Thickness:	Actual	Angle w/Vertical
Front, Upper	4.33 inches (110mm)	60 degrees
Lower	4.0 to 2.4 inches (102-61mm)	53 degrees
Sides, Front	equals 3.0 inches (76mm)	0 degrees
Rear	equals 2.0 inches (51mm)	0 degrees
Rear, Grill	equals 1.0 inches (25mm)	0 degrees
Lower	1.6 to 1.2 inches (41-30mm)	30 to 60 degrees
Top	2.25 inches (57mm)	90 degrees
Floor, Front	1.5 inches (38mm)	90 degrees
Rear	1.0 inches (25mm)	90 degrees

Turret Thickness:		
Gun Shield	4.5 inches (114mm)	30 degrees
Front	equals 7.0 inches (178mm)	0 degrees
Sides	equals 3.0 inches (76mm)	0 degrees
Rear	equals 2.0 inches (51mm)	0 degrees
Top	1.0 inches (25mm)	90 degrees

ARMAMENT

Primary: 90mm Gun M41 in Mount M87A1 in turret

Traverse: Hydraulic and manual	360 degrees
Traverse Rate: (max)	15 seconds/360 degrees
Elevation: Hydraulic and manual	+19 to −9 degrees
Elevation Rate: (max)	4 degrees/second
Firing Rate: (max)	8 rounds/minute
Loading System:	Manual
Stabilizer System:	None

Secondary:
(1) .50 caliber MG HB M2 in cupola mount on turret
(1) 7.62mm MG M73 coaxial w/90mm gun in turret
Provision for (2) .45 caliber SMG M3A1

AMMUNITION

62 rounds 90mm	8 hand grenades
600 rounds .50 caliber	
360 rounds .45 caliber	
5900 rounds 7.62mm	

FIRE CONTROL AND VISION EQUIPMENT

Primary Weapon:	Direct	Indirect
	Range Finder M17A1	Azimuth Indicator
	Periscope M32 (daylight and infrared)	M28A1
	Ballistic Drive M10A6	Elevation Quadrant
	Ballistic Computer M13B1C	M13B1
		Gunner's Quadrant
		M1A1
	Telescope M105	

Vision Devices	Direct	Indirect
Driver	Hatch	Periscope M27 (3) and Periscope M24 (1) infrared
Commander	Vision blocks (3) in cupola, (9) in vision ring, hatch	Periscope M28C (AA MG sight)
Gunner	None	Periscope M32
Loader	Hatch	None

Total Periscopes: M27 (3), M24 infrared (1), M32 (1), M28C (1)
Total Vision Blocks: (3) in M1 cupola, (9) large size in vision ring
Searchlight: AN/VSS-1, 2.2 KW, xenon white light or infrared

ENGINE

Make and Model: Continental AVDS-1790-2A
Type: 12 cylinder, 4 cycle, 90 degree vee, supercharged
Cooling System: Air   Ignition: Compression

Displacement:	1791.7 cubic inches
Bore and Stroke:	5.75 x 5.75 inches
Compression Ratio:	16:1
Net Horsepower (max):	643 hp at 2400 rpm
Gross Horsepower (max):	750 hp at 2400 rpm
Net Torque (max):	1575 ft-lb at 1750 rpm
Gross Torque (max):	1710 ft-lb at 1800 rpm
Weight:	4700 pounds, dry
Fuel: 40 cetane diesel	385 gallons
Engine Oil:	72 quarts

POWER TRAIN

Transmission: Cross-drive CD-850-6A, 2 ranges forward, 1 reverse
Single stage multiphase hydraulic torque converter
Stall multiplication: 4:1
Overall Usable Ratios: low 12.0:1   reverse: 16.9:1
high 4.3:1
Steering Control: Mechanical, steering wheel
Steering Rate: 5.7 rpm
Brakes: Multiple disc
Final Drive: Spur gear   Gear Ratio: 5.08:1
Drive Sprocket: At rear of vehicle with 11 teeth
Pitch Diameter: 24.504 inches

RUNNING GEAR

Suspension: Torsion bar
12 individually sprung dual road wheels (6/track)
Tire Size: 26 x 6 inches
10 dual track return rollers (5/track)
Dual compensating idler at front of each track
Idler Tire Size: 26 x 6 inches
Friction snubbers fitted on first 2 and last road wheels on each side
Tracks: Center guide, T97E2
Type: (T97E2) Double pin, 28 inch width, rubber chevron
Pitch: 6.94 inches
Shoes per Vehicle: 158 (79/track)
Ground Contact Length: 157.5 inches

ELECTRICAL SYSTEM

Nominal Voltage: 24 volts DC
Main Generator: (1) 24 volts, 300 amperes, gear driven by main engine
Auxiliary Generator: None
Battery: (6) 12 volts, 3 sets of 2 in series connected in parallel

COMMUNICATIONS

Radio: AN/VRC-12, 46, 47, 53, or 64 in turret bustle
AN/VRC-24 (air to ground) also may be fitted
Interphone: 4 stations plus external box

NUCLEAR, BIOLOGICAL, CHEMICAL PROTECTION

M13A1 gas, particulate filter unit system, 20 cubic feet/minute
(4) M25A1 tank protective mask

FIRE PROTECTION

(1) 30 pound carbon dioxide, fixed (2 shot)
(1) 5 pound carbon dioxide, portable

PERFORMANCE

Maximum Speed: Sustained, level road	30 miles/hour
Maximum Tractive Effort: TE at stall	75,400 pounds
Per Cent of Vehicle Weight: TE/W	70 per cent
Maximum Grade:	60 per cent
Maximum Trench:	8.5 feet
Maximum Vertical Wall:	36 inches
Maximum Fording Depth: w/o kit	48 inches
Minimum Turning Circle: (diameter)	pivot
Cruising Range: Roads	approx. 300 miles

*The surface angles and wall thicknesses of the cast hull and turret varied widely from point to point. Therefore, where indicated the data are for an equivalent thickness at 0 degrees from the vertical.

105mm GUN TANK M48A1E3

GENERAL DATA

Crew:	4	men
Length: Gun forward	366.4	inches
Length: Gun in travel position	317.1	inches
Length: Without gun	270.5	inches
Gun Overhang: Gun forward	95.9	inches
Width: Over tracks	143.0	inches
Height: Over cupola periscope	124.7	inches
Tread:	115.0	inches
Ground Clearance:	16.5	inches
Fire Height:	approx. 81	inches
Turret Ring Diameter: (inside)	85.0	inches
Weight, Combat Loaded:	108,000	pounds
Weight, Unstowed:	102,000	pounds
Power to Weight Ratio: Net	11.9	hp/ton
Gross	13.9	hp/ton
Ground Pressure: Zero penetration	12.2	psi

*ARMOR

Type: Turret, cast homogeneous steel; Hull, cast homogeneous
  steel; Welded assembly

Hull Thickness:	Actual	Angle w/Vertical
Front, Upper	4.33 inches (110mm)	60 degrees
Lower	4.0 to 2.4 inches (102-61mm)	53 degrees
Sides, Front	equals 3.0 inches (76mm)	0 degrees
Rear	equals 2.0 inches (51mm)	0 degrees
Rear, Grill	equals 1.0 inches (25mm)	0 degrees
Lower	1.6 to 1.2 inches (41-30mm)	30 to 60 degrees
Top	2.25 inches (57mm)	90 degrees
Floor, Front	1.5 inches (38mm)	90 degrees
Rear	1.0 inches (25mm)	90 degrees
Turret Thickness:		
Gun Shield	4.5 inches (114mm)	30 degrees
Front	equals 7.0 inches (178mm)	0 degrees
Sides	equals 3.0 inches (76mm)	0 degrees
Rear	equals 2.0 inches (51mm)	0 degrees
Top	0.93 inches (24mm)	90 degrees

ARMAMENT

Primary: 105mm Gun M68 in Mount M116 in turret

Traverse: Hydraulic and manual	360 degrees
Traverse Rate: (max)	15 seconds/360 degrees
Elevation: Hydraulic and manual	+ 19 to − 9 degrees
Elevation Rate: (max)	4 degrees/second
Firing Rate: (max)	7 rounds/minute
Loading System:	Manual
Stabilizer System:	None

Secondary:
  (1) .50 caliber MG HB M2 in M19 cupola mount on turret
  (1) 7.62mm MG M73 coaxial w/105mm gun in turret
  Provision for (2) .45 caliber SMG M3A1

AMMUNITION

54 rounds 105mm   8 hand grenades
900 rounds .50 caliber
360 rounds .45 caliber
5950 rounds 7.62mm

FIRE CONTROL AND VISION EQUIPMENT

Primary Weapon:	Direct	Indirect
	Range Finder M17A1	Azimuth Indicator
	Periscope M32	M28A1
	Ballistic Drive M10A4	Elevation Quadrant
	Ballistic Computer	M13A1
	M13A2	Gunner's Quadrant
	Telescope M105C	M1A1
Vision Devices	Direct	Indirect
Driver	Hatch	Periscope M27 (3) and
		Periscope M24 (1)
		infrared
Commander	Vision blocks (8)	Periscope M34 or
	in cupola, hatch	M36
Gunner	None	Periscope M32
Loader	Hatch	None

Total Periscopes: M27 (3), M24 infrared (1), M32 (1), M34 or M36 (1)
Total Vision Blocks: (8) in M19 cupola on turret top

ENGINE

Make and Model: Continental AVDS-1790-2A
Type: 12 cylinder, 4 cycle, 90 degree vee, supercharged
Cooling System: Air   Ignition: Compression

Displacement:	1791.7 cubic inches
Bore and Stroke:	5.75 x 5.75 inches
Compression Ratio:	16:1
Net Horsepower (max):	643 hp at 2400 rpm
Gross Horsepower (max):	750 hp at 2400 rpm
Net Torque (max):	1575 ft-lb at 1750 rpm
Gross Torque (max):	1710 ft-lb at 1800 rpm
Weight:	4700 pounds, dry
Fuel: 40 cetane diesel	385 gallons
Engine Oil:	72 quarts

POWER TRAIN

Transmission: Cross-drive CD-850-6A, 2 ranges forward, 1 reverse
  Single stage multiphase hydraulic torque converter
  Stall multiplication: 4:1
  Overall Usable Ratios: low 12.0:1   reverse: 16.9:1
                         high 4.3:1
Steering Control: Mechanical, steering wheel
  Steering Rate: 5.7 rpm
Brakes: Multiple disc
Final Drive: Spur gear   Gear Ratio: 5.08:1
Drive Sprocket: At rear of vehicle with 11 teeth
  Pitch Diameter: 24.504 inches

RUNNING GEAR

Suspension: Torsion bar
  12 individually sprung dual road wheels (6/track)
  Tire Size: 26 x 6 inches
  10 dual track return rollers (5/track)
  Dual compensating idler at front of each track
  Idler Tire Size: 26 x 6 inches
  Friction snubbers fitted on first 2 and last road wheels on each side
Tracks: Center guide, T97E2
  Type: (T97E2) Double pin, 28 inch width, rubber chevron
  Pitch: 6.94 inches
  Shoes per Vehicle: 158 (79/track)
  Ground Contact Length: 157.5 inches

ELECTRICAL SYSTEM

Nominal Voltage: 24 volts DC
Main Generator: (1) 24 volts, 300 amperes, gear driven by main engine
Auxiliary Generator: None
Battery: (6) 12 volts, 3 sets of 2 in series connected in parallel

COMMUNICATIONS

Radio: AN/VRC-12, 46, 47, 53, or 64 in turret bustle
  AN/VRC-24 (air to ground) also may be fitted
Interphone: 4 stations plus external box

NUCLEAR, BIOLOGICAL, CHEMICAL PROTECTION

M13A1 gas, particulate filter unit system, 20 cubic feet/minute
(4) M25A1 tank protective mask

FIRE PROTECTION

(1) 30 pound carbon dioxide, fixed (2 shot)
(1) 5 pound carbon dioxide, portable

PERFORMANCE

Maximum Speed: Sustained, level road	30 miles/hour
Maximum Tractive Effort: TE at stall	75,400 pounds
Per Cent of Vehicle Weight: TE/W	70 per cent
Maximum Grade:	60 per cent
Maximum Trench:	8.5 feet
Maximum Vertical Wall:	36 inches
Maximum Fording Depth: w/o kit	48 inches
Minimum Turning Circle: (diameter)	pivot
Cruising Range: Roads	approx. 300 miles

---

*The surface angles and wall thicknesses of the cast hull and turret varied widely from point to point. Therefore, where indicated the data are for an
  equivalent thickness at 0 degrees from the vertical.

105mm GUN TANK M48A5

GENERAL DATA

Crew:	4	men
Length: Gun forward	366.4	inches
Length: Gun in travel position	317.1	inches
Length: Without gun	270.5	inches
Gun Overhang: Gun forward	95.9	inches
Width: Over tracks	143.0	inches
Height: Over cupola periscope	120.5	inches
Tread:	115.0	inches
Ground Clearance: At escape hatch	16.5	inches
Fire Height:	78.8	inches
Turret Ring Diameter:	85.0	inches
Weight, Combat Loaded:	108,000	pounds
Weight, Unstowed:	102,000	pounds
Power to Weight Ratio: Net	11.9	hp/ton
Gross	13.9	hp/ton
Ground Pressure: Zero penetration	12.2	psi

*ARMOR

Type: Turret, cast homogeneous steel; Hull, cast homogeneous
steel; Welded assembly

Hull Thickness:	Actual	Angle w/Vertical
Front, Upper	4.33 inches (110mm)	60 degrees
Lower	4.0 to 2.4 inches (102-61mm)	53 degrees
Sides, Front	equals 3.0 inches (76mm)	0 degrees
Rear	equals 2.0 inches (51mm)	0 degrees
Rear, Grill	equals 1.0 inches (25mm)	0 degrees
Lower	1.6 to 1.2 inches (41-30mm)	30 to 60 degrees
Top	2.25 inches (57mm)	90 degrees
Floor, Front	1.5 inches (38mm)	90 degrees
Rear	1.0 inches (25mm)	90 degrees
Turret Thickness:		
Gun Shield	4.5 inches (114mm)	30 degrees
Front	equals 7.0 inches (178mm)	0 degrees
Sides	equals 3.0 inches (76mm)	0 degrees
Rear	equals 2.0 inches (51mm)	0 degrees
Top	1.0 inches (25mm)	90 degrees

ARMAMENT

Primary: 105mm Gun M68 in modified Mount M87 in turret

Traverse: Hydraulic and manual	360 degrees
Traverse Rate: (max)	15 seconds/360 degrees
Elevation: Hydraulic and manual	+ 19 to − 9 degrees
Elevation Rate: (max)	4 degrees/second
Firing Rate: (max)	7 rounds/minute
Loading System:	Manual
Stabilizer System:	None

Secondary:
**(1) 7.62mm MG M219 coaxial w/105mm gun in turret
(2) 7.62mm MG M60D on turret top
Provision for (2) .45 caliber SMG M3A1

AMMUNITION

54 rounds 105mm		8 hand grenades
360 rounds .45 caliber		
10,000 rounds 7.62mm		

FIRE CONTROL AND VISION EQUIPMENT

Primary Weapon:	Direct	Indirect
	Range Finder M17B1C	Azimuth Indicator
	Periscope M32	M28A1
	Ballistic Drive M10A6	Elevation Quadrant
	Ballistic Computer	M13B1
	M13A4	Gunner's Quadrant
	Telescope M105D	M1A1
Vision Devices	Direct	Indirect
Driver	Hatch	Periscope M27 (3) and
		Periscope M24 (1) infrared
Commander	Hatch	Periscope M17 (3)
Gunner	None	Periscope M32
Loader	Hatch	None

Total Periscopes: M27 (3), M24 infrared (1), M32 (1), M17 (3)
Total Vision Blocks: (0)
*'*Searchlight: AN/VSS-1, 2.2 KW, xenon white light or infrared

ENGINE

Make and Model: Continental AVDS-1790-2D	
Type: 12 cylinder, 4 cycle, 90 degree vee, supercharged	
Cooling System: Air Ignition: Compression	
Displacement:	1791.7 cubic inches
Bore and Stroke:	5.75 x 5.75 inches
Compression Ratio:	16:1
Net Horsepower (max):	643 hp at 2400 rpm
Gross Horsepower (max):	750 hp at 2400 rpm
Net Torque (max):	1575 ft-lb at 1750 rpm
Gross Torque (max):	1710 ft-lb at 1800 rpm
Weight:	4880 pounds, dry
Fuel: 40 cetane diesel	385 gallons
Engine Oil:	72 quarts

POWER TRAIN

Transmission: Cross-drive CD-850-6A, 2 ranges forward, 1 reverse
  Single stage multiphase hydraulic torque converter
  Stall multiplication: 4:1
  Overall Usable Ratios: low 12.0:1   reverse: 16.9:1
                         high 4.3:1
Steering Control: Mechanical, steering wheel
  Steering Rate: 5.7 rpm
Brakes: Multiple disc
Final Drive: Spur gear   Gear Ratio: 5.08:1
Drive Sprocket: At rear of vehicle with 11 teeth
  Pitch Diameter: 24.504 inches

RUNNING GEAR

Suspension: Torsion bar
  12 individually sprung dual road wheels (6/track)
  Tire Size: 26 x 6 inches
  10 dual track return rollers (5/track)
  Dual compensating idler at front of each track
  Idler Tire Size: 26 x 6 inches
  Friction snubbers fitted on first 2 and last road wheels on each side
Tracks: Center guide, T142 or T97E2
  Type: (T142) Double pin, 28 inch width, replaceable rubber pads
        (T97E2) Double pin, 28 inch width, rubber chevron
  Pitch: 6.94 inches
  Shoes per Vehicle: 156-158 (78-79/track)
  Ground Contact Length: 157.5 inches

ELECTRICAL SYSTEM

Nominal Voltage: 24 volts DC
Main Generator: (1) 24 volts, 300 amperes, gear driven by main engine
Auxiliary Generator: None
Battery: (6) 12 volts, 3 sets of 2 in series connected in parallel

COMMUNICATIONS

Radio: AN/VRC-12, 46, 47, or 64 in turret bustle
       AN/VRC-24 (air to ground) also may be fitted
Interphone: 4 stations plus external box

NUCLEAR, BIOLOGICAL, CHEMICAL PROTECTION

M13A1 gas, particulate filter unit system, 20 cubic feet/minute
(4) M25A1 tank protective mask

FIRE PROTECTION

(1) 30 pound carbon dioxide, fixed (2 shot)
(1) 2.75 pound Halon, portable

PERFORMANCE

Maximum Speed: Sustained, level road	30 miles/hour
Maximum Tractive Effort: TE at stall	75,600 pounds
Per Cent of Vehicle Weight: TE/W	70 per cent
Maximum Grade:	60 per cent
Maximum Trench:	8.5 feet
Maximum Vertical Wall:	36 inches
Maximum Fording Depth: w/o kit	48 inches
Minimum Turning Circle: (diameter)	pivot
Cruising Range: Roads	approx. 300 miles

---

* The surface angles and wall thicknesses of the cast hull and turret varied widely from point to point. Therefore, where indicated the data are for an equivalent thickness at 0 degrees from the vertical.
** The M219 can be replaced by the 7.62mm MG M240
' The AN/VSS-1 can be replaced by the AN/VSS-3 1 KW searchlight.

105mm GUN TANK M60

GENERAL DATA

Crew:	4	men
Length: Gun forward	366.5	inches
Length: Gun in travel position	320.0	inches
Length: Without gun	273.5	inches
Gun Overhang: Gun forward	93.0	inches
Width: Over tracks	143.0	inches
Height: Over cupola periscope	126.5	inches
Tread:	115.0	inches
Ground Clearance: At escape hatch	15.3	inches
Fire Height: approx.	83	inches
Turret Ring Diameter: (inside)	85.0	inches
Weight, Combat Loaded:	102,000	pounds
Weight, Unstowed:	93,500	pounds
Power to Weight Ratio: Net	12.6	hp/ton
Gross	14.7	hp/ton
Ground Pressure: Zero penetration	10.9	psi

*ARMOR

Type: Turret, cast homogeneous steel; Hull, cast homogeneous steel; Welded assembly

Hull Thickness:

	Actual	Angle w/Vertical
Front, Upper	3.67 inches (93mm)	65 degrees
Lower	5.63 to 3.35 inches (143-85mm)	55 degrees
Sides	1.41 to 2.9 inches (36-74mm)	45 to 0 degrees
Rear, Grill	equals 1.0 inches (25mm)	0 degrees
Lower	1.6 to 1.2 inches (41-30mm)	30 to 60 degrees
Top	1.41 inches (36mm)	90 degrees
Floor, Front	0.75 inches (19mm)	90 degrees
Rear	0.50 inches (13mm)	90 degrees

Turret Thickness:

Gun Shield	4.5 inches (114mm)	30 degrees
Front	equals 7.0 inches (178mm)	0 degrees
Sides	equals 3.0 inches (76mm)	0 degrees
Rear	equals 2.0 inches (51mm)	0 degrees
Top	0.93 inches (24mm)	90 degrees

ARMAMENT

Primary: 105mm Gun M68 (T254E2) in Mount M116 in turret

Traverse: Hydraulic and manual	360 degrees
Traverse Rate: (max)	15 seconds/360 degrees
Elevation: Hydraulic and manual	+ 19 to − 9 degrees
Elevation Rate: (max)	4 degrees/second
Firing Rate: (max)	7 rounds/minute
Loading System:	Manual
Stabilizer System:	None

Secondary:

#(1) .50 caliber MG M85 in M19 cupola mount on turret

(1) 7.62mm MG M73 (T197E2) coaxial w/105mm gun in turret

Provision for (2) .45 caliber SMG M3A1

AMMUNITION

57 rounds 105mm	8 hand grenades
900 rounds .50 caliber	
360 rounds .45 caliber	
5950 rounds 7.62mm	

FIRE CONTROL AND VISION EQUIPMENT

Primary Weapon:	Direct	Indirect
	Range Finder M17C	Azimuth Indicator
	Periscope M31	M28A1
	Ballistic Drive M10	Elevation Quadrant
	Ballistic Computer	M13A1
	M13A1D	Gunner's Quadrant
	Telescope M105C	M1A1
Vision Devices	Direct	Indirect
Driver	Hatch	Periscope M27 (3) and Periscope M24 (1) infrared
Commander	Vision blocks (8) in cupola, hatch	Periscope M28C (AA MG sight)
Gunner	None	Periscope M31
Loader	Hatch	None

Total Periscopes: M27 (3), M24 infrared (1), M31 (1), M28C (1)

Total Vision Blocks: (8) in M19 cupola on turret top

ENGINE

Make and Model: Continental AVDS-1790-2	
Type: 12 cylinder, 4 cycle, 90 degree vee, supercharged	
Cooling System: Air Ignition: Compression	
Displacement:	1791.7 cubic inches
Bore and Stroke:	5.75 x 5.75 inches
Compression Ratio:	16:1
Net Horsepower (max):	643 hp at 2400 rpm
Gross Horsepower (max):	750 hp at 2400 rpm
Net Torque (max):	1575 ft-lb at 1750 rpm
Gross Torque (max):	1710 ft-lb at 1800 rpm
Weight:	4527 pounds, dry
Fuel: 40 cetane diesel	385 gallons
Engine Oil:	64 quarts

POWER TRAIN

Transmission: Cross-drive CD-850-6, 2 ranges forward, 1 reverse

Single stage multiphase hydraulic torque converter

Stall multiplication: 4:1

Overall Usable Ratios: low 12.0:1   reverse: 16.9:1

high 4.3:1

Steering Control: Mechanical, steering wheel

Steering Rate: 5.7 rpm

Brakes: Multiple disc

Final Drive: Spur gear   Gear Ratio: 5.08:1

Drive Sprocket: At rear of vehicle with 11 teeth

Pitch Diameter: 24.504 inches

RUNNING GEAR

Suspension: Torsion bar

12 individually sprung dual road wheels (6/track)

Tire Size: 26 x 6 inches

6 dual track return rollers (3/track)

Dual compensating idler at front of each track

Idler Tire Size: 26 x 6 inches

** Friction snubbers fitted on first and last road wheels on each side

Tracks: Center guide, T97E2

Type: (T97E2) Double pin, 28 inch width, rubber chevron

Pitch: 6.94 inches

Shoes per Vehicle: 162 (81/track)

Ground Contact Length: 166.72 inches

ELECTRICAL SYSTEM

Nominal Voltage: 24 volts DC

Main Generator: (1) 24 volts, 300 amperes, gear driven by main engine

Auxiliary Generator: None

Battery: (6) 12 volts, 3 sets of 2 in series connected in parallel

COMMUNICATIONS

Radio: AN/VRC-12, 46, 47 or AN/PRC-25 in turret bustle

Early tanks have AN/GRC-3 thru 8 series

AN/VRC-24 (air to ground) also may be fitted

Interphone: 4 stations plus external box (AN/VIA-1 on early tanks)

FIRE PROTECTION

(1) 30 pound carbon dioxide, fixed (2 shot)

(1) 5 pound carbon dioxide, portable

PERFORMANCE

Maximum Speed: Sustained, level road	30 miles/hour
Maximum Tractive Effort: TE at stall	70,000 pounds
Per Cent of Vehicle Weight: TE/W	69 per cent
Maximum Grade:	60 per cent
Maximum Trench:	8.5 feet
Maximum Vertical Wall:	36 inches
Maximum Fording Depth:	48 inches
Minimum Turning Circle: (diameter)	pivot
Cruising Range: Roads	approx. 300 miles

---

* The surface angles and wall thicknesses of the cast hull and turret varied widely from point to point. Therefore, where indicated the data are for an equivalent thickness at 0 degrees from the vertical.

** The friction snubbers were installed during overhaul on the early production tanks. Originally, they had no shock absorbers.

Early tanks were fitted with the .50 caliber M2 MG on an external mount prior to the availability of the M85.

105mm GUN TANK M60A1

GENERAL DATA

Crew:	4	men
Length: Gun forward	371.5	inches
Length: Gun in travel position	325.0	inches
Length: Without gun	273.5	inches
Gun Overhang: Gun forward	98.0	inches
Width: Over tracks	143.0	inches
Height: Over cupola periscope	128.2	inches
Tread:	115.0	inches
Ground Clearance: At escape hatch	15.3	inches
Fire Height:	82.3	inches
Turret Ring Diameter: (inside)	85.0	inches
Weight, Combat Loaded:	105,000	pounds
Weight, Unstowed:	97,000	pounds
Power to Weight Ratio: Net	12.2	hp/ton
Gross	14.3	hp/ton
Ground Pressure: Zero penetration	11.2	psi

*ARMOR

Type: Turret, cast homogeneous steel; Hull, cast homogeneous steel; Welded assembly

Hull Thickness:	Actual	Angle w/Vertical
Front, Upper	4.29 inches (109mm)	65 degrees
Lower	5.63 to 3.35 inches (143-85mm)	55 degrees
Sides	1.41 to 2.9 inches (36-74mm)	45 to 0 degrees
Rear, Grill	equals 1.0 inches (25mm)	0 degrees
Lower	1.6 to 1.2 inches (41-30mm)	30 to 60 degrees
Top	1.41 inches (36mm)	90 degrees
Floor, Front	0.75 inches (19mm)	90 degrees
Rear	0.50 inches (13mm)	90 degrees

Turret Thickness:		
Gun Shield	5.0 inches (127mm)	60 degrees
Front	equals 10.0 inches (254mm)	0 degrees
Sides	equals 5.5 inches (140mm)	0 degrees
Rear	equals 2.25 inches (57mm)	0 degrees
Top	1.0 inches (25mm)	90 degrees

ARMAMENT

Primary: 105mm Gun M68 in Mount M140 in turret

Traverse: Hydraulic and manual	360 degrees
Traverse Rate: (max)	16 seconds/360 degrees
Elevation: Hydraulic and manual	+ 20 to − 10 degrees
Elevation Rate: (max)	4 degrees/second
Firing Rate: (max)	7 rounds/minute
Loading System:	Manual
Stabilizer System:	None

Secondary:
(1) .50 caliber MG M85 in M19 cupola mount on turret
(1) 7.62mm MG M73 or M219 coaxial w/105mm gun in turret
Provision for (2) .45 caliber SMG M3A1

AMMUNITION

63 rounds 105mm	8 hand grenades
900 rounds .50 caliber	
360 rounds .45 caliber	
5950 rounds 7.62mm	

FIRE CONTROL AND VISION EQUIPMENT

Primary Weapon:	Direct	Indirect
	Range Finder M17A1	Azimuth Indicator
	Periscope M32	M28A1
	(day & IR)	Elevation Quadrant
	Ballistic Drive M10A4	M13A3
	Ballistic Computer	Gunner's Quadrant
	M13A2	M1A1
	Telescope M105C	

Vision Devices	Direct	Indirect
Driver	Hatch	Periscope M27 (3) and Periscope M24 (1) infrared
Commander	Vision blocks (8) in cupola, hatch	Periscope M34 or M36 (day & IR)
Gunner	None	Periscope M32 (day & IR)
Loader	Hatch	Periscope M37

Total Periscopes: M27 (3), M24 infrared (1), M34 or M36 (1), M32 (1), M37 (1)
Total Vision Blocks: (8) in M19 cupola on turret top
Searchlight: AN/VSS-1, 2.2 KW, xenon white light or infrared

ENGINE

Make and Model: Continental AVDS-1790-2A	
Type: 12 cylinder, 4 cycle, 90 degree vee, supercharged	
Cooling System: Air Ignition: Compression	
Displacement:	1791.7 cubic inches
Bore and Stroke:	5.75 x 5.75 inches
Compression Ratio:	16:1
Net Horsepower (max):	643 hp at 2400 rpm
Gross Horsepower (max):	750 hp at 2400 rpm
Net Torque (max):	1575 ft-lb at 1750 rpm
Gross Torque (max):	1710 ft-lb at 1800 rpm
Weight:	4700 pounds, dry
Fuel: 40 cetane diesel	385 gallons
Engine Oil:	72 quarts

POWER TRAIN

Transmission: Cross-drive CD-850-6A, 2 ranges forward, 1 reverse
Single stage multiphase hydraulic torque converter
Stall multiplication: 4:1
Overall Usable Ratios: low 12.0:1   reverse: 16.9:1
high 4.3:1
Steering Control: Mechanical, T-bar
Steering Rate: 5.7 rpm
Brakes: Multiple disc
Final Drive: Spur gear   Gear Ratio: 5.08:1
Drive Sprocket: At rear of vehicle with 11 teeth
Pitch Diameter: 24.504 inches

RUNNING GEAR

Suspension: Torsion bar
12 individually sprung dual road wheels (6/track)
Tire Size: 26 x 6 inches
6 dual track return rollers (3/track)
Dual compensating idler at front of each track
Idler Tire Size: 26 x 6 inches
Shock absorbers fitted on first 2 and last road wheels on each side
Tracks: Center guide, T97E2
Type: (T97E2) Double pin, 28 inch width, rubber chevron
Pitch: 6.94 inches
Shoes per Vehicle: 162 (81/track)
Ground Contact Length: 166.72 inches

ELECTRICAL SYSTEM

Nominal Voltage: 24 volts DC
Main Generator: (1) 24 volts, 300 amperes, gear driven by main engine
Auxiliary Generator: None
Battery: (6) 12 volts, 3 sets of 2 in series connected in parallel

COMMUNICATIONS

Radio: AN/VRC-12, 46, 47, 53, or 64 in turret bustle
AN/VRC-24 (air to ground) also may be fitted
Interphone: 4 stations plus external box

NUCLEAR, BIOLOGICAL, CHEMICAL PROTECTION

M13A1 gas, particulate filter unit system, 20 cubic feet/minute
(4) M25A1 tank protective mask

FIRE PROTECTION

(1) 30 pound carbon dioxide, fixed (2 shot)
(1) 5 pound carbon dioxide, portable

PERFORMANCE

Maximum Speed: Sustained, level road	30 miles/hour
Maximum Tractive Effort: TE at stall	70,000 pounds
Per Cent of Vehicle Weight: TE/W	67 per cent
Maximum Grade:	60 per cent
Maximum Trench:	8.5 feet
Maximum Vertical Wall:	36 inches
Maximum Fording Depth: w/o kit	48 inches
Minimum Turning Circle: (diameter)	pivot
Cruising Range: Roads	approx. 300 miles

---

*The surface angles and wall thicknesses of the cast hull and turret varied widely from point to point. Therefore, where indicated the data are for an equivalent thickness at 0 degrees from the vertical.

152mm GUN TANK M60A2

GENERAL DATA

Crew:	4	men
Length: Gun forward	288.7	inches
Length: Gun to rear	275.3	inches
Length: Without gun	275.3	inches
Gun Overhang: Gun forward	13.4	inches
Width: Over tracks	143.0	inches
Height: Over cupola periscope	130.3	inches
Tread:	115.0	inches
Ground Clearance: At escape hatch	15.3	inches
Fire Height:	83.6	inches
Turret Ring Diameter: (inside)	85.0	inches
Weight, Combat Loaded:	114,400	pounds
Weight, Unstowed:	108,000	pounds
Power to Weight Ratio: Net	11.2	hp/ton
Gross	13.1	hp/ton
Ground Pressure: Zero penetration	12.3	psi

*ARMOR

Type: Turret, cast homogeneous steel; Hull, cast homogeneous
steel; Welded assembly

Hull Thickness:

	Actual	Angle w/Vertical
Front, Upper	4.29 inches (109mm)	65 degrees
Lower	5.63 to 3.35 inches (143-85mm)	55 degrees
Sides	1.41 to 2.9 inches (36-74mm)	45 to 0 degrees
Rear, Grill	equals 1.0 inches (25mm)	0 degrees
Lower	1.6 to 1.2 inches (41-30mm)	30 to 60 degrees
Top	1.41 inches (36mm)	90 degrees
Floor, Front	0.75 inches (19mm)	90 degrees
Rear	0.50 inches (13mm)	90 degrees

Turret Thickness:

Gun Shield	equals 11.5 inches (292mm)	0 degrees
Front	equals 11.5 inches (292mm)	0 degrees
Sides	4.75 inches (121mm)	0 degrees
Rear	2.5 inches (64mm)	0 degrees
Top	1.25 inches (32mm)	90 degrees

ARMAMENT

Primary: 152mm Gun-Launcher M162 in turret combination mount
with closed breech scavenging system

Traverse: Hydraulic and manual	360 degrees
Traverse Rate: (max)	9.1 seconds/360 degrees
Elevation: Hydraulic and manual	+ 20 to − 10 degrees
Elevation Rate: (max)	10 degrees/second
	(40 deg/sec auto. stabilized)
Firing Rate: (max)	4 rounds/minute
Loading System:	Manual
Stabilizer System:	Azimuth and elevation

Secondary:
(1) .50 caliber MG M85 in cupola mount on turret
(1) 7.62mm MG M73, M73A1, or M219 coaxial w/152mm gun-launcher
Provision for (2) .45 caliber SMG M3A1
(2) Smoke grenade launcher housings (4 tubes each) on turret

AMMUNITION

46 rounds 152mm (conventional) or	8 hand grenades
33 rounds 152mm and 13 missiles	8 M226 smoke grenade
1080 rounds .50 caliber	canisters
360 rounds .45 caliber	
5500 rounds 7.62mm	

FIRE CONTROL AND VISION EQUIPMENT

Primary Weapon:	Direct	Indirect
	Range Finder AN/VVS-1	Azimuth Indicator
	Periscope M50	M37
	(day & night)	Elevation Quadrant
	Ballistic Computer M19	M13A1
	Telescope M126	Gunner's Quadrant
	Missile Control System	M1A1
Vision Devices	Direct	Indirect
Driver	Hatch	Periscope M27 (3) and
		Periscope M24 (1)
		infrared
Commander	Vision blocks (11)	Periscope M51
	in cupola, hatch	(day & night)
Gunner	Hatch	Periscope M50
		(day & night)
Loader	Hatch	Periscope M37

Total Periscopes: M27 (3), M24 infrared (1), M51 (1), M50 (1), M37 (1)
Total Vision Blocks: (11) in cupola on turret top
Searchlight: AN/VSS-1, 2.2 KW, xenon white light or infrared

ENGINE

Make and Model: Continental AVDS-1790-2A	
Type: 12 cylinder, 4 cycle, 90 degree vee, supercharged	
Cooling System: Air Ignition: Compression	
Displacement:	1791.7 cubic inches
Bore and Stroke:	5.75 x 5.75 inches
Compression Ratio:	16:1
Net Horsepower (max):	643 hp at 2400 rpm
Gross Horsepower (max):	750 hp at 2400 rpm
Net Torque (max):	1575 ft-lb at 1750 rpm
Gross Torque (max):	1710 ft-lb at 1800 rpm
Weight:	4700 pounds, dry
Fuel: 40 cetane diesel	385 gallons
Engine Oil:	72 quarts

POWER TRAIN

Transmission: Cross-drive CD-850-6A, 2 ranges forward, 1 reverse
Single stage multiphase hydraulic torque converter
Stall multiplication: 4:1
Overall Usable Ratios: low 12.0:1   reverse: 16.9:1
high 4.3:1
Steering Control: Mechanical, T-bar
Steering Rate: 5.7 rpm
Brakes: Multiple disc
Final Drive: Spur gear   Gear Ratio: 5.08:1
Drive Sprocket: At rear of vehicle with 11 teeth
Pitch Diameter: 24.504 inches

RUNNING GEAR

Suspension: Torsion bar
12 individually sprung dual road wheels (6/track)
Tire Size: 26 x 6 inches
6 dual track return rollers (3/track)
Dual compensating idler at front of each track
Idler Tire Size: 26 x 6 inches
Shock absorbers fitted on first 2 and last road wheels on each side
Tracks: Center guide, T97E2
Type: (T97E2) Double pin, 28 inch width, rubber chevron
Pitch: 6.94 inches
Shoes per Vehicle: 162 (81/track)
Ground Contact Length: 166.72 inches

ELECTRICAL SYSTEM

Nominal Voltage: 24 volts DC
Main Generator: (1) 24 volts, 300 amperes, gear driven by main engine
Auxiliary Generator: None
Battery: (6) 12 volts, 3 sets of 2 in series connected in parallel

COMMUNICATIONS

Radio: AN/VRC-12, 46, 47, 53, or 64 in turret bustle
AN/VRC-24 (air to ground) also may be fitted
Interphone: 4 stations plus external box

NUCLEAR, BIOLOGICAL, CHEMICAL PROTECTION

M13A1 gas, particulate filter unit system, 20 cubic feet/minute
(4) M25A1 tank protective mask

FIRE PROTECTION

(1) 30 pound carbon dioxide, fixed (2 shot)
(1) 5 pound carbon dioxide, portable

PERFORMANCE

Maximum Speed: Sustained, level road	30 miles/hour
Maximum Tractive Effort: TE at stall	70,000 pounds
Per Cent of Vehicle Weight: TE/W	61 per cent
Maximum Grade:	60 per cent
Maximum Trench:	8.5 feet
Maximum Vertical Wall:	36 inches
Maximum Fording Depth: w/o kit	48 inches
Minimum Turning Circle: (diameter)	pivot
Cruising Range: Roads	approx. 280 miles

*The surface angles and wall thicknesses of the cast hull and turret varied widely from point to point. Therefore, where indicated the data are for an equivalent thickness at 0 degrees from the vertical.

105mm GUN TANK M60A1E3

GENERAL DATA

Crew:	4	men
Length: Gun forward	371.5	inches
Length: Gun in travel position	325.0	inches
Length: Without gun	273.5	inches
Gun Overhang: Gun forward	98.0	inches
Width: Over tracks	143.0	inches
Height: Over cupola periscope	129.3	inches
Tread:	115.0	inches
Ground Clearance: At escape hatch	15.3	inches
Fire Height:	82.3	inches
Turret Ring Diameter: (inside)	85.0	inches
Weight, Combat Loaded:	110,100	pounds
Weight, Unstowed:	101,100	pounds
Power to Weight Ratio: Net	11.7	hp/ton
Gross	13.6	hp/ton
Ground Pressure: Zero penetration	11.8	psi

*ARMOR

Type: Turret, cast homogeneous steel; Hull, cast homogeneous steel; Welded assembly

Hull Thickness:	Actual	Angle w/Vertical
Front, Upper	4.29 inches (109mm)	65 degrees
Lower	5.63 to 3.35 inches (143-85mm)	55 degrees
Sides	1.41 to 2.9 inches (36-74mm)	45 to 0 degrees
Rear, Grill	equals 1.0 inches (25mm)	0 degrees
Lower	1.6 to 1.2 inches (41-30mm)	30 to 60 degrees
Top	1.41 inches (36mm)	90 degrees
Floor, Front	0.75 inches (19mm)	90 degrees
Rear	0.50 inches (13mm)	90 degrees
Turret Thickness:		
Gun Shield	5.0 inches (127mm)	60 degrees
Front	equals 10 inches (254mm)	0 degrees
Sides	equals 5.5 inches (140mm)	0 degrees
Rear	equals 2.25 inches (57mm)	0 degrees
Top	1.0 inches (25mm)	90 degrees

ARMAMENT

Primary: 105mm Gun M68 w/thermal shroud in Mount M140 in turret

Traverse: Hydraulic and manual	360 degrees
Traverse Rate: (max)	16 seconds/360 degrees
Elevation: Hydraulic and manual	+ 20 to − 10 degrees
Elevation Rate: (max)	4 degrees/second
Firing Rate: (max)	7 rounds/minute
Loading System:	Manual
Stabilizer System:	Azimuth and elevation

Secondary:
- (1) .50 caliber MG M85 in M19 cupola mount on turret
- (1) 7.62mm MG M219 coaxial w/105mm gun in turret
- Provision for (2) .45 caliber SMG M3A1

AMMUNITION

63 rounds 105mm	8 hand grenades
900 rounds .50 caliber	
360 rounds .45 caliber	
5950 rounds 7.62mm	

FIRE CONTROL AND VISION EQUIPMENT

Primary Weapon:	Direct	Indirect
	Range Finder	Azimuth Indicator
	AN/VVG-2	M28A1
	Periscope XM35	Elevation Quadrant
	(day & IR)	M13A3
	Ballistic Drive M10A3	Gunner's Quadrant
	Ballistic Computer	M13A1
	XM21	
	Telescope M105D	

Vision Devices	Direct	Indirect
Driver	Hatch	Periscope M27 (3) and Periscope M24 (1) infrared
Commander	Vision blocks (8) in cupola, hatch	Periscope M36 daylight and infrared
Gunner	None	Periscope XM35 (day & IR)
Loader	Hatch	Periscope M37

Total Periscopes: M27 (3), M24 infrared (1), M36 (1), XM35 (1), M37 (1)
Total Vision Blocks: (8) in M19 cupola on turret top
Searchlight: AN/VSS-3, 1.0 KW, xenon white light or infrared

ENGINE

Make and Model: Continental AVDS-1790-2CM (RISE)	
Type: 12 cylinder, 4 cycle, 90 degree vee, supercharged	
Cooling System: Air Ignition: Compression	
Displacement:	1791.7 cubic inches
Bore and Stroke:	5.75 x 5.75 inches
Compression Ratio:	16:1
Net Horsepower (max):	643 hp at 2400 rpm
Gross Horsepower (max):	750 hp at 2400 rpm
Net Torque (max):	1620 ft-lb at 1800 rpm
Gross Torque (max):	1770 ft-lb at 2000 rpm
Weight:	4876 pounds, dry
Fuel: 40 cetane diesel	385 gallons
Engine Oil:	100 quarts

POWER TRAIN

Transmission: Cross-drive CD-850-6A, 2 ranges forward, 1 reverse
Single stage multiphase hydraulic torque converter
Stall multiplication: 4:1
Overall Usable Ratios: low 12.0:1   reverse: 16.9:1
high 4.3:1
Steering Control: Mechanical, T-bar
Steering Rate: 5.7 rpm
Brakes: Multiple disc
Final Drive: Spur gear   Gear Ratio: 5.08:1
Drive Sprocket: At rear of vehicle with 11 teeth
Pitch Diameter: 24.504 inches

RUNNING GEAR

Suspension: Torsion tube-over-bar
12 individually sprung dual road wheels (6/track)
Tire Size: 26 x 6 inches
6 dual track return rollers (3/track)
Dual compensating idler at front of each track
Idler Tire Size: 26 x 6 inches
Integral rotary hydraulic shock absorbers on first 2 and last road wheel stations
Tracks: Center guide, T142
Type: (T142) Double pin, 28 inch width, replaceable rubber pads
Pitch: 6.94 inches
Shoes per Vehicle: 160 (80/track)
Ground Contact Length: 166.72 inches

ELECTRICAL SYSTEM

Nominal Voltage: 24 volts DC
Alternator, oil cooled w/solid state regulator, 28 volts, 650 amperes, gear driven by main engine
Auxiliary Generator: None
Battery: (6) 12 volts, 3 sets of 2 in series connected in parallel

COMMUNICATIONS

Radio: AN/VRC-12, 46, 47, 53, 64, AN/GRC-125 or 160 in turret bustle
Interphone: 4 stations plus external box

NUCLEAR, BIOLOGICAL, CHEMICAL PROTECTION

(1) M13A1 gas, particulate filter unit system, 20 cubic feet/minute
(4) M25A1 tank protective mask

FIRE PROTECTION

(1) 30 pound carbon dioxide, fixed (2 shot)
(1) 5 pound carbon dioxide, portable

PERFORMANCE

Maximum Speed: Sustained, level road		30 miles/hour
Maximum Tractive Effort: TE at stall		77,000 pounds
Per Cent of Vehicle Weight: TE/W		70 per cent
Maximum Grade:		60 per cent
Maximum Trench:		8.5 feet
Maximum Vertical Wall:		36 inches
Maximum Fording Depth: w/o kit		48 inches
Minimum Turning Circle: (diameter)		pivot
Cruising Range: Roads	approx.	300 miles

*The surface angles and wall thicknesses of the cast hull and turret varied widely from point to point. Therefore, where indicated the data are for an equivalent thickness at 0 degrees from the vertical.

105mm GUN TANK M60A3

GENERAL DATA

Crew:	4	men
Length: Gun forward	371.5	inches
Length: Gun in travel position	325.0	inches
Length: Without gun	273.5	inches
Gun Overhang: Gun forward	98.0	inches
Width: Over tracks	143.0	inches
Height: Over cupola periscope	129.2	inches
Tread:	115.0	inches
Ground Clearance: At escape hatch	15.3	inches
Fire Height:	82.3	inches
Turret Ring Diameter: (inside)	85.0	inches
Weight, Combat Loaded:	114,600	pounds
Weight, Unstowed:	105,600	pounds
Power to Weight Ratio: Net	11.2	hp/ton
Gross	13.1	hp/ton
Ground Pressure: Zero penetration	12.3	psi

*ARMOR

Type: Turret, cast homogeneous steel; Hull, cast homogeneous
  steel; Welded assembly

Hull Thickness:	Actual	Angle w/Vertical
Front, Upper	4.29 inches (109mm)	65 degrees
Lower	5.63 to 3.35 inches (143-85mm)	55 degrees
Sides	1.41 to 2.9 inches (36-74mm)	45 to 0 degrees
Rear, Grill	equals 1.0 inches (25mm)	0 degrees
Lower	1.6 to 1.2 inches (41-30mm)	30 to 60 degrees
Top	1.41 inches (36mm)	90 degrees
Floor, Front	0.75 inches (19mm)	90 degrees
Rear	0.50 inches (13mm)	90 degrees
Turret Thickness:		
Gun Shield	5.0 inches (127mm)	60 degrees
Front	equals 10.0 inches (254mm)	0 degrees
Sides	equals 5.5 inches (140mm)	0 degrees
Rear	equals 2.25 inches (57mm)	0 degrees
Top	1.0 inches (25mm)	90 degrees

ARMAMENT

Primary: 105mm Gun M68E1 w/thermal shroud in Mount M140 in turret

Traverse: Hydraulic and manual	360 degrees
Traverse Rate: (max)	16 seconds/360 degrees
Elevation: Hydraulic and manual	+ 20 to − 10 degrees
Elevation Rate: (max)	4 degrees/second
Firing Rate: (max)	7 rounds/minute
Loading System:	Manual
Stabilizer System:	Azimuth and elevation

Secondary:
  (1) .50 caliber MG M85 in M19 cupola mount on turret
  (1) 7.62mm MG M240 coaxial w/105mm gun in turret
  Provision for (2) .45 caliber SMG M3A1
  (2) M239 smoke grenade launcher housings (6 tubes each) on turret

AMMUNITION

63 rounds 105mm		8 hand grenades
900 rounds .50 caliber		
360 rounds .45 caliber		
5950 rounds 7.62mm		

FIRE CONTROL AND VISION EQUIPMENT

Primary Weapon:	Direct	Indirect
	Range Finder	Azimuth Indicator
	AN/VVG-2	M28A1
	#Periscope M35E1	Elevation Quadrant
	Ballistic Drive	M13A3
	M10A3	Gunner's Quadrant
	Ballistic Computer	M1A1
	M21	
	Telescope M105D	
Vision Devices	Direct	Indirect
Driver	Hatch	Periscope M27 (3) and
		AN/VVS-2 passive
		viewer
Commander	Vision blocks (8)	# Periscope M36E1
	in cupola, hatch	
Gunner	None	#Periscope M35E1
Loader	Hatch	Periscope M37

Total Periscopes: M27 (3), AN/VVS-2 (1), M36E1 (1), M35E1 (1), M37 (1)
Total Vision Blocks: (8) in M19 cupola on turret top
Searchlight: AN/VSS-3, 1.0 KW, xenon white light or infrared

ENGINE

Make and Model: Continental AVDS-1790-2C	
Type: 12 cylinder, 4 cycle, 90 degree vee, supercharged	
Cooling System: Air Ignition: Compression	
Displacement:	1791.7 cubic inches
Bore and Stroke:	5.75 x 5.75 inches
Compression Ratio:	16:1
Net Horsepower (max):	643 hp at 2400 rpm
Gross Horsepower (max):	750 hp at 2400 rpm
Net Torque (max):	1575 ft-lb at 1750 rpm
Gross Torque (max):	1710 ft-lb at 1800 rpm
Weight:	4900 pounds, dry
Fuel: 40 cetane diesel	385 gallons
Engine Oil:	80 quarts

POWER TRAIN

Transmission: Cross-drive CD-850-6A, 2 ranges forward, 1 reverse
  Single stage multiphase hydraulic torque converter
  Stall multiplication: 4:1
  Overall Usable Ratios: low 12.0:1   reverse: 16.9:1
                         high 4.3:1

Steering Control: Mechanical, T-bar
  Steering Rate: 5.7 rpm
Brakes: Multiple disc
Final Drive: Spur gear   Gear Ratio: 5.08:1
Drive Sprocket: At rear of vehicle with 11 teeth
  Pitch Diameter: 24.504 inches

RUNNING GEAR

Suspension: Torsion bar
  12 individually sprung dual road wheels (6/track)
  Tire Size: 26 x 6 inches
  6 dual track return rollers (3/track)
  Dual compensating idler at front of each track
  Idler Tire Size: 26 x 6 inches
  Shock absorbers fitted on first 2 and last road wheels on each side
Tracks: Center guide, T142
  Type: (T142) Double pin, 28 inch width, replaceable rubber pads
  Pitch: 6.94 inches
  Shoes per Vehicle: 160 (80/track)
  Ground Contact Length: 166.72 inches

ELECTRICAL SYSTEM

Nominal Voltage: 24 volts DC
Alternator, oil cooled w/solid state regulator, 28 volts, 650 amperes, gear
  driven by main engine
Auxiliary Generator: None
Battery: (6) 12 volts, 3 sets of 2 in series connected in parallel

COMMUNICATIONS

Radio: AN/VRC-12, 46, 47 or 64 in turret bustle
  AN/VRC-24 (air to ground) also may be fitted
Interphone: 4 stations plus external box

NUCLEAR, BIOLOGICAL, CHEMICAL PROTECTION

  (1) M13A1 gas, particulate filter unit system, 20 cubic feet/minute
  (4) M25A1 tank protective mask

FIRE PROTECTION

  Automatic Halon fire extinguisher system

PERFORMANCE

Maximum Speed: Sustained, level road	30 miles/hour
Maximum Tractive Effort: TE at stall	77,000 pounds
Per Cent of Vehicle Weight: TE/W	67 per cent
Maximum Grade:	60 per cent
Maximum Trench:	8.5 feet
Maximum Vertical Wall:	36 inches
Maximum Fording Depth: w/o kit	48 inches
Minimum Turning Circle: (diameter)	pivot
Cruising Range: Roads	approx. 280 miles

---

  * The surface angles and wall thicknesses of the cast hull and turret varied widely from point to point. Therefore, where indicated the data are for an
    equivalent thickness at 0 degrees from the vertical.
  # Late M60A3s replaced the M35E1 and M36E1 image intensifier periscopes with the AN/VSG-2 tank thermal sight and the tank was designated as the
    M60A3 TTS.

FLAME THROWER TANK M67A2

GENERAL DATA

Crew:		3 men
Length: Gun forward		320.4 inches
Length: Gun in travel position		271.1 inches
Length: Without gun		270.5 inches
Gun Overhang: Gun forward		49.9 inches
Width: Over tracks		143.0 inches
Height: Over cupola periscope		121.6 inches
Tread:		115.0 inches
Ground Clearance:		16.5 inches
Fire Height:	approx.	79 inches
Turret Ring Diameter: (inside)		85.0 inches
Weight, Combat Loaded:	approx.	107,000 pounds
Weight, Unstowed:	approx.	102,000 pounds
Power to Weight Ratio: Net		12.0 hp/ton
Gross		14.0 hp/ton
Ground Pressure: Zero penetration		12.1 psi

*ARMOR

Type: Turret, cast homogeneous steel; Hull, cast homogeneous
steel; Welded assembly

Hull Thickness:	Actual	Angle w/Vertical
Front, Upper	4.33 inches (110mm)	60 degrees
Lower	4.0 to 2.4 inches (102-61mm)	53 degrees
Sides, Front	equals 3.0 inches (76mm)	0 degrees
Rear	equals 2.0 inches (51mm)	0 degrees
Rear, Grill	equals 1.0 inches (25mm)	0 degrees
Lower	1.6 to 1.2 inches (41-30mm)	30 to 60 degrees
Top	2.25 inches (57mm)	90 degrees
Floor, Front	1.5 inches (38mm)	90 degrees
Rear	1.0 inches (25mm)	90 degrees
Turret Thickness:		
Gun Shield	4.5 inches (114mm)	30 degrees
Front	equals 7.0 inches (178mm)	0 degrees
Sides	equals 3.0 inches (76mm)	0 degrees
Rear	equals 2.0 inches (51mm)	0 degrees
Top	1.0 inches (25mm)	90 degrees

ARMAMENT

Primary: Flame Thrower M7-6 in turret

Traverse: Hydraulic and manual	360 degrees
Traverse Rate: (max)	15 seconds/360 degrees
Elevation: Hydraulic and manual	+ 45 to − 12 degrees
Elevation Rate: (max)	4 degrees/second
Firing Duration: 3/4 inch nozzle	70 seconds
7/8 inch nozzle	60 seconds
Stabilizer System:	None

Secondary:
(1) .50 caliber MG HB M2 in cupola mount on turret
(1) 7.62mm MG M73 coaxial w/flame gun in turret
Provision for (1) .45 caliber SMG M3A1

AMMUNITION

365 gallons flame gun fuel   8 hand grenades
    (usable)
600 rounds .50 caliber
180 rounds .45 caliber
3500 rounds 7.62mm

FIRE CONTROL AND VISION EQUIPMENT

Primary Weapon:	Direct	Indirect
	Periscope M30C	Gunner's Quadrant M1A1
Vision Devices	Direct	Indirect
Driver	Hatch	Periscope M27 (3) and Periscope M24 (1) infrared
Commander	Vision blocks (5) in cupola, hatch	Periscope M28C (AA MG sight)
Gunner	None	Periscope M30C

Total Periscopes: M27 (3), M24 (1), M30C (1), M28C (1)
Total Vision Blocks: (5) in M1 cupola on turret top

ENGINE

Make and Model: Continental AVDS-1790-2A	
Type: 12 cylinder, 4 cycle, 90 degree vee, supercharged	
Cooling System: Air Ignition: Compression	
Displacement:	1791.7 cubic inches
Bore and Stroke:	5.75 x 5.75 inches
Compression Ratio:	16:1
Net Horsepower (max):	643 hp at 2400 rpm
Gross Horsepower (max):	750 hp at 2400 rpm
Net Torque (max):	1575 ft-lb at 1750 rpm
Gross Torque (max):	1710 ft-lb at 1800 rpm
Weight:	4700 pounds, dry
Fuel: 40 cetane diesel	385 gallons
Engine Oil:	72 quarts

POWER TRAIN

Transmission: Cross-drive CD-850-6A, 2 ranges forward, 1 reverse
    Single stage multiphase hydraulic torque converter
    Stall multiplication: 4:1
    Overall Usable Ratios: low 12.0:1   reverse: 16.9:1
                           high 4.3:1
Steering Control: Mechanical, steering wheel
    Steering Rate: 5.7 rpm
Brakes: Multiple disc
Final Drive: Spur gear   Gear Ratio: 5.08:1
Drive Sprocket: At rear of vehicle with 11 teeth
    Pitch Diameter: 24.504 inches

RUNNING GEAR

Suspension: Torsion bar
    12 individually sprung dual road wheels (6/track)
    Tire Size: 26 x 6 inches
    10 dual track return rollers (5/track)
    Dual compensating idler at front of each track
    Idler Tire Size: 26 x 6 inches
    Friction snubbers fitted on first 2 and last road wheels on each side
Tracks: Center guide, T97E2
    Type: (T97E2) Double pin, 28 inch width, rubber chevron
    Pitch: 6.94 inches
    Shoes per Vehicle: 158 (79/track)
    Ground Contact Length: 157.5 inches

ELECTRICAL SYSTEM

Nominal Voltage: 24 volts DC
Main Generator: (1) 24 volts, 300 amperes, gear driven by main engine
Auxiliary Generator: None
Battery: (6) 12 volts, 3 sets of 2 in series connected in parallel

COMMUNICATIONS

Radio: AN/VRC-12, 46, 47, 53, or 64 in turret bustle
        AN/VRC-24 (air to ground) also may be fitted
Interphone: 3 stations plus external box

NUCLEAR, BIOLOGICAL, CHEMICAL PROTECTION

Provision for (2) M8A2 gas particulate units (1 in hull, 1 in turret)

FIRE PROTECTION

(1) 30 pound carbon dioxide, fixed (2 shot)
(1) 5 pound carbon dioxide, portable

PERFORMANCE

Maximum Speed: Sustained, level road	30 miles/hour
Maximum Tractive Effort: TE at stall	75,400 pounds
Per Cent of Vehicle Weight: TE/W	70 per cent
Maximum Grade:	60 per cent
Maximum Trench:	8.5 feet
Maximum Vertical Wall:	36 inches
Maximum Fording Depth: w/o kit	48 inches
Minimum Turning Circle: (diameter)	pivot
Cruising Range: Roads	approx. 300 miles

*The surface angles and wall thicknesses of the cast hull and turret varied widely from point to point. Therefore, where indicated the data are for an
  equivalent thickness at 0 degrees from the vertical.

155mm SELF-PROPELLED GUN M53
and
8 inch SELF-PROPELLED HOWITZER M55

GENERAL DATA

Crew:		6	men
Length: Cannon in travel position, M53		382.5	inches
M55		311.4	inches
Length: Without cannon, M53		311.4	inches
M55		311.4	inches
Gun Overhang: Cannon in travel position, M53		71.1	inches
M55		0.0	inches
Width: Over fenders		141.0	inches
Height: Over AA MG mount w/o MG		136.6	inches
Tread:		110.0	inches
Ground Clearance:		18.5	inches
Fire Height:	approx.	95	inches
Turret Ring Diameter: (inside)		59.5	inches
Weight, Combat Loaded: M53	approx.	100,000	pounds
M55	approx.	98,000	pounds
Weight, Unstowed: M53	approx.	91,000	pounds
M55	approx.	90,000	pounds
Power to Weight Ratio: Net M53		14.1	hp/ton
M55		14.4	hp/ton
Gross M53		16.2	hp/ton
M55		16.5	hp/ton
Ground Pressure: Zero penetration M53		11.8	psi
M55		11.6	psi

ARMOR

Type: Turret, rolled homogeneous steel; Hull, rolled homogeneous
steel; Welded assembly

Hull Thickness:	Actual	Angle w/Vertical
Front, Upper	1.0 inches (25mm)	10 degrees
Lower	1.0 inches (25mm)	59 degrees
Sides	0.5 inches (13mm)	0 degrees
Rear	0.5 inches (13mm)	22 degrees
Top	0.5 inches (13mm)	90 degrees
Floor	0.5 inches (13mm)	90 degrees
Turret Thickness:		
Front	0.5 inches (13mm)	20 degrees
Sides	0.5 inches (13mm)	0 degrees
Rear	0.5 inches (13mm)	0 degrees
Top	0.5 inches (13mm)	90 degrees

ARMAMENT

* Primary: M53, 155mm Gun M46 (T80) in Mount M86 (T58) in turret
M55, 8 inch Howitzer M47 (T89) in Mount M86 (T58) in turret

Traverse: Hydraulic and manual	30 degrees left, 30 degrees right
Traverse Rate: 1:5 ratio	10 degrees/second
1:1 ratio	3 degrees/second
Elevation: Hydraulic and manual	+ 65 to − 5 degrees
Elevation Rate: 1:5 Ratio	7.5 degrees/second
1:1 ratio	1.5 degrees/second
Firing Rate: (max)	1 round/minute
Loading System:	Manual w/power rammer
Stabilizer System:	None

Secondary:
(1) .50 caliber MG HB M2 flexible AA mount on cupola
Provision for (1) .45 caliber SMG M3A1
Provision for (5) .30 caliber Carbine M2

AMMUNITION

10 rounds 8 inch (M55)	8 hand grenades
20 rounds 155mm (M53)	
900 rounds .50 caliber	
180 rounds .45 caliber	
900 rounds .30 caliber (carbine)	

FIRE CONTROL AND VISION EQUIPMENT

Primary Weapon:	Direct	Indirect
	Telescope M99	Panoramic Telescope M100
		Azimuth Indicator T27
		Gunner's Quadrant M1 or M1A1

Vision Devices	Direct	Indirect
Driver	Hatch	Periscope M17 (4)
Commander	Vision blocks (6) in cupola, hatch	Periscope M15A1
Gunner	None	Periscope M13 or M13B1
Cannoneers	None	None

Total Periscopes: M17 (4), M15A1 (1), M13 or M13B1 (1)
Total Vision Blocks: (6) in cupola on turret top

ENGINE

Make and Model: Continental AV-1790-5B or AV-1790-7B	
(data for AV-1790-7B)	
Type: 12 cylinder, 4 cycle, 90 degree vee	
Cooling System: Air Ignition: Magneto	
Displacement:	1791.7 cubic inches
Bore and Stroke:	5.75 x 5.75 inches
Compression Ratio:	6.5:1
Net Horsepower (max):	704 hp at 2800 rpm
Gross Horsepower (max):	810 hp at 2800 rpm
Net Torque (max):	1440 ft-lb at 2000 rpm
Gross Torque (max):	1610 ft-lb at 2200 rpm
Weight:	2581 pounds, dry
Fuel: 80 octane gasoline	380 gallons
Engine Oil:	64 quarts

POWER TRAIN

Transmission: Cross-drive CD-850-4B, 2 ranges forward, 1 reverse
Single stage multiphase hydraulic torque converter
Stall multiplication: 4:1

Overall Usable Ratios: low 13.0:1   reverse: 18.1:1
high 4.5:1

Steering Control: Mechanical, wobble stick (early vehicles), steering wheel
(late vehicles), hydraulic servo on very late equipment
Steering Rate: 5.7 rpm
Brakes: Multiple disc
Final Drive: Spur gear   Gear Ratio: 5.08:1
Drive Sprocket: At rear of vehicle with 13 teeth
Pitch Diameter: 25.038 inches

RUNNING GEAR

Suspension: Torsion bar, trailing idler
14 individually sprung dual road wheels (7/track)
Tire Size: 26 x 6 inches
6 dual track return rollers (3/track)
The rear road wheel (trailing idler) maintains and permits adjustment
of track tension
Shock absorbers fitted on first 2 road wheels on each side
Tracks: Center guide, T80E6 and T84E1
Type: (T80E6) Double pin, 23 inch width, rubber backed steel
(T84E1) Double pin, 23 inch width, rubber chevron
Pitch: 6 inches
Shoes per Vehicle: 179 (89 left track, 90 right track)
Ground Contact Length: 184 inches

ELECTRICAL SYSTEM

Nominal Voltage: 24 volts DC
Main Generator: (1) 24 volts, 300 amperes, gear driven by main engine
Auxiliary Generator: (1) 28 volts, 300 amperes, driven by auxiliary engine
Battery: (4) 12 volts, 2 sets of 2 in series connected in parallel

COMMUNICATIONS

Radio: AN/PRC-9 and AMP-AM-598/U
Interphone: AN/UIC-1, 4 stations plus extension kit

FIRE PROTECTION

(3) 10 pound carbon dioxide, fixed
(1) 5 pound carbon dioxide, portable

PERFORMANCE

Maximum Speed: Sustained, level road		35 miles/hour
Maximum Tractive Effort: TE at stall		83,200 pounds
Per Cent of Vehicle Weight: TE/W, M53		83 per cent
M55		85 per cent
Maximum Grade:		60 per cent
Maximum Trench:		8.0 feet
Maximum Vertical Wall:		42 inches
Maximum Fording Depth:		60 inches
Minimum Turning Circle: (diameter)		pivot
Cruising Range: Roads	approx.	150 miles

*The primary weapons and stowage arrangements are interchangeable between the two vehicles.

175mm SELF-PROPELLED GUN T162

GENERAL DATA

Crew:		6	men
Length: Gun in travel position		518.5	inches
Length: Without gun	approx.	319	inches
Gun Overhang: Gun in travel position	approx.	200	inches
Width: Over tracks		143.0	inches
Height: Over cupola		132.0	inches
Tread:		115.0	inches
Ground Clearance:		18.0	inches
Fire Height:	approx.	95	inches
Turret Ring Diameter: (inside)		67.6	inches
Weight, Combat Loaded:	approx.	107,000	pounds
Weight, Unstowed:	approx.	98,000	pounds
Power to Weight Ratio: Net		14.3	hp/ton
Gross		18.3	hp/ton
Ground Pressure: Zero penetration		10.1	psi

ARMOR

Type: Turret, rolled homogeneous steel; Hull, rolled homogeneous steel; Welded assembly

Hull Thickness:	Actual	Angle w/Vertical
Front, Upper	0.75 inches (19mm)	10 degrees
Lower	0.75 inches (19mm)	57 degrees
Sides	0.5 inches (13mm)	0 degrees
Rear	0.75 inches (19mm)	22 degrees
Top	0.5 inches (13mm)	90 degrees
Floor	0.5 inches (13mm)	90 degrees
Turret Thickness:		
Front	0.5 inches (13mm)	21 degrees
Sides	0.5 inches (13mm)	0 degrees
Rear	0.5 inches (13mm)	0 degrees
Top	0.5 inches (13mm)	90 degrees

ARMAMENT

Primary: 175mm Gun T181 in Mount T158 in turret

*Traverse: Hydraulic and manual	30 degrees left, 30 degrees right
Traverse Rate: 1:5 ratio	10 degrees/second
1:1 ratio	3 degrees/second
*Elevation: Hydraulic and manual	+65 to −5 degrees
Elevation Rate: 1:5 Ratio	7.5 degrees/second
1:1 ratio	1.5 degrees/second
Firing Rate: (max)	1 round/minute
Loading System:	Manual w/power rammer
Stabilizer System:	None

Secondary:
(1) .50 caliber MG HB M2 flexible AA mount on cupola
Provision for (6) .30 caliber Carbine M2

AMMUNITION

15 rounds 175mm	8 hand grenades
900 rounds .50 caliber	
900 rounds .30 caliber (carbine)	

FIRE CONTROL AND VISION EQUIPMENT

Primary Weapon:	Direct	Indirect
	Telescope T159E1	Panoramic Telescope T149E1
		Azimuth Indicator T27
		Gunner's Quadrant M1 or M1A1
Vision Devices	Direct	Indirect
Driver	Hatch	Periscope M17 (4)
Commander	Vision blocks (6) in cupola, hatch	Periscope M15A1
Gunner	None	Periscope M13 or M13B1
Cannoneers	None	None

Total Periscopes: M17 (4), M15A1 (1), M13 or M13B1 (1)
Total Vision Blocks: (6) in cupola on turret top

*Traverse and elevation were electrically powered and manual on pilot number 3.

ENGINE

Make and Model: Continental AVSI-1790-6	
Type: 12 cylinder, 4 cycle, 90 degree vee, supercharged, fuel injection	
Cooling System: Air Ignition: Magneto	
Displacement:	1791.7 cubic inches
Bore and Stroke:	5.75 x 5.75 inches
Compression Ratio:	5.5:1
Net Horsepower (max):	765 hp at 2800 rpm
Gross Horsepower (max):	980 hp at 2800 rpm
Net Torque (max):	1670 ft-lb at 2100 rpm
Gross Torque (max):	1870 ft-lb at 2400 rpm
Weight:	3050 pounds, dry
Fuel: 80 octane gasoline	420 gallons
Engine Oil:	64 quarts

POWER TRAIN

Transmission: Cross-drive XT-1400-3, 3 ranges forward, 1 reverse
Single stage multiphase hydraulic torque converter w/lock-up clutch
Stall multiplication: 3.6:1

Overall Usable Ratios: low	112:1	reverse:	121.3:1
intermediate	52.3:1		
high	24.4:1		

Steering Control: Mechanical T-bar
Steering Rate: 5.6 rpm
Brakes: Multiple disc
Final Drive: Spur gear   Gear Ratio: 4.63:1
Drive Sprocket: At rear of vehicle with 13 teeth
Pitch Diameter: 28.9 inches (estimated)

RUNNING GEAR

Suspension: Torsion bar, trailing idler
14 individually sprung dual road wheels (7/track)
Tire Size: 26 x 6 inches
6 dual track return rollers (3/track)
The rear road wheel (trailing idler) maintains and permits adjustment of track tension
Shock absorbers fitted on first 2 road wheels on each side
Tracks: Center guide, T97E1
Type: (T97E1) Double pin, 28 inch width, rubber chevron
Pitch: 6.94 inches
Shoes per Vehicle: 159 (79 left track, 80 right track)
Ground Contact Length: 190 inches

ELECTRICAL SYSTEM

Nominal Voltage: 24 volts DC
Main Generator: (1) 24 volts, 300 amperes, gear driven by main engine
Auxiliary Generator: (1) 28 volts, 300 amperes, driven by auxiliary engine
Battery: (4) 12 volts, 2 sets of 2 in series connected in parallel

COMMUNICATIONS

Radio: AN/PRC-9 and AMP-AM-598/U
Interphone: AN/UIC-1, 4 stations plus extension kit

FIRE PROTECTION

(3) 10 pound carbon dioxide, fixed
(1) 5 pound carbon dioxide, portable

PERFORMANCE

Maximum Speed: Sustained, level road	35 miles/hour
Maximum Tractive Effort: TE at stall	131,000 pounds
Per Cent of Vehicle Weight: TE/W	122 per cent
Maximum Grade:	60 per cent
Maximum Trench:	8.5 feet
Maximum Vertical Wall:	40 inches
Maximum Fording Depth:	48 inches
Minimum Turning Circle: (diameter)	pivot
Cruising Range: Roads	approx. 170 miles

MEDIUM RECOVERY VEHICLE M88

GENERAL DATA

Crew:	4	men
Length: Boom and spade in travel position	325.5	inches
Width: Over tracks	135.0	inches
Height: Over AA MG approx.	127	inches
Tread:	107.0	inches
Ground Clearance:	17.0	inches
Weight, Combat Loaded: approx.	110,000	pounds
Weight, Unstowed: approx.	106,000	pounds
Power to Weight Ratio: Net	14.8	hp/ton
Gross	17.8	hp/ton
Ground Pressure: Zero penetration	10.9	psi

ARMOR

Type: Hull and Cab, rolled and cast homogeneous steel;
Welded assembly

Hull and Cab

Thickness:	Actual	Angle w/Vertical
Front, Upper	1.25 inches (32mm)	26 degrees
Lower	1.5 inches (38mm)	47 degrees
Sides, Upper	0.75 inches (19mm)	15 degrees
Lower	1.0 inches (25mm)	0 degrees
Rear, Upper	1.0 inches (25mm)	10 degrees
Lower	1.25 inches (32mm)	0 degrees
Top	0.75 inches (19mm)	90 degrees
Floor, Front	1.5 inches (38mm)	90 degrees
Rear	1.0 inches (25mm)	90 degrees

ARMAMENT

(1) .50 caliber MG HB M2 flexible AA mount on cupola
Provision for (1) 3.5 inch Rocket Launcher M20 or M20B1
Provision for (2) .45 caliber SMG M3A1
Provision for (2) 7.62mm Rifle M14

AMMUNITION

1500 rounds .50 caliber	12 hand grenades
10 rockets 3.5 inch	
360 rounds .45 caliber	
300 rounds 7.62mm	

RECOVERY EQUIPMENT

Spade: Hydraulically operated on front of vehicle
Main Winch: 90,000 pound capacity, hydraulically operated, located below crew compartment with 200 feet of 1.25 inch cable
Boom: Hydraulically operated tubular "A" frame mounted by trunnion levers on each side of cab with a live boom lifting capacity of 50,000 pounds
Hoist Winch: Hydraulically operated, located below crew compartment with a hoisting capacity of 50,000 pounds (w/4 part 5/8 inch line)

VISION EQUIPMENT

	Direct	Indirect
Driver	Vision block (1) and hatch	Periscope M17 (3) and Periscope M24 (1) infrared
Mechanic	Vision block (1) and hatch	Periscope M17 (3)
Commander	Vision blocks (6) in cupola, hatch	
Rigger	Vision blocks (2) and hatch	Periscope M17 (1) and Periscope M24 (1) infrared

Total Periscopes: M17 (7), M24 infrared (2)
Total Vision Blocks: (10), (6) in cupola, (2) in cab front, (2) in cab rear

ENGINE

Make and Model: AVSI-1790-6A	
Type: 12 cylinder, 4 cycle, 90 degree vee, supercharged, fuel injection	
Cooling System: Air Ignition: Magneto	
Displacement:	1791.7 cubic inches
Bore and Stroke:	5.75 x 5.75 inches
Compression Ratio:	5.5:1
Net Horsepower (max):	814 hp at 2800 rpm
Gross Horsepower (max):	980 hp at 2800 rpm
Net Torque (max):	1600 ft-lb at 2100 rpm
Gross Torque (max):	1940 ft-lb at 2300 rpm
Weight:	3050 pounds, dry
Fuel: 80 octane gasoline	445 gallons
Engine Oil:	64 quarts

POWER TRAIN

Transmission: Cross-drive XT-1410-2A, 3 ranges forward, 1 reverse
Single stage multiphase hydraulic torque converter w/lock-up clutch
Stall multiplication: 3.6:1

Overall Usable Ratios:	low	112:1	reverse:	121.3:1
	intermediate	52.3:1		
	high	24.5:1		

Steering Control: Mechanical, steering wheel
Steering Rate: 5.6 rpm
Brakes: Multiple disc
Final Drive: Spur gear   Gear Ratio: 4.63:1
Drive Sprocket: At rear of vehicle with 11 teeth
Pitch Diameter: 25.000 inches

RUNNING GEAR

Suspension: Torsion bar
12 individually sprung dual road wheels (6/track)
Tire Size: 26 x 6 inches
6 dual track return rollers (3/track)
Dual compensating idler at front of each track
Idler Tire Size: 26 x 6 inches
Shock absorbers fitted on first 2 and last road wheels on each side
Tracks: Center guide, T107
Type: (T107) Double pin, 28 inch width, rubber chevron
Pitch: 7.09 inches
Shoes per Vehicle: 168 (84/track)
Ground Contact Length: 180 inches

ELECTRICAL SYSTEM

Nominal Voltage: 24 volts DC
Main Generator: (1) 24 volts, 300 amperes, gear driven by main engine
Auxiliary Generator: (1) 28 volts, 300 amperes, driven by auxiliary engine

COMMUNICATIONS

Radio: AN/GRC 3, 6, or 8 or AN/VRC 13, 14, or 15 in right sponson
Interphone: AN/UIC-1, 4 stations

FIRE PROTECTION

(8) 10 pound carbon dioxide, fixed (2 shot)
(2) 5 pound carbon dioxide, portable

PERFORMANCE

Maximum Speed: Sustained, level road	30 miles/hour
Maximum Tractive Effort: TE at stall	152,000 pounds
Per Cent of Vehicle Weight: TE/W	138 per cent
Maximum Grade:	60 per cent
Maximum Trench:	8.6 feet
Maximum Vertical Wall:	42 inches
Maximum Fording Depth:	64 inches
Minimum Turning Circle: (diameter)	pivot
Cruising Range: Roads approx.	200 miles

COMBAT ENGINEER VEHICLE M728

GENERAL DATA
Crew:	4	men
Length: Boom and dozer in travel position	347.8	inches
Length: Without boom and dozer	273.5	inches
Width: Over dozer blade	146.0	inches
Height: Over cupola periscope	128.2	inches
Tread:	115.0	inches
Ground Clearance: At escape hatch	15.3	inches
Fire Height:	82.6	inches
Turret Ring Diameter: (inside)	85.0	inches
Weight, Combat Loaded:	approx. 115,000	pounds
Weight, Unstowed:	approx. 107,000	pounds
Power to Weight Ratio: Net	11.2	hp/ton
Gross	13.0	hp/ton
Ground Pressure: Zero penetration	12.3	psi

*ARMOR
Type: Turret, cast homogeneous steel; Hull, cast homogeneous
    steel; Welded assembly

Hull Thickness:	Actual	Angle w/Vertical
Front, Upper	4.29 inches (109mm)	65 degrees
Lower	5.63 to 3.35 inches (143-85mm)	55 degrees
Sides	1.41 to 2.9 inches (36-74mm)	45 to 0 degrees
Rear, Grill	equals 1.0 inches (25mm)	0 degrees
Lower	1.6 to 1.2 inches (41-30mm)	30 to 60 degrees
Top	1.41 inches (36mm)	90 degrees
Floor, Front	0.75 inches (19mm)	90 degrees
Rear	0.50 inches (13mm)	90 degrees

Turret Thickness:		
Gun Shield	5.0 inches (127mm)	60 degrees
Front	equals 10.0 inches (254mm)	0 degrees
Sides	equals 5.5 inches (140mm)	0 degrees
Rear	equals 2.25 inches (57mm)	0 degrees
Top	1.0 inches (25mm)	90 degrees

ARMAMENT
Primary: 165mm Gun M135 in Mount M150 in turret
Traverse: Hydraulic and manual	360 degrees
Traverse Rate: (max)	40 seconds/360 degrees
Elevation: Hydraulic and manual	+ 20 to − 10 degrees
Elevation Rate: (max)	4 degrees/second
Firing Rate: (max)	approx. 2 rounds/minute
Loading System:	Manual
Stabilizer System:	None

Secondary:
(1) .50 caliber MG M85 in M19 cupola on turret
(1) 7.62mm MG M73, M219, or M240 coaxial w/165mm gun in turret
Provision for (2) .45 caliber SMG M3A1

AMMUNITION
30 rounds 165mm	12 hand grenades
600 rounds .50 caliber	
360 rounds .45 caliber	
2000 rounds 7.62mm	

FIRE CONTROL AND VISION EQUIPMENT
Primary Weapon:	Direct	Indirect
	Periscope M32C	Azimuth Indicator
	(day & IR)	M28E2
	Telescope M105F	Elevation Quadrant
		M13A3
		Gunner's Quadrant
		M1A1

Vision Devices	Direct	Indirect
Driver	Hatch	Periscope M27 (3) and
		Periscope M24 (1)
		infrared
Commander	Vision blocks (8)	Periscope M34 or
	in cupola, hatch	M36 (day & IR)
Gunner	None	Periscope M32C
		(day & IR)
Loader	Hatch	Periscope M37

Total Periscopes: M27 (3), M24 (1), M34 or M36 (1), M32C (1), M37 (1)
Total Vision Blocks: (8) in M19 cupola on turret top
Searchlight: AN/VSS-3, 1.0 KW, xenon white light or infrared

ENGINE
Make and Model: Continental AVDS-1790-2D	
Type: 12 cylinder, 4 cycle, 90 degree vee, supercharged	
Cooling System: Air Ignition: Compression	
Displacement:	1791.7 cubic inches
Bore and Stroke:	5.75 x 5.75 inches
Compression Ratio:	16:1
Net Horsepower (max):	643 hp at 2400 rpm
Gross Horsepower (max):	750 hp at 2400 rpm
Net Torque (max):	1575 ft-lb at 1750 rpm
Gross Torque (max):	1710 ft-lb at 1800 rpm
Weight:	4880 pounds, dry
Fuel: 40 cetane diesel	385 gallons
Engine Oil:	80 quarts

POWER TRAIN
Transmission: Cross-drive CD-850-6A, 2 ranges forward, 1 reverse
    Single stage multiphase hydraulic torque converter
        Stall multiplication: 4:1
        Overall Usable Ratios: low 12.0:1   reverse: 16.9:1
                high 4.3:1
Steering Control: Mechanical, T-bar
    Steering Rate: 5.7 rpm
Brakes: Multiple disc
Final Drive: Spur gear   Gear Ratio: 5.08:1
Drive Sprocket: At rear of vehicle with 11 teeth
    Pitch Diameter: 24.504 inches

RUNNING GEAR
Suspension: Torsion bar
    12 individually sprung dual road wheels (6/track)
    Tire Size: 26 x 6 inches
    6 dual track return rollers (3/track)
    Dual compensating idler at front of each track
    Idler Tire Size: 26 x 6 inches
    Shock absorbers fitted on first 2 and last road wheels on each side
Tracks: Center guide, T142
    Type: (T142) Double pin, 28 inch width, replaceable rubber pads
    Pitch: 6.94 inches
    Shoes per Vehicle: 160 (80/track)
    Ground Contact Length: 166.72 inches

ENGINEER EQUIPMENT
Boom: Turret mounted tubular "A" frame w/360 degree traverse
    Hoisting Capacity: 17,500 pounds (single line)
Winch: Planetary gear, two-speed, hydraulically operated
    Winch Capacity: 25,000 pounds w/200 feet of ¼ inch cable
Bulldozer: Hydraulically operated on front of vehicle w/146 inch
        wide blade

ELECTRICAL SYSTEM
Nominal Voltage: 24 volts DC
Main Generator: (1) 24 volts, 300 amperes, gear driven by main engine
Auxiliary Generator: None
Battery: (6) 12 volts, 3 sets of 2 in series connected in parallel

COMMUNICATIONS
Radio: AN/VRC-46 or AN/VRC-53
Interphone: AN/VIC-1, 4 stations plus external box

NUCLEAR, BIOLOGICAL, CHEMICAL PROTECTION
(1) M13A1 gas, particulate filter unit system, 20 cubic feet/minute
(4) M25A1 tank protective mask

FIRE PROTECTION
(3) 10 pound carbon dioxide, fixed (2 shot)
(1) 5 pound carbon dioxide, portable

PERFORMANCE
Maximum Speed: Sustained, level road	30 miles/hour
Maximum Tractive Effort: TE at stall	77,000 pounds
Per Cent of Vehicle Weight: TE/W	67 per cent
Maximum Grade:	60 per cent
Maximum Trench:	8.25 feet
Maximum Vertical Wall:	30 inches
Maximum Fording Depth:	48 inches
Minimum Turning Circle: (diameter)	pivot
Cruising Range: Roads	approx. 280 miles

* The surface angles and wall thicknesses of the cast hull and turret varied widely from point to point. Therefore, where indicated the data are for an equivalent thickness at 0 degrees from the vertical.

WEAPON DATA SHEETS

The following data sheets list the general characteristics of the weapons mounted on various models of the Patton and M60 series tanks. Some of the self-propelled artillery weapons which utilized the tank chassis or its components also are included. Penetration performance of the armor piercing ammunition has been omitted because many of the rounds are still subject to security restrictions. Other details of some later types of ammunition also have been left out for the same reasons.

The various dimensions listed in the data sheets are defined in the sketch below.

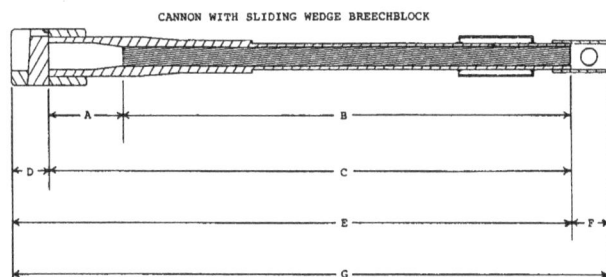

CANNON WITH SLIDING WEDGE BREECHBLOCK

A. Length of Chamber (to rifling)
B. Length of Rifling
C. Length of Bore
D. Depth of Breech Recess
E. Length, Muzzle to Rear Face of Breech
F. Additional Length, Blast Deflector, Etc.
G. Overall Length
H. Length, Breechblock and Firing Lock
I. Length of Tube
J. Length of Separable Chamber
K. Length of Tube and Chamber

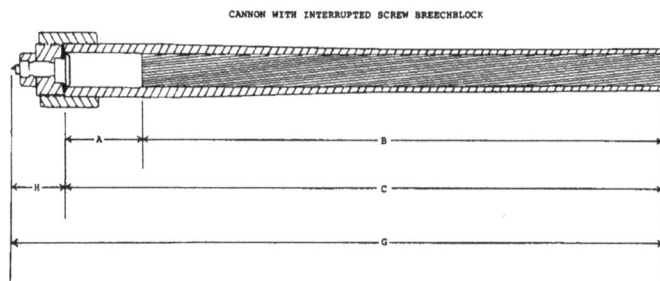

The ammunition is listed according to the official U.S. army nomenclature in use during its period of greatest service. Since this did change and sometimes needed clarification, a standard nomenclature is added in parentheses based on the following terms which are used separately and in combination.

AP	Armor piercing, uncapped
APBC	Armor piercing with ballistic cap
APCBC	Armor piercing with armor piercing cap and ballistic cap
APCR	Armor piercing, composite rigid
APDS	Armor piercing, discarding sabot
APFSDS	Armor piercing, fin stabilized, discarding sabot
HE	High explosive
HEAT	High explosive antitank, shaped charge
HESH	High explosive squash head
TP	Target practice
TPBC	Target practice with ballistic cap
TPCR	Target practice, composite rigid
MP	Multipurpose
T	Tracer

All armor plate angles in the data sheets are measured between a vertical plane and the plate surface.

The early tank guns described in this volume were fitted with single baffle muzzle brakes. These were found on all of the M46 and M46A1 tanks, but they were replaced on the production M47 and M48 series tanks by a cylindrical blast deflector. Frequently referred to in many documents as a counterweight, it served another important function in directing the muzzle blast to each side reducing target obscuration. The later T-shape blast deflectors were designed to optimize this performance and they were either cast or fabricated by welding.

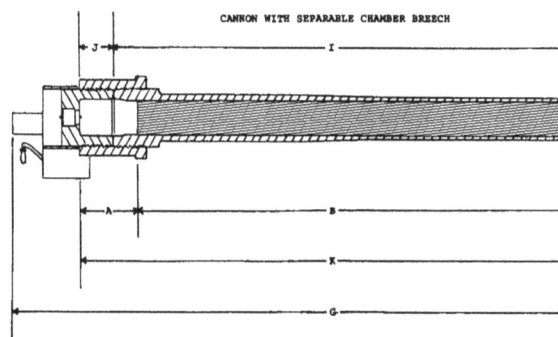

CANNON WITH INTERRUPTED SCREW BREECHBLOCK

CANNON WITH SEPARABLE CHAMBER BREECH

449

90mm Guns M3A1 and M3A2

Carriage and Mount	Medium Tank M26A1 in Mount M67A1, Medium Tanks M46 and M46A1 in Mount M73	
Length of Chamber (to rifling)	24.8 inches	
Length of Rifling	152.4 inches	
Length of Chamber (to projectile base)	20.8 inches (boat-tailed projectiles)	
Travel of Projectile in Bore	156.4 inches (boat-tailed projectiles)	
Length of Bore	177.15 inches, 50.0 calibers	
Depth of Breech Recess	9.00 inches	
Length, Muzzle to Rear Face of Breech	186.15 inches, 52.5 calibers	
Additional Length, Muzzle Brake M3E2	6.75 inches	
Overall Length	192.90 inches	
Diameter of Bore	3.543 inches	
Chamber Capacity	300 cubic inches	
Weight, Tube	approx. 1465 pounds	
Total Weight	approx. 2375 pounds	
Type of Breechblock	Semiautomatic, vertical sliding wedge	
Rifling	32 grooves, uniform right-hand twist, one turn in 32 calibers	
Ammunition	Fixed	
Primer	Percussion	
*Weight, Complete Round	AP-T M318(T33E7) Shot(APBC-T)	43.98 pounds(20.0 kg)
	APC-T M82 Projectile(APCBC/HE-T)	43.87 pounds(19.9 kg)
	HVAP-T M332 Shot(APCR-T)	32.54 pounds(14.8 kg)
	HVAP-T M304 Shot(APCR-T)	37.13 pounds(16.9 kg)
	HEAT-T T108E46 Shell(HEAT-T)	34.79 pounds(15.8 kg)
	HE M71 Shell(HE)	41.93 pounds(19.1 kg)
	WP M313 Shell(Smoke)	42.52 pounds(19.3 kg)
	HVTP-T M333 Shot(TPCR-T)	32.30 pounds(14.7 kg)
Weight, Projectile	AP-T M318(T33E7) Shot(APBC-T)	24.18 pounds(11.0 kg)
	APC-T M82 Projectile(APCBC/HE-T)	24.11 pounds(11.0 kg)
	HVAP-T M332 Shot(APCR-T)	12.44 pounds(5.7 kg)
	HVAP-T M304 Shot (APCR-T)	16.80 pounds(7.6 kg)
	HEAT-T T108E46 Shell (HEAT-T)	14.35 pounds(6.5 kg)
	HE M71 Shell(HE)	23.29 pounds(10.6 kg)
	WP M313 Shell(Smoke)	23.64 pounds(10.7 kg)
	HVTP-T M333 Shot(TPCR-T)	12.20 pounds(5.5 kg)
Maximum Powder Pressure	38,000 psi	
Maximum Rate of Fire	8 rounds/minute	
Muzzle Velocity	AP-T M318(T33E7) Shot(APBC-T)	2800 ft/sec(853 m/sec)
	APC-T M82 Projectile(APCBC/HE-T)	2800 ft/sec(853 m/sec)
	HVAP-T M332 Shot(APCR-T)	3875 ft/sec(1178 m/sec)
	HVAP-T M304 Shot(APCR-T)	3350 ft/sec(1021 m/sec)
	HEAT-T T108E46 Shell(HEAT-T)	2800 ft/sec(853 m/sec)
	HE M71 Shell(HE)	2700 ft/sec(823 m/sec)
	WP M313 Shell(Smoke)	2700 ft/sec(823 m/sec)
	HVTP-T M333 Shot(TPCR-T)	3875 ft/sec(1178 m/sec)
Muzzle Energy of Projectile, KE = ½MV²	AP-T M318(T33E7) Shot(APBC-T)	1314 ft-tons
Rotational energy is neglected and	APC-T M82 Projectile(APCBC/HE-T)	1310 ft-tons
values are based on long tons	HVAP-T M332 Shot(APCR-T)	1295 ft-tons
(2240 pounds)	HVAP-T M304 Shot(APCR-T)	1307 ft-tons
	HEAT-T T108E46 Shell(HEAT-T)	780 ft-tons
	HE M71 Shell(HE)	1177 ft-tons
	WP M313 Shell(Smoke)	1195 ft-tons
	HVTP-T M333 Shot(TPCR-T)	1270 ft-tons
Maximum Range (independent of mount)	AP-T M318(T33E7) Shot(APBC-T)	21,400 yards(19,568 m)
	APC-T M82 Projectile(APCBC/HE-T)	21,400 yards(19,568 m)
	HVAP-T M332 Shot(APCR-T)	15,700 yards(14,356 m)
	HVAP-T M304 Shot(APCR-T)	15,700 yards(14,356 m)
	HEAT-T T108E46 Shell(HEAT-T)	13,010 yards(11,896 m)
	HE M71 Shell(HE)	19,560 yards(17,886 m)
	WP M313 Shell(Smoke)	19,375 yards(17,717 m)
	HVTP-T M333 Shot(TPCR-T)	15,000 yards(13,716 m)

*Except for the HEAT-T T108E46 round, the data are for ammunition assembled with the brass M19 cartridge case (weight 11.0 pounds). These rounds also were assembled with the steel M19B1 cartridge case (weight 10.3 pounds). The T108E46 used the T27 cartridge case.

The M3A1 and M3A2 guns were identical in design. The breech ring of the M3A2 had improved metallurgical properties permitting it to be retubed five times and it was safe for use to −65 degrees Fahrenheit. The M3A1 was rated for use at −40 degrees F.

90mm Guns T119, M36(T119E1), M41(T139), and T178

Carriage and Mount	90mm Gun Tank T42 in Mount T139(T119 Gun), 90mm Gun Tanks M47, M47E1, M47E2, and M47M in Mount M78(M36 Gun), 90mm Gun Tanks T48, M48, M48A1, M48A2, and M48A3 in Mounts M87 or M87A1(M41 Gun), 90mm Gun Tank T69(T178 Gun)
Length of Chamber (to rifling)	24.4 inches
Length of Rifling	152.77 inches
Length of Chamber (to projectile base)	20.8 inches (boat-tailed projectiles)
Travel of Projectile in Bore	156.4 inches (boat-tailed projectiles)
Length of Bore	177.15 inches, 50.0 calibers
Depth of Breech Recess	9.00 inches
Length, Muzzle to Rear Face of Breech	186.15 inches, 52.5 calibers
Additional Length	7.0 inches, muzzle brake on early guns
	7.2 inches, T-shape blast deflector on late guns
Overall Length	193.2 inches w/muzzle brake
	193.4 inches w/T-shape blast deflector
Diameter of Bore	3.543 inches
Chamber Capacity	300 cubic inches
Weight, Tube	1760 pounds (M36), 1580 pounds (M41)
Total Weight approx.	2650 pounds (M36), 2370 pounds (M41)
Type of Breechblock	Semiautomatic, vertical sliding wedge
Rifling	32 grooves, uniform right-hand twist, one turn in 25 calibers
Ammunition	Fixed
Primer	Percussion

Weight, Complete Round	AP-T M318(T33E7) Shot(APBC-T)	**	43.91 pounds(19.9 kg)
	HEAT-T M431 Shell(HEAT-T)		32.25 pounds(14.6 kg)
	HE-T T91E3 Shell(HE-T)	*	36.25 pounds(16.5 kg)
	HE-T M71A1 Shell(HE-T)	#	39.54 pounds(17.9 kg)
	APERS-T XM580E1 (4100 flechettes)		41.25 pounds(18.7 kg)
	Canister M336 (1280 pellets)	*	42.50 pounds(19.3 kg)
	Canister M377 (5600 flechettes)	**	39.30 pounds(17.8 kg)
	TP-T M353(T225E1) Shot(TPBC-T)	**	43.91 pounds(19.9 kg)
Weight, Projectile	AP-T M318(T33E7) Shot(APBC-T)		24.18 pounds(11.0 kg)
	HEAT-T M431 Shell(HEAT-T)		12.75 pounds(5.8 kg)
	HE-T T91E3 Shell(HE-T)		20.25 pounds(9.2 kg)
	HE-T M71A1 Shell(HE-T)		23.57 pounds(10.7 kg)
	APERS-T XM580E1 (4100 flechettes) approx.		20 pounds(9 kg)
	Canister M336 (1280 pellets)		23.24 pounds(10.5 kg)
	Canister M377 (5600 flechettes)		20.44 pounds(9.3 kg)
	TP-T M353(T225E1) Shot(TPBC-T)		24.18 pounds(11.0 kg)
Maximum Powder Pressure	47,000 psi		
Maximum Rate of Fire	8 rounds/minute		
Muzzle Velocity	AP-T M318(T33E7) Shot(APBC-T)		3000 ft/sec(914 m/sec)
	HEAT-T M431 Shell(HEAT-T)		4000 ft/sec(1219 m/sec)
	HE-T T91E3 Shell(HE-T)		2400 ft/sec(732 m/sec)
	HE-T M71A1 Shell(HE-T)		2400 ft/sec(732 m/sec)
	APERS-T XM580E1 (4100 flechettes)		3000 ft/sec(914 m/sec)
	Canister M336 (1280 pellets)		2870 ft/sec(875 m/sec)
	Canister M377 (5600 flechettes)		2950 ft/sec(899 m/sec)
	TP-T M353(T225E1) Shot(TPBC-T)		3000 ft/sec(914 m/sec)
Muzzle Energy of Projectile, KE = ½MV²	AP-T M318(T33E7) Shot(APBC-T)		1509 ft-tons
Rotational energy is neglected and	HEAT-T M431 Shell(HEAT-T)		1414 ft-tons
values are based on long tons	HE-T T91E3 Shell(HE-T)		809 ft-tons
(2240 pounds)	HE-T M71A1 Shell(HE-T)		941 ft-tons
	APERS-T XM580E1 (4100 flechettes)		1250 ft-tons approx.
	Canister M336 (1280 pellets)		1327 ft-tons
	Canister M377 (5600 flechettes)		1230 ft-tons
	TP-T M353(T225E1) Shot(TPBC-T)		1509 ft-tons
Maximum Range (independent of mount)	AP-T M318(T33E7) Shot(APBC-T)		23,000 yards(21,031 m)
	HEAT-T M431 Shell(HEAT-T)		8,900 yards(8,138 m)
	HE-T T91E3 Shell(HE-T)		14,500 yards(13,259 m)
	HE-T M71A1 Shell(HE-T)		16,800 yards(15,362 m)
	APERS-T XM580E1 (4100 flechettes)		4,800 yards(4,389 m)
	Canister M336 (1280 pellets)		200 yards(183 m)
	Canister M377 (5600 flechettes)		440 yards(402 m)
	TP-T M353(T225E1) Shot(TPBC-T)		23,000 yards(21,031 m)

 *Assembled with the M108(T24) brass cartridge case (weight 11.0 pounds)
 **Assembled with the M108B1(T24B1) steel cartridge case (weight 10.3 pounds)
 #Assembled with the M19 brass cartridge case (weight 11.0 pounds)
 The HEAT-T M431 and APERS-T XM580E1 rounds were assembled with the M114E1 and XM200 cartridge cases respectively. The T119 gun differed from the M36(T119E1) primarily in the detailed design of the breech mechanism. The M41(T139) gun was lighter in weight than the M36 and the tube was attached to the breech ring with interrupted threads. This permitted a quick change of the tube without removing the weapon from the mount. The T178 gun was essentially the M41 mounted upside down with lugs modified to permit mounting the recoil mechanism in the forward part of the turret. In addition to the ammunition assembled with the M108 or M108B1 cartridge cases, these weapons could use any of the rounds for the lower pressure M1, M2, and M3 series of 90mm guns fitted in the M19 or M19B1 cartridge cases.

105mm Guns T140, T140E2, and T140E3

Carriage and Mount	105mm Gun Tanks T54(T140 Gun), T54E1(T140E2 Gun), and T54E2(T140E3 Gun) in Mounts T156, T157, and T174	
Length of Chamber (to rifling)	32.27 inches	
Length of Rifling	236.54 inches	
Length of Chamber (to projectile base)	28.81 inches	
Travel of Projectile in Bore	240.00 inches	
Length of Bore	268.81 inches, 65.0 calibers	
Depth of Breech Recess	9.50 inches	
Length, Muzzle to Rear Face of Breech	278.31 inches, 67.3 calibers	
Additional Length, Muzzle Brake	14.25 inches	
Overall Length	292.56 inches	
Diameter of Bore	4.134 inches	
Chamber Capacity	615 cubic inches	
Weight, Tube	approx. 3500 pounds	
Total Weight	approx. 4800 pounds	
Type of Breechblock	Semiautomatic, vertical sliding wedge	
Rifling	36 grooves, uniform right-hand twist, one turn in 25 calibers	
Ammunition	Fixed	
Primer	Percussion-electric M67	
Weight, Complete Round	AP-T T182E1 Shot(APBC-T)	72.8 pounds(33.1 kg)
	HVAPDS-T T279 Shot(APDS-T)	50.50 pounds(23.0 kg)
	HEAT-T T298E1 Shell(HEAT-T)	54.8 pounds(24.9 kg)
	TP-T T79E1 Shot(TP-T)	72.8 pounds(33.1 kg)
Weight, Projectile	AP-T T182E1 Shot(APBC-T)	35.04 pounds(15.9 kg)
	HVAPDS-T T279 Shot(APDS-T)	13.60 pounds(6.2 kg)
	HEAT-T T298E1 Shell (HEAT-T)	22.5 pounds(10.2 kg)
	TP-T T79E1 Shot(TP-T)	35.04 pounds(15.9 kg)
Maximum Powder Pressure	48,000 psi	
Maximum Rate of Fire	6 rounds/minute, manual loading(T140E2)	
Muzzle Velocity	AP-T T182E1 Shot(APBC-T)	3500 ft/sec(1067 m/sec)
	HVAPDS-T T279 Shot(APDS-T)	5100 ft/sec(1554 m/sec)
	HEAT-T T298E1 Shell(HEAT-T)	3700 ft/sec(1128 m/sec)
	TP-T T79E1 Shot(TP-T)	3500 ft/sec(1067 m/sec)
Muzzle Energy of Projectile, $KE = \frac{1}{2}MV^2$	AP-T T182E1 Shot(APBC-T)	2976 ft-tons
Rotational energy is neglected and	HVAPDS-T T279 Shot(APDS-T)	2452 ft-tons
values are based on long tons	HEAT-T T298E1 Shell(HEAT-T)	2135 ft-tons
(2240 pounds)	TP-T T79E1 Shot(TP-T)	2976 ft-tons
Maximum Range (independent of mount)	Not available	

The ammunition development was terminated for the T140 series of guns prior to completion and the test data were incomplete.

The T140 and T140E2 guns were intended for use with an automatic loader and were mounted with the vertical sliding breechblock moving up to open and down to close. The T140E3 was mounted for manual loading with the breechblock moving down to open and up to close. The ammunition for the T140 series of 105mm guns was assembled with the T43 cartridge case.

105mm Guns T254, M68(T254E2), and M68E1

Carriage and Mount	105mm Gun Tank T95E5(T254 Gun), 105mm Gun Tanks M60, M60A1, and M60A3 in Mounts M116 and M140(M68 and M68E1 Guns), 105mm Gun Tanks M48A1E1 and M48A5 in modified M87 Mounts(M68Gun)
Length of Chamber (to rifling)	24.9 inches
Length of Rifling	185.557 inches
Length of Chamber (to projectile base)	23.42 inches (APDS shot)
Travel of Projectile in Bore	187.08 inches (APDS shot)
Length of Bore	210.50 inches, 50.92 calibers
Depth of Breech Recess	8.00 inches
Length, Muzzle to Rear Face of Breech	218.50 inches, 52.85 calibers
Diameter of Bore	4.134 inches
Chamber Capacity	403 cubic inches
Weight, Tube	1534 pounds (T254), 1660 pounds (M68)
Total Weight	2475 pounds (T254), 2492 pounds (M68)
Type of Breechblock	Semiautomatic, vertical sliding wedge
Rifling	28 grooves, uniform right-hand twist, one turn in 18 calibers
Ammunition	Fixed
Primer	Electric

Weight, Complete Round	APDS-T M392A2 Shot(APDS-T)	41.0 pounds(18.6 kg)
	APFSDS-T M735 Shot(APFSDS-T)	38 pounds(17 kg)
	HEP-T M393A1 Shell(HESH-T)	46.7 pounds(21.2 kg)
	HEAT-T M456(T384E4) Shell(HEAT-T)	48.0 pounds(21.8 kg)
	APERS-T XM494E3 (5000 flechettes)	55.0 pounds(25.0 kg)
	WP-T M416 Shell(Smoke)	45.5 pounds(20.7 kg)
	TP-T M393A1 Shell(TP-T)	46.7 pounds(21.2 kg)
	TP-T M490 Shell(TP-T)	48.0 pounds(21.8 kg)
Weight, Projectile	APDS-T M392A2 Shot(APDS-T)	12.75 pounds(5.8 kg)
	APFSDS-T M735 Shot(APFSDS-T)	12.78 pounds(5.8 kg)
	HEP-T M393A1 Shell(HESH-T)	24.8 pounds(11.3 kg)
	HEAT-T M456(T384E4) Shell(HEAT-T)	22.4 pounds(10.2 kg)
	APERS-T XM494E3 (5000 flechettes)	approx. 31 pounds(14 kg)
	WP-T M416 Shell(Smoke)	25.17 pounds(11.4 kg)
	TP-T M393A1 Shell(TP-T)	24.8 pounds(11.3 kg)
	TP-T M490 Shell(TP-T)	22.4 pounds(10.2 kg)
Maximum Powder Pressure	51,540 psi	
Maximum Rate of Fire	7 rounds/minute	
Muzzle Velocity	APDS-T M392A2 Shot(APDS-T)	4850 ft/sec(1478 m/sec)
	APFSDS-T M735 Shot(APFSDS-T)	4925 ft/sec(1501 m/sec)
	HEP-T M393A1 Shell(HESH-T)	2400 ft/sec(732 m/sec)
	HEAT-T M456(T384E4) Shell(HEAT-T)	3850 ft/sec(1173 m/sec)
	APERS-T XM494E3 (5000 flechettes)	2700 ft/sec(823 m/sec)
	WP-T M416 Shell(Smoke)	2400 ft/sec(732 m/sec)
	TP-T M393A1 Shell(TP-T)	2400 ft/sec(732 m/sec)
	TP-T M490 Shell(TP-T)	3850 ft/sec(1173 m/sec)
Muzzle Energy of Projectile, KE = ½MV² Rotational energy is neglected and values are based on long tons (2240 pounds)	APDS-T M392A2 Shot(APDS-T)	2079 ft-tons
	APFSDS-T M735 Shot(APFSDS-T)	2149 ft-tons
	HEP-T M393A1 Shell(HESH-T)	990 ft-tons
	HEAT-T M456(T384E4) Shell(HEAT-T)	2302 ft-tons
	APERS-T XM494E3 (5000 flechettes)	1567 ft-tons
	WP-T M416 Shell(Smoke)	1005 ft-tons
	TP-T M393A1 Shell(TP-T)	990 ft-tons
	TP-T M490 Shell(TP-T)	2302 ft-tons
Maximum Range (independent of mount)	APDS-T M392A2 Shot(APDS-T)	40,162 yards(36,724 m)
	HEP-T M393A1 Shell(HESH-T)	
	HEAT-T M456(T384E4) Shell(HEAT-T)	8,975 yards(8207 m)
	APERS-T XM494E3 (5000 flechettes)	4,800 yards(4389 m)
	WP-T M416 Shell(Smoke)	10,400 yards(9510 m)
	TP-T M393A1 Shell(TP-T)	
	TP-T M490 Shell(TP-T)	8,975 yards(8207 m)

The T254 gun was ballistically identical with the British 105mm X15E8 or L7 cannon, but it utilized a different tube and breech design. The T254E2 weapon retained the vertical sliding breechblock of the T254, but it featured a tube that was interchangeable with that on the British gun. On the original T254E2, the U.S. tube was fitted with a concentric bore evacuator. After the British tube was adopted with its eccentric bore evacuator, the T254E2 designation was retained and the weapon was standardized as the M68. The M68E1 differed only in minor details from the M68 and it could be fitted with a mirror for a muzzle reference system. Ammunition for these weapons was assembled with the cartridge cases M115(brass), M150(brass), M150B1(steel), M148A1(brass), and M148A1B1(steel).

120mm Guns T179 and T123E6

Carriage and Mount	120mm Gun Tank T77 in Mount T169(T179 Gun) and 120mm Gun Tank T95E6 in Mount XM109(T123E6 Gun)	
Length of Chamber (to rifling)	38.05 inches	
Length of Rifling	243.95 inches	
Length of Chamber (to projectile base)	33.7 inches	
Travel of Projectile in Bore	248.3 inches	
Length of Bore	282.00 inches, 60.0 calibers	
Depth of Breech Recess	9.50 inches(T179)	9.19 inches(T123E6)
Length, Muzzle to Rear Face of Breech	291.50 inches(T179)	291.19 inches(T123E6)
Additional Length, Blast Deflector	7.25 inches(T179)	none
Overall Length	298.75 inches(T179)	291.19 inches(T123E6)
Diameter of Bore	4.7 inches	
Chamber Capacity	1021 cubic inches	
Weight, Tube	4600 pounds(T179)	2845 pounds(T123E6)
Total Weight	6280 pounds(T179)	4244 pounds(T123E6)
Type of Breechblock	Semiautomatic, vertical sliding wedge	
Rifling	42 grooves, uniform right-hand twist, one turn in 25 calibers	
Ammunition	Separated	
Primer	Percussion or percussion-electric	

Weight, Complete Round	*	AP-T M358 Shot(APBC-T)	107.31 pounds(48.8 kg)
	#	HEAT-T M469(T153E15) Shell(HEAT-T)	52.55 pounds(23.9 kg)
	**	HE-T M356(T15E3) Shell(HE-T)	89.15 pounds(40.5 kg)
	**	WP-T M357(T16E4) Shell(Smoke)	89.15 pounds(40.5 kg)
	*	TP-T M359E2(T147E7) Shot(TPBC-T)	107.31 pounds(48.8 kg)
Weight, Projectile		AP-T M358 Shot(APBC-T)	50.85 pounds(23.1 kg)
		HEAT-T M469(T153E15) Shell(HEAT-T)	31.11 pounds(14.1 kg)
		HE-T M356(T15E3) Shell(HE-T)	50.41 pounds(22.9 kg)
		WP-T M357(T16E4) Shell(Smoke)	50.41 pounds(22.9 kg)
		TP-T M359E2(T147E7) Shot(TPBC-T)	50.85 pounds(23.1 kg)
Maximum Powder Pressure		48,000 psi	
Maximum Rate of Fire		4 rounds/minute, manual loading, one loader (T123E6)	
Muzzle Velocity		AP-T M358 Shot(APBC-T)	3500 ft/sec(1067 m/sec)
		HEAT-T M469(T153E15) Shell(HEAT-T)	3750 ft/sec(1143 m/sec)
		HE-T M356(T15E3) Shell(HE-T)	2500 ft/sec(762 m/sec)
		WP-T M357(T16E4) Shell(Smoke)	2500 ft/sec(762 m/sec)
		TP-T M359E2(T147E7) Shot(TPBC-T)	3500 ft/sec(1067 m/sec)
Muzzle Energy of Projectile, KE = $\frac{1}{2}MV^2$		AP-T M358 Shot(APBC-T)	4318 ft-tons
Rotational energy is neglected and		HEAT-T M469(T153E15) Shell(HEAT-T)	3033 ft-tons
values are based on long tons		HE-T M356(T15E3) Shell(HE-T)	2184 ft-tons
(2240 pounds)		WP-T M357(T16E4) Shell(Smoke)	2184 ft-tons
		TP-T M359E2(T147E7) Shot(TPBC-T)	4318 ft-tons
Maximum Range (independent of mount)		AP-T M358 Shot(APBC-T)	25,290 yards(23,125 m)
		HEAT-T M469(T153E15) Shell(HEAT-T)	25,290 yards(23,125 m)
		HE-T M356(T15E3) Shell(HE-T)	19,910 yards(18,206 m)
		WP-T M357(T16E4) Shell(Smoke)	19,910 yards(18,206 m)
		TP-T M359E2(T147E7) Shot(TPBC-T)	25,290 yards(23,125 m)

  *With propelling charge assembly M46(T38E1) in cartridge case M109(T25)
  **With propelling charge assembly M45(T21E1) in cartridge case M109(T25)
  #With propelling charge assembly M99(T42E1) in cartridge case M111

The T179 gun was similar to the 120mm Gun M58(T123E1) in the M103 tank. However the T179 was inverted in the T169 mount for use with an automatic loader. Also, this was a rigid mount without a recoil system. The 120mm Gun T123E6 was manufactured from 160,000 psi yield strength steel as a lightweight version of the M58 gun.

152mm Gun-Launcher XM162E1 w/CBSS*

Carriage and Mount	152mm Gun Tanks M60A1E2 and M60A2	
Length of Chamber (to rifling)	10.5 inches	
Length of Rifling	94.55 inches	
Length of Chamber (to projectile base)	9 inches	
Travel of Projectile in Bore	96 inches	
Length of Tube and Chamber	105.1 inches, 17.52 calibers	
Overall Length	116.6 inches	
Diameter of Bore	6.0 inches	
Chamber Capacity	285 cubic inches	
Total Weight	1335 pounds (w/bore evacuator)	
	1230 pounds (w/o bore evacuator)	
Type of Breechblock	Semiautomatic, separable chamber, electrically operated	
Rifling	48 grooves, uniform right-hand twist, one turn in 41.2 calibers	
Ammunition	Fixed with combustible case or Shillelagh missile	
Primer	Electric	
Weight, Complete Round	MGM-51C Missile (as fired)	61.5 pounds(28.0 kg)
	MTM-51C Missile (as fired)	61.5 pounds(28.0 kg)
	HEAT-T-MP M409 Shell(HEAT-T-MP)	49.8 pounds(22.6 kg)
	HE-T XM657E2 Shell(HE-T)	50.0 pounds(22.7 kg)
	Canister M625 (10,000 flechettes)	48.0 pounds(21.8 kg)
	TP-T M411A1 Shell(TP-T)	49.8 pounds(22.6 kg)
Weight, Projectile	HEAT-T-MP M409 Shell(HEAT-T-MP)	42.8 pounds(19.5 kg)
	HE-T XM657E2 Shell(HE-T)	43.1 pounds(19.6 kg)
	Canister M625 (10,000 flechettes)	41.8 pounds(19.0 kg)
	TP-T M411A1 Shell(TP-T)	41.8 pounds(19.0 kg)
Maximum Powder Pressure	38,400 psi	
Maximum Rate of Fire	4 rounds/minute	
Muzzle Velocity	HEAT-T-MP M409 Shell(HEAT-T-MP)	2240 ft/sec(683 m/sec)
	HE-T XM657E2 Shell(HE-T)	2240 ft/sec(683 m/sec)
	Canister M625 (10,000 flechettes)	2240 ft/sec(683 m/sec)
	TP-T M411A1 Shell(TP-T)	2240 ft/sec(683 m/sec)
Muzzle Energy of Projectile, KE = ½MV²	HEAT-T-MP Shell(HEAT-T-MP)	1489 ft-tons
Rotational energy is neglected and	HE-T XM657E2 Shell(HE-T)	1499 ft-tons
values are based on long tons	Canister M625 (10,000 flechettes)	1454 ft-tons
(2240 pounds)	TP-T M411A1 Shell(TP-T)	1454 ft-tons
Maximum Range (independent of mount)	HEAT-T-MP M409 Shell(HEAT-T-MP)	9850 yards(9007 m)
	HE-T XM657E2 Shell(HE-T)	9850 yards(9007 m)
	Canister M625 (10,000 flechettes)	437 yards(400 m)
	TP-T M411A1 Shell(TP-T)	9850 yards(9007 m)

The M409, M625, and M411A1 rounds were assembled with the M157 combustible case and the M189 charge. The XM657E2 round was assembled with the XM157 combustible case and the XM190 charge.

*CBSS (closed breech scavenging system) With this equipment, the bore evacuator was not required, but it was still fitted on many of the earlier gun-launchers.

155mm Gun M46(T80)

Carriage and Mount	155mm Self-Propelled Gun M53(T97) in Mount M86(T58)	
Length of Chamber (to rifling)	44.03 inches	
Length of Rifling	230.56 inches	
Length of Chamber (to projectile base)	36.275 inches	
Travel of Projectile in Bore	238.315 inches	
Length of Bore	274.59 inches, 45.0 calibers	
Length, Breechblock and Firing Lock	17.18 inches(early T80)	16.96 inches(M46)
Length, Muzzle to Rear of Firing Lock	291.77 inches(early T80)	291.55 inches(M46)
Additional Length, Muzzle Brake	13.0 inches(early T80)	none(M46)
Overall Length	304.8 inches(early T80)	291.55 inches(M46)
Diameter of Bore	6.102 inches	
Chamber Capacity	1640 cubic inches	
Weight, Tube	5825 pounds	
Total Weight	7350 pounds(early T80)	7195 pounds(M46)
Type of Breechblock	Interrupted screw, carrier hinged, opening to right side	
Rifling	48 grooves, uniform right-hand twist, one turn in 25 calibers	
Ammunition	Separate loading	
Primer	Percussion, percussion-electric, or electric	
Weight, Complete Round	HE M101 Shell(HE), Supercharge	126.73 pounds(57.6 kg)
	HE M101 Shell(HE), Normal	116.08 pounds(52.8 kg)
	WP M104 Shell(Smoke), Supercharge	126.23 pounds(57.4 kg)
	WP M104 Shell(Smoke), Normal	115.58 pounds(52.5 kg)
	* HD M104 Shell(Gas), Supercharge	126.23 pounds(57.4 kg)
	* HD M104 Shell(Gas), Normal	115.58 pounds(52.5 kg)
Weight, Projectile	HE M101 Shell(HE)	95.10 pounds(43.2 kg)
	WP M104 Shell(Smoke)	94.60 pounds(43.0 kg)
	* HD M104 Shell(Gas)	94.60 pounds(43.0 kg)
Maximum Powder Pressure	40,000 psi	
Maximum Rate of Fire	1 round/minute	
Muzzle Velocity	HE M101 Shell(HE), Supercharge	2800 ft/sec(853 m/sec)
	HE M101 Shell(HE), Normal	2100 ft/sec(640 m/sec)
	WP M104 Shell(Smoke), Supercharge	2800 ft/sec(853 m/sec)
	WP M104 Shell(Smoke), Normal	2100 ft/sec(640 m/sec)
	* HD M104 Shell(Gas), Supercharge	2800 ft/sec(853 m/sec)
	* HD M104 Shell(Gas), Normal	2100 ft/sec(640 m/sec)
Muzzle Energy of Projectile, $KE = \frac{1}{2}MV^2$	HE M101 Shell(HE), Supercharge	5168 ft-tons
Rotational energy is neglected and	HE M101 Shell(HE), Normal	2907 ft-tons
values are based on long tons	WP M104 Shell(Smoke), Supercharge	5141 ft-tons
(2240 pounds)	WP M104 Shell(Smoke), Normal	2892 ft-tons
	* HD M104 Shell(Gas), Supercharge	5141 ft-tons
	* HD M104 Shell(Gas), Normal	2892 ft-tons
Maximum Range (independent of mount)	HE M101 Shell(HE), Supercharge	25,715 yards(23,456 m)
	HE M101 Shell(HE), Normal	18,605 yards(17,012 m)
	WP M104 Shell(Smoke), Supercharge	25,715 yards(23,456 m)
	WP M104 Shell(Smoke), Normal	18,605 yards(17,012 m)
	* HD M104 Shell(Gas), Supercharge	25,715 yards(23,456 m)
	* HD M104 Shell(Gas), Normal	18,605 yards(17,012 m)

* As its designation indicated, the HD M104 was loaded with a persistent gas, in this case 11.70 pounds of distilled mustard gas (HD) and Burster M6.

In addition to its standard ammunition, the M46 gun could fire the HE M107 shell and the BE M116 shell(red, green, or yellow smoke) designed for the 155m howitzer. However, these rounds were restricted to the use of the normal propelling charge only.

165mm Guns M57(T156) and M135(XM135)

Carriage and Mount	Combat Engineer Vehicle M728(T118E1) in Mount M150(M135 Gun) and Combat Engineer Vehicle M102(T39E2) in Mount T186(M57 Gun)
Length of Chamber (to rifling)	5.0 inches
Length of Rifling	90.1 inches
Length of Chamber (to projectile base)	4.1 inches
Travel of Projectile	91.0 inches
Length of Tube and Chamber	95.1 inches, 14.6 calibers
Depth of Breech Recess	9.4 inches
Length, Muzzle to Rear Face of Breech	104.5 inches, 16.1 calibers
Diameter of Bore	6.5 inches
Chamber Capacity	750 cubic inches
Weight, Tube	875 pounds (M57)
Total Weight	1425 pounds(M57) 1465 pounds(M135)
Type of Breechblock	Semiautomatic, oblique sliding (M57) Manual, oblique sliding (M135)
Rifling	36 grooves, uniform right-hand twist, one turn in 15 calibers
Ammunition	Fixed
Primer	Electric
Weight, Complete Round	HEP M123A1 Shell(HESH) 67.60 pounds(30.7 kg) TP M623 Shell(TP) 67.60 pounds(30.7 kg)
Weight, Projectile, w/o handle	HEP M123A1 Shell(HESH) 65.0 pounds(29.5 kg) TP M623 Shell(TP) 65.0 pounds(29.5 kg)
Maximum Powder Pressure	3600 psi (M57) 5100 psi (M135)
Maximum Rate of Fire approx.	2 rounds/minute
Muzzle Velocity	HEP M123A1 Shell(HESH) 850 ft/sec(259 m/sec) TP M623 Shell(TP) 850 ft/sec(259 m/sec)
Muzzle Energy of Projectile, $KE = \frac{1}{2}MV^2$ Rotational energy is neglected and values are based on long tons (2240 pounds)	HEP M123A1 Shell(HESH) 326 ft-tons TP M623 Shell(TP) 326 ft-tons
Maximum Range (independent of mount)	HEP M123A1 Shell(HESH) 1000 yards(914 m) TP M623 Shell(TP) 1000 yards(914 m)

The perforated M104 cartridge case was threaded to the projectile base and remained attached to it when the round was fired. The HEP M123A1 shell contained 35 pounds of Composition A3 plastic explosive with base detonating fuze M62A2.

175mm Gun T181

Carriage and Mount	175mm Self-Propelled Gun T162 in Mount T158	
Length of Chamber (to rifling)	58.4 inches	
Length of Rifling	355.0 inches	
Length of Chamber (to projectile base)	46.5 inches	
Travel of Projectile in Bore	366.93 inches	
Length of Bore	413.4 inches, 60.0 calibers	
Length, Breechblock and Firing Lock	13.1 inches	
Length, Muzzle to Rear of Firing Lock	426.5 inches	
Additional Length	none	
Overall Length	426.5 inches	
Diameter of Bore	6.89 inches	
Chamber Capacity	2275 cubic inches	
Weight, Tube	13,670 pounds	
Total Weight approx.	15,100 pounds	
Type of Breechblock	Interrupted screw, carrier hinged, opening to right side	
Rifling	48 grooves, uniform right-hand twist, one turn in 20 calibers	
Ammunition	Separate loading	
Primer	Electric	
Weight, Complete Round	HE T203E3 Shell(HE)	194.3 pounds(88.3 kg)
Weight, Projectile	HE T203E3 Shell(HE)	146.5 pounds(66.6 kg)
Maximum Powder Pressure	36,000 psi	
Maximum Rate of Fire	1 round/minute	
Muzzle Velocity	HE T203E3 Shell(HE)	2850 ft/sec(869 m/sec)
Muzzle Energy of Projectile, KE = ½MV² Rotational energy is neglected and values are based on long tons (2240 pounds)	HE T203E3 Shell(HE)	8249 ft-tons
Maximum Range (independent of mount)	HE T203E3 Shell(HE) approx.	30,000 yards(27,432 m)

8 inch Howitzer M47(T89)

Carriage and Mount	8 inch Self-Propelled Howitzer M55(T108) in Mount M86(T58)	
Length of Chamber (to rifling)	35.20 inches	
Length of Rifling	164.80 inches	
Length of Chamber (to projectile base)	26.17 inches	
Travel of Projectile in Bore	173.83 inches	
Length of Bore	200.00 inches, 25.0 calibers	
Length, Breechblock and Firing Lock	16.74 inches	
Length, Muzzle to Rear of Firing Lock	216.74 inches	
Additional Length, Muzzle Brake	17.0 inches(early T89)	none(M47)
Overall Length	233.74 inches(early T89)	216.74 inches(M47)
Diameter of Bore	8.00 inches	
Chamber Capacity	1485 cubic inches	
Weight, Tube	4777 pounds	
Total Weight	6392 pounds(early T89)	6120 pounds(M47)
Type of Breechblock	Interrupted screw, carrier hinged, opening to right side	
Rifling	64 grooves, uniform right-hand twist, one turn in 25 calibers	
Ammunition	Separate loading	
Primer	Percussion, percussion-electric, or electric	
Weight, Complete Round	HE M106 Shell(HE) max. M2 charge	228.8 pounds(104.0 kg)
	GB M246 Shell(Gas) max. M2 charge	227.8 pounds(103.5 kg)
	VX M246 Shell(Gas) max. M2 charge	227.8 pounds(103.5 kg)
Weight, Projectile	HE M106 Shell(HE)	200.0 pounds(90.9 kg)
	GB M246 Shell(Gas)	199.0 pounds(90.5 kg)
	VX M246 Shell(Gas)	199.0 pounds(90.5 kg)
Maximum Powder Pressure	33,000 psi	
Maximum Rate of Fire	1 round/minute	
Muzzle Velocity	HE M106 Shell(HE) max. M2 charge	1950 ft/sec(594 m/sec)
	GB M246 Shell(Gas) max. M2 charge	1950 ft/sec(594 m/sec)
	VX M246 Shell(Gas) max. M2 charge	1950 ft/sec(594 m/sec)
Muzzle Energy of Projectile, $KE = \frac{1}{2}MV^2$	HE M106 Shell(HE) max. M2 charge	5272 ft-tons
Rotational energy is neglected and	GB M246 Shell(Gas) max. M2 charge	5246 ft-tons
values are based on long tons	VX M246 Shell(Gas) max. M2 charge	5246 ft-tons
(2240 pounds)		
Maximum Range (independent of mount)	HE M106 Shell(HE) max. M2 charge	16,008 yards(14,638 m)
	GB M246 Shell(Gas) max. M2 charge	16,008 yards(14,638 m)
	VX M246 Shell(Gas) max. M2 charge	16,008 yards(14,638 m)

The GB M246 and VX M246 shells were loaded with 14.5 pounds of non-persistent or persistent gases respectively. The M83 Burster contained 7 pounds of Composition B4 explosive.

REFERENCES AND SELECTED BIBLIOGRAPHY

Books and Manuscripts

Beckhoff, R., "History of the M48A2 Tank", Chrysler Corporation, November 1956

Davis, V.M., "A Summary Analysis—The Problem of Defeating the Antivehicular Mine", U.S. Army Engineer Research and Development Laboratories, Fort Belvoir, Virginia, January 1966

Dunstan, Simon, "Vietnam Tracks", Presidio Press, Novato, California, 1982

Hay, Lieutenant General John H. Jr., "Vietnam Studies: Tactical and Materiel Innovations", Department of the Army, Washington, D.C., 1974

Palmer, Dave Richard, "Summons of the Trumpet", Presidio Press, Novato, California, 1978

Starry, General Donn A., "Vietnam Studies: Mounted Combat in Vietnam", Department of the Army, Washington, D.C., 1978

Tillotson, Geoffrey, "M48, Modern Combat Vehicles: 4", Ian Allan, Ltd., London, 1981

————, "The Super M60 Battle Tank", Teledyne Continental Motors, General Products Division, Muskegon, Michigan, undated

Reports and Official Documents

"Army Equipment Development Guide", U.S. Army, Fort Monroe, Virginia, 29 December, 1950

"Army Equipment Development Guide", Department of the Army, Washington, D.C., 3 May 1954

"Engineering Concepts on Medium Recovery Vehicle T88", Ordnance Tank-Automotive Command, Centerline, Michigan, December 1954

"M60A1E3 Combat Tank, Characteristics and Description Book", prepared by Chrysler Corporation for the Project Manager M60 Tanks, January 1973

"Notes on Development Type Materiel, 90mm Gun T119", Watervliet Arsenal, New York, August 1950

"Notes on Development Type Materiel, 90mm Gun T119E1", Watervliet Arsenal, New York, February 1952

"Notes on Development Type Materiel, 90mm Gun T139", Watervliet Arsenal, New York, October 1951

"Notes on Development Type Materiel, 90mm Gun T139E1", Watervliet Arsenal, New York, May 1952

"Notes on Development Type Materiel, 105mm Gun T140E2", Watervliet Arsenal, New York, March 1953

"Notes on Development Type Materiel, 105mm Gun T140E3", Watervliet Arsenal, New York, October 1953

"Notes on Development Type Materiel, 105mm Gun T254E2", Watervliet Arsenal, New York, March 1959

"Notes on Development Type Materiel, 120mm Gun T123E6", Watervliet Arsenal, New York, August 1958

"Notes on Development Type Materiel, 152mm Gun-Launcher, XM81", Watervliet Arsenal, New York, March 1961

"Notes on Development Type Materiel, Engineer Armored Vehicle T39", Detroit Arsenal, Centerline, Michigan, 15 September 1951

"Notes on Development Type Materiel, Engineer Armored Vehicle T39E2", Detroit Arsenal, Centerline, Michigan, 30 June 1955

"Notes on Development Type Materiel, 90mm Gun Tank T42", Detroit Arsenal, Centerline, Michigan, 1 June 1951

"Notes on Development Type Materiel, 105mm Gun Tank T54 with Automatic Ammunition Loading Equipment", Detroit Arsenal, Centerline, Michigan, 31 August 1956

"Notes on Development Type Materiel, 120mm Gun Tanks T57 and T77, Trunnion Mounted Turrets with Automatic Loading Equipment", Detroit Arsenal, Centerline Michigan, 31 August 1956

"Notes on Development Type Materiel, Tank, Combat, Full Tracked, 105mm Gun M48A1E1", Detroit Arsenal, Centerline, Michigan, January 1960

"Notes on Development Type Materiel, Self-Propelled 155mm Gun T97", Detroit Arsenal, Centerline, Michigan, 15 February 1952

"Notes on Development Type Materiel, Self-Propelled 8 inch Howitzer T108", Detroit Arsenal, Centerline, Michigan, 27 June 1952

"Notes on Development Type Materiel, Self-Propelled 175mm Gun T162", Detroit Arsenal, Centerline, Michigan, 31 August 1954

"Report of the Army Ground Forces Equipment Review Board", Washington, D.C., 20 June 1945

"Report of the War Department Equipment Board" (Stilwell Board), Washington, D.C., 19 January 1946

"Studies of Recovery Vehicles, Combat Engineer Vehicles, Armored Vehicle Launched Bridges", Ordnance Tank-Automotive Command, Centerline, Michigan, 29 January 1957

"Technical Manual TM5-5420-200-12, Launcher, M48A2 Tank Chassis, for Transporting Bridge, Armored Vehicle Launched, Scissoring Type, Class 60", Department of the Army, Washington, D.C., March 1962

"Technical Manual TM5-5420-202-10, Launcher, M60A1 Tank Chassis, for Transporting Bridge, Armored Vehicle Launched, Scissoring Type, Class 60", Department of the Army, Washington, D.C., April 1966

"Technical Manual TM5-5420-203-14, Bridge, Armored Vehicle Launched, Scissoring Type, Aluminum, 60 Foot Span", Department of the Army, Washington, D.C., October 1972

"Technical Manual TM9-500 Ordnance Equipment Data Sheets", Department of the Army, Washington, D.C., September 1962

"Technical Manual TM9-718A Tank, 90mm Gun, M47", Department of the Army, Washington, D.C., January 1952

"Technical Manual TM9-718B Tank, 90mm Gun, T48", Department of the Army, Washington, D.C., August 1952

"Technical Manual TM9-1000-213-35 105mm Gun Cannon M68(T254E2), Combination Gun Mount M116, and Caliber .50 Machine Gun Tank Commander's Cupola M19(T9)", Department of the Army, Washington, D.C., 5 August 1960

"Technical Manual TM9-1300-203 Artillery Ammunition for Guns, Howitzers, Mortars, and Recoilless Rifles", Department of the Army, Washington, D.C., April 1967

"Technical Manual TM9-1375-200 Demolition Materials", Department of the Army, Washington, D.C., 28 January 1964

"Technical Manual TM9-2320-222-10 Recovery Vehicle, Full Tracked, Medium M88", Department of the Army, Washington, D.C., April 1966

"Technical Manual TM9-2350-200-12 Tank, Combat, Full Tracked, 90mm Gun M47", Department of the Army, Washington, D.C., October 1958

"Technical Manual TM9-2350-215-10 Tank, Combat, Full Tracked, 105mm Gun M60", Department of the Army, Washington, D.C., June 1960

"Technical Manual TM9-2350-215-10 Tank, Combat, Full Tracked, 105mm Gun M60 and M60A1 and Tank Gunnery Trainer XM30", Department of the Army, Washington, D.C., September 1962

"Technical Manual TM9-2350-220-10 Vehicle, Combat Engineer, Full Tracked, T118E1", Department of the Army, Washington, D.C., 1965

"Technical Manual TM9-2350-224-10 Tank, Combat, Full Tracked, 90mm Gun M48A3", Department of the Army, Washington, D.C., July 1963

"Technical Manual TM9-2350-232-10 Tank, Combat, Full Tracked, 152mm Gun-Launcher M60A2", Department of the Army, Washington, D.C., 13 April 1973

"Technical Manual TM9-2350-256-10 Recovery Vehicle, Full Tracked, Medium M88A1", Department of the Army, Washington, D.C., March 1977

"Technical Manual TM9-2350-258-10 Tank, Combat, Full Tracked, 105mm Gun M48A5", Department of the Army, Washington, D.C., June 1976

"Technical Manual TM9-3031 90mm Gun M41(T139) and Combination Gun Mount T148 for 90mm Gun Tank M48", Department of the Army, Washington, D.C., 4 January 1955

"Technical Manual TM9-7012 Tank, 90mm Gun M48", Department of the Army, Washington, D.C., August 1954

"Technical Manual TM9-7022 Tank, Combat, Full Tracked, 90mm Gun M48A2", Department of the Army, Washington, D.C., March 1958

"Technical Manual TM9-7220 Self-Propelled 8 inch Howitzer T108", Department of the Army, Washington, D.C., May 1955

"U.S. Army Tank Modernization Program", U.S. Army Materiel Command, Washington, D.C. December 1968

"Trunnion Mounted Turret with Automatic Loading Mechanism for 105mm Gun Tank T54E1", Final Report by United Shoe Machinery Corporation for Detroit Arsenal, Centerline, Michigan, February 1957

Abrams, General Creighton W. Jr., 6
Add-on stabilization system (AOS), 199-200
Ad Hoc Group on Armament for Future
    Tanks or Similar Combat Vehicles
    (ARCOVE), 149, 178
Advanced Radar-directed Gun Air Defense
    System (ARGADS), 263
Aircraft Armaments, Inc., 102, 125, 134
Airscrew Howden, Ltd., 402-403
ALCO Products, Inc., 55, 59, 73, 119
American Locomotive Company (ALCO), 55,
    59, 73, 119
Ammunition Stowage
    Chrysler K tank, 197
    Elimination of ammunition stowage
        above turret ring, 401
    M47 tank, 53
    M47-M tank, 79
    M48A1E1 tank, 220
    M48A3 tank, 224-225
    M48A5 tank, 239
    M53 SPG, 256
    M55 SPH, 256
    M60 tank, 161
    M60A1 tank, 173
    M60A1E3 tank, 204
    M60A2 tank, 192
    T39 CEV, 289
    T39E2 CEV, 292
    T42 tank, 39-41
    T54 tank, 127
    T54E1 tank, 131
    T54E2 tank, 139
    T69 tank, 49
    T77 tank, 143
Antiaircraft Tanks, 263-271
    General Electric DIVAD proposal,
        30mm SPG, 264
    M42 40mm SPG, 263
    Raytheon DIVAD proposal, 35mm
        SPG, 264
    Sergeant York, 40mm SPG, 268, 408
    Sperry DIVAD proposal, 35mm SPG,
        265
    XM246 (General Dynamics DIVAD)
        35mm SPG, 265-268
    XM247 (Ford Aerospace DIVAD) 40mm
        SPG, 267-271
ARCOVE, 149, 178
Armored Brigade, U.S.
    194th, 372
Armored Cavalry Assault Vehicle (ACAV),
    383-384
Armored Engineer Vehicle T39, 14
Armored top loading air cleaner (ATLAC),
    210
Armored Vehicle Launched Bridge (AVLB),
    287, 304-309
    AVLB on M46 tank chassis, 304-305
    AVLB on M48 tank chassis, 306
    AVLB on M48A2C tank chassis, 307-308
    AVLB on M48A5 tank chassis, 309
    AVLB on M60A1 tank chassis, 308-309,
        396
    Folding bridge towed by M48 tank, 305
Armor, special
    Active (explosive) Israeli Blazer, 400-401
    Applique, 124, 217
    Grill, 125
    Siliceous cored, 123-124, 152, 156,
        158, 168
Army Equipment Development Guide, 9,
    261, 330

Army Ground Forces Equipment Review
    Board, 9, 255
Army, Republic of Vietnam (ARVN)
    20th Tank Regiment, 397-398
Automatic loader, 47-51, 126-127, 131, 143,
    145
Beirut, Lebanon, 351, 355
Berres, Colonel John, 263
Birmingham Fabricating Company, 331
Blast deflector, 55, 59
Blazer, active armor, 400-401
Bore evacuator, 12, 16, 119, 152, 189
Bowen-McLaughlin-York Company (BMY),
    77, 227, 230, 273-274, 276, 280
Bradley, Clifford, 179
Bulldozers, Tank Mounting, 283-285, 292,
    299
    M3, 283, 289
    M3E1, 283
    M6, 283, 292
    T18, 283-284
    T18E1, 283-284
    M8, 284
    M8A1 (M8E1), 284
    M8A3, 284-285
    M9, 284-285, 299
Bureau of the Budget (BOB), 152
Cadillac Gage Company, 49, 51, 79, 114,
    134, 143, 225
    Gun control system, 51, 74, 79, 114, 131,
        134, 143, 202, 225
Call, Colonel William, 85
Cannon
    30mm General Electric GAU-8, 263
    30mm Mauser F gun, 263
    35mm Oerlikon gun, 263, 265, 268
    35mm Sperry gun, 263
    37mm Sperry Vigilante gun, 263
    40mm Bofors L/70 gun, 263, 267-268
    75mm M2 or M3 guns, 83
    76mm T91 gun, 32, 126
    76mm T94 gun, 32
    3 inch T98 gun, 12
    3 inch 70 caliber Navy AA gun, 12
    90mm M3 gun, 12, 58
    90mm M3A1 gun, 12, 16, 26, 34, 450
    90mm T54 gun, 12
    90mm T119 gun, 33-34, 40, 85
    90mm M36 (T119E1) gun, 58-59, 72,
        75, 85, 451
    90mm M41 (T139) gun, 47-48, 85, 96,
        126-127, 222, 451
    90mm T139E1 gun, 72
    90mm T178 gun, 48, 451
    90mm T208E9 smooth bore gun, 154-155,
        198
    105mm T5E2 gun, 126
    105mm T140 gun, 126-127, 452
    105mm T140E2 gun, 126, 130-131, 452
    105mm T140E3 gun, 134, 452
    105mm T210 smooth bore gun, 198
    105mm T254 gun, 152-153, 155, 453
    105mm T254E2 gun, 153-155, 453
    105mm T254E3 gun, 153
    105mm M68 (T254E2) gun, 75, 79, 153,
        161, 178, 202, 219-222, 238, 453
    105mm French gun (AMX 30 tank), 75
    105mm L7 British gun, 75, 403-404
    105mm Rheinmetall smooth bore gun, 404
    120mm M58 (T123E1) gun, 153-155
    120mm T123E6 gun, 153-155, 454
    120mm T179 gun, 143, 454
    120mm Delta gun, smooth bore, 198

    120mm Rheinmetall smooth bore gun, 404
    152mm XM81 gun-launcher, 178-181
    152mm XM81E5 gun-launcher, 179
    152mm XM81E6 gun-launcher, 179
    152mm XM81E10 gun-launcher, 180
    152mm XM81E13 gun-launcher, 181
    152mm XM150 gun-launcher, 190,
        196-197
    152mm XM162 gun-launcher, 181, 455
    152mm gun-launcher, conventional
        ammunition, 178, 181, 189-190, 195
    155mm M46 gun, 256, 258, 456
    6.5 inch Mk I demolition gun, British,
        287-288
    165mm M57 (T156) demolition gun,
        287-288, 292, 297-298, 457
    165mm M135 (XM135) demolition gun,
        298-299, 457
    175mm T181 gun, 261-262, 458
    8 inch M47 howitzer, 256, 258, 459
    20 pounder, British gun, 33
    Ex 20 pounder (105mm) British gun, 152
    Non-recoil guns, 72, 143
    Quick change tube, 72, 85
    Rigid mount guns, 72, 143
    Thermal shroud for gun tubes, 202, 208
Chain-link fence as a defense against
    shaped charge rounds, 31, 381, 397
Chemical alarm system, 215
Chemical, biological, radiological (CBR)
    equipment, 215, 223, 230
Chinese Forces, 25
Chrysler Corporation, 55, 79, 85-86, 90, 92,
    102-103, 119, 157, 168, 196, 199,
    234, 300, 308
Chrysler España, 79
Colby, Brigadier General Joseph M., 52
Closed breech scavenging system (CBSS), 189
Combat Engineer Vehicles (CEV), 283-302
    T31 demolition tank, 286
    T39 pioneer tank, 286-291, 304
    T39E1 engineer armored vehicle, 292
    M102 (T39E2) CEV, 292-296, 299
    T118 CEV, 297-299
    M728 (T118E1) CEV, 298-302, 322, 393, 448
    AVRE, British, 286
    Combined TRV/CEV proposal
        concepts, 297
    Sherman, 286
Combustible case ammunition, 178, 181,
    189-193, 195-197
Compact turret design, 179-180
Constant or controlled pressure hydraulic gun
    control system, 49, 51, 74, 79,
    114, 131, 134, 143, 202, 225
Crismon, Major Fred, 215, 371-372
Crombez, Colonel Marcel G., 26
Crocodile, British flame thrower, 247
Cupolas, 19, 40, 53, 62, 73, 102-103, 106,
    130-131, 156-158, 192-193,
    197, 199, 238
    AAI model 30, 102-103, 106
    AAI model 108, 73, 125, 127, 134
    Chrysler low silhouette cupola,
        102-103, 197
    Interim cupola, 130-131, 157
    Israel Defence Force cupola, 239
    Low silhouette cupola with pop-up hatch,
        216, 239
    M1 cupola, 102-103, 106, 124-125, 157,
        220, 225, 227-228, 238
    M19 cupola, 158, 170, 199, 216, 228, 299
    T6 cupola, 158-160

T9 cupola, 157
Vision block ring, 227-230, 238
XM60 full vision cupola, 156-157
Dart missile, 297
Division air defense (DIVAD) gun system, 263-271
Divisions, U.S.
  1st Cavalry, 357-358, 387
  1st Infantry, 381
  1st Marine, 27-29, 368, 417, 420
  2nd Armored, 352-353, 359
  2nd Marine, 368
  3rd Armored, 208, 356, 360
  3rd Marine, 228, 370, 373-380
  4th Armored, 356, 362-363
  4th Infantry, 334-335
  5th Infantry, 396
  25th Infantry, 31, 382, 388
Driver's hatches on M48 and M48A1 Tanks, 98
Electric amplidyne gun control system, 47, 74, 114, 131, 134
Electric-hydraulic gun control system, 61, 92, 114, 225
Engines
  AGT-1500, 215, 346
  AOS-895, 34
  AOS-895-3, 42, 100
  AV-1790 series, 11-13, 16-17, 21, 24, 55, 75, 100, 311, 402
  AV-1790-1, 11-13
  AV-1790-3, 13, 24
  AV-1790-5, 16, 24
  AV-1790-5A, 16, 21, 24
  AV-1790-5B, 16-17, 21, 24, 55, 75, 100, 311
  AV-1790-7, 100, 311
  AV-1790-7B, 100, 311
  AV-1790-7C, 100, 311
  Avco-Lycoming 650, 346-347
  AVCR-1790-1, 211
  AVCR-1790-1B, 216
  AVDS-1790 series, 77, 79, 150, 152, 158, 162, 174, 200-201, 211, 220, 223, 280
  AVDS-1790-P, 150
  AVDS-1790-1, 150
  AVDS-1790-2, 158, 162, 211, 220
  AVDS-1790-2A, 77, 174, 223
  AVDS-1790-2C, 79, 201
  AVDS-1790-2D, 200
  AVDS-1790-2DR, 280
  AVDS-1790-5A, 211
  AVDS-1790-7A, 211
  AVI-1790-8, 107-108, 117, 276, 278
  AVSI-1790-6, 261, 276
  AVSI-1790-6A, 278-280
  Ford GAF, 10, 12
  Garrett GT601, 346-348
  GM 12V71TA, 402
  MTU MB-837, 75, 403-404
  RISE engine, 200-202
  Steam power plant, 72
  Variable compression ratio, 211
Escape hatch improvement on the M60A1, 211
Exhaust deflector for gun travel lock, 92
Fire control equipment
  Ballistic computer T30, 94-95
  Ballistic computer T32, 127, 136
  Ballistic computer T34, 131
  Ballistic computer M13B1, 225
  Ballistic computer XM16C, 170
  Ballistic computer XM19, 193
  Ballistic computer XM21, 208
  Coincidence type range finder, 119, 170
  Direct sight telescope, 33, 93-95, 127, 131, 136, 145

Fire control system, phase I, 94-95
Fire control system, phase II, 94
Fire control system, phase III, 94
Fire control system, phase IV, 94
Infrared fire control, 156-157, 178-179, 227
Laser range finder, 190, 200, 208
M47 tank fire control system, 61
M48, M48A1 tank fire control system, 94-95
M48A1E1 tank fire control system, 220
M48A2C tank fire control system, 119
M48A3 tank fire control system, 225
M60 tank fire control system, 160, 220
M60E1 tank fire control system, 170
M60A1 tank fire control system, 172
M60A1E3 tank fire control system, 204
M60A2 tank fire control system, 190, 192-193
M60AX tank fire control system, 217-218
Muzzle reference system, 215
Passive sight, 202, 210
Periscope M20 (T35), 40, 49, 55, 58, 94-95
Periscope M21 (T39), for flame thrower, 250
Periscope M28 (T42), cupola mount, 102
Periscope M28C, cupola mount, 159
Periscope M31, 225
Periscope M35E1, 210
Periscope XM30, for flame thrower, 252
Periscope XM34, 170
Periscope XM36, 170
Periscope XM50, 193
Periscope XM51, 193
Range Finder M13A1, 119
Range Finder M17 (M13A1E1), 119
Range Finder M17C, 170
Range Finder M17A1, 171
Range Finder M17B1C, 225
Range Finder, short base, 190
Range Finder T41, 40-41, 55
Range Finder T46E1, 94-95
Range Finder T46E2, 51
Range Finder T46E3, 127, 136
Range Finder T50, 131, 145
Shock mounts in T54E2 tank, 136
Solid state computer, 200
Stereoscopic range finder, 33, 41
Tank thermal sight (TTS), 210
Telescope M76E5, 288
Telescope M83, 12
Telescope M105, articulated, 225
Telescope M105C, articulated, 170
Telescope M105F, articulated, 299
Telescope XM126, articulated, 193
Wind sensors, 190, 208
Flame thrower M7-6 (E28-30R1), main armament, mechanized, 248
Flame thrower M7A1-6, main armament, mechanized, 252
Flotation Devices, 310-313
  T8 for M26 tank, 310-311
  T15 for M47 tank, 310-312
  T44 for M48 tank, 312-313
Food Machinery and Chemical Corporation (FMC), 274
Ford Aerospace and Communciations Corporation, 178, 264, 267-268
Ford Motor Company, 10, 12, 90
Fording Equipment, 311, 322
  Deep water for M47 tank, 311
  Deep water for M48 (T48) tank, 312
  Deep water for M48A2 tank, 313
  Deep water for M48A3 tank, 313
  Deep water for M60 tank, 313-315
  Deep water for medium recovery vehicle

M88, 321
  Underwater for M60 tank at Fort Knox, 316
  Underwater for M60, M60A1 tanks, 202, 316-320
  Underwater for CEV M728, 322
Fort Churchill, Canada, 150
Friction snubbers, 77, 112, 169, 220
General Dynamics Corporation, 264-268, 336
General Electric Company, 263-264
General Motors Corporation, Detroit Diesel-Allison Division, 11, 34, 90, 211, 215, 261, 278, 280, 325, 402-403
Germany, 32, 64-66, 75, 351-353, 356, 359-360, 362-365, 368, 371, 403-406, 408
  Berlin, 32, 364
Grill armor, 125
Halon fire extinguisher system, 215
Herringbone formation, 383
Howell, Lieutenant Colonel Martin D., 384
Hurt, Lieutenant Colonel Robert D. III, 417
Infrared suppression, 108-109, 345
International Harvester Company, 51
Iran, 79, 402
Israel, 399-401
Israel Military Industries, 79
Italy, 75, 80
Iwo Jima, 245
Japan, 25, 248
Jettisonable fuel tanks, 107, 119, 273-274
Johnson, Ron, 127, 146
Joint Coordinating Committee on Ordnance, 149
Kiljan, Second Lieutenant Gary, 369
Kopicki, Captain Chester F., 330
Korea, 17, 25-26, 28-30, 47, 64, 242, 248, 263, 330, 357-358, 369-370, 407
  Chipyong-ni, 26
  Pusan, 25, 242
Korean War, 17, 25-26, 35, 52, 248, 330
Leaflet II searchlight tank, 29
Le Tourneau, R. G., Inc., 325
Littlefield, Jacques, 127, 146
Lockley Machine Company, 331
Loop, Lieutenant Colonel James, 335, 392-394
Machine guns
  Antiaircraft, 40, 51, 97, 130, 143, 159
  Blister, 33
  Coaxial, 33, 40, 48, 85, 92, 96, 116, 127, 130, 134, 143, 157, 197, 288
  Cupola, 40, 102-103, 106, 124-125, 127, 130-131, 156-157, 159, 170, 192, 197, 199, 220, 299
  Fender, 45
  M2 HB .50 caliber, 45, 125, 159, 220, 242, 278, 289
  M37 .30 caliber, 157
  M60D 7.62mm, 239, 242
  M73 (T197E2) 7.62mm, 157, 160, 174, 220, 254, 299
  M85 (T175E2) .50 caliber, 159, 170, 192, 199
  M219 7.62mm, 174, 208-209
  M240 7.62mm, 208-209
  M1919A4 .30 caliber, 45
  MAG-58 7.62mm Belgian, 208-209
  Ranging .50 caliber, 157
  T175 .50 caliber, 157
McGaw, Colonel Charles D., 369
McLemore, Captain Dwight, 364
Mine clearance equipment, 287, 294, 323-339
  Birmingham roller, 330-333
  Daisey Mae point charge demolition snake, 338
  Demolition snake, 287, 294, 337-339

ENSURE 202 roller, 333-335
ERDL model I roller, 330-331
ERDL model II roller, 330-331
Excavators, 323-327
Excavator, heavy pusher type, 325
Excavators, jet type, 323-324
Fargo Express rotating hoe, 326
Flail, 287, 323
Galloping Ghost self-propelled
    exploder, 330
Giant Viper, British flexible line
    charge, 338-339
High Herman heavy roller, 328, 330
Kopicki, 330
Larruping Lou light roller, 328-330
Lockley roller, 330-332
M157 projected line charge, 337
M173 projected line charge, 338
M174 practice projected line charge, 338
Mountain Lion triple projected line
    charge, 338-339
Peter Pan tank mounted mine plow,
    325-326
Peter Rabbit (Wild Bill) point charge
    projector, 339
PT54 and KMT5 Soviet rollers, 336
ROBAT, 339
Tank mounted mine clearing roller
    (TMMCR), 336
Track width mine plow, 326-327
Minneapolis Honeywell, 72, 114
Motor-und-Turbinen Union (MTU),
    75, 403-404
Muzzle brakes, 12, 16, 55, 256
Muzzle reference system, 215
Napco Industries, Inc., 402-403
National Waterlift, 216
Nomenclature for tanks, 35
North Atlantic Treaty Organization
    (NATO), 52
North Korean Army, 25
Nuclear, biological, chemical (NBC)
    equipment, 215, 223, 230
Okinawa, 245
Oilgear gun control system, 61, 92, 114, 225
Operations or Exercises
    Atlanta, 383
    Badger Tooth, 390
    Bonded Item, 367-368
    Cedar Falls, 386
    Certain Forge, 363
    Fisher, 393
    Fortress Lightning, 370
    Hickory, 381
    Incinerator, 379
    Junction City, 387
    Paul Revere IV, 384
    Pershing, 390
    Pipestone Canyon, 395
    Piranha, 376
    Solid Shield, 366
    Starlight, 373
    Utah Mesa, 396
Oscillating turret, 47, 126, 130-133, 143-145
Oto Melara, 79-80
Pacific Car and Foundry Company, 255-256,
    261-262, 174
Pakistan, 75, 79, 399
Patton, General George S. Jr., 5, 14
Personnel heater exhaust extensions, 92
Questionmark III conference, 149
Radosevich, Lieutenant Wilbert, 383
Rapid railway destructor, 303
Raytheon Company, 264
Regiments, U.S.
    2nd Armored Cavalry, 371
    4th Cavalry, 381, 383, 392, 416

8th Cavalry, 358
11th Armored Cavalry, 334-335, 381,
    383-385, 387, 396, 417
32nd Armor, 382, 386
34th Armor, 382, 386
35th Armor, 355
40th Armor, 357, 364, 414
69th Armor, 360-361, 364, 390, 395, 416
72nd Armor, 369-370
77th Armor, 397
Renk, 211, 216
Republic of Korea (ROK) Army, 25
Research Designing Service, 311
Retrofit of M48 series tanks, 219
Rheem Manufacturing Company, 47,
    126-127, 130, 143
Searchlights, 26, 29, 64-65, 122, 157, 166-167,
    173, 179, 190, 202, 225
Self-Propelled Artillery, 255-262
    M40 155mm GMC, 255
    M43 8 inch HMC, 255
    M53 (T97) 155mm SPG, 255-259, 261, 445
    M55 (T108) 8 inch SPH, 255-259, 261, 445
    M55E1 8 inch SPH, 257, 260
    T146 240mm HMC, 262
    T147 8 inch GMC, 262
    T162 175mm SPG, 261-262, 446
Semiautomatic loader in K tank, 196-197
Senior Officers Materiel Review Board
    (SOMRB), 199
Shillelagh missile MGM 51C, 190-193, 195, 419
Shillelagh weapon system, 178-181, 234
Shock mount for fire control system, 136
Short range air defense (SHORAD)
    study, 263
Siliceous cored armor, 123-125, 152, 156,
    158, 168
Smoke Grenade launchers
    M239, 208, 210
    XM176E1, 192-193
Sparkman and Stephens, Inc., 312
Sperry Products Company, 303
Sperry Rand Corproation, 264
Stabilizers, 9, 33, 40, 58, 72, 192, 197,
    199-200, 211
Starry, General Donn A., 5, 383
Stilwell Board, 9, 32, 255
Stilwell, General Joseph W., 9
Suspensions
    Advanced torsion bar (ATB) suspension,
        206-207
    Aluminum road wheels, 158, 211
    Flat track suspension, 47
    Horizontal volute spring suspension
        (HVSS), 30
    Hybrid suspension, 207
    Hydropneumatic suspension system
        (HSS), 205-207, 216
    M46 tank, 13, 22
    M47 tank, 55, 62
    M47-M tank, 77
    M48 tank, 85, 92, 101
    M48A1E1 tank, 220
    M48A2 tank, 112
    M48A3 tank, 227
    M53-M55 self-propelled artillery, 258
    M60 tank, 158, 163
    M60A1 tank, 174
    M60A1E3 tank, 202
    M60E1 tank, 169
    Rotary shock absorbers, 198, 201-202,
        206-207
    T40 tank, 13
    T42 tank, 42-43
    T48 tank, 85, 92, 101
    T48E2 tank, 112
    T162 self-propelled gun, 261-262

Track tension idler, 13, 16, 77, 92,
    119, 220
Tube-over-bar (TOB) suspension, 198,
    201-202, 205
Tank Battalions, U.S.
    1st Marine, 31, 368, 389, 395, 417
    2nd Marine, 366-368
    3rd Marine, 228, 373-380, 388-389
    6th, 25-26
    29th, 352
    57th, 352-353
    64th, 25-26
    710th, 354
    713th, 245
    759th, 353
Tank Recovery Vehicles (TRV), 272-282
    Combined TRV/CEV proposal, 297
    M31 (T2), 272
    M32 (T5), 272
    M74 (T74), 272
    M88 (T88) medium recovery vehicle,
        274-279, 321, 387, 394, 447
    M88A1 (M88E1) medium recovery
        vehicle, 280-282
    Modified M48 tank as an interim recovery
        vehicle, 275
    T88 medium recovery vehicle concepts,
        272-274
    T88E1 medium recovery vehicle, 275
Tanks, British
    Centurion, 72
Tanks, Chinese
    T59, 398
Tanks, French
    AMX 30, 404
Tanks, German
    Leopard, 200
Tanks, Israeli
    Blazer, active armor, 400-401
    M48A2, 399
    M48A2 (105), 399
    M48A3, 400
    M48A4, 400
    M60, 400
    M60A1, 400-401
    Mine roller, 401
    Track width mine plow, 401
Tanks, Soviet
    PT76, 149, 395, 397-398
    T34/85, 25
    T54, 397-398
Tanks, Swiss
    Pz 61, 75
Tanks, U.S.
    Abrams main battle tank, 5, 6, 215, 409
    Airborne reconnaissance/airborne assault
        vehicle (AR/AAV), 179
    Chrysler K tank, 196-198
    Crocodile flame thrower tank, 247
    General Patton II, 59
    Flame thrower based on the M26 tank,
        245-247
    M1 Abrams 105mm gun tank, 5, 6,
        215, 409
    M3 medium tank, 83, 272
    M4 series medium tank, 5, 10, 25, 33, 35,
        83, 245, 272, 286-287, 351
    M4A3 medium tank, 26, 272, 292
    M26 medium tank, 5, 10-11, 13-14, 16-17,
        25-26, 52, 58, 245, 286-287, 310,
        323-324
    M26 medium tank, modernized, 12,
        14, 32
    M26E1 medium tank, 12
    M26E2 medium tank, 11-12
    M26A1 medium tank, 16
    M41 (T41E1) 76mm gun tank, 14, 100,

126, 398
M42B1 flame thrower tank, 245, 248
M42B3 flame thrower tank, 245, 248
M46 medium tank, 10-31, 33, 42, 52-53,
    55, 58, 92, 283, 287-289, 292,
    303-305, 310, 325, 351, 422
M46 medium tank (New), 17
M46A1 medium tank, 10-31, 55, 100, 310, 422
M46E1 medium tank, 53
M47 90mm gun tank, 5, 17, 42, 52-80, 83,
    92, 100, 248, 283, 292, 294, 310-312,
    325, 351, 399, 402, 425
M47E1 90mm gun tank, 72
M47E2 90mm gun tank, 73
M47E modernized by Chrysler España, 79
M47-M modernized by BMY, 77-79, 426
M47RKM modernized in Israel, 79
M48 90mm gun tank, 72, 83-108, 149,
    248, 283-284, 292, 306, 312, 346,
    399, 402, 428
M48 105mm gun tank modernized in
    Germany for Turkey, 404
M48C 90mm gun tank, 103, 306
M48A1 90mm gun tank, 103-107, 219,
    222-223, 234, 242, 283-284, 346,
    351, 400, 402, 429
M48A1 tank with applique armor installed
    in Germany, 406
M48A1E1 105mm gun tank, 219-223, 435
M48A1E2 90mm gun tank, 222
M48A1E3 105mm gun tank, 234-237, 437
M48A2 90mm gun tank, 114-122,
    150-155, 157-158, 252, 284,
    313, 351, 434
M48A2C 90mm gun tank, 119-122, 125,
    150, 158, 307-308, 313, 351, 393, 434
M48A2C 105mm gun tank, 122
M48A2E1 90mm gun tank, 150-152, 158
M48A2G tank modernized in Germany,
    404-406
M48A3 90mm gun tank, 5, 6, 119,
    222-234, 242, 254, 284-285, 313,
    333-335, 373-377, 379-392, 394-399
M48A3 (Mod B) 90mm gun tank,
    227-234, 238, 436
M48A3E1 105mm gun tank, 238
M48A3 tank modified in Korea, 407
M48A4 105mm gun tank, 234-237, 400
M48A5 105mm gun tank, 238-242,
    264-265, 267-268, 309, 346, 370-371,
    400, 408, 418, 438
M48A5PI 105mm gun tank, 239-242
M51 proposed new tank, 152
M60 105mm gun tank, 5, 6, 77, 156-168,
    171, 199, 219, 234, 284-285, 298,
    313-316, 326-327, 351, 418, 439
M60E1 105mm gun tank, 155, 168-170,
    174, 179
M60E2 152mm gun tank, 179
M60E3 152mm gun tank, 179
M60A1 105mm gun tank, 5, 75, 171-177,
    179, 196, 199-202, 211, 215-216,
    242, 284-285, 298-299, 308-309,
    316-320, 326-327, 351, 408-409, 440
M60A1 (PI) 105mm gun tank, 199
M60A1 (AOS) 105mm gun tank, 200
M60A1 (RISE) 105mm gun tank, 201
M60A1 (RISE) (PASSIVE) 105mm gun
    tank, 202
M60A1E1 152mm gun tank, 180-188, 234
M60A1E2 152mm gun tank, 188-197, 419
M60A1E3 105mm gun tank, 201-205, 208, 442
M60A2 152mm gun tank, 5, 188-197, 408, 441
M60A3 105mm gun tank, 5, 79,
    208-215, 408, 443
M60A3 (TTS) 105mm gun tank, 210-215
M60AX 105mm gun tank, 216-218

M67 (T67) flame thrower tank,
    248-252, 254
M67A1 flame thrower tank, 252-253
M67A2 (M67E1) flame thrower tank, 254,
    377-378, 380, 420, 444
M103 (T43E1) 120mm gun tank, 143,
    153-155
Main battle tank (MBT) 70, 6, 196,
    206, 408
MBT (MR), 179
Patton, 5, 6, 25, 26
Patton 47, 59
Patton 48, 93
Pershing, 5, 10, 14, 25-26, 245, 351
Retrofit of M48 series tanks, 219
Sheridan, 179
Sherman, 5, 10, 25, 33, 35, 83, 245, 272,
    286-287, 351
T35 flame thrower tank, 245-247
T37 light tank, 32-33, 83
T40 90mm gun tank, 13-16, 34-35, 42, 287
T41 76mm gun tank, 32, 35, 52, 83
T42 90mm gun tank, 14, 17, 32-47, 52-53,
    72, 83, 248, 423
T43 120mm gun tank, 35, 52, 83, 85, 283
T48 90mm gun tank, 42, 59, 83-93,
    126, 312, 427
T48E1 90mm gun tank, 108
T48E2 90mm gun tank, 108-114
T48E2 tank with turret from the T54E2,
    141-142
T48E3 90mm gun tank, 114
T52 searchlight tank, 26, 29
T54 105mm gun tank, 126-129, 131, 141, 430
T54E1 105mm gun tank, 126,
    130-133, 141, 431
T54E2 105mm gun tank, 134-142, 432
T57 120mm gun tank, 143
T66 flame thrower tank, 248
T69 90mm gun tank, 47-51, 424
T77 120mm gun tank, 143-146, 433
T87 90mm gun tank, 47, 149
T95 tank series, 6, 131, 141, 149-150,
    152-158, 168, 179, 206, 274-275,
    297-298
XM60 weapon system, 152-157
XM66 concept studies, 180
XM735 105mm gun tank, 234
XM736 105mm gun tank, 238
XM803 152mm gun tank, 6, 408
Task Force Crombez, 26
Taylor, General Maxwell D., 149
Teledyne Continental Motors (TCM), 10, 16,
    34, 150, 158, 174, 210-211, 216,
    261, 402
Telfare, Sergeant First Class Nathaniel, 343
Television in tanks, 345
Thermal shroud for gun tube, 202, 208
Top loading air cleaner (TLAC), 199-200
Tracks
    T80E6, 42
    T84E1, 42, 62
    T95, 42
    T97, 200
    T97E2, 158
    T142, 200
    23 inch wide on M53 and M55,
    256-257
Training Devices, 340-344
    Driver trainer for M48A2 tank, 342
    Evasive target tank (ETT), 340, 344
    Laser training device, 343
    Lightweight turret trainer, 340
    M30 turret trainer, 341
    M30A1 turret trainer, 341
    M34 driver trainer, 342
    M37 turret trainer, 342

    Simfire device, 343
    T13 turret trainer, 340
    T18 turret trainer, 340
    T20 turret trainer, 341
    Telfare device, 343
Transmissions
    CD-500, 34
    CD-500-3, 42
    CD-850 series, 11-13, 16-17, 21, 42, 55,
        75, 77, 107-108, 150, 158, 220,
        223, 402-403
    CD-850-1, 11-13
    CD-850-2, 13
    CD-850-3, 16-17, 21
    CD-850-4, 17, 21, 42, 55, 75, 107-108, 403
    CD-850-4B, 108
    CD-850-6, 150, 158, 220
    CD-850-6A, 77, 223
    Cross drive, 11-13, 16, 21, 26, 34, 107
    Renk RK-304, 211, 216
    X-700, 211
    X-1100, 211, 215
    XT-500, 46-47
    XT-1400, 107-108, 150, 276
    XT-1400-2, 278-279
    XT-1400-3, 261
    XT-1410-4, 280
United Nations Forces, 25
United Shoe Machinery Corporation,
    127, 130
Unit Rig and Equipment Company, 307
Utz, Chester C., 85
Vaughan, Russell P., 370-371, 406-407
Vehicle engine exhaust smoke system
    (VEESS), 210
Vickers gun control system, 74
Vietnam, 225, 227-228, 254, 263, 308,
    333-335, 373-398
    Anke, 395
    An Lao river, 387
    A Shau valley, 396
    Ben Dong, 383
    Ben Het Special Forces Camp, 395
    Bien Hoa, 381, 384
    Binh Son, 380
    Bong Son, 388, 390
    Can Lo, 397
    Chu Lai, 375
    Cu Chi, 382
    Da Nang, 373-374, 376, 379, 389
    Dong Ha, 380, 397
    Dong Ha river, 398
    Gia Ray, 383
    Hoa Long, 377
    Hue, 391
    Hung Nghiz, 396
    Khe Sanh, 388, 392, 394
    Khong Nhon, 379
    Lai Khe, 387
    Phu Bai, 377
    Pleiku, 384
    Quang Tri Province, 390, 393, 397
    Rockpile, 397
    Saigon, 383
    Suoi Cat ambush, 383-385
Vision block ring, 227-230, 238
War Department Equipment Review Board
    (Stilwell Board), 9, 32, 255
Westmoreland, General William, 373
Williams, Joseph, 83, 179
XM13 missile (Shillelagh), 178-179
XM60 armament tests, 152-155
XM60 ranging machine gun, 157
York, Sergeant Alvin C., 268
Yuba Manufacturing Company, 72
Yuma test rig, 150
Zaloga, Steven, 217-218

E P B M We hope you enjoyed this title
from Echo Point Books & Media

Before Closing this Book, Two Good Things to Know

1. Buy Direct & Save

Go to www.echopointbooks.com to see our complete list of titles. We publish books on a wide variety of topics—from spirituality to auto repair.

Buy direct and save 10% at www.echopointbooks.com

DISCOUNT CODE: EPBUYER

2. History Buff? Tank Lover? We've got you covered!

Echo Point Books & Media is proud to announce the release of new, top quality reprints of R.P. Hunnicutt's tank books. Finally back in print at an affordable price!

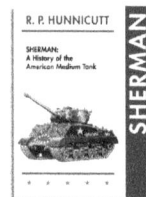

Echo Point Books & Media is the official publisher of the R.P. Hunnicutt tank series. Easily the most comprehensive series of reference books on the developmental history of American military vehicles, Hunnicutt's 10-volume compendium of tank information is an absolute must-have for every serious military and history buff's bookshelf. These books contain a wealth of detailed information including:

o   Important facts and figures

o   Thousands of photographs

o   Hundreds of detailed line drawings and diagrams

View our entire catalog of military history books at:
www.echopointbooks.com

Follow us to keep up with new releases in the Hunnicutt Series!

f echopointbooks   **y** @EPBM   **P** echopointbooks

www.ingramcontent.com/pod-product-compliance
Lightning Source LLC
Chambersburg PA
CBHW050237220326
41598CB00047B/7436